Structure and Function in Cell Signalling

Structure and Function in Cell Signalling

John Nelson
Queen's University, Belfast, UK

John Wiley & Sons, Ltd

Telephone (+44) 1243 779777

Email (for orders and customer service enquiries): cs-books@wiley.co.uk
Visit our Home Page on www.wileyeurope.com or www.wiley.com

Other Wiley Editorial Offices

John Wiley & Sons Inc., 111 River Street, Hoboken, NJ 07030, USA

Jossey-Bass, 989 Market Street, San Francisco, CA 94103-1741, USA

Wiley-VCH Verlag GmbH, Boschstr. 12, D-69469 Weinheim, Germany

John Wiley & Sons Australia Ltd, 42 McDougall Street, Milton, Queensland 4064, Australia

John Wiley & Sons (Asia) Pte Ltd, 2 Clementi Loop #02-01, Jin Xing Distripark, Singapore 129809

John Wiley & Sons Canada Ltd, 6045 Freemont Blvd, Mississauga, Ontario, L5R 4J3, Canada

Wiley also publishes its books in a variety of electronic formats. Some content that appears in print may not be available in electronic books.

Library of Congress Cataloging-in-Publication Data

Nelson, John, 1953 Feb. 27-
Structure and function in cell signalling / John Nelson.
p. ; cm.
Includes bibliographical references and index.
ISBN 978-0-470-02550-5 (alk. paper) – ISBN 978-0-470-02551-2 (alk. paper)
1. Cellular signal transduction. I. Title.
[DNLM: 1. Signal Transduction–physiology. 2. Amino Acid Sequence–physiology. 3. Intercellular Signaling Peptides and Proteins–physiology. 4. Intracellular Signaling Peptides and Proteins–physiology. QU 375 N427s 2008]
QP517.C45N45 2008
571.7′4–dc22 2007040695

British Library Cataloguing in Publication Data

A catalogue record for this book is available from the British Library

ISBN 9780470025505 HB
9780470025512 PB

Typeset in 10.5/12.5 pt Times by Thomson Digital, Noida, India
Printed and bound in Spain by Grafos SA, Barcelona

The cover image shows the superimposition of two insulin receptor kinase structures. The two structures are aligned on the β-sheet region between residues 990 and 1080 (grey). The α-helices are shown as cylinders, coloured orange for the inactive structure and yellow for the active kinase. The A-, P- and C-loops are shown as solid ribbons coloured violet, red and green, with the darker shades being for the inactive kinase. The substrate and ATP analogue are shown as ball-and-stick, coloured by atom type, while the two magnesium atoms are shown as maroon spheres. An electrostatic contour surrounds the peptide substrate. This diagram was prepared using InsightII.

Contents

Acknowledgments

I dedicate this book to my wife Debbie, my daughter Alex, my son Jamie, my mother Bettie and the memory of my father James.

My special thanks go to Professor Marc Chabre for his invaluable and instructive comments particularly in the area of G protein coupling and 'affinity-shift'. Thank you, Marc. I very much enjoyed our exchange of emails and your robust but entertaining critique. And thank you to Dan Cassel for suggesting that I contact Marc in the first place, and to Maeli Waelbroeck for helpful discussions on the thermodynamics of 7-pass receptor coupling. Thank you too, to all the other scientists who took the time to answer my queries and to my colleagues for their help and suggestions, especially Neil McFerran for assistance with superimposition of IRK and Flt structures.

Finally, I (and many others, I am sure) gratefully acknowledge the generosity of Roger Sayle, who wrote RasMol and made it freely available. Thanks also to Herbert Bernstein who maintains and curates the cross-platform updates of RasMol as an open source project. For more information please visit www.rasmol.org and the 2 following references:

Sayle, R. and James Milner-White, E. (1995) RasMol: Biomolecular graphics for all. *Trends in Biochemical Sciences (TIBS)*, **20**(9): 374.

Bernstein, H.J. (2000) Recent changes to RasMol, recombining the variants. *Trends in Biochemical Sciences (TIBS)*, **25**(9): 453–455.

Preface

To the non-expert, cell signalling is daunting. Even for a specialist, straying from one area of signalling to an unfamiliar one can be somewhat testing.

The scientific literature now features signal transduction maps that, at first sight, appear to be bewildering collections of coloured blobs connected by circuitous lines and arrows. This is an undeniably efficient way to contain the burgeoning knowledge base, but without an understanding of the underlying fundamentals, such maps are as inscrutable as electronic circuits are to those ignorant of the function of resistors and capacitors. Fortunately, just like electronic circuits, cell signalling pathways are constructed from a limited number of types of components that rely upon a small number of discrete mechanisms of action. Indeed, this is exactly the analogy that Rodbell used when proposing that a 'transducer' (*G protein*) laid between the multiple information 'discriminators' (*receptors*) and the signal 'amplifier' (*effector enzyme*).

My aim is to break signalling down into common elements and activities – the 'nuts and bolts' of cellular information exchange. And, as we shall see, signalling is really just a series of recognition events that lead to either simple binding and conformational change or catalysis.

In order to explore adequately the molecular details of transduction components and their interactions, I have focused on 7-pass and single-pass receptor signalling. Unfortunately, space and time constraints did not permit coverage of neural and visual signalling; similarly, ion channels are not covered in detail, and adhesion- and cytokine signalling are not discussed.

What is in the book...

The book begins with basic principles, including a historical account of key discoveries and their impact, a simplified treatment of classical mathematical methods of quantitation, and a survey of structural domains and linear peptide signatures of the commonest transduction molecules. To give the book a narrative, I have focused on metabolic signalling and the control of cell cycle progression because all the key players (G proteins, 2nd messengers, kinases, adaptors) are present and these pathways, being the longest studied, are relatively well understood. As you shall see, however, understanding of even these venerable pathways and networks is still in a partial state of flux. Finally, I have done my best to clarify the alternative nomenclatures, numberings and terminologies that have emerged from the diverse specialities that now exist within cell signalling.

I have deliberately used the freeware molecular rendering programme RasMol throughout to encourage readers to try it for themselves (simple instructions are in the Appendix). Just remember: successful use of molecular graphics depends upon being able to find your bearings in a (perhaps unfamiliar) linear sequence by relating it to conserved motifs and known three dimensional structural homologues. Furthermore, one must always bear in mind that the sequence numbering in coordinate files from an X-ray or NMR study is not always the same as the native sequence numbering because proteins are often truncated to aid crystallisation. Lastly, always read the source paper before attempting to examine the structural model.

For the past half century, our understanding of metabolic and signalling pathways has been built from *in vitro* measurements of the activities of individual components isolated from homogenised cells, the behaviour of the entire pathway being inferred mathematically by summation. Recently, however, modern techniques increasingly allow whole pathways to be monitored in single living cells. The familiar paradigms of metabolic pools and 'control point' enzymes (derived, *in vitro*, from kinetic assays of individual glycolytic enzymes) has been challenged by 'metabolic control theory' developed from global analysis of glycolytic flux in living whole cells. Similarly, the paradigm of signal amplification by soluble protein kinase cascades has been modified by the discovery of 'scaffolding proteins', which hold kinases and their downstream kinase substrates in one-to-one cassette-like complexes. New real-time receptor binding assay methodologies (surface plasmon resonance, microcalorimetry) increasingly allow measurements of on- and off-rate constants (k_1 and k_2) of native ligands to be obtained – something previously difficult to measure using labelled ligands.

The 21st century is likely to be dominated by the study of the cell physiological functions of protein-to-protein interactions as they occur in whole cells. Both metabolic and signal transduction research are in a sense coming full circle to the realisation that one only really gets a true picture of how pathways work by looking at them in their entirety *and* in their natural environment: the complex, crowded and elastic milieu of the living cell. We have taken the clock to pieces, now we are putting it back together again.

John Nelson
April 2007

1 The components and foundations of signalling

Endocrine glands produce hormones that act as *first messengers*, informing other tissues about the external environment, overall energy status, etc., on a strictly 'need-to-know' basis. Tissues that do not need to respond to a given first messenger may lack a functional receptor or pathway for that signal. First messengers (except steroids) cannot enter the cell, and so elaborate ways of transducing the signal to the intracellular compartments must be employed. Some receptors couple with an effector enzyme that produces a *second messenger* inside the cell. These second messengers are almost invariably charged or membrane bound and therefore cannot leave the cell. The final stages in the transduction of the signal usually involve the activation of downstream protein kinase enzymes, often by direct binding of the second messenger. Downstream kinases are mostly serine/threonine kinases and they activate (or inactivate) target proteins by phosphorylation.

The biological response that a cell makes to a received signal can also vary. cAMP-inducing catabolic signals such as adrenaline activate glycolysis in the heart but have an opposite effect in the liver. This is due to the differential effects of PKA phosphorylation on the different isoforms of an effector enzyme that either makes or breaks down the second messenger, fructose-2,6-bisphosphate. In most tissues, insulin is strictly anabolic, inducing target cells to incorporate glucose into macromolecular stores. In the heart, however, insulin uniquely stimulates glycolysis.

Structure and Function in Cell Signalling by John Nelson

A multicellular animal (*metazoan*) faces a major problem not shared by unicellular forms (*protozoans and prokaryotes*) namely, how to integrate, organise and control the dynamics of the diverse collection of differentiated tissues and organs that make up a body. The most obvious means of control, of course, is the central nervous system: one wills a skeletal muscle to move, and it moves. However, much of the signalling between tissues is done entirely unconsciously and it is this unconscious signalling that controls vital processes such as growth, reproduction, food and energy usage, host defence and immunity, fight or flight responses and many other day-to-day functions that one tends to take for granted. Some of these latter signals are initiated by the autonomic nervous system; others are entirely hormonal with no conscious or unconscious neuronal input.

In its widest definition, *cell signalling* encompasses the generation and transmission of a signal (in the form of a blood borne first messenger, for example), the reception of the signal (by the target cell's receptor) and the propagation of that signal (via second messengers, for example) within the receiving cell. *Signal transduction* is the process whereby an extracellular signal is converted (or *transduced*) into a different form (or forms) of intracellular signal. Transduction of the signal often results in amplification and is frequently a multi-step process. Crucially, these processes are time-limited. In other words, a signal is only allowed to be transduced for a certain length of time before it is turned off.

1.1 Definition of terms used

1.1.1 First messengers

A first messenger is an extracellular molecule that is recognised by target cells because they possess receptors that bind the molecule specifically. This is how cells can communicate with each other – often over long distances. Cells that should not respond to a particular first messenger will lack the matching receptor. However, signal transduction in responsive cells can be still be attenuated by either down-regulating the number of receptors per cell, up-regulating an enzyme that inactivates the first messenger, or desensitising the downstream intracellular transduction pathway. There are hundreds of distinct first messengers ranging in size from large glycosylated polypeptide hormones such as gonadotrophin, through polypeptides, tripeptides, cholesterol-like steroids to amino acid derivatives (Figure 1.1). Smaller messengers are the neurotransmitter amine acetyl choline and even smaller paracrine 'hormones': the gaseous hormones, nitric oxide and hydrogen sulphide.

1.1.2 Glands and types of secretion

Glands can be divided roughly into *exocrine* and *endocrine*. Exocrine glands secrete into the gastrointestinal tract or onto the surface of the body (Figure 1.2). Exocrine

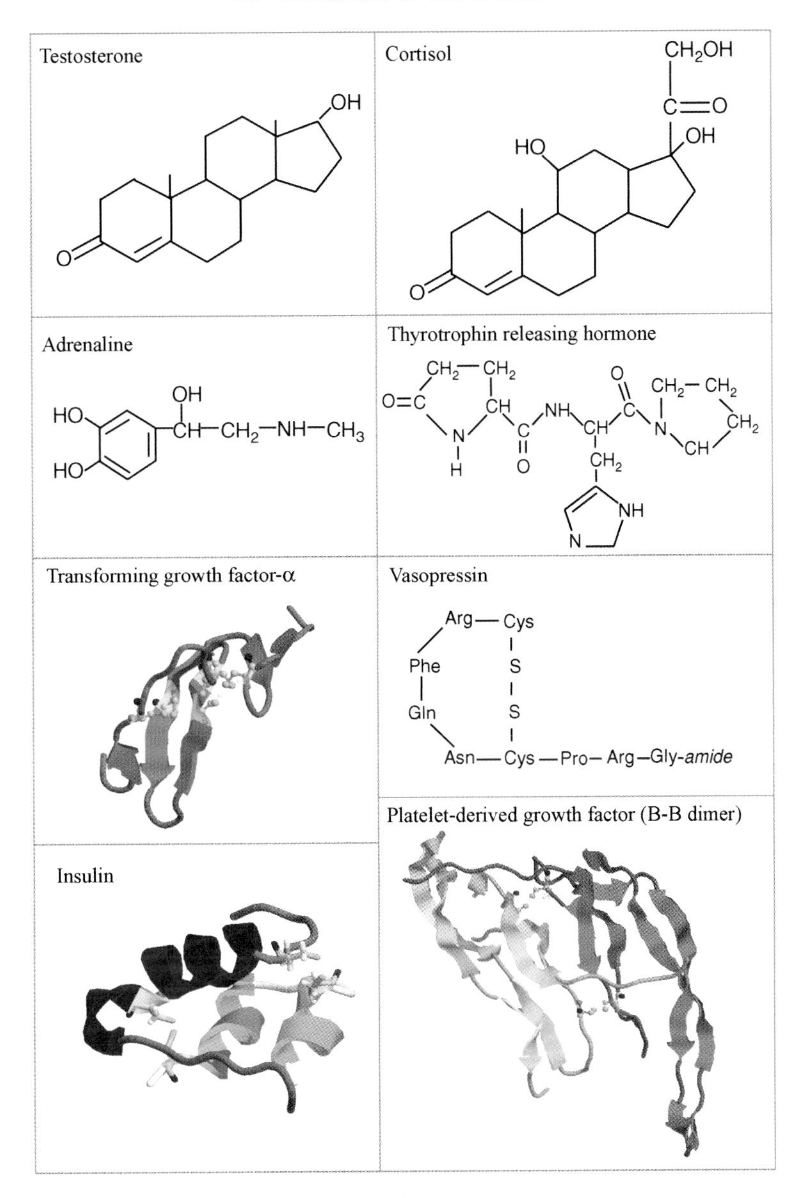

Figure 1.1 First messengers

secretions generally play a digestive or protective role and are not involved in signalling. Endocrine glands, on the other hand, secrete informational molecules into the blood system. A few organs, such as the pancreas, contain both exocrine and endocrine compartments. Not all first messengers travel through the blood; some act locally or even self-stimulate via autocrine secretion (see Section 10.3) (Figure 1.3).

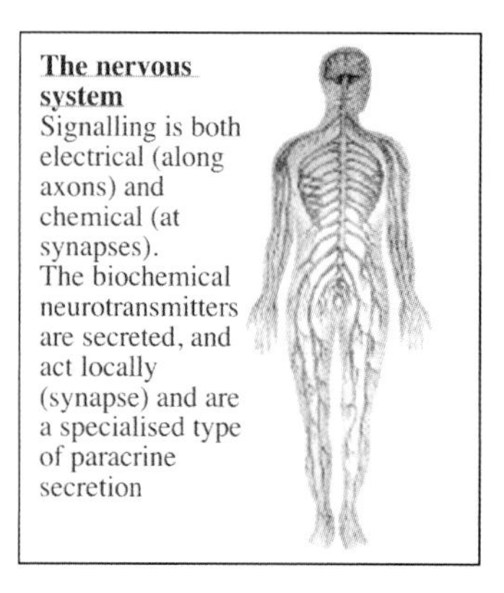

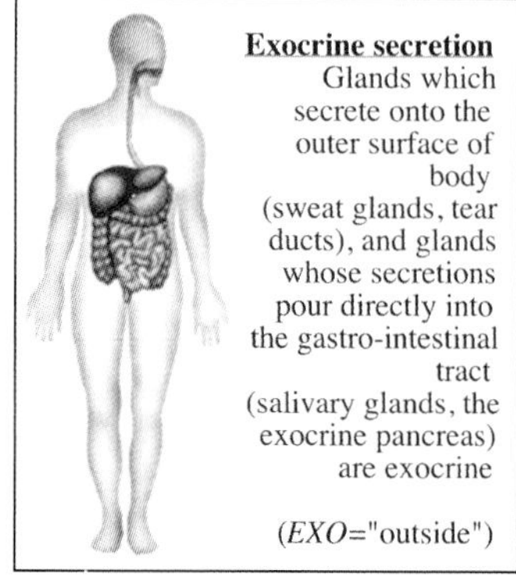

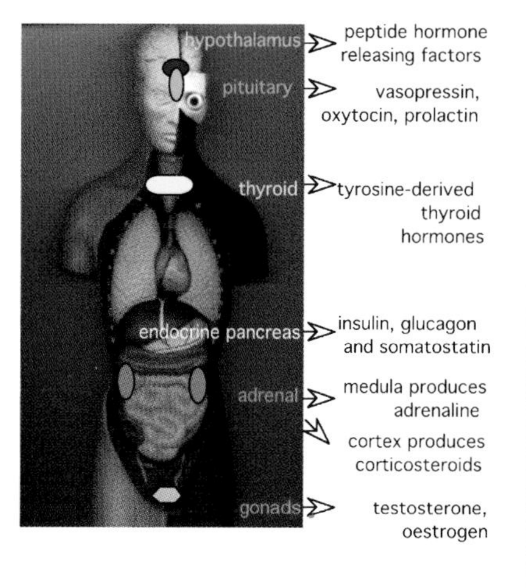

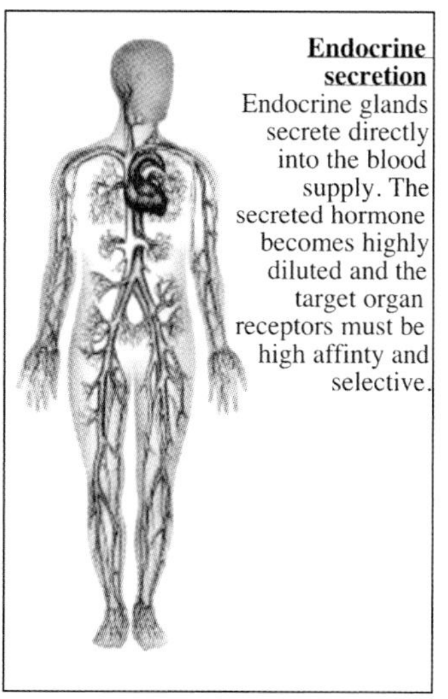

Figure 1.2 Glands

1.1.3 Ligands

All first messengers are ‘ligands’, which simply means: a molecule that binds to a given receptor protein in a specific and saturable manner. Second messengers are also ligands, but for intracellular receptors.

1.1.4 Agonists

Agonists are ligands that cause a biological response when they bind to their cognate receptor. They include:

- the natural hormone (or first messenger);

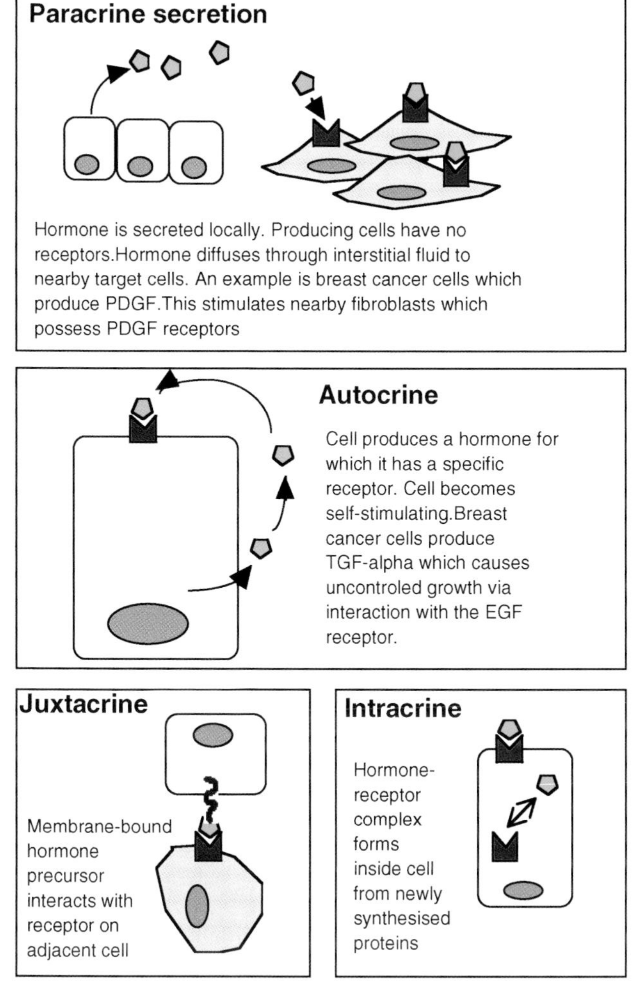

Figure 1.3 Informational molecules do not all travel through the circulation

- synthetic analogues (or drugs) which produce a similar response when interacting with the receptor in question – for example, adrenaline is the natural agonist, and the analogue isoproterenol an artificial agonist, of the β-adrenergic receptor.

1.1.5 Antagonists

Antagonists are also ligands, which bind specifically and saturably to a given receptor but without producing a cellular response. Thus, antagonists occupy the receptor, effectively blocking the activating effects of the natural first messenger. Almost all antagonists are human-made. Very few natural ligands are true receptor antagonists.

Table 1.1 Ligand types

Conventional agonists	Full agonist	At full receptor occupancy, elicits full biological response
	Partial agonist	Even at saturation, only a partial biological response is elicited‡
	Inverse agonist	Binds to, and stabilises the inactive form of the receptor (or converts R* to R)§
Antagonists	Competitive antagonist	Binds to the same site on the receptor as the agonist and blocks its effects
	Noncompetitive antagonist	Binds to a different site on the receptor than the agonist and reduces its effects

‡ Critically depends on system tested
§ See Chapter 2

Blocking receptor pathways in nature almost always comes about by indirect antagonism from another pathway.

It is likely that most polypeptide antagonists bind unproductively to the same site on the receptor that would be occupied by the native agonist. However, synthetic catecholamine ligands fall into many subcategories (Table 1.1). Undoubtedly, this multiplicity was discovered because of the huge research effort inspired by the clinical importance of these receptors as drug targets, the ease of synthesis of such small ligands, and the availability of simple *in vitro* screens. Whether receptors for large complex ligands such as glycoproteins also display such diverse behaviour is open to question. However it is possible given that insulin & IGF-1 bind to different domains of InsR Ectodomain (see Chapter 2, Section 2.3.3).

1.1.6 Receptors for first messengers

The receptors we are concerned with are large proteins that 'receive' information in the form of first messengers. 'Reception' is by direct binding of the hormone to a specialised ligand-binding site on the receptor. Other 'receptors' can act both as information receivers and attachment factors such as the integrins that bind cells to basement membranes and participate in platelet activation. In the broadest sense, the term 'receptor' also includes *non-informational* receptors such as endocytic or scavenger receptors. These receptors (the liver's asialoglycoprotein receptor is an example) serve a waste-disposal role in the body, mopping-up aged proteins from the circulation – they do not 'receive' information.

First messenger receptors fall into two broad categories: soluble or membrane-spanning. Intracellular soluble receptors are members of the 'zinc-coordinating domain' superfamily of transcription factors. Being transcription factors, they are largely nuclear in location. They include receptors for lipophilic hormones and include oestrogen-, androgen-, and progesterone-receptors, as well as thyroid hormone receptors and those for retinoids and vitamin D3 analogues.

Membrane-spanning receptors are multi-subunit, single-pass, or 7-pass (Figure 1.4). The multi-subunit receptors are ligand-gated ion channels that allow specific ions to cross the membrane when agonist-occupied, through a central pore that closes when the

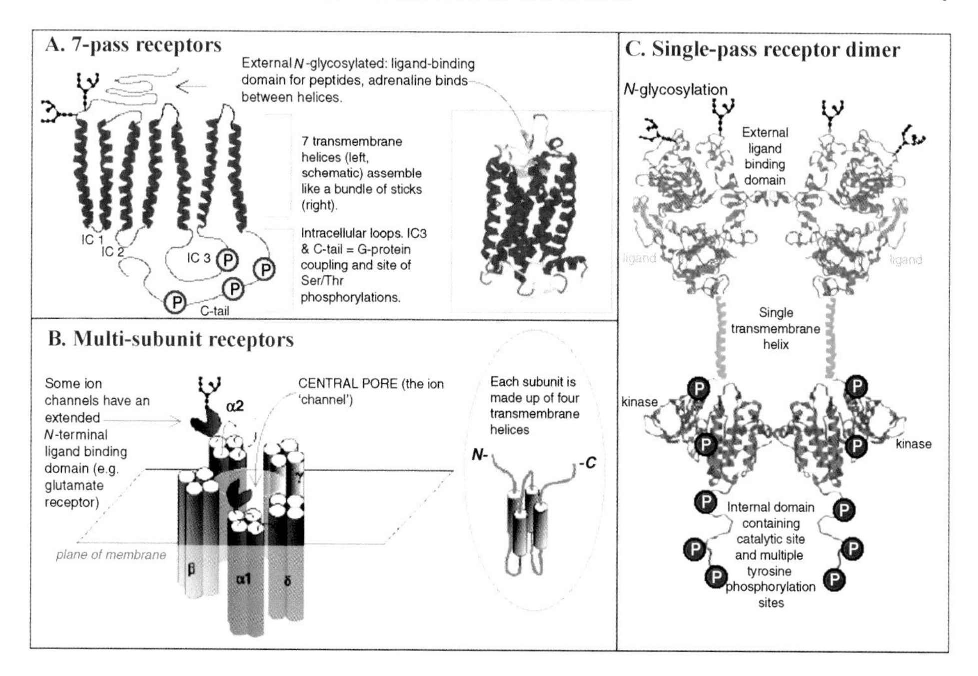

Figure 1.4 Schematic cartoons of some membrane-spanning receptor types

receptor is unoccupied. These are a numerous and diverse group of fast-acting switches that are typically made up of four or more subunits that are usually (but not always) products of more than one gene. The nicotinic acetyl choline receptor, for example, is made up of five subunits: two ligand-binding α-subunits and one each of β-, γ- and δ-subunits. The central pore is Na^+-selective and, when activated, extracellular sodium ions are allowed into the cell, causing local depolarisation of the membrane and activation of voltage-gated channels.

Unfortunately, space does not permit a detailed discussion of these receptor types and only the IP3 multi-subunit receptor will be mentioned in the text.

Receptors that have a single transmembrane helix, I shall term 'single pass'. These are either catalytic or non-catalytic. Non-catalytic single-pass receptors rely on recruitment of cytoplasmic proteins (often tyrosine kinases) to transduce their signals. These include the 'tumour necrosis factor' receptor and the 'cytokine receptors'such as those for the interleukins. Again, space does not permit a detailed discussion of non-catalytic single-pass receptors.

Catalytic single-pass receptors are predominantly tyrosine kinases and can be phylogenically connected through catalytic domain homologies (see Appendix 1). A smaller group are serine/threonine kinases, including the growth-inhibitory 'transforming growth factor-β' receptor.

The receptor tyrosine kinase class of single-pass receptors play major roles in all aspects of metazoan life and, being the best understood, are one focus of the present text. Receptor tyrosine kinases, activated by polypeptide 'growth factor'-type ligands, phosphorylate specific intracellular substrates, and are characterised by themselves becoming tyrosine-phosphorylated. They dimerise when ligand-activated and phosphorylate each other in the first steps in a signal transduction programme.

The evolutionary more ancient 7-pass receptors are non-catalytic, phylogenetically linked and the single largest receptor superclass. They often become serine/threonine-phosphorylated after signalling has peaked.

1.1.7 Second messengers

Second messengers are confined within the cell that synthesised them because all are either charged (and thus impermeant to the cell membrane) or are incorporated into the inner leaflet of the plasma membrane. For example, although sometimes secreted by social amoeba, cAMP release from a cell is only possible when the membrane is ruptured, and extracellular cAMP is thus a sign of cellular damage.

1.1.7.1 A few types of second messenger

Second messengers are generally smaller than first messengers and there are far fewer individual types (Figure 1.5). Discrimination in cellular information processing is concentrated upon the large diversity of hormone receptors and extracellular cognate ligands. Once the signal is transduced into the cell, the options are reduced. Many different receptors may be capable of stimulating production of a common second messenger.

Second messengers are produced by 'effector enzymes' downstream of extracellular receptors. The effector enzymes responsible for the synthesis of second messengers are

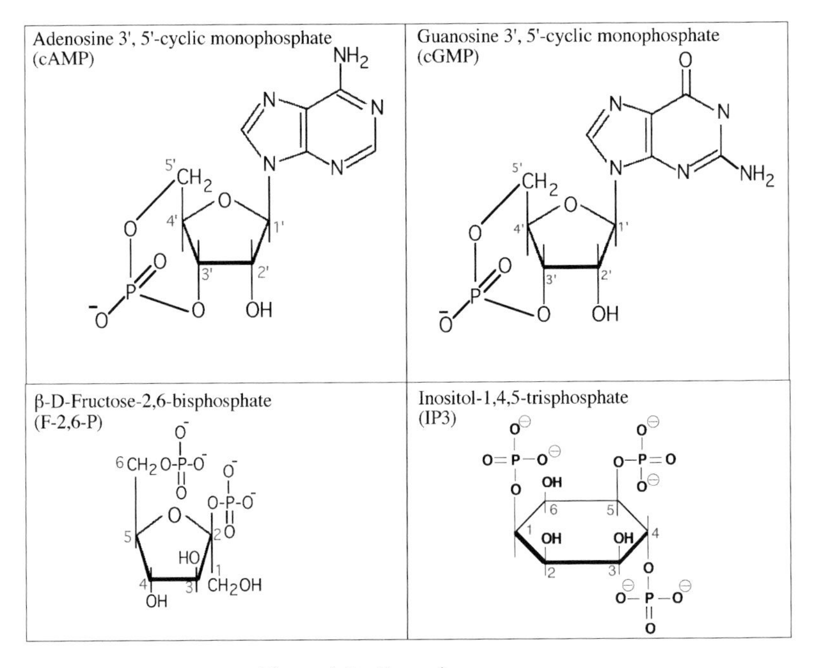

Figure 1.5 Second messengers

under tight control, only being active for short bursts. Crucially, transduction systems have the means to destroy the second messengers or sequester them when the signal needs to be terminated.

An archetypal system is represented by the effector enzyme adenylyl (or adenylate) cyclase that produces cAMP, which is in turn broken down by the signal-terminating enzyme phosphodiesterase.

1.1.8 Soluble second messengers

cAMP The cyclic nucleotides cAMP and cGMP (Figure 1.5) are produced by cyclase enzymes that use ATP or GTP as substrates. 'Adenyate' or adenylyl cyclase enzymes are a diverse family of membrane-spanning proteins. In mammals they have a 'split' catalytic that is re-united upon stimulation by a variety of upstream signals.

Calcium Calcium release from intracellular stores in the endoplasmic reticulum lumen is controlled by ligand-gated ion channels that are receptors for inositol-1,4, 5-trisphosphate (IP3). Free calcium is maintained at a low level (around 150 nM) in the cytosol of resting cells against a high extracellular concentration of 1–3 mM in plasma (Figure 1.6). This low level is maintained by highly active calcium pumps that either expel the ion to the extracellular space, or channel it into internal storage in the endoplasmic reticulum. However, when IP3 binds to its receptors in the endoplasmic reticulum membrane cytosolic calcium rises rapidly to around 1 μM. IP3 receptors are calcium-selective channels that open to release the ion into the cytosol. Other sources of calcium come from cell surface 'store-operated' channels that permit extracellular calcium to refill the internal stores as well as contributing to certain types of signalling. Finally, membrane depolarisation downstream of plasma membrane ligand-gated sodium channels (like the nicotinic acetyl choline receptor) also causes external calcium entry via voltage-gated calcium channels.

Much of calcium's effects are mediated by binding to the ubiquitous calcium-binding protein, calmodulin, various forms of calcium-dependent protein kinase (PKC) and the multi-subunit phosphorylase kinase.

Fructose-2,6-bisphosphate Fructose-2,6-bisphosphate (F-2,6-P) is produced by phosphofructokinase-2 (PFK-2). It should not be confused with fructose-1,6-bisphosphate (F-1,6-P). F-1,6-P is a key intermediate in the energy-harvesting glycolysic pathway, whereas F-2,6-P plays no part in energy-generation but instead acts solely as a second messenger.

1.1.9 Membrane-bound second messengers

Like soluble second messengers, the key to signalling lipids is that they normally are absent, or at a very low level, in the resting cell membrane. Phosphatidyl inositol-4,5-bisphosphate (PIP2) is an important lipid substrate for a variety of effector enzymes.

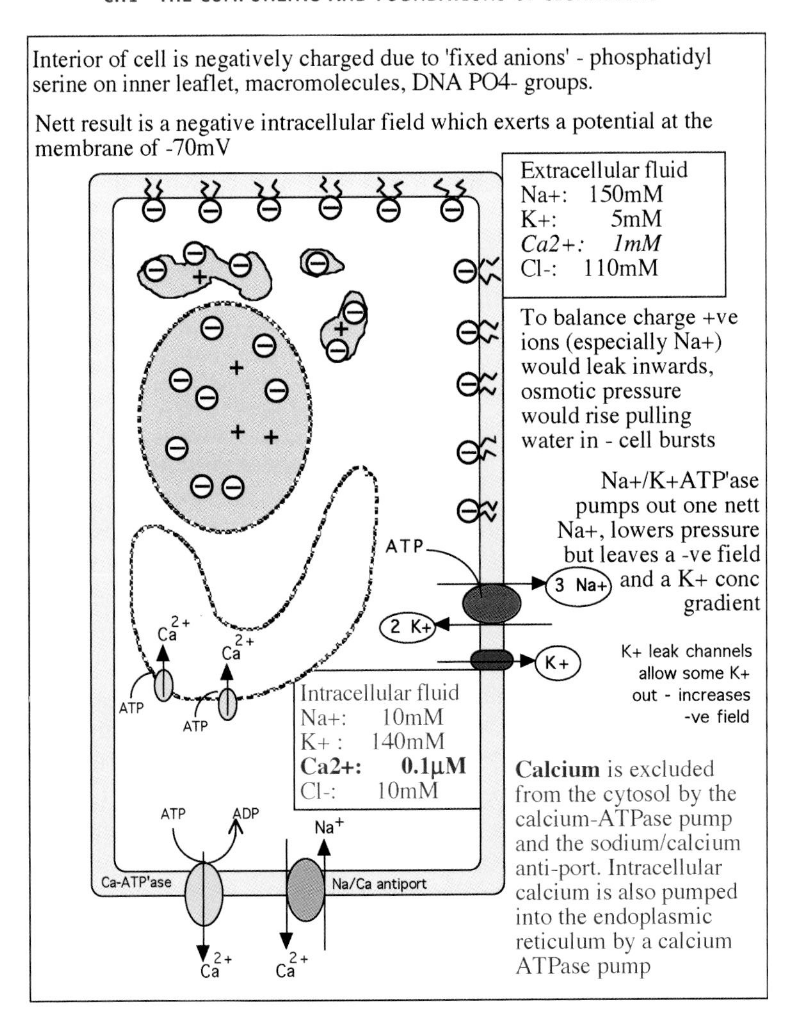

Figure 1.6 Summary of ionic homeostasis

Phospholipase C signals – diacylglycerol and IP3 Diacylglycerol (DAG) is the product of PIP2 hydrolysis by signal-activated phospholipase C (PLC). PIP2 distribution is asymmetric: it is only found on the inner leaflet of the plasma membrane. PLC cleavage of PIP2 yields two second messengers – the soluble hexitol IP3 and the membrane-bound DAG – which are produced simultaneously (Figure 1.7). There is a family of these signal-activated phospholipases, including phospholipase A and phospholipase D, each with a characteristic cleavage each site in PIP2.

PI-3-kinase signals – PIP3 Phosphatidyl inositol-3,4,5-trisphosphate (PIP3) is the product of the enzymic activity of an important family of signal-activate lipid kinases that also use PIP2 as a substrate. Like DAG, PIP3 is practically absent in the plasma membrane of resting cells.

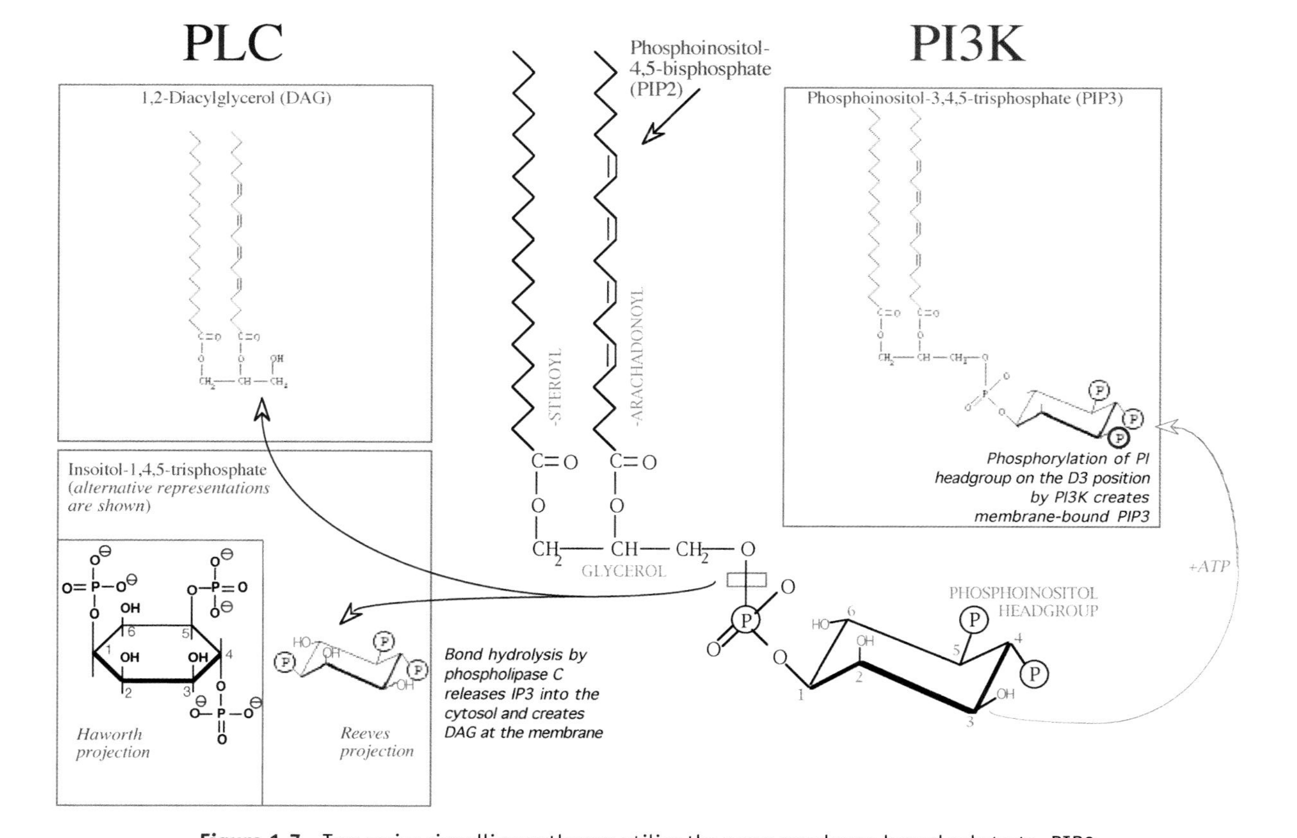

Figure 1.7 Two major signalling pathways utilise the same membrane bound substrate, PIP2

Phosphoinositide-3-kinase (PI-3-kinase) family members are downstream effectors of both single-pass and 7-pass receptors and all produce the activating 3-phosphorylation of the PI headgroup.

The PI-3-kinase family is divided into three classes[1]. Class 1 PI-3-kinases use PIP2 as a substrate and produce PIP3 in response to signals from single-pass receptor tyrosine kinases or heterotrimeric G protein β/γ subunits.

Class 1A are heterodimers consisting of a large, 110 kDa, catalytic subunit (p100α, p110β or p110δ) and a smaller regulatory subunit that processes the upstream signal. The catalytic subunits possess Ras-binding activity. Class 1A regulatory subunits, of which several splice variants exist, are of two basic types: p85 and p55. Both types contain a tandem pair of SH2 domains on either side of an interaction domain that binds the catalytic subunit. The smaller p55 subunits have lost an *N*-terminal regulatory region present in p85 that consists of an SH3 domain, polyproline segments and a domain resembling a Rho/Rac-GAP (see Chapter 3).

Class 1B PI-3-kinases are typified by the p110γ catalytic subunit. This class of PI-3-kinases do not interact with SH-2 containing adaptors but instead have a unique binding partner (p101). Class 1B types are activated by free β/γ subunits released from activated G proteins and it is the p101 regulatory subunit that mediates this activation[2].

Like the Class 1A p110α catalytic subunits, p110γ also has a Ras-binding domain and it has been shown that PI-3-kinase may act as an effector of Ras. Binding of activated Ras<GTP> increases PI-3kinase activity by 8 to 20-fold[2].

Class II PI-3-kinases do not use PIP2 as a substrate, but instead use phosphatidyl inositol-4-phosphate. They are >200 kDa and contain a calcium-binding C2 domain. Class III types use phosphatidyl insositol only.

1.2 Historical foundations

Fortune favours the prepared mind

Louis Pasteur

I include the following brief notes on early experimental milestones for three reasons. First, early serendipitous findings led to protocols still in use today and it is worth knowing where some of these odd reagents and methods came from. Second, these founding episodes illustrate the passion, persistence, good and bad luck, controversy and sheer bloody-mindedness that was (and still is) part-and-parcel of the field. Third, the way that hypotheses were set up, and then knocked down on the road to clarity, is a good exemplar of the scientific method. Even after a century, some important areas of signalling are still mysterious and many paradoxes remain unresolved.

1.2.1 When did the discipline of cell signalling begin?

One could argue that the discipline was born out of the furious scientific debates ignited by the work of the Russian physiologist, Ivan Petrovich Pavlov, and his discovery of the 'conditioned reflex'. Pavlov was the first to reveal the workings of the autonomic nervous system in his famous demonstration that dogs, conditioned to expect food at the sound of a bell, would salivate simply at the sound, even without food being presented. He predicted that the autonomic nervous system would be found to be responsible for all aspects of control of digestion: In other words, (unconscious) neuro-electrical control solely governed gastrointestinal secretions, in particular initiation of exocrine pancreatic secretion after a meal, which he had shown was dependent upon vagal nerve stimulation.

1.2.2 The discovery of 'hormones' – Bayliss and Starling, 1902

In a lab in University College London, on the afternoon of Wednesday, 16th January 1902, the English physician Ernest Henry Starling and his physiologist colleague William Maddock Bayliss proved Pavlov wrong and in the process discovered *secretin*, the first known signalling molecule. Having dissected all nervous tissue from the exposed jejunum of an anaesthetised dog, leaving only connected blood vessels, they found that merely stimulating the denervated jejunum with dilute hydrochloric acid prompted pancreatic secretion even though there was no neuronal connection between the two tissues. Pavlov's theory held this to be impossible.

At this discovery, Starling declared it to be a 'chemical reflex' and immediately improvised a protocol that proved the signal from the intestine had arrived at the pancreas through the blood, rather than the nervous system. Starling excised a section of the jejunum, treated it with HCl, filtered the resulting mucosal secretion and injected the filtrate into the jugular vein of the anaesthetised dog. A friend, Charles J. Martin, who 'happened to be present' recounts that 'after a few moments the pancreas responded by a much greater secretion than had occurred before'. Our witness further remarked that 'it was a great afternoon'.

In 1902, Bayliss and Starling named the jejunal factor 'secretin' and reasoned correctly that it exists, stored in the jejunal mucosa, as an inactive prohormone: *prosecretin*. In 1905, Starling introduced the term 'hormone' (from the Greek *hormao*, meaning 'arousing, initiating or exciting agent') to describe blood-borne substances (chemical first messengers) that initiate such 'chemical reflexes'.

This novel hypothesis could not be accepted without experimental replication. But, in Russia, Starling's chemical stimulation experiment did not work and in London, Pavlov's vagal nerve stimulation could not be reproduced. After some contention, honour was satisfied – experimental protocols were at fault. Neuronal stimulation could not be demonstrated in London because the dog was treated with morphine prior to anaesthesia, whereas Pavlov's group discovered that they had destroyed the secretin by 'over-neutralising' the HCl extract. Both were correct. Neuronal and hormonal control of gastrointestinal secretions are alternative, complimentary systems. Thus began the discipline of 'endocrinology'[3].

Bayliss and Starling are generally credited with the discovery of the first hormone, but the 'suprarenal gland' hormone, *adrenaline*, was actually discovered earlier (in 1901) and its total synthesis achieved in 1904[3]. Starling appeared unaware of this at the time of his lectures of 1905. The 'suprarenal' gland (meaning *above the kidney*) is now known as the adrenal gland.

In the following decades, many more hormones were discovered but most were peptides whose structures would not be elucidated until the 1950s. Identification of their receptors would have to wait until the 1970s.

1.2.3 The discovery of insulin and the beginning of endocrine therapy – Banting and Best, 1921

Insulin was discovered and first purified by Frederick Grant Banting and Charles Best in the labs of Dr. J.J.R. MacLeod at the University of Toronto in 1921, and later shown to rescue patients with diabetes in early 1922. Much later, in 1951, insulin was the first polypeptide to be sequenced (by Fred Sanger and Hans Tuppy) and, later still, the first to be synthesised in the lab.

1.2.4 Peptide sequencing – Fred Sanger, 1951

It is impossible to exaggerate Fred Sanger's importance in modern biochemistry. Although Sanger was not directly involved in signalling or molecular genetics, these fields of research would not exist without the sequencing techniques that Sanger developed; first peptide sequencing in the 1950s and later oligonucleotide sequencing in the 1970s. In recognition of these revolutionary pieces of work, he was awarded two Nobel Prizes.

Sanger did more than provide the world with sequencing techniques. He also dispelled the prevailing notions of what made up proteins. At the beginning of the 1950s, proteins were considered to be mixtures of substances lacking unique structure, whose amino acid compositions were not directly encoded, but rather 'managed' by genes. However, clues that specific amino acids had singular roles to play in proteins were emerging, in particular the finding that a single amino acid substitution in haemoglobin could give rise to sickle cell anaemia. Sanger chose insulin as a target for his first foray into protein sequencing because it was medically important and was available in large quantities and high purity[4]. His technique depended on *N*-terminal labelling with fluorodinitrobenzene (FDNB), which results in covalent addition of dinitrophenyl (DNP) groups to any free amino groups. DNP-labelled peptides are yellow and can be easily identified after separation by chromatography. Proteolytic cleavage combined with partial acid hydrolysis of the peptide (which more-or-less randomly cleaves peptide bonds, yielding cleavage products with new amino termini) and specific proteolytic cleavage, the peptide sequence could be reconstructed from the amino acid compositions of overlapping sequences. Here again, a spin-off was the first proof that the proteases Sanger used (trypsin and chymotrypsin, primarily) were both irreversible and truly specific: *trypsin* cleaving after basic amino acids only (lysine, arginine), and *chymotrypsin* cleaving after aromatic amino acids only (tyrosine and phenylalanine).

Although Sanger's peptide sequencing was largely replaced by automated Edman *N*-terminal degradation – only possible by the later development of high performance liquid chromatography – trypsin and chymotrypsin are still used to fragment proteins for identification by mass spectrometry. Information available from the various genome projects allows one to identify the trypsin cleavage sites in all known proteins. Each protein will yield predictable and unique trypsin cleavage fragments with unique molecular weights and such patterns can be used to identify the protein (as long as post-translational modifications, such as phosphorylation, are taken into account).

1.2.5 Discrimination of beta- and alpha-adrenergic responses – Ahlquist, 1948

Although adrenaline (also known as 'epinephrine') was long known to have neurotransmitter and hormone-like activities, its diversity of functions was not properly understood until the work of Raymond Ahlquist[5]. Before this seminal work (sadly ignored for five years after its publication), adrenaline was thought to act through either 'excitatory' or 'inhibitory' receptors. By careful analysis of tissue-specific responses to adrenaline, and related agonists and antagonists, Ahlquist confirmed that two distinct receptors existed, but realised that the division into 'excitatory' or 'inhibitory' was too simplistic.

He gave the name 'beta-adrenotrophic' to the receptor mediating adrenaline's 'inhibitory' effects on blood vessels (vasodilation) and uterus (relaxation) as well as its 'excitatory' effect on the heart (myocardial stimulation). The 'alpha-adrenotrophic' receptors, then, were those responsible for 'excitatory' effects on blood vessels (vasocontraction) and uterus (muscular contraction) as well as the 'inhibitory' effect on intestine (relaxation).

Alquist's proof lay in the fact that the alpha responses to adrenaline (both 'excitatory' and 'inhibitory') could be blocked by addition of the antagonist ergotoxine. Yet this alpha-blocking agent had no effect upon beta type responses. Furthermore, the rank order of potency of the known agonists differed between the two types.

Although such receptors remained 'hypothetical structures or systems' until the 1970s, the pharmacological distinction between alpha and beta responses greatly assisted the development of adrenergic-targeted drugs such as 'beta-blockers' (such as propanolol) and laid the groundwork for the later subdivision of adrenergic receptors into the multiple forms known today.

1.2.6 'Acrasin' = cAMP – the ancient hunger signal

In 1947, J.T. Bonner discovered 'acrasin', a name he gave to a diffusable soluble chemical that caused the usually solitary amoeba of the slime mould *Dictyostelium discoideum* to swarm together to form a temporary multi-cellular slug in times of 'famine'. Some 20 years later, it was realised that 'acrasin' was in fact cAMP[6]. Social amoebae such as *Dictyostelium* are the only organisms known to secrete cAMP deliberately, using it as both extracellular first messenger as well as intracellular second messenger[7]. In all other organisms, cAMP is locked within the cell that produces it and acts solely as an intracellular second messenger, only escaping as a result of cellular damage.

1.3 Early milestones in signal transduction research

1.3.1 Cell-free experiments and the discovery of cAMP – Sutherland, Rall and Berthet, 1957

Arguably, the most important early milestone in signal transduction was the discovery by Earl Sutherland of the first 'second messenger': the cyclic nucleotide cAMP[8]. Almost as important, however, was the group's success in working with homogenised cells and purified subcellular fractions, thus disproving a prevailing consensus that cells must be intact for hormonal action to be manifest.

It is important to realise that in the early 1950s the cell was thought of as a mechanism too complex to be disassembled. To quote Sutherland:

> When I first entered the study of hormone action, some 25 years ago, there was a widespread feeling among biologists that hormone action could not be studied meaningfully in the absence of organised cell structure.[9]

Sutherland, and colleagues Jacques Berthet and Theodore Rall, began a series of experiments in November of 1955 aimed at understanding how the hormones adrenaline and glucagon activate the liver enzyme, glycogen phosphorylase[10]. They were favoured by complementing expertise and a measure of good luck. As Theodore Rall comments, 'I have lost count of how many wrong ideas got us to do the right experiments'. Perhaps the most significant happenstance was that Earl Sutherland's laboratory used dogs, rather than rats, as a source of tissues. This was fortunate because his experiments might not have worked otherwise, but unfortunate because most other labs used liver from young male rats.

Their work was greatly simplified by the fact that cAMP is heat-stable. Their final piece of good luck was that the ATP they used was impure – it contained small amounts of GTP. Without this contaminant, their experiments would not have worked, for reasons that would not become clear for another 20 years.

1.3.2 Fluoride – a stimulator of G proteins

By the early 1950s it was known that treating slices of dog liver with adrenaline caused glycogen to be broken down into glucose because the enzyme *glycogen phosphorylase* (GP) was somehow activated. Glucagon or adrenaline could activate GP in liver slices, but not in homogenised liver. Perhaps something liberated by homogenisation was inhibiting the reaction . . . .

Sutherland was able to isolate activated GP from liver slices and show that it could be inactivated by an enzyme that later proved to be a *serine/threonine protein phosphatase*. The phosphatase was found to be inhibited by the *fluoride* ion and so this was routinely added to their incubation mixtures in the expectation that this would allow the GP activation to proceed in homogenates. Sensibly, they always included fluoride-free controls[9]. Had they not done so, the hormonal effects would actually have been masked, because fluoride is capable of *directly* stimulating the G proteins involved in adrenaline and glucagon signalling. This is discussed further in Section 1.4.4.

1.3.3 ATP and subcellular fractionation

Taking a cue from the work of Krebs and Fischer, Sutherland's team were able to restore hormonal sensitivity to homogenates by adding ATP and Mg^{2+}. This suggested that the system might be amenable to dissection. Initial experiments on centrifuged liver homogenate cells suggested that the whole process of adrenaline activated glycogen breakdown occurred in the supernatant (i.e., the *cytosol*). However, when more stringently separated cytosolic fractions were prepared, activity was lost. But when a small portion of the particulate fraction (i.e., the spun-down membrane pellet) was added back, adrenaline responsiveness returned[11].

1.3.4 Heat-stable factor – cAMP

Sutherland and co-workers then found that they could do the experiment in two stages (Figure 1.8). The purified membrane pellet was separated (from the cytosol) and incubated in buffer containing adrenaline and ATP. It could be boiled yet would still retain the ability to activate glycogenolysis when added back to the reserved cytosol. The 'boiled stuff' contained something that stimulated glycogen breakdown (and glycogen phosphorylase activation) in the cytosol. They knew that the active substance was heat-stable and therefore not a protein. They also realised it was not adrenaline or ATP, because neither could stimulate glycogen turnover in purified cytosol.

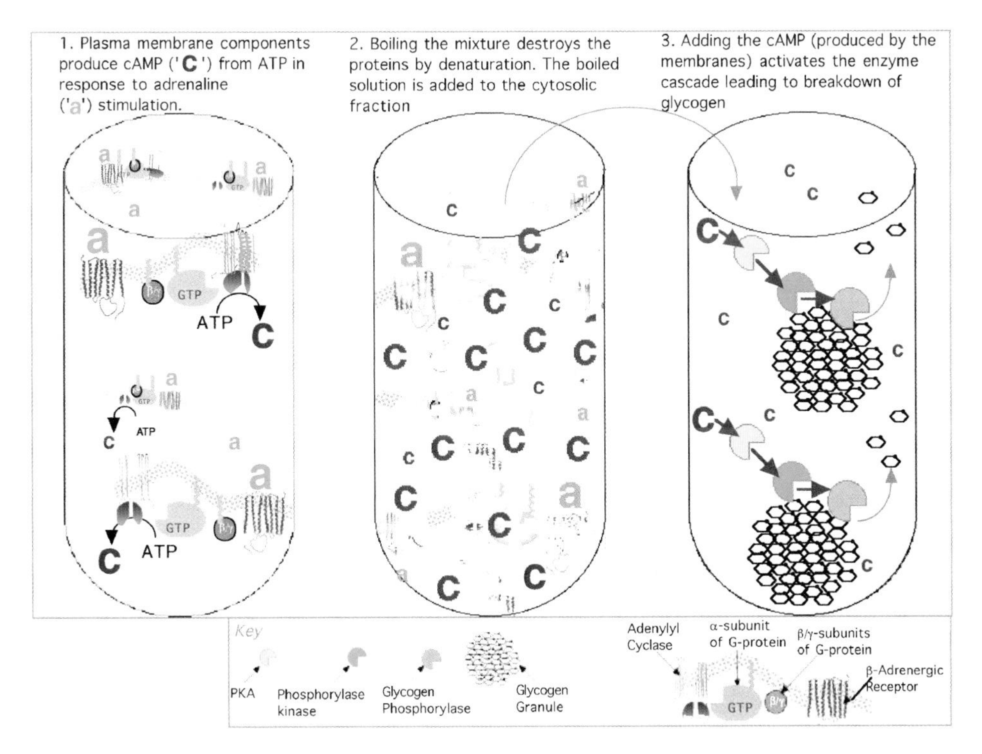

Figure 1.8 Earl Sutherland's crucial experiment

It was not possible to interpret these experiments fully at first because receptors were not understood, protein kinases were unknown and G proteins undreamt-of. However, we know now that the 'particulate' fraction contains the plasma membranes with their embedded adrenaline receptors, adenylyl cyclase and the stimulatory G protein, Gs. Incubating these membranes with adrenaline and ATP stimulates the first steps in the transduction pathway leading to the production of cAMP. The purified cytosol has no receptors, but instead contains the soluble components of the pathway: cAMP-dependent protein kinase (PKA), phosphorylase kinase (PhK), glycogen phosphorylase and the glycogen granules. Adding the cAMP (produced by the receptor-G protein-adenylyl cyclase coupling that occurred during the membrane incubation) to the cytosol activates PKA, and then PhK and finally glycogen phosphorylase itself.

The hunt was now on to find the structure of this heat-stable second messenger. It took little time to purify, crystallise, and then solve the structure of adenosine 3′, 5′-cyclic monophosphate (Figure 1.2). Unfortunately the hypothesis that this cyclic nucleotide was a universal hormonal mediator met with initial scepticism when the findings were announced.

1.3.5 The problem with rats

One serious objection (that persisted right up to the award of Sutherland's Nobel laureate) was that many labs simply could not demonstrate adrenaline-stimulated, cAMP-mediated, activation of PhK. Unfortunately, these labs were all using the most common of lab animals, rats. And they were always young males (females and older males being used primarily for breeding). It has since become clear where the problems lay: young male rat liver contains no responsive beta-type adrenaline receptors, instead adrenaline acts solely through hepatic alpha-1 receptors that stimulate PhK through calcium mobilisation alone. During their lifetimes, however, all rats do have hepatic receptors that can activate PhK via cAMP generation – female rats express both alpha and beta-adrenergic receptors, as do males that are either very young or post-mature[12] (see also Chapter 5, Section 5.7.2).

1.3.6 The discovery of hormonally regulated protein kinases – phosphorylase kinase, serine phosphorylation and Ca^{2+} – Krebs and Fischer, 1958–1968

Serendipity also played a role in the next step in unravelling how glycogen breakdown can be initiated by either hormonal (adrenaline, glucagon) or electrical (neuronal) stimulation. Strangely, filter paper had a large part to play...

In the 1950s, Edwin Krebs and Ed Fischer were working in the laboratories of Carl and Gerty Cori, assisting in purification of glycogen phosphorylase. The Coris were interested in the mechanism by which muscle glycogen phosphorylase *b* (the inactive form) could be converted to glycogen phosphorylase *a* (the active from) as a result of neuronal stimulation. All that was known at the time was that supra-physiological concentrations of 5′-AMP could activate the enzyme – a finding of Krebs that remained unpublished for some time – but this allosteric modulation by AMP was not considered likely to explain activation *in vivo*. Krebs and Fischer were tasked with crystallising the *a*-form of the enzyme using a reliable purification scheme worked out by the Coris.They were unsuccessful. Their extract contained only impure *b*-form that refused to crystallise. They soon realised that they had not followed the Coris' protocol exactly: instead of 'clarifying' the muscle homogenate by filtration through filter paper, they had centrifuged it.

When they repeated the Coris' procedure exactly (with filtration through paper instead of centrifugation), they obtained active phosphorylase-*a*, which crystallised readily. A few more experiments provided two more clues. First, leaving the homogenate sitting around for too long ('ageing') prevented the conversion of *b* to *a*, even using the filtration method. Second, if the filter paper was stringently washed prior to use, the filtration method did not cause conversion even when fresh muscle homogenate was used[13].

1.3.7 Discovery of calcium as activator of phosphorylase kinase

Since the centrifugation step yielded only *b*-form, it was concluded that muscle contained predominantly inactive glycogen phosphorylase-*b*, and that the filter paper (in the Coris' protocol) had artefactually activated it. The activating contaminant (which could be washed out of the filter paper) turned out to be *calcium* and the essential component destroyed by ageing the homogenate proved to be *ATP*.

Krebs and Fischer identified a 'converting enzyme' that could be separated from glycogen phosphorylase and discovered that this enzyme, when provided with ATP and calcium, could convert phosphorylase-*b* to the *a*- form. ATP, at the same time, was converted to ADP. It was quickly concluded that this was a phosphotransferase step, glycogen phosphorylase being activated by the covalent addition of a phosphoryl group. The 'converting enzyme' was dubbed *phosphorylase kinase*, the first protein kinase discovered. They also identified the serine residue of GP whose phosphorylation led to the activation.

Phosphorylase kinase (PhK) uses ATP as a co-substrate to phosphorylate, and activate, its protein substrate GP. Further, Krebs and Fischer had established that PhK could itself exist in inactive and active forms, interconverted by loss or addition of calcium. This explained how neuronal stimulation induced glycogen breakdown in skeletal muscle – through release of calcium.

1.3.8 cAMP-dependent protein kinase

Further work in Kreb's lab established that highly purified PhK could be alternatively activated by a combination of ATP and cAMP. The activation of PhK was this time due to its serine phosphorylation, and at first this alternative activation mechanism was thought to be due to PhK phosphorylating itself (i.e., autophosphorylation). This turned out to be incorrect. Instead a minor protein contaminant was responsible. This 'phosphorylase kinase *kinase*' proved to be a distinct protein kinase that was activated by cAMP and was capable of phosphorylating other proteins as well as PhK. It was termed 'cAMP-dependent protein kinase' and is almost universally referred to as *protein kinase A* or PKA.

1.4 The discovery of receptors and G proteins

1.4.1 Radioligand receptor assays prove receptors are discrete entities

As a philosophical concept, the 'receptor' has a long history, but in the first half of the 20th century there was no expectation that receptors might turn out to be simple proteins. Rather, the term was shorthand for a cellular 'mechanism' that produced a biological endpoint after hormonal stimulus. By the late 1960s and early 1970s, receptor binding assays were being performed with radiolabelled hormones on cells and extracts and this allowed the *Scatchard* equation and other mathematical modelling (see Chapter 2) to be used to quantify receptors directly, rather than relying upon indirect assays of down-stream effects (such as phosphorylase activation and glycogenolysis). But even as late as 1973, the prominent pharmacologist Ahlquist, who developed the concept of separate alpha and beta adrenergic receptors, stated: 'To me they are an abstract concept conceived to explain observed responses of tissues produced by chemicals of various structure...' (see Reference 14).

1.4.2 Oestrogen receptor directly detected by radioligand binding assays – Jensen and Gorski, 1962

The first receptor to be extracted and assayed by radioligand binding assays was the oestrogen receptor in 1962. This work of Toft and Gorski followed ground-breaking research by Elwood Jensen who first used radiolabelled oestradiol to follow the fate of the hormone in rats, noting that it accumulated in target tissues (uterus and vagina) but not in non-target tissues (muscle, kidney, liver), and that it was *chemically unaltered*. This disproved the notion that oestrogens were metabolised to somehow provide energy for the biological response provoked (i.e., growth stimulation of breast cancer)[15].

Jenson's work also provided the first indications that receptors may be simple proteins. Sucrose density ultracentrifugation analysis gradients of [^{3}H]-oestradiol-bound oestrogen receptors showed a discrete '8S' band – 'S' stands for Svedberg units, an indirect measure of molecular mass, named after the Swedish chemist and Nobel Laureate, Theodor Svedberg.

1.4.3 Purification of the β-adrenergic receptor – Caron and Lefkowitz, 1976

In 1970, Lefkowitz was the first to use a radioactive ligand to label and assay a *membrane* receptor directly. The ligand was radio-iodinated adrenocorticotrophic hormone (ACTH) and the receptor was from adrenal gland membranes[16]. Iodination became a standard way of labelling peptides and proteins, whereas small ligands such as catecholamines and steroids are usually tritiated. In 1976, the β-adrenergic receptor was first purified to homogeneity and was shown to exhibit stereospecific binding to agonists and antagonists[17]. The final proof that the receptor, G protein and adenylyl cyclase were separate proteins came with the demonstration that a fully functional adrenergic-activated adenylyl cyclase could be constructed by reconstitution of the three isolated proteins into artificial phospholipid vesicles[18].

1.4.4 The discovery of G proteins. Guanine nucleotides, fluoride and aluminium – Gilman and Rodbell, 1971–1983

The magical 'contaminant' in Sutherland's ATP was re-visited after previously reliable experiments ceased to work when pure ATP analogue was used. At the same time, an important discovery was made because glass test tubes were substituted with plastic ones...

Alfred G. Gilman and Martin Rodbell worked independently; Gilman in Sutherland's old lab at Case Western Reserve University (by this stage run by Theodore Rall) and Rodbell at NIH, Bethesda.

Rodbell's group found the first clue that a 'transducer' may couple multiple receptors to the single effector enzyme, adenylyl cyclase (AC)[19]. Rodbell had shown that multiple hormones (including adrenaline, ACTH, TSH, LH, secretin, and glucagon) could all activate AC in fat cells. His key insight was that they were not additive in effect when applied in combinations. This argued against Sutherland's guess that each receptor had its own individual AC activity and the assumption that the receptor and AC may actually be a single entity (Figure 1.9).

1.4.5 Magnesium

Sutherland had earlier shown that the fluoride ion could stimulate AC independently of hormones[20] and Rodbell found that both ACTH and fluoride were able to stimulate AC in fat cell membranes in a MgATP-dependent manner. However, fluoride stimulation of AC exhibited Mg^{2+}-dependence with a Hill coefficient of 2 (not 1 as expected), suggesting two sites for Mg^{2+} action. One site must be AC (which used MgATP to produce cAMP), so what might the other be?

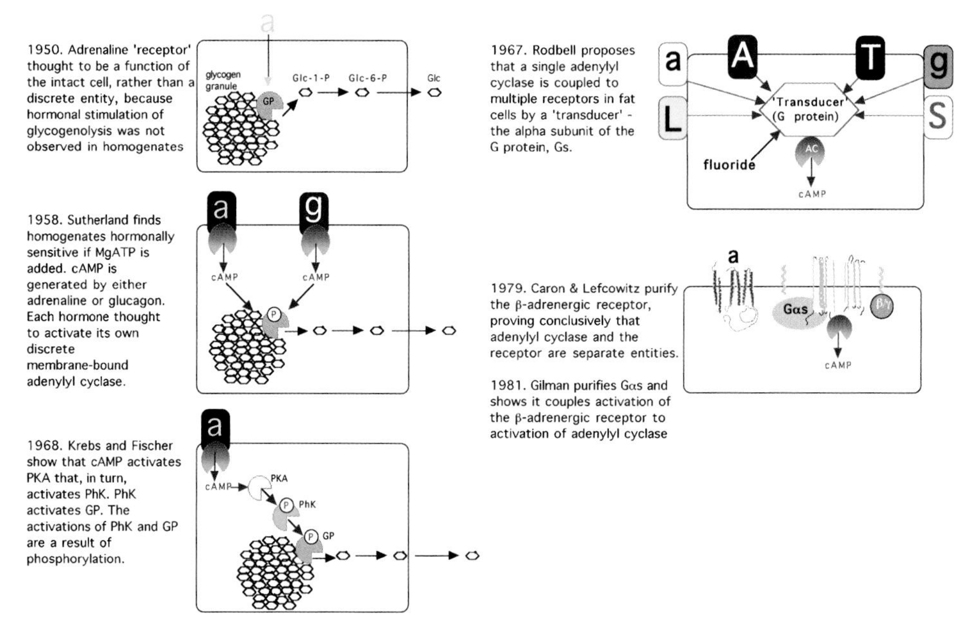

Figure 1.9 The evolution of ideas of how hormones work

1.4.6 High and low glucagon affinities

Rodbell's group found that direct binding of ^{125}I-glucagon to liver membranes did not match the kinetics of AC activation. In AC assays, glucagon raised cAMP levels within seconds of addition to liver membranes, and was quickly reversed when the hormone was removed. In binding assays, however, ^{125}I-glucagon remained bound to the membranes even after removal of free label. The label could not be displaced even with extensive washing. The problem lay in the protocols: the ^{125}I-glucagon binding assay was carried out in simple buffer, whereas the AC assay contained MgATP. Adding MgATP to the ^{125}I-glucagon assay produced a dramatic change in binding: ^{125}I-glucagon binding was easily and quickly reversed after the simple removal of ^{125}I-glucagon.

1.4.7 GTP (contaminant of ATP) lowers 7-pass receptor affinity

Rodbell knew that commercial ATP preparations were contaminated with other nucleotides. After testing a variety of candidates, he found that very low concentrations of GTP mimicked his contaminated ATP – the presence of GTP effectively lowered the affinity of the receptor for glucagon ($^{+}$GTP → fast off-rate; $^{-}$GTP → slow off-rate).

Conclusively, a pure synthetic ATP analogue (App(NH)p) did not support AC activation by glucagon unless GTP was also added[21].

Lefkowitz's group confirmed this affinity shift in adrenaline receptors and further showed that the affinity-lowering effects of GTP applied only to adrenergic receptor *agonists*; the affinity of *antagonists* was the same whether GTP was present or not[22]. It is worth noting that in both studies, GTP and GDP were equipotent in causing agonist affinity shifts. Nevertheless, such affinity-lowering effects on G protein coupled receptors are much more pronounced if non-hydrolysable GTP analogues are used, such as: guanyl-5′-imidodiphosphate (Gpp(NH)p) or guanosine-5′-*O*-(3-thiotriphosphate) (GTPγS).

As we shall see, this affinity-lowering effect of GTP analogues towards 7-pass receptors and their agonists is a general phenomenon that came to be used to identify such G protein coupled receptors *in vitro* and to predict agonist versus antagonist behaviour *in vivo*. It is important to note that 'agonist affinity-shift' is only observable in isolated plasma membranes and is not found in whole cell receptor assays, nor in solubilised receptor preparations. Frustratingly, the mechanism behind the agonist affinity-lowering effects of guanine nucleotides is still a matter of controversy (Chapter 2).

1.4.8 GTP analogues and adenylyl cyclase activation

Artefactual AC-stimulating effects of GDP were encountered in some early studies. This proved to be due to the AC assay conditions[23], which included ATP-regenerating systems to maintain an effective level of ATP (AC substrate) against a high background of nucleoside triphosphatase activity. Unfortunately this had the unintentional side effect of converting GDP into GTP.

Significantly, there is evidence that something like this can occur in nature. G protein-bound GDP can be converted to GTP by the action of nucleoside diphosphate kinase (NDPK), an enzyme that associates with G protein β/γ subunits. Unusually, NDPK phosphorylates the β subunit on a histidine residue and this high-energy phosphate group is subsequently transferred to GDP, providing a receptor-independent means of activating G proteins. This unusual activation pathway plays a part in the aetiology of congestive heart failure[24].

The mechanism of G protein activation was clarified when adenylyl cyclase assays were performed using 5′-adenylylimidodiphosphate (App(NH)p) as a substrate instead of the ATP regenerating systems. Salomon and Rodbell clearly showed that, in these clean assay conditions, only GTP or Gpp(NH)p could support glucagon stimulation of AC activity. GDP had no effect. In the presence of Gpp(NH)p, the effects of glucagon were much longer lasting than with GTP. Further work showed that GTP activation of AC was reversible because the nucleotide is hydrolysed by the G protein, which then reverts to the inactive GDP-bound state[25]. Gpp(NH)p and GTPγS are both non-hydrolysable and so produce persistent AC-activating effects that resemble the effects of cholera toxin (see below).

1.4.9 cAMP toxicity and clonal mutants of S49 cells

The next stage in the isolation of G proteins is rather confusing, if told as it unfolded, so I intend to spare you the brain-churning interpretations that Gilman's team must have gone through.

The key tool was the lymphoma cell line, 'S49', which is killed when excess cytosolic cAMP is generated. Wild-type S49 cells contain both β-adrenergic receptors and a functional adenylyl cyclase system. Chronic adrenaline treatment thus selects mutants that do not make cAMP in response to adrenergic stimulation. Two clones were isolated that were resistant to killing by adrenaline: *cyc*$^-$ and *unc* mutants. Both cell lines express adrenergic receptors but are nevertheless resistant to adrenaline-induced cell death.

Since *cyc*$^-$ cells could not produce cAMP in response to fluoride ion, it was assumed that they simply lacked AC[26]. In a failed attempt to restore activity, AC was detergent-extracted from AC+ve cells then incubated with *cyc*$^-$ cell preparations. With hindsight, we can see that this would never have worked because AC is an integral membrane protein and cannot re-integrate into a native plasma membrane. The same can be said for the receptor. The best one can do is construct an artificial membrane vesicle around the protein using pure phospholipids.

Eventually, it was found that adrenaline-responsive AC activity could be restored to *cyc*$^-$ plasma membranes by adding extracts from wild type S49 or the B82 cell line (that was capable of cAMP production but lacked β-adrenergic receptors). B82 plasma membranes were dissolved in detergent then mildly heat-treated (37°C, 30 minutes) to destroy AC activity. When the heat-inactivated membrane extracts were incubated with *cyc*$^-$ membranes, activation in response to both adrenergic agonist and fluoride cells was restored. Whatever the restorative factor was, it could not be adenylyl cyclase, and it was obviously not β-adrenergic receptor. As it transpired, the heat-resistant factor was the alpha subunit (Gαs) of the stimulatory G protein Gs, which is the only component of the AC system missing from *cyc*$^-$ cells (it had first been thought that AC was absent).

This works because the G protein is a peripheral membrane protein, only associating through a covalently bonded fatty acid chain that (unlike an integral protein's trans-membrane helix) can easily re-integrate into native membranes.

The S49 *unc* mutant is interesting from our point of view because, although able to produce cAMP when stimulated by fluoride, it is unresponsive to adrenaline, even though the cell line possesses β-adrenergic receptors. The *unc* cell line *does* express Gαs but the protein is defective because it contains a point mutation near the *C*-terminus. An arginine at Gαs position 389 (6 amino acids from *C*-terminus; number includes start methionine, see Table 9.2) is substituted with a proline. This Arg → Pro substitution prevents receptor coupling, a finding we shall come back to in later chapters.

The guanine nucleotide responsive factor, Gs, was purified to homogeneity by Gilman's group in 1980[27]. Reconstituted into phospholipid vesicles, the purified protein was able to mediate the activation of AC activity by fluoride and it was found that the protein dissociated into α- and β/γ-subunits after binding GTP (with stoichiometry of one GTP-to-one α-subunit. Incidentally, they had also unwittingly purified the

AC-inhibitory G protein, Gi, without immediately realising it (see Reference 28) – in fact, Gi-containing fractions from their gel filtration columns languished in a freezer for some time before finally being identified by labelling with pertussis toxin (discussed later).

1.4.10 Aluminium is needed for fluoride activation of G proteins

The final piece of serendipity did not reveal its full importance for another ten years. Here is an extract from Gilman's Nobel speech:

> Additional work on the mechanism of activation of Gs by fluoride provided surprises and even amusement. The effect of fluoride, observable when experiments were performed in glass test tubes ... was lost ... when experiments were done in plastic test tubes.
>
> Paul Sternweis ... purified the coactivator from ... aqueous extracts of disposable glass test tubes. A metal seemed to be involved, and neutron-activation analysis revealed Al^{3+} as the culprit.

So fluoride ion is not enough, the real inorganic stimulator of Gs-proteins is aluminium fluoride.

1.4.11 Use of bacterial toxins

At this point it is worth noting that the work of Rodbell and Gilman was paralleled and greatly assisted by the findings of many others: pre-eminently Pfeuffer and Cassel, who respectively demonstrated that GTP binding[29] and hydrolysis[30] accompanied adenylyl cyclase activation.

Gilman group's final achievement of Gs purification was by gel filtration with monitoring for AC activation in the eluted fractions, but the scheme depended partly upon labelling the Gαs subunit with a bacterial toxin excreted by *Vibrio cholerae*.

In 1978, Cassel and Pfeuffer[31], and simultaneously Gill and Meren[32], first produced evidence of a distinct guanine nucleotide binding protein by the use of cholera toxin, which (as they correctly guessed) is an ADP-ribosylating enzyme. The use of the toxin was also prompted by a number of observations including their findings that (a) cholera toxin could activate AC while at the same time inhibiting a GTPase activity downstream of adrenaline binding and (b) a GTP-binding protein was indispensable for the fluoride activation of AC[33].

Cholera toxin (CTX) is an enzyme that specifically transfers ADP-ribose from NAD^+ to the αs subunit of Gs – a reaction known as 'ADP-ribosylation' – and this was used by Cassel and Pfeuffer to radiolabel the alpha subunit of the AC-stimulating G protein. Radioactivity from the substrate ([^{32}P]-labelled NAD^+) is transferred to an arginine of Gαs by the catalytic activity of CTX. Thus labelled, it was easy to estimate the G protein's molecular weight following electrophoresis, blotting and autoradiography.

This provided a simple and reliable means to identify Gαs and, in the longer term, led to cholera and pertussis toxin (PTX) labelling being used as a generic method for discriminating between AC-stimulating, and AC-inhibiting G proteins in widely different species. Perhaps more importantly, investigations into the effects of the two toxins led to new insights into G protein mechanisms. Both toxins are exquisitely specific: CTX only labels a single arginine found in an equivalent position in all Gαs subunits (Arg201 of Gαs), whereas PTX labels a single *C*-terminal cysteine residue similarly conserved in all Gαi subunits. The effects of both modifications and their relevance to molecular details of Gα coupling mechanisms are discussed in Chapter 8.

Pertussis toxin is also an ADP-ribosylating enzyme but it is specific for the alpha subunit of the AC-inhibiting G protein, Gαi. PTX was used by Gilman's group to radio-label, and hence identify and purify, the 41 kDa AC-inhibitory G protein from the side fractions kept from their earlier Gs purifications[34].

By 1986, Lefkowitz's group had purified the alpha-2 adrenergic receptor and a consensus model of opposing adrenergic mechanisms of AC regulation had emerged.

1.4.12 The calcium signal

The ability of adrenergic agonists to stimulate PhK in young male rat liver was not fully understood until the discovery by Michael Berridge of a third type of G protein in 1984[35]. The G protein was at first referred to as 'Np', later 'Nq' and finally was given the name 'Gq'. Activation of the AC-inhibitory Gi protein can, in some instances, also lead to calcium mobilisation but this is pertussis-toxin sensitive. In the case of Gq activation, the calcium release is insensitive to both cholera and pertussis.

1.5 cAMP pathways

1.5.1 A simple mammalian signalling pathway – F-2,6-bisP as a second messenger

Fructose-2,6-bisphosphate (F-2,6-bisP) is a side-product of glycolysis that is not used for energy production, but instead acts as a second messenger. It was discovered in 1980 and was soon shown to be a widespread metabolic indicator in eukaryotes, and under hormonal control in certain mammalian tissues.

The directional flux of glycolysis versus gluconeogenesis is subject to moment-to-moment control by energy-sensing allosteric enzymes but these local controls can be over-ridden by signals (hormones) sent from remote endocrine organs.

An understanding of how the two levels of control work is contained in a classical model of metabolic pools and control points in glycolysis (Figure 1.10):

- Many enzymatic conversions are easily reversible, others (usually those requiring energy [ATP] or releasing energy [ATP]) are not.

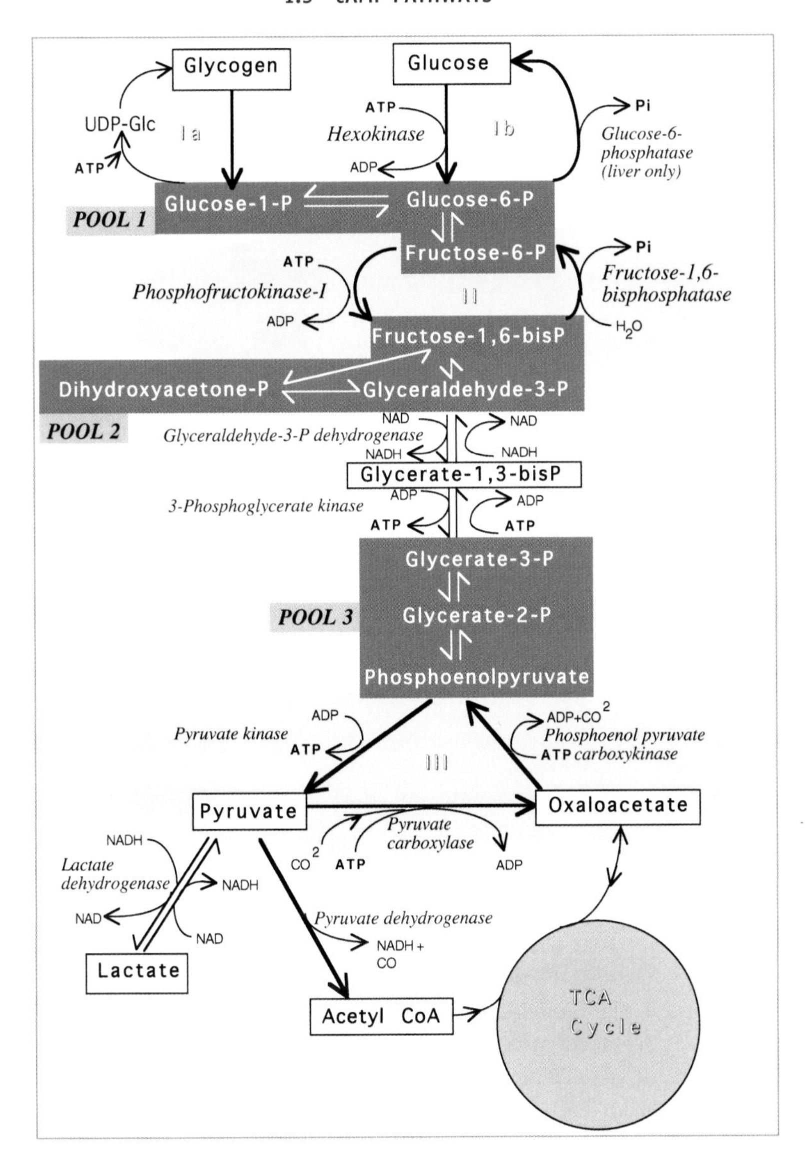

Figure 1.10 Metabolic control points in glycolysis & gluconeogenesis

- Reactions that cannot be reversed represent *control points*. This is where, for example, glycolysis uses one enzyme, but the reverse gluconeogenic step uses a different enzyme.

- Metabolites that can be interconverted by reversible reaction (using the same single enzyme and only responding to changes in substrate/product levels) are said to be in *metabolic pools*.

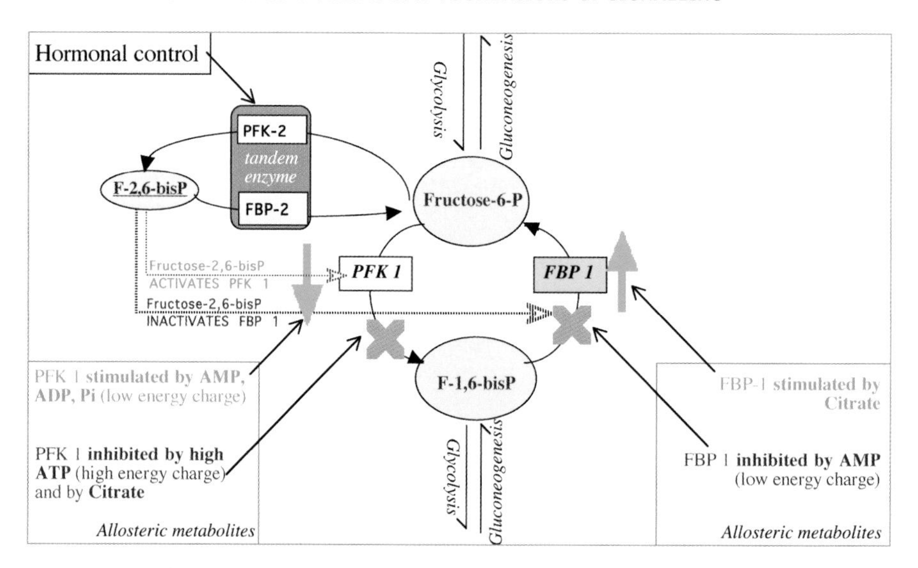

Figure 1.11 Allosteric and hormonal control of PFK-1 and FPB-1

- Control points between pools are theorised to be responsible for the overall direction of the metabolic flux.

- The important control point enzymes are *allosteric*, with active sites that can be modulated by molecules binding at non-overlapping sites ('allosteric sites') – substrate turnover may be either stimulated or inhibited (Figure 1.11).

Glycolysis should operate when energy and carbon skeleton supplies are limited – low [ATP] and [citrate]; high [ADP] and [AMP]. The pathway may be reversed, when energy supplies and carbon skeleton building blocks are high, so as to store energy in macromolecules (polysaccaharides such as glycogen, triglycerides, proteins). The 'anabolic' direction of flux is then gluconeogenic. Gluconeogenesis should only operate when energy supplies and carbon skeleton supplies are elevated – i.e., when [ADP] and [AMP] are low and [ATP] and [citrate] are high.

1.5.2 PFK-1 and FBP-1

A primary control point is 'control point II' where fructose-6-phosphate (F-6-P) is phosphorylated to fructose-1,6-bisphosphate (F-1,6-bisP) by phosphofructokinase-1 (PFK-1) or de-phosphorylated by fructose-1,6-bisphosphatase (FBP-1). Both enzymes are allosteric and respond in opposite ways to the same metabolic indicators, such as ATP, AMP and citrate – low energy charge indicators stimulate PFK-1, but inhibit FBP-1; high energy charge indicators inhibit PFK-1, but stimulate FBP-1 (Figure 1.11).

1.5.3 PFK-2/FBP-2 – a 'tandem' enzyme

A second F-6-P kinase/phosphatase system exists that controls the phosphorylation of position 2 of F-6-P. F-6-P can be 'diverted' from mainstream metabolism by being phosphorylated by phosphofructokinase-2 (PFK-2) to give fructose-2,6-bisphosphate (F-2,6-bisP), which can be de-phos-phorylated back to F-6-P by a FBP-2 activity. Whereas PFK-1 and FBP-1 are separate proteins, PFK-2 and FBP-2 activities turned out to be due to a single polypeptide, a so-called 'tandem enzyme'.

F-2,6-bisP acts as an alternative allosteric modulator of PFK-1 and FBP-1, stimulating PFK-1 and inhibiting FBP-1 (Figure 1.11). And, because the 'tandem enzyme' that produces or destroys it, is itself under hormonal control, F-2,6-bisP is effectively a second messenger in certain tissues. Its tissue-specificity is also instructive in that the same hormone/receptor combination can elicit very different signalling outcomes.

PFK-2/FBP-2 is the product of four genes in humans and there are many splice variants. The 'liver' isozyme comes as three isoforms from the same gene (*PFKFBP1*); the 'heart' isozyme is found as two isoforms from the same gene (*PFKFBP2*); the 'brain' gene (*PFKFBP3*) produces two isoforms; the testes gene (*PFKFBP4*) produces an isozyme that, unlike the others, is tissue-specific[36]. The 'liver' gene produces the ***L*** isoform in liver cells but in skeletal muscle, the same gene produces the ***M*** isoform, which differs from the ***L*** form at the *N*-terminal region. The ***L*** isoform has an *N*-terminal 32 residue regulatory region (including a PKA phosphorylation site); in the ***M*** isoform, this is replaced with an unrelated sequence lacking phosphorylation sites (Table 1.2).

1.5.4 Control of PFK-2/FBP-2 by phosphorylation – liver

The single PKA phosphorylation site of the liver ***L*** isoform is the main point of control. When stimulated by glucagon or adrenaline, the cognate hepatocyte 7-pass receptor is activated, couples with the stimulatory G protein (Gs), and causes its activation by GDP dissociation and GTP binding. The activated αs subunit then couples with AC to produce

Table 1.2 Organisation of PFK-2/FBP-2 isozyme domains

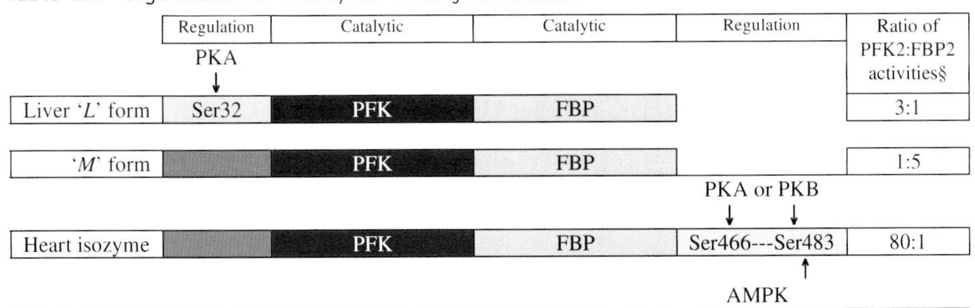

	Regulation	Catalytic	Catalytic	Regulation	Ratio of PFK2:FBP2 activities§
	PKA ↓				
Liver '*L*' form	Ser32	PFK	FBP		3:1
'*M*' form		PFK	FBP		1:5
				PKA or PKB ↓ ↓	
Heart isozyme		PFK	FBP	Ser466---Ser483	80:1
				↑ AMPK	

§Hue, L & Rider, M.H. (1987) Role of fructose 2,6-phosphate in the control of glycolysis in mammalian tissues. *Biochem. J.*, **245**: 313-324.

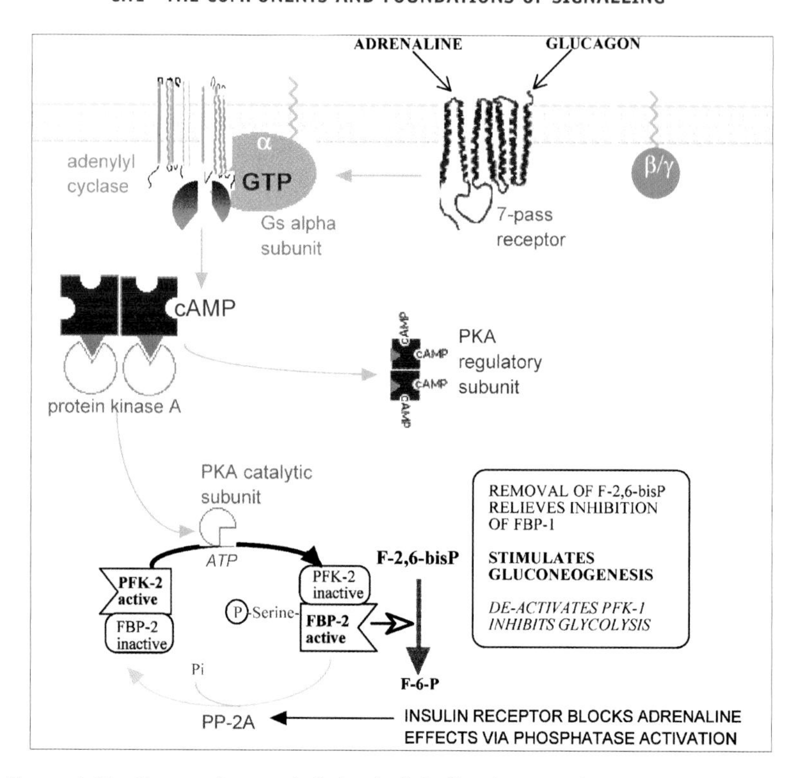

Figure 1.12 Hormonal control of glycolysis in liver by PFK-2/FBP-2 phosphorylation

cAMP. cAMP binds to the regulatory subunits of PKA, causing dissociation of the active catalytic subunits. The freed PKA phosphorylates the ***L*** tandem enzyme on Ser32 and this has two effects: (1) The PFK-2 activity is inhibited and (2) the FBP-2 activity is stimulated. The consequences of glucagon / adrenaline stimulation is that the second messenger F-2,6-bisP disappears, PFK-1 is inactivated by its loss and the inhibition of FBP-1 is relieved. This effectively switches the pathway from glycolysis to gluconeogenesis. This is just one of the ways in which the liver responds to the body's need for glucose export in starvation or stress (Figure 1.12). In the liver, insulin blocks adrenaline effects by activating a phosphdiesterase, destroying cAMP and potentially allowing de-phosphorylation to occur (Chapter 9, Section 9.8). The tandem enzyme activity ratio can be reversed by de-phosphorylation of Ser32 by protein phosphatase-2A.

1.5.5 Control of PFK-2/FBP-2 by phosphorylation – heart

The heart isozyme is controlled by phosphorylations of its *C*-terminal regulatory domain and its hormonal control presents a most unusual example of insulin and adrenaline signals producing the same outcome. The heart is highly adaptable in terms of energy requirements and will happily use low energy fuels like lipids or lactate, but when glucose is plentiful after a meal or when fight-or-flight stimulus requires extra energy, the heart switches to glycolysis. Similarly, in anaerobic conditions the heart also must switch to glycolysis because the TCA cycle shuts down.

Downstream of adrenaline, PKA phosphorylates Ser466 and Ser483 of the heart tandem enzyme and this has the effect of activating the PFK-2 activity with little effect upon the FBP-2 activity, which is already low[36]. The net effect of PKA activation, then, is an increase in glycolytic flux. Interestingly, insulin has the same effect of increasing glycolysis when glucose is abundant. Insulin activates a serine/threonine kinase (related to PKA) called PKB (see Chapter 9). PKB (or another insulin-regulated kinase) phosphorylates the same two sites on the heart PFK-2/FBP-2 enzyme with the same outcome of increasing glucose usage via glycolysis (Figure 1.13). Recently, it has been found that this phosphorylation by PKB leads to the phospho-PFK-2/FBP-2 being bound by a 14-3-3 protein, an event that contributes to the activation[37]. 14-3-3 proteins are phospho-serine binding proteins (see Chapter 3, Section 3.7.1).

The 'Pasteur effect' describes the shift from energy production by the TCA cycle to glycolysis in anaerobic conditions. Underlying this effect is an increase in 5′-AMP, which stimulates a 5′-AMP-dependent protein kinase (AMPK). AMPK can also phosphorylate the heart isozyme on Ser466, again with the same outcome: glycolysis is stimulated.

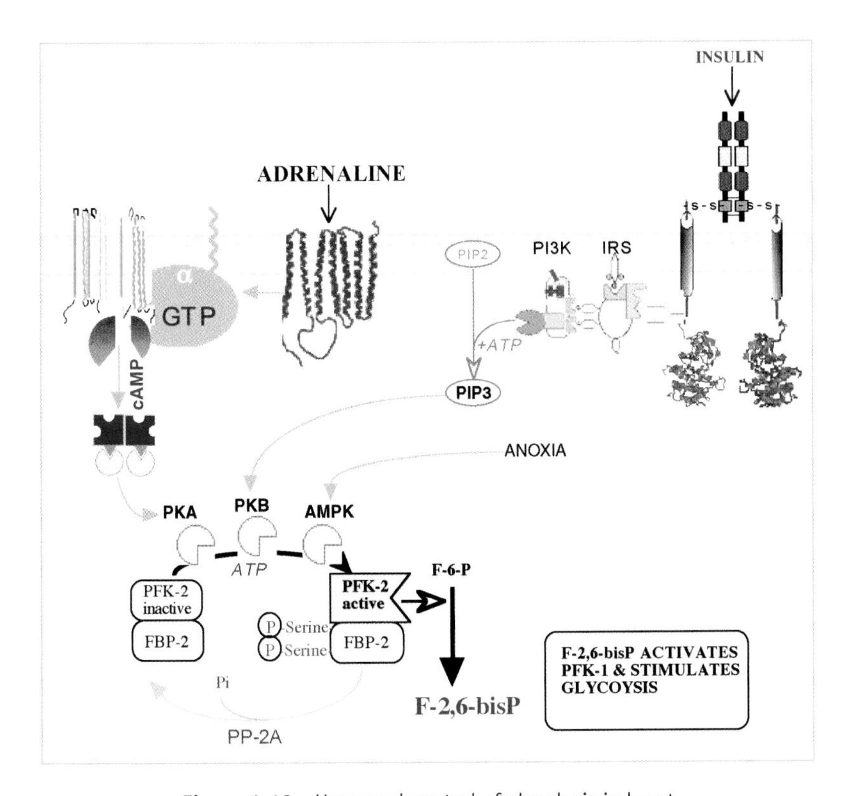

Figure 1.13 Hormonal control of glycolysis in heart

1.5.6 F-2,6-bisP in tumours

Many types of cancer cells produce large amounts of lactate even in aerobic conditions – the 'Warburg effect'. PFK-2 isozymes are often highly active in cancer cells and levels of F-2,6-bisP are high. It is theorised that this may contribute to the dominance of anaerobic glycolysis in cancer cells[38], although a more direct explanation is that at least some cancers have a defective glycerol phosphate shuttle and are forced to use lactate (rather than the mitochondrion) to regenerate NAD^+.

1.6 cAMP: ancient hunger signal – primitive signalling in amoebazoans and prokaryotes

1.6.1 Slime moulds

Being part-time multicellular animals, slime moulds appear to be positioned somewhere between unicellular eukaryotes and metazoans in terms of complexity; as such, social amoebae are often termed *amoebazoans*.

The most striking feature of slime moulds is their ability to circumvent starvation by swarming together to form a slug that can migrate and differentiate into a fruiting body. This process is activated and guided by a single molecule but, since the *Dictyostelium* genome has been completely deciphered, this seemingly simple system has proved to be much more complex than expected. The following is a summary.

When food runs out, individual *Dictyostelium* amoebae in the soil begin to secrete bursts of cAMP every six minutes. cAMP acts as a *chemoattractant*. Nearby cells sense the stimulus via cell surface 7-pass cAMP receptors (cAR) receptors and begin to move up the concentration gradient towards the source of cAMP. Such directional migration towards a chemoattractant is known as *chemotaxis*. At the same time, these responding amoebae begin to secrete cAMP themselves, setting up concentric waves of migration towards a central collecting point. cAMP is produced by an *adenylyl cyclase*. The on-off six minute pulses are achieved by a negative feedback mechanism that desensitises the receptor by destroying the extracellular cAMP long enough for the cell's cAMP receptor to recover, sense the cAMP concentration gradient, and then begin secreting more of its own cAMP. The destruction of cAMP is caused by an *extracellular phosphodiesterase* (PDE) (Figure 1.14).

The culmination of all this chemotaxis signalling is the formation of multicellular motile 'slug' made up of as many as 100,000 amoebae. The slug migrates towards light and once it has found a suitable position on the surface it begins to *differentiate* into stalk cells and spore cells. The spore cells become desiccated and eventually are dispersed by the wind, to germinate into individual amoebae if conditions are right, or remain dormant until conditions improve. The stalk cells, however, are sacrificed in the process.

The most remarkable fact is that the whole process of sporulation (including swarming, differentiation and eventual germination) is controlled by a single molecule, cAMP, and key types of pathway components shared with humans.

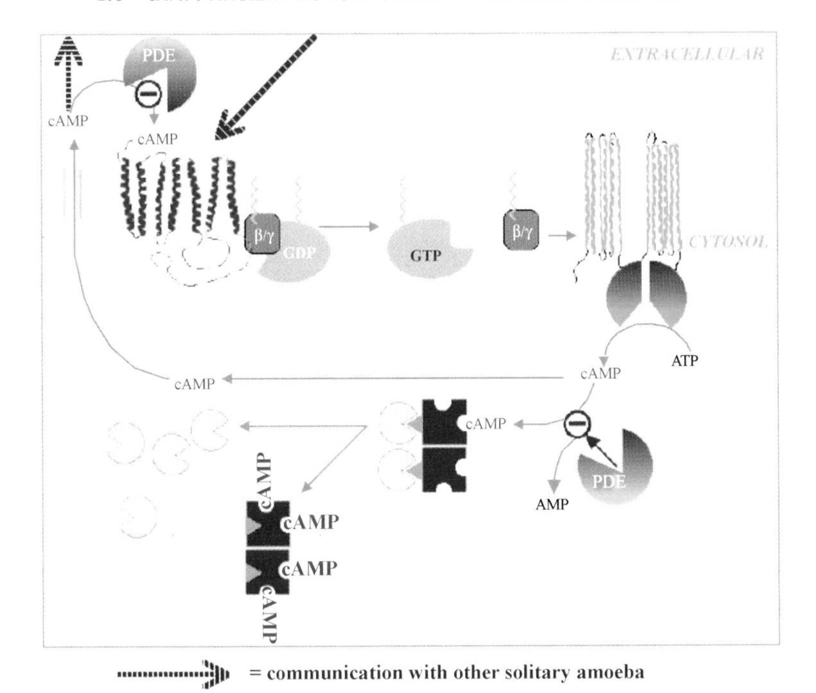

Figure 1.14 cAMP signal transduction in slime mould

The *Dictyostelium* genome is now fully sequenced and although the genes for the components of the chemotaxis/differentiation pathways are known, few of the gene products' activities have been biochemically characterised.

Dictyostelium has six chromosomes with an estimated 12,500 genes. There are at least 55 genes for 7-pass G protein-coupled-receptors (GPCRs) and these can be grouped into those resembling mammalian GPCR families 2, 3 and 5 (see Chapter 5, Section 5.1) and the unrelated family of cAMP receptors (cAR1, cAR2, cAR3 and cAR4) and cAMP receptor-like GPCRs (CRLs) that are unique to slime moulds[39]. The genome contains 14 heterotrimeric G protein alpha subunits, but only two Gβ and a single Gγ gene are present[40]. The G protein alpha subunits (Gα1–Gα12) are all most closely related to the mammalian Gαi family[41]. In addition, there are at least 40 monomeric G proteins, including members of Ras/Rap, Rac, Rab, Ran and Arf families; the effector pathways of Ras (MAP kinase or Erk) are also present, as are the regulators of Ras (GEFs and GAPs)[42].

The genome codes for three adenylyl cyclases that are involved in *Dictyostelium's* development programme and their activities are opposed by the activities of both extracellular and cytosolic PDE enzymes.

A major difference in slime mould GPCR signalling, compared to metazoans, is the pre-eminence of G protein β/γ subunits in activation of *Dictyostelium* adenylyl cyclase. Whereas *all* mammalian adenylyl cyclases are activated by Gαs subunits[43], in slime mould the Gα subunits have no such effect on the enzyme.

The membrane-bound, Gβ-coupled adenylyl cyclase A (ACA) of *Dictyostelium* is homologous with mammalian adenylyl cyclases, sharing the same structure of *N*-terminal 6-pass transmembrane domain, followed by a cytosolic catalytic domain, separated by a second 6-pass transmembrane domain from a second, *C*-terminal, catalytic domain that makes up the other half of the split catalytic site. ACA controls the chemotaxis and aggregation stage of multicellular development. A second cyclase (ACG) is G protein-independent and controls spore development. It consists of a single transmembrane domain and a single catalytic domain and is controlled by osmolarity[7]. The third adenylyl cyclase is cytosolic and is responsible for correct spore development. This enzyme is referred to as either 'ACB'[44] or 'ACR'[45].

ACA is coupled to cAR1 via the heterotrimeric G protein Gα2β/γ but, unlike mammalian adenylyl cyclases, ACA is activated solely by the β subunit (Figure 1.14). This appears reminiscent of type II and IV mammalian adenylyl cyclases, which are also activated by β/γ. However, activation of mammalian adenylyl cyclases by β/γ is conditional, being dependent upon the presence of an activated Gαs subunit, which must also bind simultaneously to AC II at an alternative site to that occupied by β/γ[43] (see Chapter 8, Section 8.6).

Their developmental programme involves not only metazoan-like signalling molecules that are absent in unicellular forms (SH2 domains, for example), but it also retains ancient signal systems absent in metazoans. A good example of the latter, prokaryote-like system is the His/Asp phosphorelay used by *Dictyostelium* to control RegA, one of its phosphodiesterases[46]. Like RegA, ACR also contains a prokaryotic response regulator domain[46].

1.6.2 cAMP and *E. Coli*

In 1965, Makman and Sutherland[47] made the surprising discovery that cAMP was present in the cytoplasm of the enteric bacterium, *E. coli*, and that the concentration of the cyclic nucleotide increased as growth slowed down when glucose was depleted or removed from the medium. Under these conditions, the bacteria are forced to utilise less favoured fuel sources like lactose, which are more costly to metabolise and consequently less energy efficient. 'Catabolite repression' was, at the time, a relatively well-understood phenomenon due to the work of Jacob and Monod who discovered that for *E. coli* to metabolise lactose, a whole new set of genes needed to be transcribed – the *Lac* operon. The *Lac* operon contains three 'structural' genes encoding β-galactosidase, thiogalactoside transacetylase, and a transporter (galactoside permease). It was first thought that a catabolite of glucose was responsible for repressing the transduction of these genes, and that when glucose was used up, the 'catabolite' disappeared and the 'repression' was lifted. It later transpired that this was the opposite of what was happening, but the name 'catabolite repression' stuck and survives today.

In the absence of glucose, *E. coli* needs to find alternatives. It does this by sensing the background turnover of other sugars present in the environment. Lactose, for example, is a substrate for one of the *Lac* operon enzymes (β-galactosidase) that is present in very

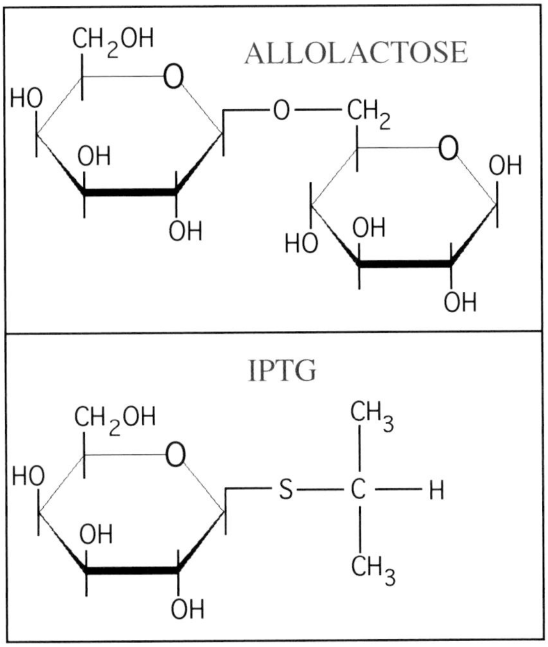

Figure 1.15 Lac operon inducers

low amounts even in the presence of glucose. β-Galactosidase normally hydrolyses lactose to galactose plus glucose, but is also capable of catalysing the production of small amounts of *allolactose*. Allolactose binds to the *Lac* repressor protein, altering its conformation such that it can no longer bind to the DNA of the operator site upstream of the structural genes of the operon. De-binding of the *Lac* repressor protein allows RNA polymerase to bind weakly, but fully productive binding is only achieved if cAMP is also present *at high levels*. That is because cAMP binds to the transcription factor CAP (catabolite activator protein). CAP is a primordial cAMP receptor that bears a remarkable structural resemblance to the main vertebrate cAMP receptor, the R-subunits of PKA (see Chapter 4). It is worth noting that a structural analogue of allolactose is in wide use today: *isopropylthiogalactoside* (IPTG) (see Figure 1.15) is used as a non-hydrolysable inducer of the *Lac* operon in artificial protein expression systems.

References

1. Vanhaesebroeck, B., Leevers, S.J., Panayotou, G. and Waterfield, M.D. (1997) Phosphoinositide 3-kinases: a conserved family of signal transducers. *TIBS*, **22**: 267–272.
2. Pacold, M.E., Suire, S., Persic, O., Lara-Gonzalez, S., Davis, C.T., Walker, E.H., Hawkins, P.T., Stephens, L., Eccleston, J.F. and Williams, R.L. (2000) Crystal structure and functional analysis of Ras binding to its effector phosphoinositide 3-kinase γ. *Cell*, **103**: 931–943.

3. Henriksen, J. H. and Schaffalitzky De Muckadell, O.B. (2000) Secretin, its discovery, and the introduction of the hormone concept. *Scand. J. Clin. Lab. Invest.*, **60**: 463–472.
4. Stretton, A.O.W. (2002) The first sequence: Fred Sanger and insulin. *Genetics*, **162**: 527–532.
5. Ahlquist, R.P. (1948) A study of the adrenotropic receptors. *American Journal of Physiology*, **153**: 586–600.
6. Bonner, J.T. (1999) The history of the cellular slime moulds as a 'model system' for developmental biology. *J. Biosci.*, **24**: 7–12.
7. Schaap, P. (2002) Survival by cAMP in social amoebae: an intersection between eukaryote and prokaryote signalling systems. *Microbiology Today*, **29**: 136–138.
8. Sutherland, E.W. and Rall, T.W. (1958) Fractionation and characterization of a cyclic adenine ribonucleotide formed by tissue particles. *J. Biol. Chem.*, **233**: 1077–1091.
9. Sutherland, E.W. (1971) Studies on the mechanism of hormone action. *Nobel Lecture*.
10. Rall, T.W., Sutherland, E.W., and Berthet, J. (1957) The relationship of ephinephrine and glucagon to liver phosphorylase. *J. Biol. Chem.*, **224**: 463–475.
11. Gallagher, G.L. (1990) Getting the message across. *The Journal of NIH Research*, **2**: 77–78.
12. Katz, M.S., Dax, E.M. and Gregerman, R.I. (1993) Beta adrenergic regulation of rat liver glycogenolysis during aging. *Experimental Gerontology*, **28**: 329–340.
13. Krebs, E.G. (1992) Protein phosphorylation and cellular regulation, I. *Nobel Lecture*.
14. Lefkowitz, M. (2004) Historical review: a brief history and personal retrospective of seven-transmembrane receptors. *TRENDS in Pharamacological Sciences*, **25**: 413–422.
15. Jensen, E.V. and Jordan, V.C. (2003) The estrogen receptor: a model for molecular medicine. *Clinical Cancer Research*, **9**: 1980–1989.
16. Lefkowitz, R.J., Roth, J., Pricer, W. and Pastan, I. (1970) ACTH receptors in the adrenal: specific binding of ACTH-125I and its relation to adenyl cyclase. *Proc. Natl. Acad. Sci. USA*, **65**: 745–752.
17. Caron, M.C. and Lefkowitz, M. (1976) Solubilization and characterisation of the β-adrenergic receptor binding sites of frog erythrocytes. *J. Biol. Chem.*, **25**: 2374–2384.
18. Cerione, R.A., Sibley, D.R., Codina, J., Benovic, J.L., Winslow, J., Neer, E.J., Birnbaumer, L., Caron, M.G. and Lefkowitz, R.J. (1984) Reconstitution of a hormone-sensitive adenylate cyclase system. The pure beta-adrenergic receptor and guanine nucleotide regulatory protein confer hormone responsiveness on the resolved catalytic unit. *J. Biol. Chem.*, **259**: 9979–9982.
19. Rodbell, M. (1994) Signal transduction: evolution of an idea. *Nobel Lecture*.
20. Sutherland, E.W., Rall, T.W. and Menon, T. (1962) Adenyl cyclase I. Distribution, preparation, and properties. *J. Biol. Chem.*, **237**: 1220–1227.
21. Rodbell, M., *et al.* (1971) The glucagon-sensitive adenylyl cyclase system in plasma membranes of rat liver IV-V. *J. Biol. Chem.*, **246**: 1872–1876; 1877–1882.
22. Lefkowitz, R.J., Mullikin, D. and Caron, M.G. (1978) Regulation of β-adrenergic receptors by guanyl-5′-imidodiphosphate and other purine nucleotides. *J. Biol. Chem.*, **252**: 4686–4692.
23. Salomon, Y., Len, M.C., Londos, C., Rendell, M. and Rodbell, M. (1975) The hepatic adenylate cyclase system. I. Evidence for transition states and structural requirements for guanine nucleotide activation. *J. Biol. Chem.*, **250**: 4239–4245.
24. Lutz, S., Hippe, H.J., Niroomand, F. and Wieland, T (2004) Nucleoside diphosphate kinase-mediated activation of heterotrimeric G proteins. *Methods in Enzymology*, **390**: 403–418.
25. Salomon, Y. and Rodbell, M. (1975) Evidence for specific binding sites for guanine nucleotides in adipocyte and hepatocyteplasma membranes. A difference in the fate of GTP and guanosine 5′-(β,γ-imino)triphosphate. *J. Biol. Chem.*, **250**: 7245–7250.
26. Ross, E.M., Howlett, A.C., Ferguson, K.M. and Gilman, A.S. (1978) Reconstitution of hormone-sensitive adenylate cyclase activity with resolved components of the enzyme. *J. Biol. Chem.*, **253**: 6401–6412.

27. Northup, J. K., Sternweis, P. C., Smigel, M. D., Schleifer, L. S., Ross, E. M., and Gilman, A. G. (1980) Purification of the regulatory component of adenylate cyclase. *Proc. Natl. Acad. Sci. U.S.A.*, **77**: 6516–6520.
28. Gilman, A.G. (1994) G proteins and regulation of adenylyl cyclase. *Nobel Lecture*.
29. Pfeuffer, T and Helmreich, E.J.M. (1975) Activation of pigeon erthrocyte membrane adenylate cyclase by guanylnucleotide analogues and separation of a nucleotide binding protein. *J. Biol. Chem.*, **250**: 867–876.
30. Cassel, D. and Selinger, Z. (1976) Catecholamine-stimulated GTPase activity in turkey erythrocyte membranes. *Biochim. Biophys. Acta*, **452**: 538–551.
31. Cassel, D. and Pfeuffer, T. (1978) Mechanism of cholera toxin action: covalent modification of the guanyl nucleotide-binding protein of the adenylate cyclase system. *Proc. Natl. Acad. Sci. USA*, **75**: 2669–2673.
32. Gill, D.M. and Meren, R. (1978) ADP-ribosylation of membrane proteins catalysed by cholera toxin: basis of the activation of adenylate cyclase. *Proc. Natl. Acad. Sci. USA*, **75**: 3050–3054.
33. Pfeuffer, T. (1977) GTP-binding proteins in membranes and the control of adenylate cyclase activity. *J. Biol. Chem.*, **252**: 7224–7234.
34. Bokoch, G.M., Katada, T., Northup, J.K., Hewlett, E.L., and Gilman, A.G. (1983) Identification of the predominant substrate for ADP-ribosylation by islet activating protein. *J. Biol. Chem.*, **258**: 2072–2075.
35. Berridge, M.J. and Irvine, R.F. (1984) Inositol trisphosphate, a novel second messenger in cellular signal transduction. *Nature*, **312**: 315–321.
36. Rider, M.H., Bertrand, L., Vertommen, D., Michels, P.A., Rousseau, G.S. and Hue, L. (2004) 6-Phosphofructo-2-kinase/fructose-2,6-bisphosphatase: head-to-head with a bifunctional enzyme that controls glycolysis. *Biochem. J.*, **381**: 561–579.
37. Pozuelo Rubio, M., Peggie, M., Wong, B.H.C., Morrice, N. and MacKintosh, C. (2003) 14-3-3s regulate fructose-2,6-bisphosphate levels by binding to PKB-phosphorylated cardiac fructose-2,6-bisphosphate kinase/phosphatase. *EMBO J.*, **22**: 3514–3523.
38. Hue, L. and Rider, M.H. (1987) Role of fructose 2,6-phosphate in the control of glycolysis in mammalian tissues. *Biochem. J.*, **245**: 313–324.
39. Williams, J.G., Noegel, A.A. and Eichinger, L. (2005) Manifestations of multicellularity: *Dictyostelium* reports in. *Trends in Genetics*, **21**: 392–398.
40. Eichinger, L. *et al.* (2005) The genome of the social amoeba *Dictyostelium discoideum. Nature*, **435**: 43–57.
41. Brzostowski, J.A., Johnson, C. and Kimmel, A.R. (2002) Gα-mediated inhibition of developmental signal response. *Current Biology*, **12**: 1199–1208.
42. Wilkins, A and Install, R.H. (2001) Small GTPases in *Dictyostelium*: lessons from a social amoeba. *Trends in Genetics*, **17**: 41–48.
43. Hurley, J.H. (1999) Structure, mechanism, and regulation of mammalian adenylyl cyclase. *J. Biol. Chem.*, **274**: 7599–7602.
44. Kim, H.J., Chang, W.T., Meima, M., Gross, J.D. and Schaap, P. (1998) A novel adenylyl cyclase detected in rapidly developing mutants of *Dictyostelium. J. Biol. Chem.*, **273**: 30859–30862.
45. Soderbom, F., Anjard, C., Iranfar, N., Fuller, D. and Loomis, W.F. (1999) An adenylyl cyclase that functions during late development of *Dictyostelium. Development*, **126**: 5463–5471.
46. Thomason, P. and Kay, R. (2000) Eukaryotic signal transduction via histidine-aspartate phosphorelay. *J. Cell Sci.*, **113**: 3141–3150.
47. Markham, R.S. and Sutherland, E.W. (1965) Adenosine 3′,5′-phosphate in *Escherichia coli* grown in continuous culture. *J. Biol. Chem.*, **240**: 1309–1314.

2

Enzymes and receptors – quantitative aspects

Classically, enzymes are assayed at 'steady state' – during the initial phase of turnover of substrate – by monitoring product generation. Receptors are instead assayed at equilibrium. Scatchard analysis of equilibrium binding assays yields straight-line graphs where receptors are single independent binding sites, but gives curved graphs where there are multiple independent sites, or cooperativity. Similarly, displacement of tracer radioligands may display monophasic (one site) dissociation or biphasic (two site) dissociation. Such deviations from linearity are not fully understood, but are becoming more easily explicable with reference to new structural and functional information. New methods allow real time monitoring of ligand binding and de-binding (surface plasmon resonance detectors, for example) and this may provide a route to a future resolution of such problems.

2.1 Enzyme steady state assays – Michaelian enzymes

A *simple* enzyme obeys Michaelis-Menten kinetics, because each and every molecule in a pure sample of that enzyme will have the same unalterable, intrinsic properties: each will share the same affinity for the substrate and each will be limited to the same maximum rate at which it can turnover substrate. At low substrate concentrations, an individual enzyme will turnover substrate slowly; at high concentrations it works faster. But there is a limit to its turnover rate. As the concentration of substrate rises, the enzyme approaches its theoretical maximum velocity. After this point, further increases in

Structure and Function in Cell Signalling John Nelson

substrate concentration make no difference to the rate. If we are assaying a sample of many molecules of that enzyme, the maximum velocity observed, ***Vmax***, cannot be further increased because all the catalytic centres are filled and working at their upper limit of turnover. The enzyme is then said to be *saturated*.

2.1.1 How are enzymes assayed?

Kinetic measurements cannot easily be performed on an individual enzyme molecule. Instead we use samples containing a population of enzyme molecules, which may be more, or less, pure. A sample might be a wet weight of liver homogenate, a milligram of protein from semi-purified cell extract or a microgram of purified enzyme. Obviously the observed rate for such samples depends not only upon the concentration of substrate it is exposed to but also upon the number of enzyme molecules present. For this reason, kinetic analysis requires that enzyme concentration be kept constant in each incubation while the substrate concentration is varied.

2.1.2 Steady state

Please note that in the following, square brackets indicate a concentration term.

- (a) Classically, enzymes are not measured at equilibrium but instead are assayed during the initial period of incubation only, which is when a 'steady state' concentration of enzyme-substrate complex (ES) exists. An *individual* enzyme-substrate complex is a transient entity, so the steady state ES compartment is actually made up of a constant flux of substrate passing through on its way to become product, opposed by ES dissociation back to free enzyme (E) and free substrate (S) (see Panel 1).

- (b) The rate of ES formation is determined by an intrinsic rate constant of the enzyme (k_1), and varies with both the substrate concentration [S] and the enzyme concentration [E]. The observed rate is equal to k_1 multiplied by [E] multiplied by [S].

- (c) ES can dissociate to E and S again and this reaction is governed by a second intrinsic rate constant (k_2). The rate of ES breakdown to E + S is equal to [ES] multiplied by (k_2).

- (d) The other way ES breaks down is the formation of product (P) and free enzyme. This is considered a 'one-way street' during the steady state – in other words, product does not revert to substrate during this initial phase. This 'irreversible' step is governed by a third intrinsic rate constant (k_3) and the rate of product formation is equal to [ES] multiplied by (k_3).

  Note, that $E + S \rightarrow ES$ is a second order reaction (that is, one whose rate is dependent upon either a single concentration squared, or upon the mathematical

Panel 1.

(a) $E + S \underset{k_2}{\overset{k_1}{\rightleftharpoons}} ES \xrightarrow{k_3} E + P$

(b) velocity of ES formation $= k_1.[E].[S]$

(c) velocity of S de-binding $= k_2.[ES]$

(d) velocity of ES catalytic breakdown $= k_3.[ES]$

(e) $v = k_{cat} \cdot [ES]$

(f) $E + S \underset{k_2}{\overset{k_1}{\rightleftharpoons}} ES \xrightarrow{catalysis} EP \xrightarrow{release} E + P$

Panel 2.1

product of two concentrations, in this case: [E] and [S]). The forward second order rate constant k_1 has units of $M^{-1}s^{-1}$. The other reactions are first order, being dependent upon a single reactant concentration. k_2 and k_3 have units of s^{-1}, meaning 'events per second'.

- (e) k_3 is the catalytic rate constant and is often referred to as k_{cat} (strictly speaking this is an approximation because k_{cat} is the maximum turnover rate when substrate is saturating may be determined by several rate constants other than k_3 alone). The velocity (v) that one measures at any given substrate concentration is governed by this constant and the steady state level of ES complex achieved at that substrate concentration.

- (f) The above treatment is a simplification because there is at least one step missing. Catalysis occurs within the active site. So ES is actually converted to an enzyme-product complex (EP) first, and this is followed by release of product after catalysis.

For *Michaelian enzymes*, the release of product from the enzyme-product complex (EP) is very rapid and *the rate-limiting step is the catalytic step* (ES → EP). The release step is therefore ignored in the following simple treatment.

Returning to the simple Michaelis-Menten treatment, it is clear that if the concentration of ES could be measured directly there would be a simple route to determine k_{cat} (k_3). However, in most cases this is impractical: only [S] is known, and only product concentration is available for measurement. Often the product is coloured or fluorescent while the substrate is not. The question is how do we determine k_{cat} when only product can be measured?

The constancy of the concentration of [ES] is due to the rate of its formation being equal to the overall rate of its breakdown. Therefore,

$$k_1.[E].[S] = (k_2 + k_3).[ES] \tag{i}$$

This equation can be rearranged (see Equation (i) in Panel 2).

2.1.3 K_M – the Michaelis-Menten constant

The Michaelis-Menten constant (K_M) was introduced to simplify the preceding equation. This is the ratio between the sum of ES breakdown rate constants and the rate of its formation.

Substituting K_M for the denominator in Equation (i) simplifies Equation (iii). The rest of the derivation of the 'Michaelis-Menten' plot is summarised in Panel 2.

2.1.4 *Vmax* is reached when the enzyme becomes saturated

The maximum velocity of reaction is only approached when substrate concentration is very much higher than K_M, and then

$$\frac{[S]}{K_M + [S]} \text{ is almost } = 1 \tag{ix}$$

and *at saturating concentrations of substrate*, the observed velocity $v = Vmax$, and virtually all of the enzyme is in the form of enzyme substrate complex. That is, $[E_T] = [ES]$. So, in that case, equation (vii) becomes

$$Vmax = kcat.[E_T] \tag{x}$$

Substituting (x) into (viii) gives the *Michaelis-Menten equation*

$$v = Vmax.\frac{[S]}{K_M + [S]}$$

Plotting observed rates against substrate concentration gives a rectangular hyperbola that approaches saturation (Figure 2.1). Note that saturation is *never* reached but must be extrapolated from the data.

Panel 2.

(i) $[ES] = \dfrac{[E].[S]}{(k_2 + k_3)/k_1}$

(ii) $K_M = \dfrac{k_2 + k_3}{k_1}$

Substituting (ii) for the denominator in (i) gives (iii)

(iii) $[ES] = \dfrac{[E].[S]}{K_M}$

Note that the total concentration of enzyme $[E_T]$ is split between free enzyme and that complexed with substrate (iv)

(iv) $[E_T] = [E] + [ES]$

The total substrate concentration is likewise split between free and complexed concentrations. However, the experiment is set up so that substrate (at whatever concentration in the range used) is in a huge excess over enzyme and effectively the free substrate concentration is not depleted during the initial rate. So, [S] is effectively equal to $[S_T]$.

Substitute (iv) into (iii) to give (v)

(v) $[ES] = \dfrac{([E_T]-[ES]).[S]}{K_M}$

Solving for [ES] gives (vi)

(vi) $[ES] = [E_T] \cdot \dfrac{[S]}{K_M + [S]}$

Recall that the catalytic step is the rate-limiting one governed by k_3 or 'k_{cat}'.

(vii) $v = k_{cat}.[ES]$

Substitute [ES] in (vii) for that in (vi) to get (viii)

(viii) $v = k_{cat}.[E_T]\dfrac{[S]}{K_M + [S]}$

(ix) $\dfrac{[S]}{K_M + [S]}$ is almost = 1

(x) $Vmax = k_{cat}.[E_T]$

Finally, substitute (x) into (viii) to get the Michaelis-Menten equation

$$v = Vmax.\frac{[S]}{K_M + [S]}$$

Panel 2.2

2.1.5 What does the K_M mean?

The K_M value for an enzyme has two meanings:

- K_M is the substrate concentration at which the enzyme is half saturated and reaction velocity is half maximal (see Figure 2.1). If the initial velocity, v is half $Vmax$ then

$$\frac{Vmax}{2} = Vmax.\frac{[S]}{K_M + [S]} \ldots \textit{divide across by Vmax} \ldots \frac{1}{2} = \frac{[S]}{K_M + [S]}$$

$$\textit{solve for } K_M \ldots 2.[S] = K_M + [S] \quad \textit{Thus, } K_M = [S] \textit{at half Vmax}$$

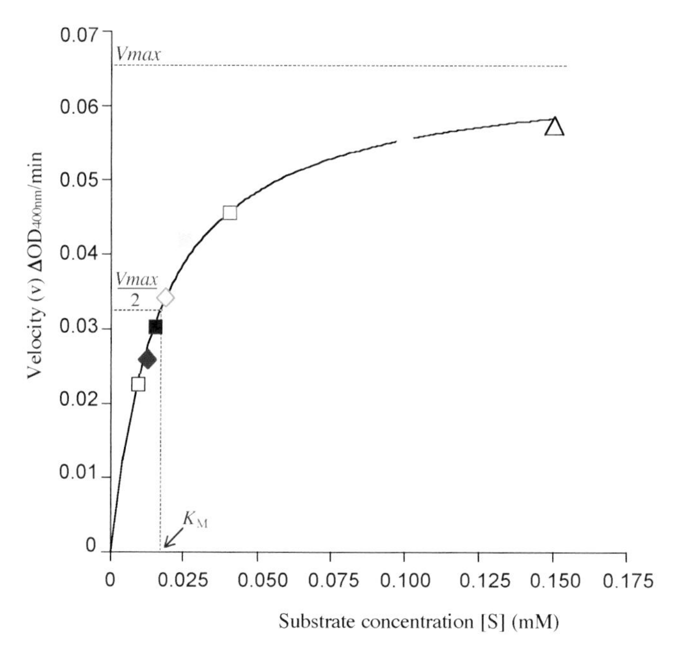

Figure 2.1 Michaelis-Menten plot of alkaline phosphatase enzymic activity

- K_M is the ratio of rate constant for ES formation versus those for ES breakdown. If the rate constant for product formation (k_3) is much smaller than that for ES $\rightarrow$ E + S dissociation (k_2), then $K_M \approx k_2/k_1$. In this case only (strictly speaking) $K_M = K_{EQ}$, which is the equilibrium or dissociation constant for substrate-to-enzyme binding E + S $\rightleftharpoons$ ES

In many cases, and to a close approximation, K_M is a measure of the affinity the enzyme has for its substrate and is *expressed in units of Molarity*. The lower the K_M value, the higher is the affinity. To look at it another way, enzymes with low K_M require lower substrate concentrations to achieve 1/2 *Vmax* than those with high K_M.

From the Michaelis-Menten plot, *Vmax* can be obtained by extrapolation using computerised non-linear curve fitting (e.g., Graphpad Prism) and K_M can be obtained from the curve at 1/2*Vmax* (Figure 2.1). In the days before personal computers, this was done visually and was therefore inaccurate. For this reason, linearised re-arrangements of the Michaelis-Menten equation were introduced. The most familiar are

Lineweaver-Burke plot

$$\frac{1}{v} = \frac{1}{Vmax} + \frac{K_M}{Vmax} \cdot \left(\frac{1}{[S]}\right)$$

Eadie-Hoftsee plot

$$\frac{v}{[S]} = \frac{Vmax}{K_M} - v.\left(\frac{1}{K_M}\right)$$

The data in Figure 2.1 have been re-plotted as linear plots, as shown in Figure 2.2A and B. The experimental details, raw data and calculations are included in Appendix 1 so that you can try this out for yourselves. Note that the data points are colour coded to illustrate the order of data in each analysis.

2.1.6 Non-Michealian enzymes – G proteins and Ras

Enzymes that display cooperativity are non-Michaelian. They are exceptional in that their intrinsic activities can be altered by the progressive binding of substrate (*homotropic*) or the binding of an unrelated modulator molecule (*heterotropic*). Homotropic cooperativity is almost exclusively found in multimeric proteins. In simple terms, binding of substrate to one subunit alters the conformation of both it and the other subunits so that the next substrate binds more tightly (positive cooperativity) or less tightly (negative cooperativity). Such propagated conformational change is termed allosteric: *allosteric* means 'differently shaped'.

Homotropic cooperativity only involves the active site. Heterotropic regulators, however, bind at an alternative binding site on the enzyme. Although such allosteric modulators do not bind directly to the active site, they nevertheless affect its conformation either positively or negatively so as to cause stimulation or inhibition of the enzyme's activity. Co-operativity can be detected by Hill Plot analysis (see Appendix).

2.1.7 Non-Michaelian enzymes – cooperativity and allostery

It is important to realise that the catalytic mechanism is quite different in GTPases such as monomeric Ras and the heterotrimeric *G proteins*. In these peculiar enzymes *product release is the rate-limiting step*. In Gα subunits, the dissociation of GDP from the guanine nucleotide binding cleft/catalytic site is the rate-limiting step, rather than the hydrolytic step of GTP $\rightarrow$ GDP. Added to that, G protein nucleotide binding and GTPase activity is under the control of accessory proteins that variously upregulate or downregulate both hydrolysis and product release. Thus, these 'GTPases' display kinetics that are very far from being 'Michaelian'.

2.2 Receptor equilibrium binding assays

Receptor assays differ from enzyme assays in one major respect: receptors are measured at equilibrium. However, in many ways simple receptor assays have some analogy with Michaelian enzyme kinetic assays. Ligand (L) binds to receptor (R) in a reversible manner that is dependent on forward and reverse rate constants as well as the concentrations of ligand [L], receptor [R] and the ligand-receptor complex [LR].

$$R + L \underset{k_2}{\overset{k_1}{\rightleftharpoons}} LR$$

2.2.1 Equilibrium

As with enzymes, certain conditions need to be met in order for a meaningful assay to be performed. The ligand needs to be present in sufficient amounts so that its concentration

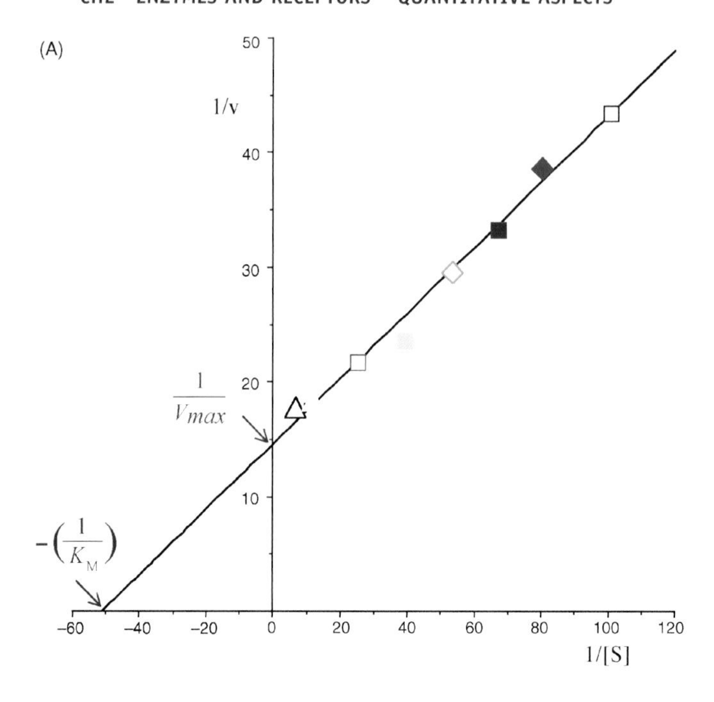

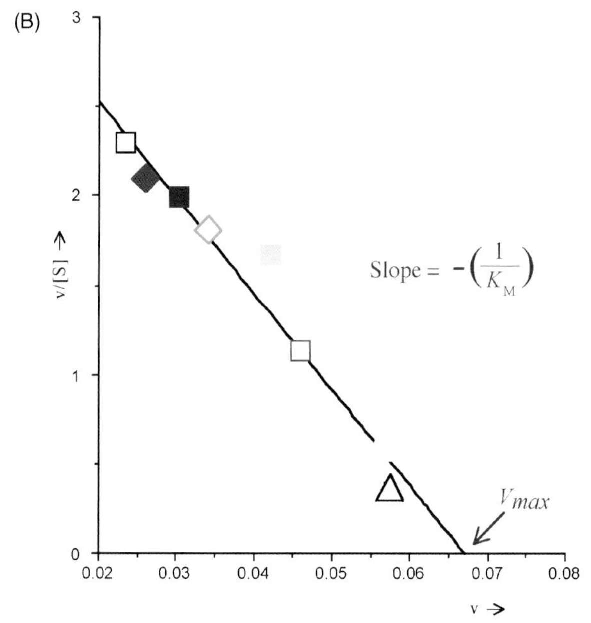

Figure 2.2 (A) Lineweaver-Burke plot of alkaline phosphatase assay; (B) Eadie-Hoftsee plot of alkaline phosphatase assay

does not change appreciably when binding occurs. In practice, this means that the receptor concentration is very much lower than the ligand's. A second condition is that the binding process must be allowed to go to equilibrium before being assayed. As a general rule, receptor binding reaches equilibrium in around 30 minutes to an hour at 37°C, or two hours at room temperature, or overnight at 4°C. Finally, the receptor in question is assumed to have a single binding site and all receptors in a sample have the same intrinsic affinity as for ligand.

Binding and de-binding is faster at higher temperatures and equilibrium state is reached more quickly. Nevertheless, the equilibrium concentration of the complex reached will be the same for all temperatures from 4°C to 37°C. One caveat is that at 4°C, receptors that were already occupied by an endogenous ligand prior to assay will not be labelled because ligand exchange is inefficient at low temperatures.

The derivation of formulae for receptor assays is similar to, but simpler than, enzyme kinetics (see Panel 3). Formation of LR is dependent upon the concentrations of L and R and on the forward rate constant k_1 (the 'on-rate') and the rate of its formation is k_1 multiplied by [L] multiplied by [R]. The dissociation of LR is governed by k_2 (the 'off-rate') and the concentration of LR. The dissociation rate is equal to k_2 multiplied by [LR]. Equilibrium is achieved when the rate of formation of LR is equally balanced by the rate of dissociation. Or:

$$k_1.[\mathrm{L}].[\mathrm{R}] = k_2.[\mathrm{LR}]$$

Thus, receptor assays only have to deal with two rate constants. The derivation of a saturation curve analogous to the Michaelis-Menten is shown in simplified form in Panel 3.

2.2.2 K_D – the dissociation constant

The dissociation constant (K_D) of receptors is analogous to the Michaelis-Menten constant of simple enzymes in that it is a measure of the intrinsic affinity the receptor has for the ligand.

2.2.3 *Bmax* – the maximum binding capacity is a 'count' of the receptors in a sample

Maximum binding of ligand in a receptor sample occurs as ligand approaches high, saturating concentrations of ligand. For simple, single site binding this means that all receptor molecules are occupied (with a single ligand each). Therefore, at saturating ligand concentrations:

$$[\mathrm{R_T}] = Bmax \qquad \text{(vii)}$$

In other words the concentration of receptor is equal to the concentration of bound ligand when the receptor is saturated. Substituting Equation (vii) into Equation (vi) (Panel 3) gives:

$$[\mathrm{LR}] = Bmax.\frac{[\mathrm{L}]}{K_D + [\mathrm{L}]} \qquad \text{(viii)}$$

Panel 3.

(i) $$[LR] = \frac{[L].[R]}{k_2 / k_1}$$

(ii) $$K_D = \frac{k_2}{k_1}$$

Substituting (ii) for the denominator in (i) gives (iii)

(iii) $$[LR] = \frac{[L].[R]}{K_D}$$

The initial or total receptor concentration [RT] is split between free receptor [R] and that complexed with ligand [LR]. Thus...

(iv) $$[R_T] = [R] + [LR]$$

Substitute (iv) into (iii) to give (v)

(v) $$[LR] = \frac{([R_T]-[LR]).[L]}{K_D}$$

Solve for [LR]...

(vi) $$[LR] = [R_T]\cdot\frac{[L]}{K_D + [L]}$$

At saturating ligand concentrations...

(vii) $$[R_T] = Bmax$$

Substitute (vii) into (vi)...

(viii) $$[LR] = Bmax.\frac{[L]}{K_D + [L]}$$

Panel 2.3

As you can see, this takes the same form as the Michaelis-Menten equation and similarly produces a hyperbolic curve of [LR] *versus* [L].

In papers 'dealing with hormone-receptor interactions, the LR complex is often referred to as 'bound ligand' (B) or 'specifically bound' (SB), reflecting the fact that only the ligand concentration can be assayed because it is the only component that is labelled. By the same token, L is usually referred to as 'free ligand' or F.

$$[SB] = Bmax.\frac{[F]}{K_D + [F]}$$

2.2.4 The meaning of K_D

You will probably recognise that this last equation resembles the Michaelis-Menten equation and produces a similar rectangular hyperbolic plot of [SB] *versus* [F]. By the same argument used for enzyme analysis, K_D can be shown to be the concentration required to reach 1/2*Bmax*, or half-saturation. Thus, K_D is a measure of the affinity of the interaction, and *the higher the* K_D, *the lower the affinity*. A receptor with a high K_D requires high ligand concentrations to reach half-saturation. A high affinity receptor reaches half-saturation at relatively low ligand concentrations.

Again, visually extrapolating the saturation curve (Figure 2.3) gives a very rough approximation of *Bmax*, and thus K_D can be obtained. This is accurately done by computerised fitting, but in earlier times was made more accurate by the use of a linear rearrangement. The commonest linear plot is that introduced by Scatchard in 1954 (originally employed to measure radioactive calcium binding to Ca^{2+}-binding proteins).

The results in Figure 2.3 are from a receptor assay of soluble oestrogen receptors. Again, the protocols and raw data are provided in Appendix 1 with an explanation of the experimental problem of 'non-specific' binding.

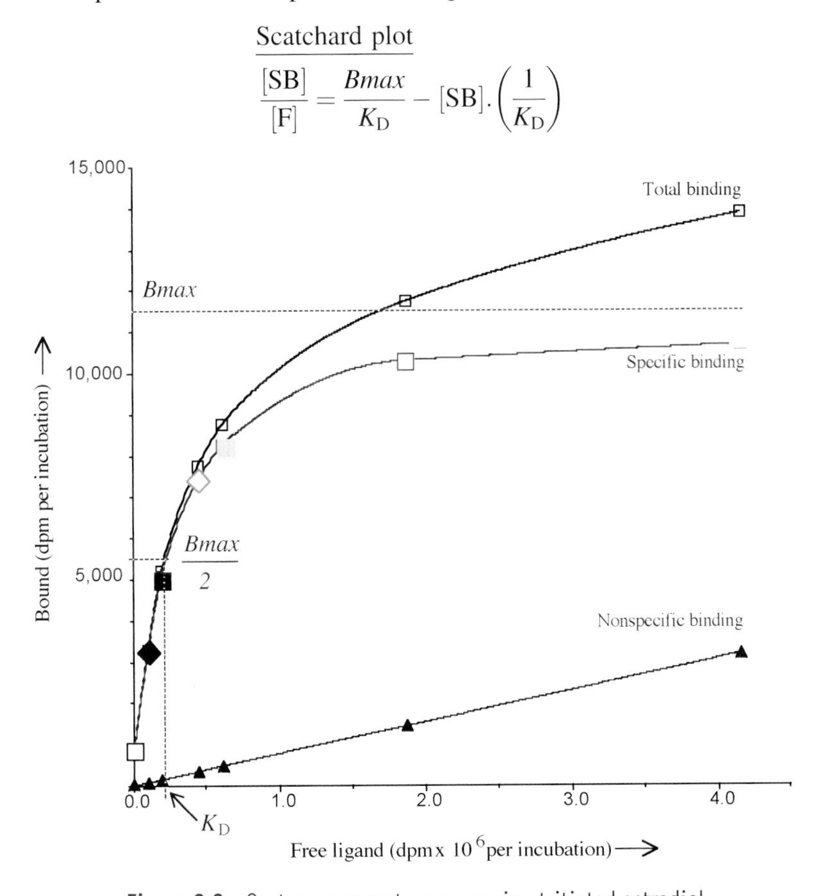

Figure 2.3 Oestrogen receptor assay using tritiated ostradiol

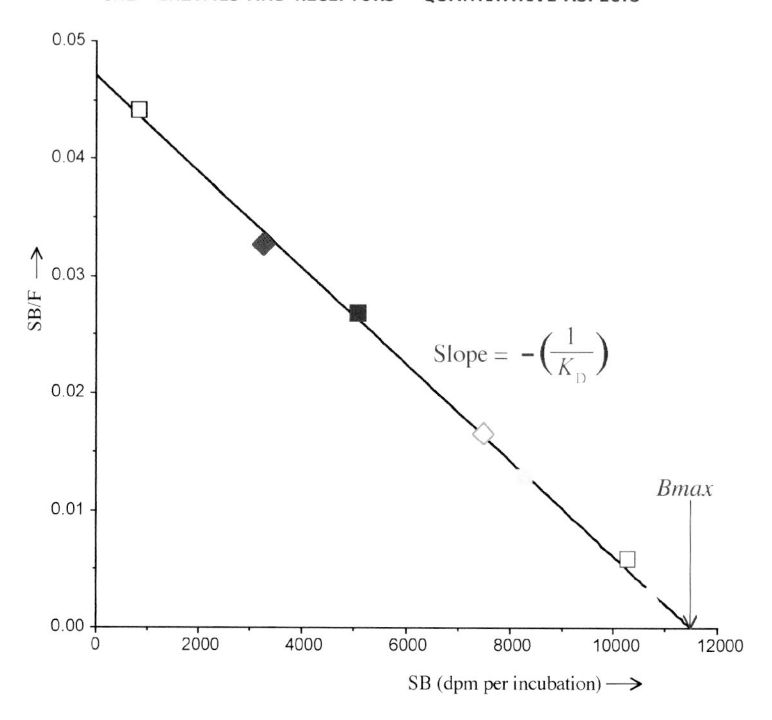

Figure 2.4 Oestrogen receptor assay - Scatchard plot

Alternative arrangements of the same Scatchard equation are

$$[\mathrm{SB}] = \mathrm{Bmax}.\frac{[\mathrm{F}]}{K_\mathrm{D} + [\mathrm{F}]} \; OR \ldots \frac{[\mathrm{SB}]}{[\mathrm{F}]} = \frac{(\mathrm{Bmax} - [\mathrm{SB}])}{K_\mathrm{D}}$$

Looking at the Scatchard plot equation, one can see that it resembles the Eadie-Hoftsee plot for enzyme kinetics. Plotting [SB]/[F] against [SB] gives a linear graph with the form

$$\mathrm{y} = (-\mathrm{mx}) + \mathrm{c}$$

having a negative slope of $-1/K_\mathrm{D}$ and an intercept on the [SB] axis of *Bmax*.

The data from the saturation curve have been re-plotted as a Scatchard plot (see Figure 2.4 and Appendix 1).

2.2.5 Displacement assays

Looking at the worked example of saturation binding analysis, you will appreciate that large quantities of radioligand are required to even approach saturation values.

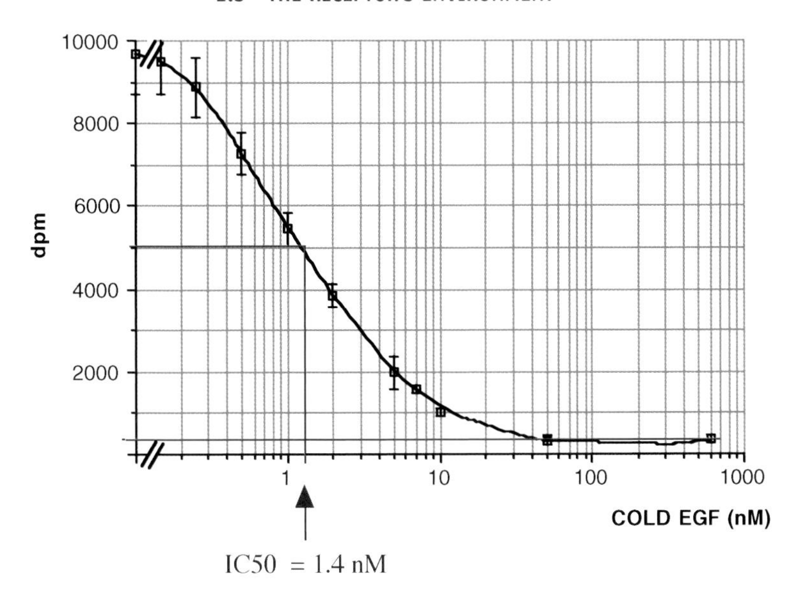

Figure 2.5 EGF receptor assay using the tracer displacement method

Radioactively labelled compounds are exceedingly expensive. If one is only interested in the relative affinity of one ligand compared to another, a cheaper alternative is to use 'tracer' amounts of radioligand in a 'competition' assay (also referred to as 'displacement'). Displacement assays are unable to measure *Bmax*, and K_D values are usually an approximation. An EGF receptor displacement (or 'tracer') assay is shown in Figure 2.5 (data in Appendix 1). Note that the 'x-axis' is logarithmic. This semi-log graph gives a sigmoidal ('S-shaped') curve – halfway between the inflection points is roughly equivalent to half-saturation and the concentration of cold ligand required to achieve this is referred to as the inhibitory concentration for 50% saturation (IC_{50}). Often this is used as a very rough approximation of the K_D.

2.3 The receptor's environment

Cell surface receptors often display very different ligand-binding characteristics depending upon whether they are assayed in whole cells, isolated plasma membranes, or as highly purified native or recombinant proteins. Cellular modulators of receptor function may be lost during membrane purification, often resulting in isolated membrane-bound receptors with apparently different properties than whole cell receptors. Examples of such well-known regulators that modulate affinity include ions, guanine nucleotides, soluble protein kinases and phosphatases. Moreover, the isolated protein, whether native or recombinant, may need to be incorporated into phospholipid vesicles in order to bind ligand – of course, this artificial lipid milieu cannot adequately mimic all of the features of the native bilayer.

7-pass and single-pass receptors can be removed from membranes by dissolving them out of the lipid bilayer with detergent. Some 7-pass receptors that have been solubilised are found to bind ligand with full retention of stereospecificity, even in solution (i.e., when not incorporated into artificial phospholipid vesicles)[1]. However this is not often possible – the choice of detergent appears crucial, with digitonin being most often successful. Interestingly, such soluble 7-pass receptors display monophasic high affinity binding whereas the isolated membrane-bound form displays bi-phasic high- and low-affinity binding. Whole cell assays of the same receptor give monophasic low affinity binding (dealt with in detail later in the chapter).

2.3.1 Heterogeneity of binding sites – positive cooperativity

2.3.1.1 Positive cooperativity

The concept of cooperativity emerged from studies of the binding of oxygen to haemoglobin. Haemoglobin is tetrameric and exhibits positive cooperativity in its uptake of oxygen compared with non-cooperative binding by monomeric myoglobin. The first haemoglobin monomer binds the first oxygen with relatively low affinity, but its binding causes a conformational change in the monomer that is transmitted to the other monomers such that each subsequent binding is made easier – effectively, the affinity for oxygen increases in the empty sites after the first binding event. The saturation of four sites with oxygen, or the loss of all four oxygens, occurs over a very narrow range of O_2 partial pressures and is 'switch-like'. Myoglobin, on the other hand, is monomeric, non-cooperative and binds or releases oxygen over a much wider range, which is more smoothly graded (see Figure 2.6).

Ligand-gated ion channels are prototypical of cooperative multi-protein receptor systems that act as 'hair trigger' sensors.

Classic positive cooperativity is (almost) exclusively seen in multimeric proteins, which contain at least two individual subunits that possess functionally identical ligand binding sites. The (two or more) binding sites each begin with an identical basal ligand affinity, but the first ligand binding event changes the conformation/orientation of its ligated subunit and that is transmitted to the empty subunit, which also changes conformation, giving its binding site a temporarily higher affinity for the next ligand, and so on. Ligand-gated ion channels need this switch-like response because their signal transduction depends only upon *how long* the central pore remains open (or shut) – *time is important, not how 'open' the channel is*[2].

IP3 receptor The IP3 receptor is a tetramer of four identical subunits, each possessing a single IP3 binding site. The 'bundle' of subunits straddles the endoplamic reticulum membrane and forms a central pore (a calcium channel) that can open when adequately stimulated by ligand. When only one or two subunits are occupied, the channels stays closed – it only opens when three or all subunits are occupied[3]. Thus, there is a very sharp discrimination made between IP3 levels that will permit opening or

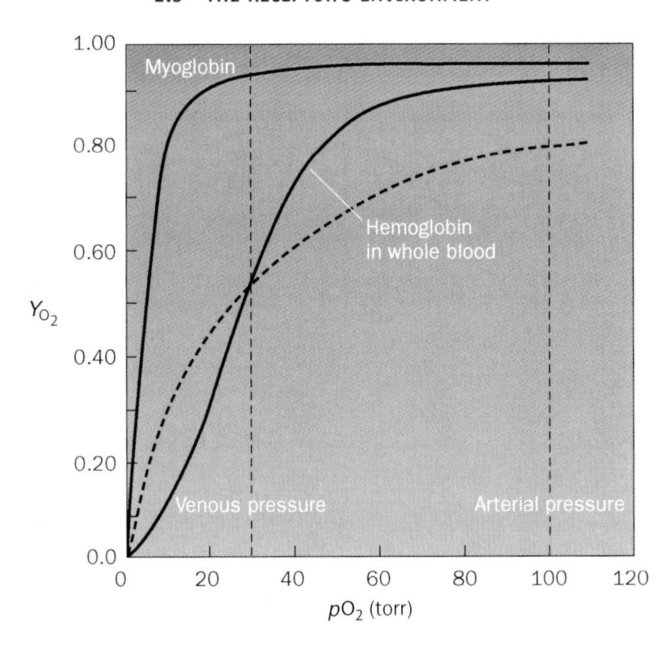

Figure 2.6 Oxygen dissociation curves of Mb and of Hb in whole blood. The normal sea level values of human arterial and venous values are indicated. The dashed curve is a hyperbolic curve with the same as Hb (26 torr). From Voet D. and Voet J. G., *Biochemistry* (Third Edition); Copyright © 2004 by Donald Voet and Judith G. Voet, reprinted with permission of John Wiley & Sons, Inc.

closure of the pore. Without cooperativity, the channel would probably display varying stages of 'openness' at almost all levels of IP3. These receptors are subject to other modulating signals including feedback from calcium and phosphorylation.

2.3.2 Binding site heterogeneity – two site models versus negative cooperativity

With some single-pass receptors, radioligand binding assays produce well-behaved data that yield linear Scatchard plots, indicating that the receptor sites are all identical and independent. Examples include the growth hormone receptor (a non-catalytic single-pass cytokine-like receptor)[4] and the PDGF receptor[5]. However, other such receptors display complex binding patterns that plot as non-linear Scatchard graphs – with very few exceptions, these are 'upwardly concave' (Figure 2.7). This is not the behaviour expected of the 'single independent site' model implicitly assumed in Scatchard analysis. This inconvenient behaviour is incompletely understood and has been the subject of considerable debate over the decades. The interpretation of these binding curves can be somewhat arbitrary without additional evidence to favour one model over another. Fortunately, advances in our knowledge of receptor-modulating cytoplasmic

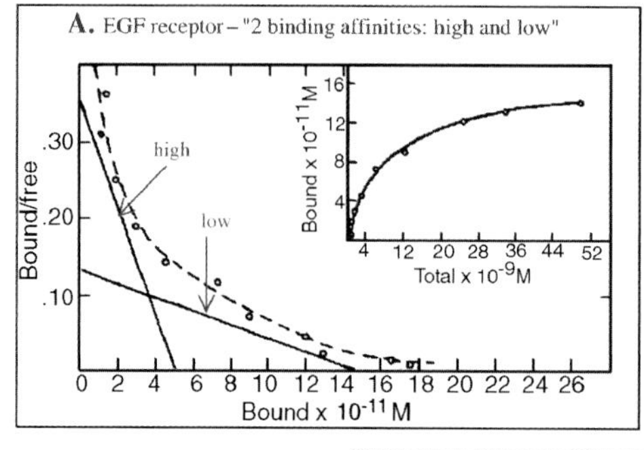

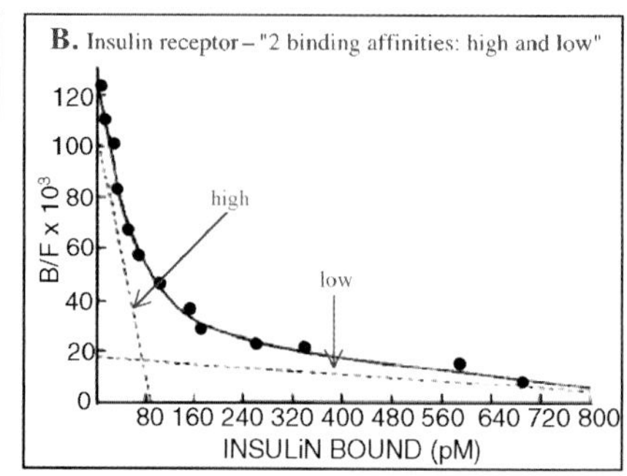

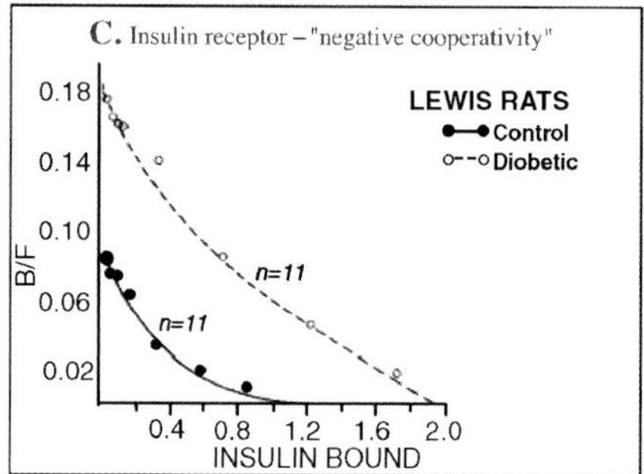

Figure 2.7 Curvilinear Scatchard plots of ligand-binding to single-pass receptors. (A) Reproduced with kind permission from Livneh, E., Prywes, R., Kashles, O., Reiss, N., Sasson, I., Mory, Y., Ullrich, A. and Schlessinger, J. (1986) Reconstitution of human epidermal growth factor receptor and its deletion mutants in cultured hasmter cells. *J. Biol. Chem.*, **261**: 12490–12497. (B) Reproduced with kind permission from Rubin, C.S., Hirsch, A., Fung, C. and Rosen, O.M. (1978) Development of hormone receptors and hormone responsiveness in vitro. *J. Biol. Chem.*, **253**: 7560–7578. (C) Reproduced with kind permission from Davidson, M.B. and Kaplan, S. (1977) Increased insulin binding by hepatic plasma membranes from diabetic rats. *J. Clin. Invest.*, **59**: 22–30

interactions and recent detailed structural information on how ligands dock with receptors have clarified previous interpretations of biphasic radioligand binding results, at least in some cases.

Upwardly-concave 'curvilinear' Scatchard plots have been repeatedly and uniformly obtained from radioligand binding assays on both insulin receptors and EGF receptors from a variety of sources when whole cells or native membrane preparations are used.

All single pass receptors dimerise to create a holoreceptor with two binding surfaces, and it is this arrangement that gives them the *potential* to 'cooperate'. The question is: to what extent are curved Scatchards a sign of cooperativity or heterogeneity? Frequently cited sources of artefactual curvilinearity are: (a) non-specific binding not being fully eliminated; (b) internalisation of the radioligand (in whole cell assays) creating a non-dissociable subset of receptors; or (c) endogenous ligand pre-occupying a subset of receptors (low affinity due to the need for ligand exchange). Nevertheless, none of these artefactual arguments explains how so many well-controlled studies in different systems get the same result – upwardly-concave Scatchard plots for insulin or EGF binding.

Taken at face value, and with nothing else to go on, upwardly-*concave* curvilinear Scatchard plots may be legitimately explained in two main ways:

- ***A.*** There are two (or more) pre-existing independent sites, each with an intrinsically different ligand affinity ('receptor heterogeneity'); binding at one site does not affect the other.

- ***B.*** There is a single site with a single basal affinity that is subject to negative cooperativity during ligand binding, giving rise to two (or more) observed affinities. At one theoretical extreme, this may mean a continuously varying spectrum of (non-discrete) affinities.

Simple equilibrium binding assays cannot discriminate between these two possible scenarios. Indeed the same InsR binding data set successfully fits mathematically into both models[6]. Note that the appearance of a 'dog leg' in the curve (perhaps suggestive of two sites) is an illusion of scaling. These curves are asymptotes, and remain microscopically so if one zooms in on (seemingly) linear sections.

If ***A.*** were true, there should be two subset of receptors in basal conditions. During assay, the ratio of the two types should not change with increasing ligand occupancy. Proof of ***A.*** would lie in the ability to purify separately, and study the two types of receptor.

If ***B.*** were true, the initial binding events should be high affinity and the later ones low affinity. The binding sites should be the same in basal conditions, with the low affinity site only observable when ligand-occupancy approaches some threshold level. In other words, binding at the 'empty site' is made less avid by initial binding at the 'filled site'.

Both models imply that there would be a biphasic dissociation of radioligand from receptor when ligand is removed and the preparation diluted in ligand-free buffer (slow off-rate = high affinity; fast off-rate = low affinity). However, it is argued that biphasic dissociation from negatively cooperative receptors (***A.***) should be sensitive to presence of unlabelled ligand[7].

2.3.3 Negative cooperativity of the insulin receptor or two site model

InsR radioligand receptor binding assays give upwardly *concave* curvilinear Scatchard plots. These have most often been adduced as evidence of negative cooperativity rather than independent site heterogeneity. In early studies the non-linearity was cited as representing two sites (high or low affinity) but there has been debate about whether they are two independent sites or a manifestation of negative cooperativity between the monomeric halves of the InsR holoreceptor[6]. Nonetheless, we now know that the insulin receptor is a good candidate for such cooperative behaviour, being a persistent dimer and having a bivalent ligand. Insulin-binding 'bridges' (or 'crosslinks') the two α-chain ectodomains by virtue of the fact that it has two distinct binding surfaces that pair up with two alternate binding domains on each ectodomain. In a sense, the single insulin ligand allows the two receptor halves to 'communicate' their occupancy status.

The curvilinear InsR Scatchard plots are clearly not an experimental artefact but something unique to insulin – a counter-intuitive proof of their authenticity is that under identical conditions, IGF-I gives linear Scatchard plots whereas insulin produces upwardly-concave curves in membrane assays[8]. Furthermore, in uniform conditions the dissociation of IGF-I from a soluble dimerised IGF-IR ectodomain is monophasic whereas dissociation of insulin from the same IGF-IR ectodomain dimer is biphasic (two off-rates)[9]. This suggests that it is the ligand, not the receptor, which determines this effect. Indeed, insulin and IGF-I binding sites are in different regions of the receptors' α-chains – a primary site for insulin is centred on the 'L1' domain whereas the primary IGF-I site is the cysteine-rich domain[10].

Decades of research (beginning with photoaffinity labelling, followed by site-directed mutagenesis and later receptor domain swapping) has painstakingly identified the regions of insulin and the receptor that are needed for full ligand binding activity with associated negative cooperativity. To summarise, the 'classic' insulin binding site is centred on the L1 domain whereas a second site has been identified as comprising the L2 domain and part of the *C*-terminal portion of the α-chain, namely the fibronectin domain 0 (Fn0)[11]. The domain structure of InsR is dealt with in Chapter 9.

Site-directed mutagenesis has identified residues needed for the interaction of insulin and InsR. Insulin's critical residues roughly map to opposite ends of the molecule. The *C*-terminal portion of insulin's B-chain binds (mostly) hydrophobically to 'site 2' of InsR (the 'cooperative site' that is encoded by exon 2 – 'L1'); a smaller interaction site centred on insulin's A-chain makes salt-bridge contacts with 'site 1', the product of InsR exons 6 and 7 – 'L2' and Fn0[7,12]. Electron microscopy (EM) has visualised two distinct conformational states of InsR: unoccupied and insulin-occupied. The unoccupied form of the ectodomain is in an open state with the two monomer halves splayed apart; in the occupied form, the ectodomain is closed around a single gold-labelled insulin molecule. Guided by the mutational and receptor binding studies, crystal structures of the sub-domains of InsR/IGF-IR have been carefully mapped onto the EM structure model of the ligand-occupied InsR and measurements of interacting residues are consistent with two distinct sites (see Figure 2.8)[12]. This construct illustrates how prescient the early models of negative cooperativity were[7,13].

A highly simplified scheme is presented in Figure 2.8 based on the above-cited models of insulin-InsR interaction[7,10,11,13]. The first ligand binding events are bridging (or 'crosslinking') and this accounts for the high affinity. Insulin binds to one InsR monomer first, probably through long range electrostatic effects between 'site 1' of insulin and the L2/Fn0 domains, immediately followed by binding to the other monomer through insulin's opposite face, 'site 2', which binds predominantly hydrophobically.

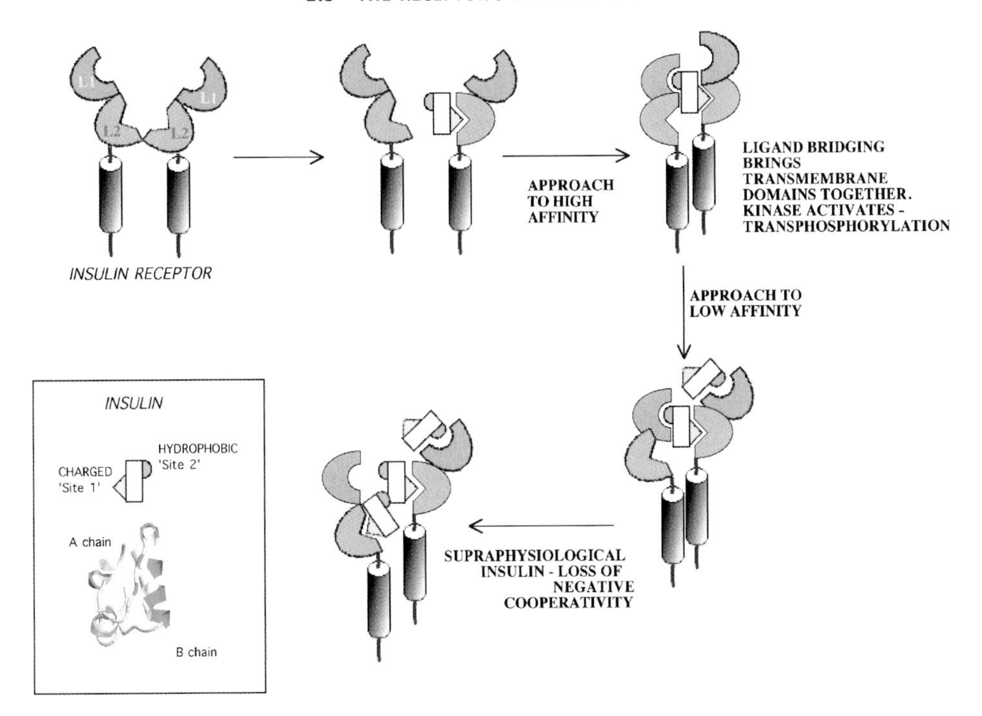

Figure 2.8 Simplified model of insulin receptor ligand binding and negative cooperativity

This bridging action creates a high affinity site that brings the monomers closer together and is speculated to cause lateral movements of the transmembrane domains such that the InsR tyrosine kinase domains are brought close enough to *trans*-autophosphoryate. The ability to create this novel bi-partite site is dependent upon the monomers being persistently held together by the two interchain disulphide bonds.

The bridging effect of low concentration insulin obviously leaves two empty sites – a 'site 1' binding region on one monomer and a 'site 2' on the other. Binding at the left-over 'site 1' region at higher insulin concentrations is thus rendered low affinity because the initial bridging ligand precludes formation of a second bridge. However, both 'site 1' and 'site 2' binding regions on InsR overlap[12]. So, when insulin levels are high enough, the initially bound insulin will begin to be displaced (i.e., an insulin-accelerated off-rate). At super high concentrations, the acceleration of dissociation by cold insulin (negative cooperative effect) is lost, presumably because the bridging is suppressed by overwhelming one-to-one interactions of ligands with both sites.

2.3.4 Site heterogeneity of the EGF receptor – independent two site model?

Like InsR, EGFR assays on whole cells and isolated membranes give upwardly *concave* curvilinear Scatchard plots (for example, Figure 2.7). However with EGFR, these have classically been interpreted to represent true receptor heterogeneity. Usually, this is taken to mean that there are likely to be two independent EGF binding sites: one high affinity, the other low affinity. In other words, in a cell (or its isolated plasma membranes) there are actually two populations of EGFR: the majority (>80%) of the membrane-bound EGFRs are low affinity, with a smaller number of them displaying high affinity. The absence of negative cooperativity has been recently confirmed by ingenious pulse-chase experiments, wherein the dissociation rate of labelled EGF from receptors was unaffected by addition of unlabelled EGF[14].

Attempts to separate and study these native entities physically have been somewhat inconclusive, but there are several ways to modulate their ratios, which strongly suggests that they are independent realities.

Treatment of cells with phorbol esters such as PMA can selectively eliminate the high affinity component of EGFR (see Figure 2.9)[15,16]. PMA activates conventional forms of protein kinase C (PKC) and this leads to phosphorylation of threonine654 in the cytoplasmic juxtamembrane region of EGFR and subsequent selective down-regulation of the high-affinity state. In contrast, sphingosine treatment shifts the EGFR population to a high affinity state (Figure 2.9)[17].

The EGF receptor exists as a monomer in the absence of ligand. EGF binds to the receptor monomer through two regions L1 and L2 and this leads to dimerisation with another occupied EGFR with the result that the activated homodimer contains two EGF ligands. Although the interaction of the two binding sites within the individual monomer with a single EGF looks suspiciously like a positive cooperative[18], it appears that high affinity binding depends upon formation of a dimer and is suggested to be dependent upon prior exposure of one half to EGF[19,20]. Again, this appears to suggest positive cooperativity.

In contrast to membrane-inserted EGFR, detergent-solubilised native EGFR gives linear Scatchard plots, at least when dissolved at a low concentration. But more concentrated solutions of solubilised native EGFR, or its ectodomain, display evidence of positive cooperativity – i.e., Hill plots (see Appendix 1) with a gradient 1.7[21]. Obviously, in highly concentrated solutions, EGFR is more likely to collide and homodimerise (in 3-dimensional space) and it is argued that this high concentration may mimic a calculated concentration of EGFR in (2-dimensional) native membranes.

Unfortunately, at the present time there is simply no easy way to explain the behaviour of membrane inserted native EGFR without resorting to a hypothetical cellular or membrane modulator (or 'external site') that the ligand-occupied receptor binds to and is thereby stabilised in a high affinity state[18]. This 'external site' has not been identified, but is thought to bind to the cytoplasmic side of the receptor. It is tempting to speculate whether this may have something to do with the reversible interaction of the cytoplasmic juxtamembrane region of EGFR with the lipid bilayer proposed by McLaughlin and co-workers to control its tyrosine kinase autoinhibition and activation[22]. I hasten to add that this is mere speculation on my part and not an hypothesis advanced by the authors, who confined their model to the receptor's enzymic activity alone (discussed in Chapter 6, Section 6.7).

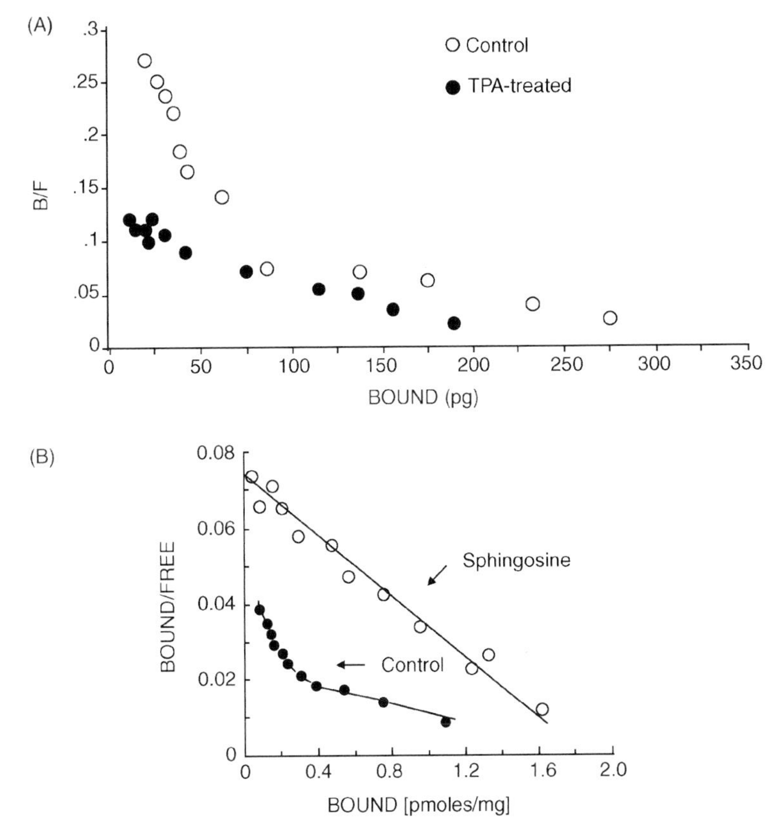

Figure 2.9 Differential effects of (A) phorbol ester or (B) sphingosine on high affinity of component of EGFR. Reproduced from Decker, S.J., Ellis, C., Pawson, T. and Velu, T. (1989) Effects of substitution of threonine 654 of the epidermal growth factor receptor on epidermal growth factor-mediated activation of phospholipase C. *J. Biol. Chem.*, **265**: 7009–7015. Davis, R.J., Girones, N. and Faucher, M. (1988) Two alternative mechanisms control the interconversion of functional states of the epidermal growth factor receptor. *J. Biol. Chem.*, **263**: 5373–5379, 1988 with permission from the American Society for Biochemistry and Molecular Biology

2.4 Guanine nucleotides and the agonist 'affinity-shift' of 7-pass receptors

Therapeutic drugs directed at adrenergic receptors (and other pharmacologically interesting 7-pass receptors) are often characterised by radioligand 'displacement' binding assays performed upon plasma membranes purified from cells or tissues. These assays employ a tracer amount of radiolabelled neutral ('competitive') antagonist – 'displacement' of the label by increasing concentrations of unlabelled agonist or antagonist gives an estimate of their respective affinities for the receptor (see Section 2.2.5). A signature characteristic of 7-pass receptors in native plasma membranes is that antagonists bind with a single affinity but agonists often display bi-phasic displacement that can be modelled as being two-site: (i) a high affinity component that ranges from a low level to as high as 92% of total binding; and (ii) a low affinity component that makes up the rest (see Figure 2.10). This bi-phasic agonist binding can be reduced to a mono-phasic low affinity curve by the addition of non-hydrolysable GTP analogues, $GTP\gamma S$ or Gpp(NH) p, either of which can cause a shift in affinity from high to low[23]. So one can sort antagonists from agonists because the former are unaffected by $GTP\gamma S$, while the latter shift affinity in its presence.

2.4.1 The ternary complex 'equilibrium' model

The classic paradigm holds that the agonist-receptor-$G\alpha\beta/\gamma$ 'ternary complex' is the high affinity form whereas the agonist-receptor complex is low affinity. The high agonist affinity form is thought to be disrupted by addition of non-hydrolysable GTP analogues because they trigger uncoupling of $G\alpha$ from the agonist-occupied receptor, leaving the latter in the uncoupled, low affinity state. There has been a common assumption that the GDP-occupied G protein is the form that is found in the high affinity ternary complex, but that does not explain why *both* GDP and GTP have also been observed to produce agonist affinity shifts in 7-pass receptors[24,25,26,27].

The classical ternary complex model[23,28] (Figure 2.10) has been extended and elaborated over the years to account for the apparent diversity of ligand activities observed in membrane assays and bioassays as well as for the finding that constitutively activated 7-pass receptor mutants can activate G proteins in the absence of agonist. Later versions of this model[29] are exceedingly complex and contain much that is theoretical (for a critique, see Reference 2). In the simple form of the model, the receptor (R) exists in two states that are in equilibrium with one another. In the absence of the hormone or agonist ('A'), low agonist-affinity 'R' predominates, but a small proportion is in the high agonist-affinity 'R*' state. R is a biologically inactive conformation but R* is in an active conformation that can couple to a heterotrimeric G protein that is GDP-occupied: $G\alpha<GDP>\beta/\gamma$. Neutral antagonists bind to both R and R* with the same affinity but agonists, which bind preferentially to R*, tend to stabilise this activated form. The result is that the equilibrium is shifted in favour of R* by agonist binding, and the rise in [AR*] is further stabilised by its coupling with $G\alpha<GDP>\beta/\gamma$ (see Figure 2.10). Dissociation of GDP from this ternary complex and replacement with GTP is accompanied by uncoupling of the high affinity complex. $G\alpha<GTP>$ is free to activate its effector enzyme (E) and uncoupled R* reverts to the low affinity R state.

With uniform materials (single membrane source, constant level of protein expression) and well-controlled conditions (unvarying membrane isolation and assay procedures), agonist affinity shift membrane assays are pharmacologically predictive in screening compound libraries. For example, they can identify (and partially characterise) antagonists more quickly than complex biological endpoint assays such as animal experiments. The extended ternary model is most frequently used to aid the

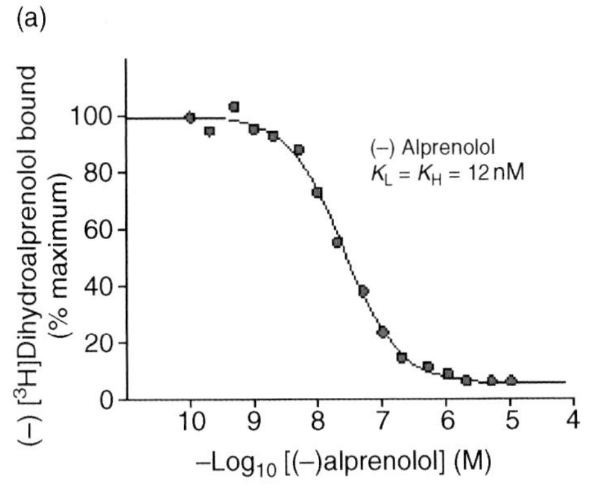

(a) Displacement of tracer radioligand (tritiated antagonist, dihydroalprenolol) by unlabelled antagonist (alprenolol) is monophasic and unaffected by guanine nucleotides.

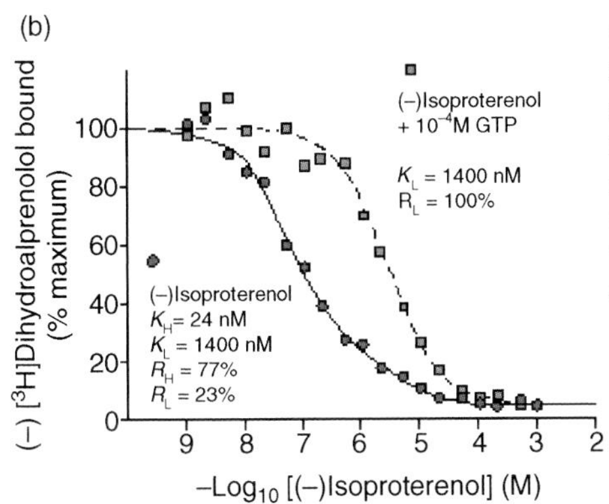

(b) Displacement of tracer radioligand by agonist (isoproterenol) is bi-phasic in the absence of guanine nucleotides, and resolves into two components: a subset of receptors with high agonist affinity and another with low affinity. Addition of GTP 'shifts' the high affinity receptors to low affinity - binding is now monophasic.

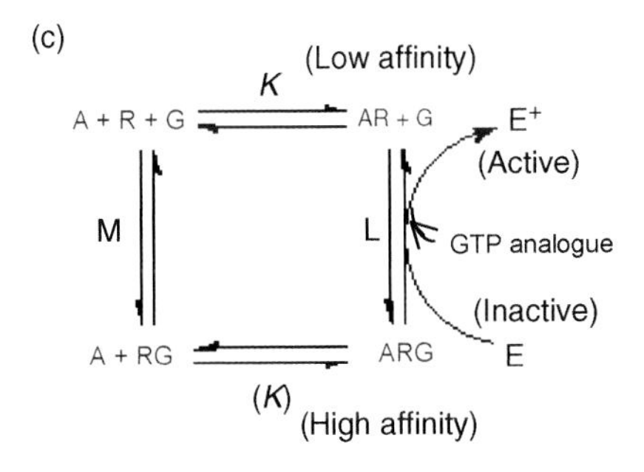

(c) The classical ternary complex equilibrium model postulates that the receptor exists in at least four different pools (free, liganded, unliganded but G-coupled and the ternary complex); interchange between pools is governed by four dissociation constants (K, L, K', M). The ternary complex (ARG) is the high agonist affinity form that is disrupted by addition of guanine nucleotide (usually in the form of non-hydrolysable GTP analogues like GTP-γ-S). 'E' = effector enzyme.

Figure 2.10 The ternary complex equilibrium model cap

classification of ligands on the basis of observed affinity shift, among other endpoints. However, despite its success there are experimental findings that question the model's relevance to how the system actually works *in vivo*.

If the ARGα<GDP>β/γ complex is the high agonist-affinity form, how can one account for the fact that GDP can also cause an affinity shift just like GTP? High agonist affinity is only observable in washed or re-constituted membranes that contain both receptor and cognate G protein and, most importantly, the assay must be carried out in the complete absence of soluble guanine nucleotides. In guanine nucleotide-free conditions, how does the putative high agonist-affinity ARGα<GDP>β/γ complex survive if GDP is continuously dissociating from it? As Colquhoun *et al.* point out, 'the GTP-hydrolysis cycle is inherently *not* an equilibrium process, and some sort of (quasi-) steady-state treatment is likely to be essential, as a minimum'[2].

2.4.2 The 'empty pocket' form of Gα

Questions like these and others have lead to the proposal that the high affinity agonist binding observed in nucleotide-free membrane assays, represents agonist-occupied receptors that are coupled to G proteins in the empty-pocket form (Gα<EMPTY>βγ), the GDP having dissociated during the assay period. There is recent support for this suggestion in the work of Kobilka's group[26], who have shown that a recombinant protein of βAR fused with Gαs exhibits bi-phasic agonist binding in the absence of guanine nucleotide, whereas the recombinant βAR alone, displays low affinity binding only. These experiments were done by expressing the proteins in insect cells (Sf9 cells) that do not contain G proteins capable of coupling to mammalian 7-pass receptors. When affinity-shift assays were done on isolated plasma membranes from the respective clones, it was found that addition of GTPγS, GTP or GDP abolished high-affinity binding of the βAR-Gα's fusion receptor but GTPγS had no effect on the affinity of the singly-expressed βAR[26].

2.4.3 The thermodynamic 'catalytic' or 'kinetic' model

An alternative model has been proposed that appears better able to account for the high agonist-affinity state and the affinity-lowering effects of GDP. This was first proposed by Marc Chabre to account for the kinetics of the rhodopsin-transducin cycle[30,31] and later by Magali Waelbroeck to describe the kinetics of muscarinic receptor G-coupling[27,32]. Indeed, Chabre has successfully modelled the rhodopsin-transducin cycle using simple Michaelis-Menten kinetics (for his analysis, the reader is referred to Reference 33). The following treatment is much simplified.

One should think of the 7-pass receptor as a kind of catalyst, in that a single agonist occupied receptor can activate several G proteins, one-by-one, in turn. The bonds made and broken in the conformational changes of the G protein are not covalent, so this is not classically 'enzymatic'. Nevertheless, large changes in free energy accompany the coupling/uncoupling of G proteins to receptors, the dissociation of GDP, the association of GTP, and its ultimate hydrolysis back to GDP. The thermodynamic catalytic model recognises that important steps in the activation cycle of 7-pass receptors are very far from equilibrium and, being practically irreversible, their kinetics are more akin to enzymic catalysis. For example, the activated receptor acts as a 'guanine nucleotide exchange factor' when it couples to a GDP-occupied G protein. Although there is no making or breaking of covalent bonds, the agonist-occupied receptor nevertheless induces significant conformational change in the G protein, particularly in the 'switch' regions (see Chapter 8). In other words, the activated receptor acts like an isomerase, contributing high levels of binding energy to re-model the GDP-binding site such that GDP is expelled from Gα. In living cells, this is practically an irreversible step because *in vivo* the direction of flux is further strengthened by the immediate binding of GTP to the α-subunit, which causes a further conformational change in the α-subunit. This causes loss of the β/γ subunit and de-coupling of the receptor. The disassembled G protein can no longer interact with a receptor and so is not in equilibrium with any of the receptor pools. The energy barrier leading to GDP dissociation is lower (Figure 2.11 dashed line) in the presence of the agonist than in its absence (Figure 2.11 solid line) because agonist binding causes an activating conformational change in the receptor that happens more rarely in the absence of agonist.

The thermodynamic model posits that the high-affinity agonist-binding state of 7-pass receptors is the nucleotide-free form of the ternary complex, i.e., AR**G*α<EMPTY>$\beta\gamma$. The following presents a summary of the model and supporting evidence. (Note that many intermediate steps are left out for simplicity.)

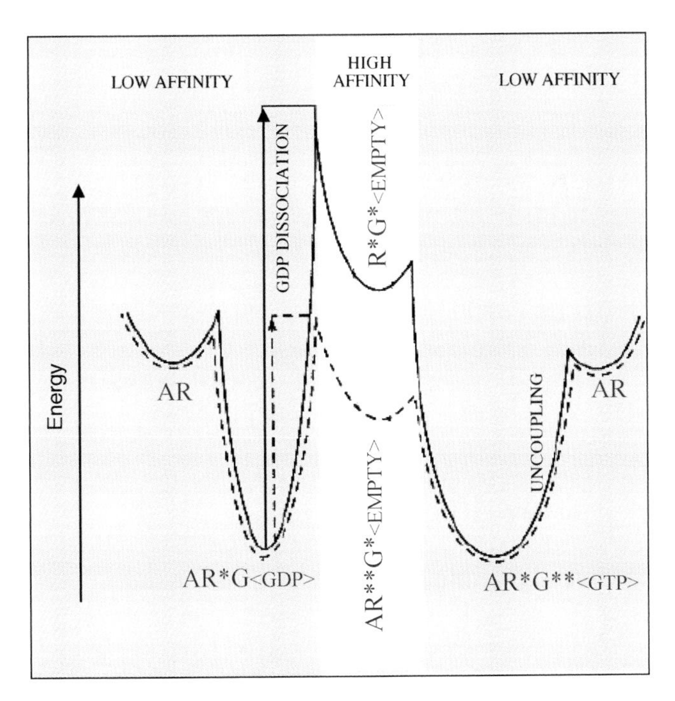

Figure 2.11 A thermodynamic model of ligand binding and G-protein coupling, and indicating free energy changes between each state

(**A**) 7-pass receptors that bind soluble, diffusible ligands have two natural binding sites: one is an agonist-binding site (that is located in the extracellular-domain or TM-region), and the other is an intracellular Gα<GDP>βγ-coupling site (in the *C*-terminal tail and IC3 loop of the receptor).

The ligand-free receptor oscillates between two conformations:

- R (inactive – the intracellular coupling site has no affinity for Gα<GDP>βγ)
- R* (active – the intracellular coupling site has high affinity for Gα<GDP>βγ)

In the absence of agonist, R is largely predominant. A small proportion of R* may account for possible constitutive activity of receptor. Agonist affinities of R and R* are not separately measurable. Upon agonist binding, all the occupied receptors are in the active AR* conformation. It is often assumed that the ligand binds only to R* and stabilises its active conformation, but kinetic arguments suggest that the agonist binds as well to the predominant R conformer and converts it to R*.

A third conformation may be observed for the agonist-occupied receptor upon its coupling to the G protein and following its ensuing release of GDP from the bound Gα:

$$\mathrm{AR}^* + \mathrm{G}\alpha <\mathrm{EMPTY}> \beta\gamma \rightarrow \mathbf{AR^{**}\ G^{*}}\ \alpha <\mathbf{EMPTY}> \beta\gamma + \mathrm{GDP}$$

In this AR**G* α<EMPTY>βγ complex, the receptor has high affinities for both agonist and the G protein. The GDP that was stochiometrically bound to Gα is released into the assay buffer where it dilutes to a concentration below its affinity for the complex. The AR**G* α<EMPTY>βγ complex is stable in isolation as long as GTP is excluded from the system, for example by the washing procedures involved in the isolation of plasma membranes.

(**B**) In the successive complexes, the heterotrimeric G protein's α-subunit exists in three states and each has a different conformation (crystal and NMR structures are available for the three conformations of monomeric G proteins, see Chapter 7):

- Gα<GDP>βγ (Gα has no affinity for R, but has a high affinity for both AR* and GDP)
- AR**G*α<EMPTY> (G*α has a low affinity for GDP, but has high affinity for both AR** and GTP)
- G**α<GTP> (G**α has no affinity for either AR* or β/γ, but has high affinity for both effector enzyme and GTP)

2.4.4 Constitutive signalling in the absence of ligand

It is widely accepted that 7-pass receptors can weakly signal, constitutively. To produce this basal ligand-free signalling, the receptor spontaneously switches conformation from R to the rarer R* state. The more constitutively active the receptor is, the higher the percentage of time spent in the R* state.

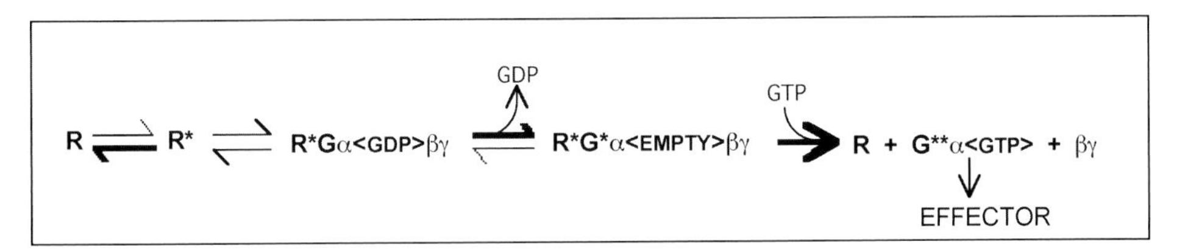

Figure 2.12 Constitutive signalling - agonist affinity of the receptor is invisible

Agonist binding to membrane bound receptor in absence of G protein This system (Figure 2.12) is, of course, not open to interrogation with radioligand. However, in membrane assays of receptors expressed in cells *without* cognate G protein (e.g., Sf9 insect cells), agonist binding to a 7-pass receptor is overwhelmingly mono-phasic and low affinity. Even the higher constitutively-signalling mutant receptors show mono-phasic low affinity agonist binding in the absence of G protein. So, although agonist changes the receptor's conformation from R to R* and therefore shifts the equilibrium in favour of AR*, the agonist affinity is unchanged. This indicates that the agonist affinity is similar for the R and R* states.

2.4.5 The effect of limiting concentrations of G protein on agonist binding

Preparation of plasma membranes for receptor assay typically entails repeated centrifugation. This effectively washes out any guanine nucleotides that are not already bound to G proteins. When GTP is omitted (and its synthesis from GDP is also suppressed) during the receptor assay, the limiting amount of Gα<GDP>βγ available for coupling results in a proportion of agonist-occupied receptors building-up in a high affinity (low off-rate) state. The reason for this is that as binding and coupling proceed, Gα<GDP>βγ quickly runs out. Without GTP, occupied receptors that have coupled and catalysed GDP release become trapped in a nucleotide-free complex with G (see Figure 2.13). The rest of the occupied receptors, having had no chance to couple, remain in a low affinity state.

Gα<GDP>βγ cannot be recreated from Gα<EMPTY>βγ because there is not enough GTP. Adding unlimited GTP to such a system allows G protein recycling, there is then a steady supply of Gα<GDP>βγ, and the high affinity sites no longer accumulate. In the presence of GTP, this low agonist-affinity pool presumably contains AR and AR* only.

2.4.6 Agonist binding in membrane preparations where the cognate G protein is in unlimited supply

When G protein is unlimited, GTP suppression will cause a higher (or even total) accumulation of agonist-receptor in the high affinity state. Again, this is reversed by adding GTP, which induces mono-phasic low affinity binding. With excess G protein and GTP, AR**G*α<EMPTY>βγ is transitory and undetectable.

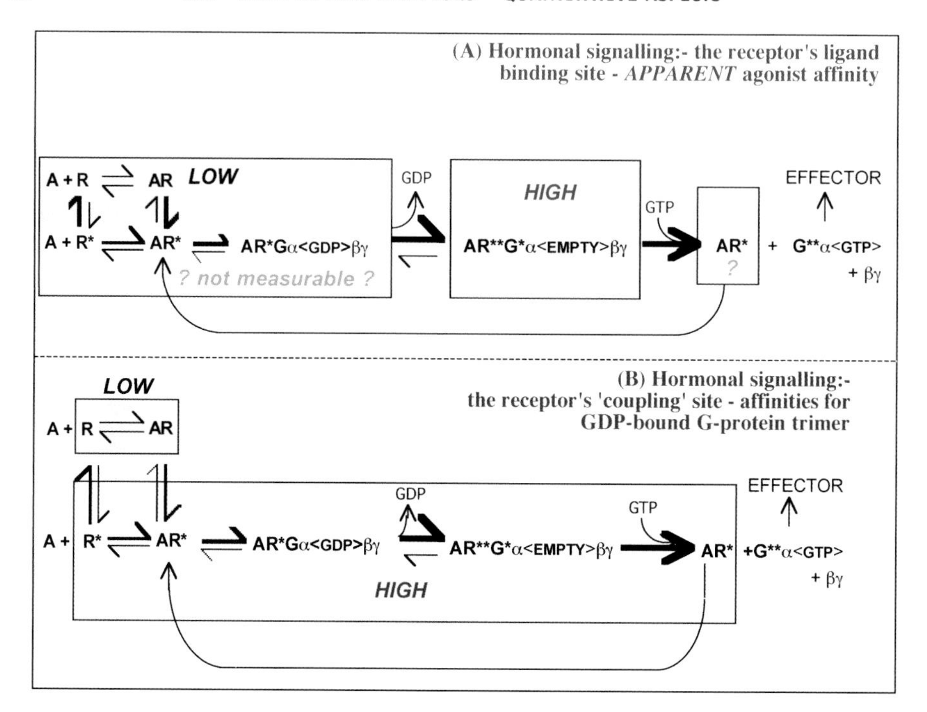

Figure 2.13

Support for the aforementioned effects of varying G protein:receptor ratios upon apparent agonist affinity ratios has been obtained experimentally[34,35].

2.4.7 *In vivo* GTP versus GDP concentrations

One final problem with the ternary equilibrium model is the oft-quoted assertion that (*in vivo*) the dissociation of GDP from the guanine-binding pocket is immediately followed by occupancy with GTP because 'GTP is at a 30- (or sometimes 10-) fold higher concentration in the cytoplasm'. If this latter statement were true, an equilibrium model could easily explain how de-coupling is favoured after ligand binding. However, there is little evidence for such a large difference in concentrations. A single supporting reference *estimates* that the cellular concentration of GTP is at least 30-fold higher than GDP; these measurements were made in frog oocytes[36].

More often, the tissue concentrations of GTP and GDP are found to be close to equal ($\approx$ 1-2·5 mM)[37,38,39,40]. With equal concentrations, GTP and GDP would be expected to compete equally as they have the same on-rate, and a purely equilibrium binding process would be expected to produce very sluggish de-coupling. G<GTP> would be continuously reverting to G<GDP> because of the high levels of GDP present. The corollary would be that, as well as being sluggish, the absolute amount of activation with equimolar GDP background should be half that found in its absence. This is not what is found[33].

The 'kinetic' or 'thermodynamic' model considers the step after GTP binds to be irreversible. In other words, after each G<GTP> is formed, it is completely removed from the 'equilibrium'– because of the 'catalytic' effect of GTP-binding, there is a continual forward flux that is not so badly affected by an equimolar background of GDP. For example, the maximum rate of transducin activation by rhodopsin is 1 · 2 ms in the complete absence of GDP but only increases modestly to 2·4 ms when GTP and GDP are equimolar[33].

The crucial point is that, although it takes longer, a single flash of light achieves the same level of transducin activation whether GDP is absent or is equimolar with GTP[33].

As mentioned earlier, the agonist affinity-shift assay has been intensively utilised to 'sort' ligands into various functional categories. Despite the contradictory interpretations of the affinity-shift mechanism, the two models do not necessarily disagree with its classification outcomes.

Very recent research has proven the existence of metastable activated-receptor/ G*α<EMPTY>βγ complex in the form of an activated, detergent-solubilised rhodopsin bound to nucleotide-free $G\alpha_T$ (transducin)[41]. The α-subunit in the trapped complex is structurally distinct from both the nucleotide-occupied and the *uncoupled* nucleotide-free forms of the α-subunit. The receptor-bound 'empty pocket' Gα is conformationally dynamic, whereas isolated 'empty pocket' Gα has a stable conformation. This proves that it is the activated receptor that prepares the subunit for GTP loading (acting as a 'guanine nucleotide exchange factor'), because it is the interaction of the activated receptor with the G protein that causes its profound conformational change rather than of the simple loss of GDP alone. An upsurge of interest in the nucleotide-free forms of heterotrimeric G protein α-subunits and monomeric G proteins has flowed from new structural studies. Their interactions with guanine nucleotide exchange factor (like 7-pass receptors or mSos) are the subject of intense research and, in the coming years, more light will no doubt be shed on these key intermediate states.

References

1. Caron, M.C. and Lefkowitz, M. (1976) Solubilization and characterisation of the β-adrenergic receptor binding sites of frog erythrocytes. *J. Biol. Chem.*, **25**: 2374–2384.
2. Colquhoun, D. (1998) Binding, gating, affinity and efficacy: the interpretation of structure-activity relationships for agonists and of the effects of mutating receptors. *British Journal of Pharmacology*, **125**: 924–947.
3. Meyer, T., Holowka, D. and Styer, L. (1988) Highly cooperative opening of calcium channels by inositol 1,4,5-trisphosphate. *Science*, **240**: 653–656.
4. Smal, J., Closset, J., Hennen, G. and de Meyts, P. (1985) Receptor-binding and down-regulatory properties of 22000-Mr human growth hormone and its natural 20000-Mr variant on IM9 human lymphocytes. *Biochem. J.*, **225**: 283–289.
5. Kelly, J.D., Haldeman, B.A., Grant, F.J., Murray, M.J., Seifert, R.A., Bowen-Pope, D.F., Cooper, J. A. and Kazlauskas, A. (1991) Platelet-derived growth factor (PDGF) stimulates PDGF receptor subunit dimerisation and intersubunit *trans*-phosphorylation. *J. Biol. Chem.*, **266**: 8987–8992.
6. Mamounas, M., Gervin, D. and Englesberg, E. (1989) The insulin receptor as a transmitter of a mitogenic signal in Chinese hamster ovary CHO-K1 cells. *Proc. Natl. Acad. Sci. USA*, **86**: 9294–9298.

7. De Meyts, P. (1994) The structural basis of insulin and insulin-like growth factor-I receptor binding and negative co-operativity, and its relevance to mitogenic versus metabolic signalling. *Diabetologia*, **37**: S135–S148.
8. Ney, D.M., Huss, D.J., Gillingham, M.B., Kritsch, K.R., Dahly, E.M., Talamantez, J.L. and Adamo, M.L. (1999) Investigation of insulin-like growth factor (IGF)-I and insulin receptor binding and expression in jejunum of parenterally fed rats treated with IGF-I or growth hormone. *Endocrinology*, **140**: 4850–4860.
9. Jansson, M., Hallen, D., Koho, H., Andersson, G., Berghard, L., Heidrich, J., Nyberg, E., Uhlen, M., Kordel, J. and Nilsson, B. (1997) Characterisation of ligand binding of a soluble human insulin-like growth factor receptor variant suggests a ligand-induced conformational change. *J. Biol. Chem.*, **272**: 8189–8197.
10. Surinya, K.H., Molina, L., Soos, M.A., Brandt, J., Kristensen, C., and Siddle, K. (2002) Role of insulin receptor dimerization domains in ligand binding, cooperativity, and modulation by anti-receptor antibodies. *J. Biol. Chem.*, **277**; 16719–16725.
11. Hao, C., Whittaker, L. and Whittaker, J. (2006) Characterization of a second ligand binding site of the insulin receptor. *Biochem. Biophys. Res. Commun.*, **347**: 334–339.
12. Ottensmeyer, F.P., Beniac, D.R., Luo, R.Z-T. and Yip, C.C. (2000) Mechanism of transmembrane signaling: insulin binding and the insulin receptor. *Biochemistry*, **39**: 12103–12112.
13. Schaffer, L. (1994) A model for insulin binding to the insulin receptor. *Eur. J. Biochem.*, **221**: 1127–1132.
14. Wilkinson, J.C., Stein, R.A., Guyer, C.A. Beechem, J.M. and Staros, J.V. (2001) Real-time kinetics of ligand/cell surface receptor interactions in living cells: binding of epidermal growth factor to the epidermal growth factor receptor. *Biochemistry*, **40**: 10230–10242.
15. Decker, S.J., Ellis, C., Pawson, T. and Velu, T. (1990) Effects of substitution of threonine 654 of the epidermal growth factor receptor on the epidermal growth factor-mediated activation of phospholipase C. *J. Biol. Chem.*, **265**: 7009–7015.
16. Bao, J., Alroy, I., Waterman, H., Schejter, E.D., Brodie, C., Gruenberg, J. and Yarden, Y. (2000) Threonine phosphorylation diverts internalized epidermal growth factor receptors from a degradative pathway to the recycling endosome. *J. Biol. Chem.*, **275**: 26178–26186.
17. Davis, R.J., Girones, N. and Faucher, M. (1988) Two alternative mechanisms control the interconversion of functional states of the epidermal growth factor receptor. *J. Biol. Chem.*, **263**: 5373–5379.
18. Klein, P., Mattoon, D., Lemmon, M.A. and Schlessinger, J. (2004) A structure-based model for ligand binding and dimerisation of EGF receptors. *Proc. Natl. Acad. Sci. USA*, **101**: 929–934.
19. Hurwitz, D.R., Emanuel, S.L., Nathan, M.H., Sarver, N., Ullrich, A., Felder, S., Lax, I. and Schlessinger, J. (1991) EGF induces increased ligand binding affinity and dimerisation of soluble epidermal growth factor (EGF) receptor extracellular domain. *J. Biol. Chem.*, **266**: 22035–22043.
20. Boni-Schnetzler, M. and Pilch, P. (1987) Mechanism of epidermal growth factor receptor autophosphorylation and high-affinity binding. *Proc. Natl. Acad. Sci. USA*, **84**: 7832–7836.
21. Lemmon, M.A., Bu, Z., Ladbury, J.E., Zhou, M., Pinchasi, D., Lax, I., Engelman, D.M. and Schlessinger, J. (1997) Two EGF molecules contribute additively to stabilization of the EGFR dimer. *EMBO J.*, **16**: 281–294.
22. McLaughlin, S., Smith, S.O., Hayman, M.J. and Murray, D. (2005) An electrostatic engine model for autoinhibition and activation of the epidermal growth factor receptor (EGFR/ErbB) family. *J. Gen. Physiol.*, **126**: 41–53.
23. De Lean, A., Stadel, J.M. and Lefkowitz, R.J. (1980) A ternary complex model explains the agonist-specific binding properties of the adenylate cyclase-coupled β-adrenergic receptor. *J. Biol. Chem.*, **255**: 7108–7117.

24. Rodbell, M., Krans, M.J., Pohl, S.L. and Birnbaumer, L. (1971) The glucagon-sensitive adenylyl cyclase system in plasma membranes of rat liver IV-V. *J. Biol. Chem.*, **246**: 1872–1876; 1877–1882.
25. Lefkowitz, R.J., Mullikin, D. and Caron, M.G. (1978) Regulation of β-adrenergic receptors by guanyl-5′-imidodiphosphate and other purine nucleotides. *J. Biol. Chem.*, **252**: 4686–4692.
26. Seifert, R., Wenzel-Seifert, K., Lee, T.W., Gether, U., Sanders-Bush, E. and Kobilka, B.K. (1998) Different effects of $G_s\alpha$ splice variants on β_2-adrenoreceptor-mediated signalling. *J. Biol. Chem.*, **273**: 5109–5116.
27. Waelbroeck, M. (2001) Activation of guanosine 5′-[γ-^{35}S]thio-triphosphate binding through M_1 muscarinic receptors in transfected Chinese hamster ovary cell membranes: 1. Mathematical analysis of catalytic G-protein activation. *Molecular Pharamcology*, **59**: 875–885.
28. Lefkowitz, M. (2004) Historical review: a brief history and personal retrospective of seven-transmembrane receptors. *TRENDS in Pharamacological Sciences*, **25**: 413–422.
29. Christopoulos, A. and Kenakin, T. G (2002) Protein-coupled receptor allosterism and complexing. *Pharmacol Rev.*, **54**: 323–374,
30. Chabre, M., Bigay, J., Bruckert, F., Bornacin, F., Deterre, P., Pfister, C. and Vuong, T.M. (1988) Visual signal transduction: the cycle of transducin shuttling between rhodopsin and cGMP phosphodiesterase. *Cold Spring Harb. Symp. Quant. Biol.*, **53**(1): 313–324.
31. Bornancin, F. Pfister, C. and Chabre, M. (1989) The transitory complex between photoexcited rhodopsin and transducin. Reciprocal interaction between the retinal site in rhodopsin and the nucleotide site in transducin. *European Journal of Biochemistry*, **184**: 687–698.
32. Waelbroeck, M. (1999) Kinetics versus equilibrium: the importance of GTP in GPCR activation. *TiPS*, **20**: 477–481.
33. Bruckert, F., Chabre, M. and Vuong, T.M. (1992) Kinetic analysis of the activation of transducin by photoexcited rhodopsin. *Biophys. J.*, **63**: 616–629.
34. Chabre, M. (1987) Receptor-G-protein precoupling: neither proven nor needed. *TINS*, **10**: 355.
35. Florio, V.A. and Sternweis, P.C. (1989) Mechanism of muscarinic receptor action on Go in reconstituted phospholipid vesicles. *J. Biol. Chem.*, **264**: 3909–3915.
36. Trahey, M. and McCormick, F. (1987) A cytoplasmic protein stimulates normal N-ras p21 GTPase, but does not affect oncogenic mutants. *Science*, **238**: 542–545.
37. Leitner, G.C., Neuhauser, M., Weigel, G., Kurze, S., Fischer, M.B. and Hocker, P. (2001) Altered intracellular purine nucleotides in gamma-irradiated red blood cell concentrates. *Vox Sanguinis*, **81**: 113–118.
38. Jagodzinski, P., Lizakowski, S., Smolenski, R.T. Slominska, E.M., Goldsmith, D., Simmonds, A. and Rutkowski, B. (2004) Mycophenolate mofetil treatment following renal transplantation decreases GTP concentrations in mononuclear leukocytes. *Clinical Science*, **107**: 69–74.
39. DeAzerdo, F.A., Lust, W.D. and Passoneau, J.V. (1984) Light induced change in energy metabolites, guanine nucleotide and guanylate cyclase within frog retinal layer. *J. Biol. Chem.*, **256**: 2731–2735.
40. Robinson, W.E. and Hagins, W.A. (1979) GTP hydrolysis in intact rod outer segments and the transmitter cycle in visual excitation. *Nature*, **280**: 398–400.
41. Abdulaev, N.G., Ngo, T., Ramon, E., Brabazon, D.M., Marino, J.P. and Ridge, K.D. (2006) The receptor-bound ‘empty pocket’ state of the heterotrimeric G-protein α-subunit is conformationally dynamic. *Biochemistry*, **45**: 12986–12997.

3

Modules and motifs in transduction

A great leap in the understanding of cellular signal transduction pathways came with the realisation that 'motifs' and 'modules' are conserved within the structures of diverse (and often unrelated) proteins involved in signalling:

- motifs are linear peptide sequences
- modules are 3-dimensional folded domains

Although a few of these motifs and modules are also found in structural and transport proteins, most are unique to signalling molecules.

Motifs are short, contiguous, linear stretches of amino acids that act, for example, as address tags for retention or translocation of the protein, or serve as target sequences for covalent modification (such as glycosylation or phosphorylation).

Modules are transposable protein domains – tightly folded discrete 3-dimensional polypeptide structures – that have been inserted ('transposed') into unrelated proteins during evolution. Individual modules are typically globular and have *N*- and *C*-termini close together, ensuring minimum disruption to the overall structure/function of the acceptor protein. The same module may be found in quite unrelated types of proteins that are involved in signal transduction. Certain types of modules are important monitors of the phosphorylation status of their specific targets, binding or de-binding depending upon presence or absence of a phosphoryl group (resulting from protein kinase or phosphatase activation, respectively). SH2 and PTB domains bind to phosphorylated tyrosine residues in specific sequence motifs, and 14-3-3

Structure and Function in Cell Signalling John Nelson

proteins bind to phosphoserine- or phosphothreonine-containing motifs, with forkhead-associated domains being specific for phosphothreonine. Other modules target proteins to specific cellular locations, often in a conditional manner. PH domains bind to the plasma membrane inner leaflet because they are specific for phosphatidyl inositides – in many cases they bind only the 3-phosphorylated form (PIP3) and are thus translocated in response to activation of phosphatidyl inositol-3-kinase. Still other modules (EF hands, C1 and C2 domains) respond to changes in levels of cytosolic calcium.

3.1 Src homology domains

Understanding of signalling transduction was greatly informed by studies of the retroviral oncoprotein vSrc produced by the Rous ***sarc***oma virus. The virus causes sarcoma in chickens by expression of this single protein – a host protein that has been incorporated into the viral gemone in a corrupted form. The chick vSrc gene product is a peripheral plasma membrane protein that has unregulated tyrosine kinase activity. In fact, Src was the first protein tyrosine kinase to be identified. It has a normal cellular counterpart in the proto-oncogene cSrc and is closely related in sequence to a number of other cytosolic and peripheral non-receptor tyrosine kinases that make up the 'Src family' of protein tyrosine kinases. Currently, there are 11 known members: Src, Yes, Lck, Hck, Lyn, Frk, Fgr, Fyn, Srm, Blk, and Brk[1]. The Src family share catalytic domain homology with the large number of protein tyrosine kinases found in metazoan genomes. Protein tyrosine kinases are not found in lower organisms such as prokaryotes, yeast and plants but instead appear to be a hallmark of animals and presumably such tyrosine kinases evolved recently from a common ancestor.

Beyond the kinase domain, these Src family proteins were also found to share structural similarities with other quite unrelated proteins. In particular, three other regions of Src protein were found to be homologous with discrete regions in a large number of other signalling molecules. Src gave its name to these modules and motifs.

3.1.1 Src-homology-1 (SH1) region represents the tyrosine kinase domain

SH1 is a rarely used term for the ≈ 300 amino acid modules that represent Src's catalytic protein tyrosine kinase domain. Tyrosine kinase domains are found in a large family of Src-related tyrosine kinases (both cytosolic and peripheral membrane types), as well as in more distantly related proteins of the single-pass receptor class of receptor tyrosine kinases (RTKs), serine/threonine kinases (STKs) and mixed-function kinases (STYKs).

At this point it is worth drawing a distinction between 'modules' and other domains. The protein kinase domain is not, strictly speaking, a module because kinases are clearly related and have descended from a common ancestral kinase through lineal succession.

Modules, on the other hand, appear to have been acquired through horizontal gene transfer.

Both serine/threonine- and tyrosine-protein kinase domains share the same overall structure and contain regions of sequence homology in a common pattern of 12 subdomains. The detailed structures of these protein kinases are discussed in Chapter 4.

3.1.2 Src-homology-2 (SH2) modules are phosphotyrosine-binding domains

The sole function of SH2 domains is to bind to phosphorylated tyrosine residues in specific polypeptide motifs. This 'docking' behaviour allows SH2-containing proteins to form non-covalent complexes with tyrosine-phosphorylated proteins and is the basis for 'signal transduction particle' formation at activated growth factor receptor dimers. Large multi-protein complexes can be built up in this way. Some SH2 proteins have no catalytic activity and function solely as 'adaptors' – much like Lego building blocks.

Although the majority of SH2 domains are monospecific and bind only to phosphotyrosine-containing peptide motifs, there are notable exceptions: some SH2 domains bind to specific tyrosine motifs whether phosphorylated or not. The EAT-2 protein of natural killer cells (consisting of a single SH2 domain) binds to a tyrosine motif in the *C*-terminal tail of a membrane-bound virus receptor called SLAM, regardless of whether it is phosphorylated or not[2]. Still other pTyr-binding SH2 domains can also bind (as an alternative ligand) the phosphate groups of phosphatidylinositol lipid headgroups. PI-3-kinase regulatory subunit and PLCγ both contain PIP3-binding SH2 domains (see Chapter 6).

SH2 domains are true modular domains and are therefore found in a much more diverse set of proteins than those containing SH1 domains. SH2 domains are ≈ 100 amino acid modules and they share seven regions of high homology with one another. In particular, there is an absolutely conserved arginine-containing sequence motif: Phe.Leu.Val.Arg ('FLVR'). The basic arginine sidechain of the FLVR motif (equivalent to Arg179 of human c-Src) is responsible for coordinating the phosphate of pTyr, and the phosphoamino acid is also contacted by a second arginine in the first α-helix in the SH2 domain (Arg159 of human c-Src). The crystal structure of the SH2 domain of v-Src bound to phosphotyrosine-containing peptides was first solved in 1992[3].

Further SH2 module structures soon followed. A particularly interesting example is the adaptor protein 'Shc' (for '***S****rc-****h****omology 2, and* ***c****ollagen-homology'), which has both a *C*-terminal SH2 domain and an *N*-terminal PTB domain (see below) separated by a GlyPro-rich linker containing a phosphorylatable tyrosine. Shc thus exhibits two modes of phosphotyrosine-binding as well as being an SH2 target in itself. The gallery of SH2 structures allow a canonical core domain to be delimited

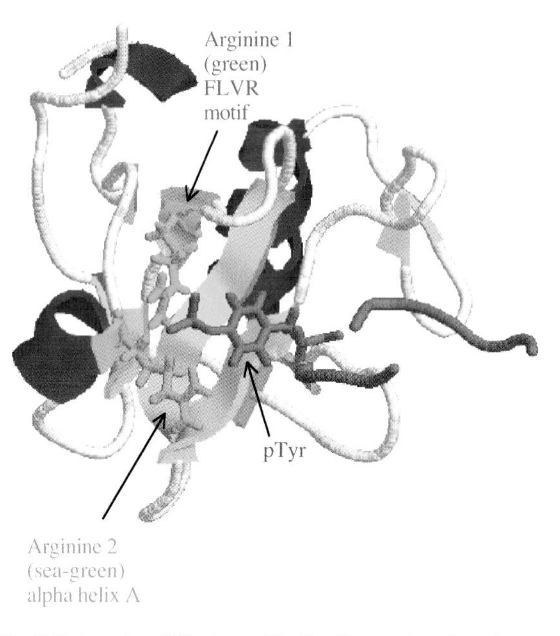

Figure 3.1 The SH2 domain of Shc bound to T cell receptor phosphopeptide (in grey)

with the phosphopeptide-binding module consisting of a central 3-stranded anti-parallel β-sheet sandwiched between two α-helices.

In all cases examined, the arginine residue of the SH2 domain's Phe.Leu.Val.Arg (FLVR) conserved sequence motif in β-strand-2 is completely buried in the hydrophobic core and forms an ion pair with the phosphate group of the phosphotyrosine. Figure 3.1 shows the Shc SH2 domain complexed with a pTyr-peptide from the T cell receptor. A second arginine in the first alpha helix also binds pTyr via an amino-aromatic interation with the amino acid's ring structure and a salt bridge with the phosphate[3,4]. In the SH2 structures from Shc, Lck, PLC-γ1 and Syp, the phosphopeptide is extended and binds across the central β-sheet at right angles to the strands with the pTyr binding to one side of the sheet and the residues *C*-terminal of the pTyr being recognised by loops on the opposite side of the sheet. Behind the peptide-binding surface are two short parallel β-strands and the core structure is further elaborated in some examples of the domain, such as that of Shc[5].

Individual SH2 domains have distinct specificities for the peptide motif adjacent to the phosphotyrosine (pTyr) target, particularly the amino acid sequence immediately *C*-terminal of the pTyr. So, a given SH2 module will only bind to a protein that contains that motif. Examples of consensus sequence targets for a number of SH2 proteins[6] are shown in Table 3.1.

The normal form of the viral Src oncogene, cSrc, presents a striking example of auto-inhibition that depends, in part, upon its SH2 domain binding to a regulatory motif found at its own extreme *C*-terminus. In inactive cSrc, the *C*-terminal regulatory motif contains a phosphotyrosine. An X-ray crystal structure of full length inactive c-Src was obtained[7], and this shows its SH2 domain forming an intrachain bond to its own

Table 3.1 Ligand preferences among SH2 domains

	P^0	P^{+1}	P^{+2}	P^{+3}
Shc	pTyr	Glu/Φ	Xxx	Met/Φ
Src	pTyr	Glu	Glu	Ile
Vav	pTyr	Met	Glu	Pro
$P85^{PI3K}$	pTyr	Met/Φ/ Glu	Xxx	Met
Grb2	pTyr	Gln/Φ	Asn	Gln/Φ
Csk	pTyr	Thr/Ala	Lys/Arg/ Gln/Asn	Met/Φ/ Arg
Csk (optimal sequence: Src A-loop Tyr416)	pTyr	Thr	Ala	Arg

C-terminal pTyr (PDB file: 1FMK). This inhibitory SH2-pTyr intrachain bond is reinforced by a similarly inhibitory intrachain bond involving the SH3 domain of Src (see next section).

3.1.3 Src-homology-3 (SH3) modules are polyproline-binding domains

The SH3 module is found in many signalling molecules, but is also common in cytoskeleton-associated proteins such as spectrin. The size of the module varies: spectrin SH3 domain is ≈ 60 amino acids; the SH3 domain from PI-3-kinase regulatory subunit is ≈ 80 amino acids; the SH3 domain of Src encompasses 140 amino acids. The structure of the SH3 domain from chicken α-spectrin was first published in 1992[8]. SH3 core structures contain a single α-helix and a compact β-barrel in which five anti-parallel β-strands form two sheets with β-strand-2 shared between them (see Figure 3.2). The binding site is a hydrophobic surface on one side of the β-barrel with shallow binding pockets for the individual residues of the SH3 target ligand, which is almost invariably a polyproline type II (PPII) helix. Two loops (the 'RT'- and 'n-Src'-loops) flanking the SH3 module's β-barrel provide further specific interactions with the ligand, increasing both the specificity and strength of the interaction[9].

Note that, as with the SH2 domain, the *N*- and *C*-termini are close together, allowing insertion into different protein sequences without overall disturbance to the 'host' protein. This is a common feature of all transposable modules.

Many SH3 domains bind to cytoskeletal elements (actin) in the region of the plasma membrane – this may aid localisation of signalling enzymes to their membrane-bound substrates (see PLC-γ). Other SH3 domains bind cytosolic proteins (for example, Grb2 binding to mSos).

3.1.3.1 SH3 ligand specificity

SH3 domains bind predominantly to proline strings, that is: left-handed type II polyproline (PPII) helices in which there are three residues per turn with the proline

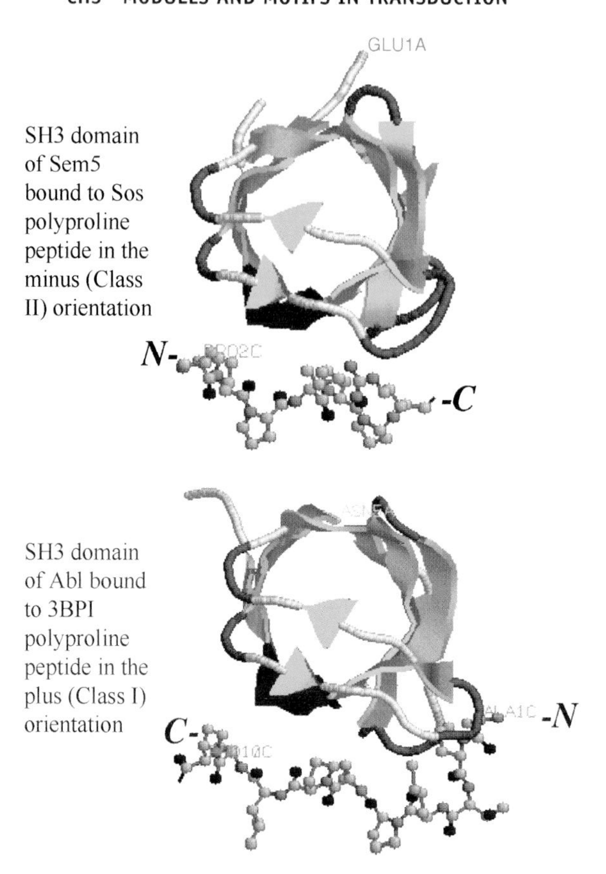

Figure 3.2 SH3 domains bind polyproline helices in either N → C or C → N orientations (PDB file: 1SEM)

at P_0 being co-planar with the proline at P_3 (Table 3.2). The structure is stabilised (in the same way that the proline-rich collagen helix is stabilised) by steric repulsion between the pyrrolidone rings of the prolines. In effect, the PPII helix is the most energetically favourable way in which the rings can keep out of each other's way. The minimum SH3-binding sequence requirement appears to be just PxxP but, crucially, specificity arises from non-proline amino acids in the flanking and internal sequence motif. Although a high proline content is needed to stabilise the structure properly, non-proline residues can also adopt a PPII type helix[10] and this fact is key to the unique autoinhibition of Src-like tyrosine kinases (see Section 3.2.1).

Interestingly, the PPII helix exhibits pseudosymmetry and, although not being a true palindrome, SH3 ligand motifs can be divided into two types depending upon the direction in which they are 'read'. Native ligands for a selection of SH3 domains are shown in Table 3.2. The entries in black are 'Class I' peptide ligands that bind in the 'plus' orientation *N*-to-*C*. The entries in red are 'Class II' ligands and bind in the 'minus' *C*-to-*N* orientation (hence red sequences are reversed). Although native ligands

Table 3.2 SH3 domain ligand consensus sequences

SH3 domain		Native binding sequences								*ligand origin*
					Class I orientation					
		P_{-3}	P_{-2}	P_{-1}	P_0	P_1	P_2	P_3		
Abl	*N*-	M	P	P	**P**	L	P	**P**	-*C*	3BP1
Fyn	*N*-	Y	P	P	**P**	P	V	**P**	-*C*	3BP2
PI-3-kinase	*N*-	R	K	L	**P**	P	R	**P**	-*C*	RLP1
Src	*N*-	R	P	L	**P**	V	A	**P**	-*C*	PI3K (93-99)*
Grb2 (*N*-SH3)	*N*-	R	P	L	**P**	C	T	**P**	-*C*	c-Cbl
					Class II orientation					
PI-3-kinase	*C*-	R	S	**P**	V	P	**P**	A	-*N*	Dynamin
Src	*C*-	K	P	**P**	L	A	**P**	A	-*N*	PI3K (310-304)*
Sem5/Grb2 (*C*-SH3)	*C*-	R	P	**P**	V	P	**P**	P	-*N*	Sos

*sequence numbering from human p85 regulatory subunit of PI-3-Kinase (isoform 1, NCBI accession number NP 852664)
Information in the Table is taken from Lim, W.A. *et al.* (1994) and Sparks, A.B. *et al.* (1996).

are 'pseudosymmetric' and thus exclusive in orientation, truly palindromic synthetic (non-native) ligands have been constructed and shown to bind in both orientations. For example, PI3K SH3 domain binds RSLRPLPPLPPRPXX in either Class I (RSL[RPLPPLP]PRPXX) or Class II (RSLRPL[PPLPPR]PXX) orientation[11].

Consensus SH3 binding sites have been derived by panning for ligands in phage-display libraries in which the PXXP sequence is fixed and both flanking and the two internal residues are randomised[11,12]. The consensus sequences (Table 3.2) are in good agreement with the currently known natural ligands. The core binding region for an SH3 ligand is optimally **R**PL**P**PL**P** (Class I) or Φ**PP**L**P**X**R** (Class II) where Φ = a hydrophobic residue and X = any amino acid; both sequences are written *N*-to-*C*. The two proline rings (PxxP) slot between the rings of two aromatic residues (Trp and Tyr) on the SH3 surface rather like gear teeth in a ratchet[1]. Note that the arginine of the core ligand, although not strictly conserved, is thought to strengthen association with cognate SH3 domains through a charge-stabilised H-bond with a conserved glutamate in the RT-loop of the module; more importantly, perhaps, this ligand arginine determines in which orientation the PPII helix is read[11].

The 'n-Src loop' of the SH3 is also important in determining specificity of ligand interaction. c-Src is ubiquitous but two alternative splice variant isoforms are also present. These are only in neuronal cells where they are expressed at high levels. These isoforms contain either 6- or 11-amino acid inserts in the n-Src loop, and are referred to as Src-NI and Src-NII, respectively[13,14]. The additional insert in the n-Src loop leads to a disturbance in its SH3 domain with the result that n-Src is more active (the auto-inhibitory intrachain SH3-to-PPII linker bond is weak) and less able to interact with the SH3 target motifs to which c-Src naturally binds, such as dyanmin and synapsin[15]. Figure 3.2 shows the structure of the *C*-terminal SH3 domain of Sem-5 bound to a mSos-derived polyproline peptide: PPPVPPRRR (from PDB file: 1SEM)[16]. Sem-5 is the *C. elegans* homologue of human Grb2 ('***G***rowth factor ***r***eceptor ***b***inding protein-2' – a SH2.SH3.SH3-only adaptor), and corresponds to 'Drk' in *Drosophila.* In this structure, the mSos polyproline peptide binds in the minus (i.e., Class II) orientation. The other structure in Figure 3.2 is the SH3 domain of the non-receptor tyrosine kinase Abl (Abl is not a Src family member) bound to a polyproline peptide (APTMPPPLPP) from 3BP1. 3BP1 is a GTPase activating protein that has 'breakpoint cluster region homology' (referred to as 'Bcr-homology'). Such 'BcrH' domains are often flanked by proline-rich motifs that are potential SH3 binding ligands[17]. The proline helix ligand sequence of 3BP1 is bound to Abl in the plus (Class I) orientation. The Abl SH3 structure is from PDB file: 1ABO[18].

3.1.4 Src-homology-4 (SH4) motif and Src 'unique domain'

Much of Src's modular architecture and function is preserved in the other family members. All Src kinase types are thought to exist in a similar auto-inhibited state with the same intrachain SH2-pTyr527 and SH3-PPII linker bonds in place. Furthermore, Src knockouts are less severe than might be expected because there is redundancy in the family – that is to say, other ubiquitously expressed Src-like kinases can take over and compensate for the missing activity. Where the family members differ is in the extreme *N*-terminal SH4/unique domain, which is the most divergent in sequence between family members.

3.1.4.1 SH4 myristoylation site (membrane-targetting)

SH4 is a rarely used term for the extreme *N*-terminal motif of Src that becomes glycine-myristoylated during processing[19].

SH4 of human c-Src proto-oncogene M^{1}GSNKSKPK

SH4 of Rous Sarcoma v-Src oncogene M^{1}GSSKSKPK

This SH4 motif is similar to a selection of other 9-amino acid consensus sequences found in a variety of other unrelated proteins that are also myristoylated by the enzyme: *N*-myristoyltransferase. However, aside from the glycine at position 2, there is little global consensus amongst these non-SH4 myristoylation motifs[19].

Src family kinases are synthesised on free ribosomes and these proteins would not, therefore, associate with membranes unless fatty acylated. This is in contrast to all integral membrane proteins, which are synthesised in the rough endoplasmic reticulum where they are co-translationally membrane-inserted and efficiently transported (membrane-bound) in vesicles to fuse with cell surface membranes.

Myristoylation of non-integral membrane proteins also occurs co-translationally. The start methionine is first removed then the new *N*-terminus (i.e., the α-amino group of

Gly2) becomes fatty acylated through an amide linkage with 14-carbon myristic acid. This saturated fatty acid modification allows mature Src to associate with the membrane, peripherally, by insertion of the lipophilic myristoyl function into the inner leaflet of the plasma membrane.

Myristoylation alone is barely able to maintain membrane localisation; indeed many myristoylated proteins are soluble and mainly localised to the cytoplasm. For stable membrane association, myristoylation must be augmented somehow. Note that the suggested presence of a membrane-bound receptor protein that is specific for myristoylated peptides has been discounted[19].

Many membrane associated myristoylated proteins augment the signal motif with stretches or patches of polybasic amino acids. In the case of Src, it has been shown that the presence of the three lysines in its SH4 motif is essential to reinforce membrane association (probably by electrostatic attraction to negatively charged phospholipid head groups)[20]. This is probably further strengthened by adjacent basic residues in the following Src unique domain, in particular a triplet of arginines preceding a serine phosphorylation site for PKA (Ser17)[21].

Such augmenting polybasic patches need not be contiguous with the myristoyl motif. The '***m***yristoylated ***a***lanine-***r***ich ***C***-***k***inase ***s***ubstrate' (MARCKS), for example, depends upon a polybasic patch that is around 150 amino acids distant from the myristoylated glycine in the primary sequence[22].

Other myristoylated proteins employ a different strategy. In many myristoylated proteins, myristoylation is followed by palmitoylation – the addition of this second hydrocarbon chain enhances the association with the membrane. In these proteins myristoylation is a prerequisite signal for palmitoylation on a cysteine near the myristoylated glycine. For example, the Src family members Lck, Hck, Fgr and Fyn are both myristoylated and palmitoylated and this acts as an address that redirects these Src family members to calveolae[23]. Src, which lacks Cys3, is not palmitoylated. Most heterotrimeric G protein α-subunits are also myristoylated and palmitoylated at the *N*-terminus[19].

SH4 of Hck	MGCMKSKFL
SH4 of Lyn	MGCIKSKRK
N-terminus of Gαi1	MGCTLSAED
N-terminus of Gαz	MGCRQSSEE

3.1.4.2 The 'myristoyl-electrostatic switch' – translocation and/or activation

The membrane targetting by the myristoyl group can be blocked or reversed in certain proteins by serine phosphorylation. Such a 'myristoyl-electrostatic switch' is best characterised in the MARCKS protein that, like Src, relies upon basic amino acids to aid membrane association through electrostatic attraction to negatively charged acidic phospholipids of the inner leaflet. Unlike Src, MARCKS polybasic domain is in the middle of the sequence. This basic patch is essential for both membrane localisation and MARCKS function[22].

N-terminus of MARCKS M^{1}GAQFSKTA

Basic phosphorylation domain of MARCKS K^{152}KKKKRF<u>S</u>FKK<u>S</u>FKL<u>S</u>GFSFKKNKK

MARCKS acts downstream of PKC in control or modulation of calmodulin localisation. In unstimulated cells, MARCKS is membrane associated via phosphatidylserine (and PIP2), and also binds and sequesters calmodulin at the inner cell surface. MARCKS also binds actin. In stimulated cells, MARCKS becomes multiply serine phosphorylated and both it, and calmodulin (to which it binds), translocate from the outer membrane to cytoplasmic structures – MARCKS-bound calmodulin may thus act as a dormant reservoir in unstimulated cells[24]. Significantly, in activated cells, the release from the membrane appears to be a two-step process triggered by activated PKC, which serine phosphorylates the basic domain on three serines (red, underlined). The rapid initial phosphorylation (of Ser167 and/or Ser170) causes calmodulin release first and the slower phosphorylation of the remaining site(s) increases the negative charge enough to neutralise the basic region and cause release of MARCKS into the cytoplasm[21,22,24]. The *basic domain* is often referred to as the *MARCKS effector domain*.

It must be emphasised that not all myristoylated proteins are membrane bound. Indeed the first well-documented case of eukaryotic protein myristoylation was the catalytic subunit of PKA. PKA C-subunits are completely soluble (at least when released from their regulatory subunits), and free C-subunits exhibit no membrane binding but distribute between cytosol and nucleus. Interestingly, this modification was discovered because the PKA C-subunit is resistant to Edman sequencing, suggesting a blocked *N*-terminus. Usually the *N*-terminal amino group is formylated, acetylated or converted to pyroglutamate, but PKA was found to be myristoylated[25]. The function of this was only recently examined (early studies used non-myristoylated PKA). Classically, all PKA C-subunits were thought to be able to diffuse to the nucleus when released from their regulatory subunits, but a proportion of PKA C-subunits are de-amidated on an asparagine directly *C*-terminal of the myristoylated glycine. The postranslational change of Asn → Asp introduces a negative charge that appears to exclude such modified C-subunits from entry to the nucleus after activation[26].

In the non-receptor tyrosine kinase Abl (related to, but not a member of, the Src family), an autoinhibitory SH3 bond forms with the SH2-kinase linker PPII (as with Src) but Abl lacks the *C*-terminal regulatory pTyr527 of Src. Instead, its SH2 domain is blocked by being bound to the kinase C-lobe. In contrast to Src, the myristoyl group of Abl plays a key role in autoinhibition (as well as determining Abl's translocation between cytosol and nucleus). Abl does not appear to associate with membranes through the myristosyl group, which is buried in the active site of the auto-inhibited enzyme[27]. The myristoyl inhibition of Abl kinase appears to be relieved by phosphorylation of a tyrosine in the SH2-kinase linker PPII, a modification that breaks the intrachain SH3 bond, allowing Abl to *trans*-autophosphorylate on its A-loop tyrosine to achieve full activation.

3.1.4.3 Src 'unique domain' – a 'myristoyl-electrostatic switch?

The SH4 and unique domain of Src are contiguous and cooperate in controlling Src localisation. PKA phosphorylates Ser17 in the centre of the basic *N*-terminal segment.

$$M^{1}GSN\underline{\mathbf{K}}S\underline{\mathbf{K}}P\underline{\mathbf{K}}DASQ\underline{\mathbf{RRR}}\,\boxed{pS}L$$

In murine fibroblasts, this PKA-dependent phosphorylation appears to switch Src from being a proliferative signal transducer to one with anti-proliferative effects, which are apparently transduced through the monomeric G protein, Rap1[28]. These anti-proliferative effects of Rap1 are cell type-specific because in human embryonic kidney cells, c-Src mediates crosstalk from β-adrenergic receptors that leads to activation of the Erk mitogenic pathway through Rap1 activation. Again, phosphorylation of Ser17 of Src by PKA is an essential step; this is underlined by the finding that a Ser17-to-Asp17 mutant (the negatively charged sidechain mimicking serine phosphorylation) can constitutively activate Rap1 in the same cells[29]. Phosphorylation of Ser17 by PKA, downstream of PDGF receptor activation, is also reported to cause c-Src to dissociate from the membrane, presumably through the introduction of negative charges into the basic SH4/unique domain region[30] (discussed further in Chapter 7, Section 7.7.4).

3.1.4.4 Src 'unique domain' – a protein interaction domain

Src family members' functions are to some extent redundant, as has been demonstrated in multiple knockouts, but some functions are unique to individual members. Some specificity is gained by differential expression between tissues. Although many members are ubiquitous, specificity still remains. The divergence in unique domain sequences among Src family members might be expected to explain such divergence in their functions; indeed this is what is beginning to emerge.

One of Src's neuronal functions is the regulation of synaptic glutamate receptors. The unique domain of Src is essential for its interaction with, and upregulation of, neuronal *N*-methyl-D-aspartate (NMDA, glutamate-class) receptors. It is perhaps surprising that the interaction is facilitated by a mitochondrially encoded protein, NADH dehydro genase subunit 2 (ND2), which functions outside of the mitochondrion. The region 40–58 of Src's unique domain binds to ND2, which acts as a bridging adaptor between Src and the NMDA receptor complex[31].

3.1.4.5 Immune receptor interactions with Src family unique domains

The Src family member, Lyn, also uses its unique domain as a protein-selective interaction domain, in this case to bind to IgE receptors in immune cells[32]. Significantly, this close interaction of Lyn kinase with the receptor leads to the latter's tyrosine phosphorylation. A second member of the Src family, Fyn, is also involved in immune signalling and again uses its *N*-terminal SH4/unique domain (first 10 amino acids) to associate with the cytoplasmic portions of the CD3-zeta subunit of the T-cell receptor. This association is specific to Fyn as neither Src nor Lck is able to associate with CD3-zeta[33].

3.1.5 The C-terminal Src regulatory motif and Src family autoinhibition

3.1.5.1 SH2 to pTyr416 binding

cSrc has a regulatory motif in its *C*-terminal tail that consist of a tyrosine phosphorylation sequence, which is a substrate for ***C***-terminal ***S***rc ***k***inase (Csk). Csk is an SH2-containing (non-receptor) tyrosine kinase whose function is to downregulate Src by inhibiting its kinase activity. Csk-SH2 recognises activated Src through the signature of its phosphorylated activation loop – the sequence context of Src's A-loop pTyr416 is an optimal ligand for Csk-SH2. Csk thus binds to active Src and phosphorylates tyrosine527 in Src's *C*-terminal tail. This *C*-terminal phosphotyrosine motif (although not an optimal sequence – see Panel 3.1) is recognised by Src's own SH2 domain, with the result that Src SH2 binds to its own regulatory motif intramolecularly.

Note that the numbering of Src residues in the literature is somewhat confusing due to human and avian forms being different sizes and, of course, because of the presence of neuronal isoforms. By convention, the inhibitory tyrosine in the regulatory tail is usually referred to as Tyr527 as this is the position it *would* occupy in the viral oncogene product v-Src. v-Src is a truncated version of normal chicken c-Src. It is because v-Src actually lacks the regulatory tyrosine that it is constitutively active. Panel 3.1 shows human numbering versus that based upon avian v-Src.

3.1.5.2 SH3 bonding to SH2-kinase linker sequence

The inhibitory effect of the intramolecular bonds between the pTyr527 motif and the SH2 domain is reinforced by a second set of intramolecular bonds between the SH3 domain and the proline-containing linker peptide that runs between the catalytic domain and the SH2 domain (Figure 3.3). The linker peptide does not contain a classic PxxP binding motif for SH3 domains (Table 3.2). Nevertheless, in the inhibited structure, the linker does adopt a PPII-like helix that binds in the Class II minus orientation[7].

The net result of these interactions is a restrained assembly of the three domains that holds the active site in an inhibited conformation in which the activation loop (A-loop) blocks the active site cleft – the activating A-loop residue (Tyr416, vSrc numbering) is buried, and the C-helix is displaced (see Figure 3.3). This inhibited structure (PDB file: 2SRC) has both SH2 and SH3 domains packed against the rear of the catalytic domain, which is in a 'closed' or 'restrained' conformation[34]. However, when tyrosine527 is dephosphorylated, cScr adopts an 'open' active conformation (see Figure 3.4; PDB file 1Y57). The role of kinase subdomains in auto-inhibitory states is discussed further in Chapter 4.

HUMAN pp60^{cSrc} proto-oncogene product

```
1   MGSNKSKPKD ASQRRRSLEP AENVHGAGGG AFPASQTPSK PASADGHRGP SAAFAPAAAE
61  PKLFGGFNSS DTVTSPQRAG PLAGGVTTFV ALYDYESRTE TDLSFKKGER LQIVNNTEGD
121 WWLAHSLSTG QTGYIPSNYV APSDSIQAEE WYFGKITRRE SERLLLNAEN PRGTFLVRES
181 ETTKGAYCLS VSDFDNAKGL NVKHYKIRKL DSGGFYITSR TQFNSLQQLV AYYSKHADGL
241 CHRLTTVCPT SKPQTQGLAK DAWEIPRESL RLEVKLGQGC FGEVWMGTWN GTTRVAIKTL
301 KPGTMSPEAF LQEAQVMKKL RHEKLVQLYA VVSEEPIYIV TEYMSKGSLL DFLKGETGKY
361 LRLPQLVDMA AQIASGMAYV ERMNYVHRDL RAANILVGEN LVCKVADFGL ARLIEDNEYT
421 ARQGAKFPIK WTAPEAALYG RFTIKSDVWS FGILLTELTT KGRVPYPGMV NREVLDQVER
481 GYRMPCPPEC PESLHDLMCQ CWRKEPEERP TFEYLQAFLE DYFTSTEPQY QPGENL 536
```

human numbering (plus the start methionine)

1-9 = SH4 domain

10-83 = Src unique domain

84-145 = SH3 domain

151-248 = SH2 domain

270-523 = SH1 Catalytic domain

307-318 = "C-helix"

Activation loop = 407-435

SH2-SH1 linker = 249-269. *PPII-like helix = 252-256*

524=536 = Regulatory domain

activating tyrosine = Tyr419

inhibitory tyrosine = Tyr530

T = insertion point of NI and NII splice variant sequences of neuronal Src isoforms

---

Rous Sarcoma viral Src oncogene

```
1   MGSSKSKPKD PSQRRRSLEP PDSTHHGGFP ASQTPNTTAA PDTHRTPSRS FGTVATEPKL
61  FGDFNTSDTV TSPQRAGALA GGVTTFVALY DYESWIETDL SFKKGERLQI VNNTEGNWWL
121 AHSVTTGQTG YIPSNYVAPS DSIQAEEWYF GKITRRESER LLLNPENPRG TFLVRESETT
181 KGAYCLSVSD FDNAKGLNVK HYKIRKLDSG GFYITSRTQF SSLQQLVAYY SKHADGLCHR
241 LTNVCPTSKP QTQGLAKDAW EIPRESLRLE VKLGQGCFGK VWMGTWNGTT RVAIKTLKPG
301 TMSPEAFLQE AQVMKKLQHE KLVQLYAVVS KEPIYIVIEY MSKGSLLNFL KGEMGKYLRL
361 PQLVDMAAQI ASGMAYVERM NYVHRDLRAA NILVGENLVC KVADFGLARL IEDNEYTARQ
421 GAKFPIKWTA PEAALYGRFT IKSDVWSFGI LLTELTTKGR VPYPGMGNGE VLDRVERGYR
481 MPCPPECPES LHDLMCQCWR RDPEERPTFE YLQAQLLPAC VLEVAE 526
```

chick numbering (plus the start methionine)

1-9 = SH4 domain

10-80 = Src unique domain

81-143 = SH3 domain

148-245 = SH2 domain

267-514 = SH1 Catalytic domain

304-315 = "C-helix"

Activation loop = 404-432

SH2-SH1 linker = 246-267. *PPII-like helix = 249-253*

515=526 = Regulatory domain

activating tyrosine = Tyr416

inhibitory tyrosine (absent), would be = Tyr527

Panel 3.1 Src numbering

Taking the human sequence and converting it to chicken c/v-Src numbering by setting the signature tyrosines to 416 and 527, gives the following…

Human cellular Src proto-oncogene product with chick numbering

```
1                                                 MGSN¹KS KPKDASQRRR SLEPAENVHG
24   AGGGAFPASQ TPSKPASADG HRGPSAAFAP AAAEPKLFGG FNSSDTVTSP QRAGPLAGGV
084  TTFVALYDYE SRTETDLSFK KGERLQIVNN TEGDWWLAHS LSTGQTGYIP SNYVAPSDSI
144  QAEEWYFGKI TRRESERLLL NAENPRGTFL VRESETTKGA YCLSVSDFDN AKGLNVKHYK
204  IRKLDSGGFY ITSRTQFNSL QQLVAYYSKH ADGLCHRLTT VCPTSKPQTQ GLAKDAWEIP
264  RESLRLEVKL GQGCFGEVWM GTWNGTTRVA IKTLKPGTMS PEAFLQEAQV MKKLRHEKLV
324  QLYAVVSEEP IYIVTEYMSK GSLLDFLKGE TGKYLRLPQL VDMAAQIASG MAYVERMNYV
384  HRDLRAANIL VGENLVCKVA DFGLARLIED NEYTARQGAK FPIKWTAPEA ALYGRFTIKS
444  DVWSFGILLT ELTTKGRVPY PGMVNREVLD QVERGYRMPC PPECPESLHD LMCQCWRKEP
504  EERPTFEYLQ AFLEDYFTST EPQYQPGENL⁵³³
```

1-6 = SH4 domain
7-80 = Src unique domain
81-143 = SH3 domain
148-245 = SH2 domain
267-520 = SH1 Catalytic domain
304-315 = "C-helix"
Activation loop = 404-432
SH2-SH1 linker = 246-267. *PPII-like helix = 249-253*
515=526 = Regulatory domain
activating tyrosine = Tyr416
inhibitory tyrosine = Tyr527
T = insertion point of NI and NII splice variant sequences of neuronal Src isoforms

Panel 3.1 (*Continued*)

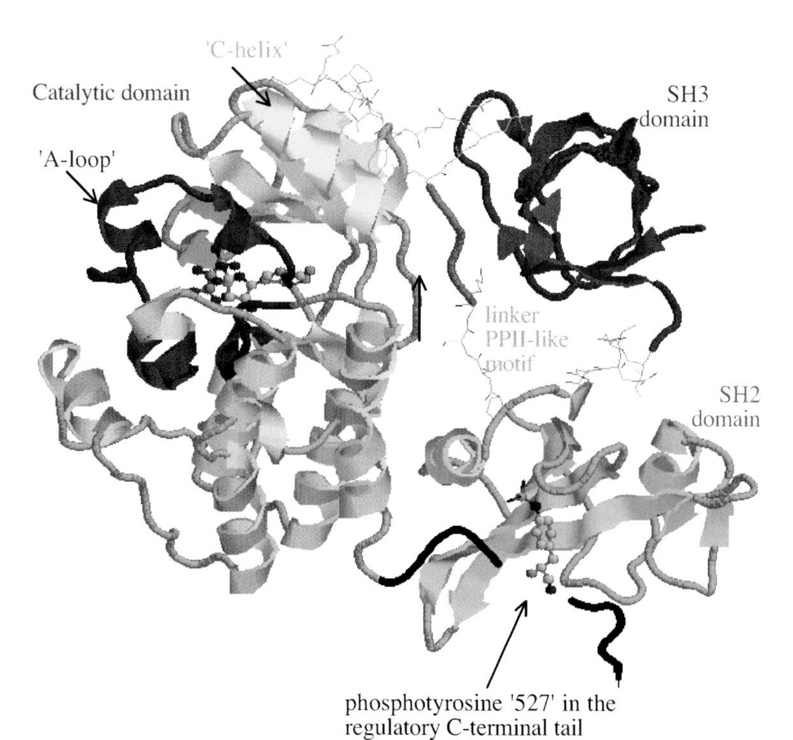

Figure 3.3 cSrc in an inactive conformation (PDB file: 2SRC)

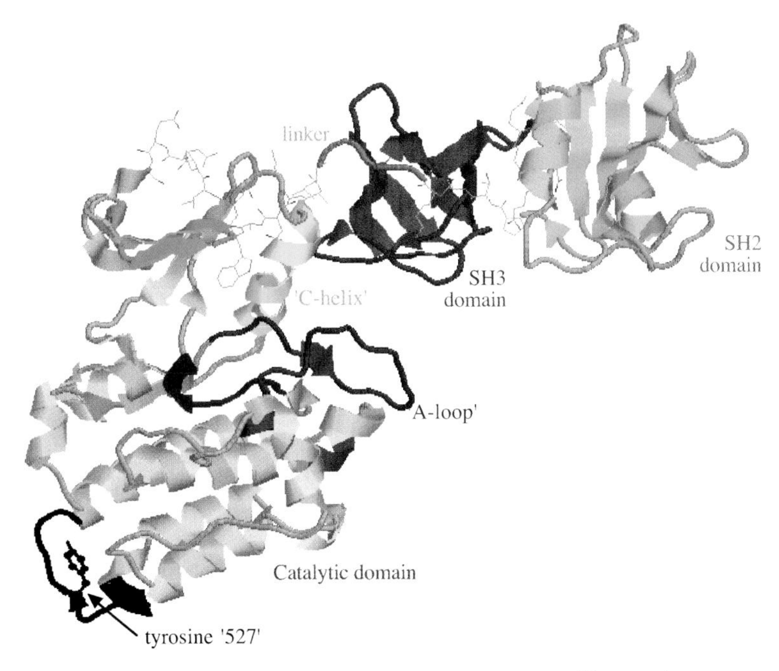

Figure 3.4 cSrc in an active conformation (PDB file: 1Y57)

3.2 PH superfold modules: PH-, PTB- and PDZ-domains

Pleckstrin ***h***omology (PH) domains share a similar structural fold with a ***p***hospho***t***yrosine ***b***inding module (PTB) quite distinct from the SH2 module. A PH-like fold is also found in PDZ domains and this common structure has become known as the 'PH superfold'[35,36]. PDZ modules are named after the first three of these proteins to be identified: ***P***ost-synaptic density protein-95, ***D***iscs large protein and ***Z***onula occludens-1 protein[37].

Despite the common 3D folded structures of PH, PTB and PDZ domains, and the fact that the three modules exhibit some mechanistic overlap, their primary amino acid sequences are unrelated[35]. Therefore these three module groups are not true domain 'families' – there is no evidence of conserved motifs derived from a common ancestor, such as the FLVR motif found in all SH2 modules. Indeed, even within the group of known PTB domains there is a distinct lack of sequence consensus. Thus, in many cases it is impossible to detect these domains by primary sequence homology; instead, their identification depends upon structural similarity.

The PH superfold contains a core structure consisting of a 7-stranded β-sandwich with a *C*-terminal α-helix lying along one edge. The central core is further elaborated in some structures. The three 'flavours' of PH superfold module may have converged functionally during evolution despite their lack of primary sequence homology. Many such PH-superfolds, for example, appear to bind acidic phospholipids in addition to primary peptidyl ligands and vice versa[35].

3.2.1 PH domains – phosphoinositide lipid-binding modules, or Gβ/γ-interacting modules

A typical PH domain consists of a 7-stranded 'β-sandwich' made up of two anti-parallel β-sheets capped with a *C*-terminal α-helix, although the structure may be elaborated in some forms by additional β-strands[38]. Although PH domains were originally identified as phosphotidyl inositide binders, some PH domains bind to peptide motifs on proteins, in particular β/γ-subunits of heterotrimeric G proteins[39].

Phosphoinositide-binding PH domains vary in their specificity for the phosphorylation status of the inositol head group, from promiscuous to specific. For example, the PH domain of the PLCγ1 form of phosphoinositide-specific phospholipase C (PLC) is highly specific for phosphatidyl inositol-(3,4,5)-trisphosphate (PtdIns(**3**,4,5)P_3 PIP3), and its binding to PIP3-enriched membranes can be displaced by competition with the soluble inositol-(1,**3**,4,5)-tetrakisphosphate (IP4) head group. The PLCδ1 PH domain, on the other hand, is selective for PtdIns(4,5)P_2 (PIP2) and its lipid-binding is dissociated with inositol-(1,4,5)-trisphosphate (IP3), whereas protein kinase B (PKB, also known as Akt) contains a PH domain that binds to the D3-phosphorylated PtdIns(**3**,4) P_2[40]. Note that the PtdIns molecules with the D3 phosphoryl group are the product of activating signals from PI-3-kinases, whereas PIP2 is the preferred substrate for all forms of PLC (see Chapter 1). It is interesting to note that the PH domain of PLCδ first allows membrane binding of the protein, localising it to its substrate, and then releases it from the membrane as levels of IP3 product compete for PH-binding to membrane-bound PIP2. Its PH domain specificity can be distinguished from that of its catalytic site because the former can also bind PIP3[41].

PH domains that bind β/γ subunits of heterotrimeric G proteins are typified by the ***β-a***drenergic ***r***eceptor ***k***inase (βARK), a serine/threonine kinase that de-activates the adrenergic receptor after ligand binding (see Chapter 5). The PH domain of βARK can also bind PIP2[41].

3.2.2 PTB domains – phosphotyrosine binding modules

PTB domains serve a function similar to that of SH2 domains in that they bind to specific phosphotyrosine-containing motifs and, like SH2 domains, they also employ conserved arginines to coordinate and charge-neutralise the negatively charged phosphotyrosine residue[35]. However, PTB domains differ from SH2 in both being larger ($\approx$ 160 amino acids) and in having a preferred specificity for amino acid motifs immediately *N*-terminal of the target pTyr (in SH2 domains, the amino acids *C*-terminal of the pTyr are contextually more important)[42].

As mentioned earlier, the Shc adaptor protein has a PTB domain at its *N*-terminus and an NMR structure (PDB file: 1SHC) of this module shows a mode of binding quite different from that of the SH2 domain, despite both modules' common use of arginines in phosphotyrosine recognition (Figure 3.5). In PTB binding, the otherwise extended phosphotyrosine peptide adopts a β-turn just before the pTyr[35,43], a ligand

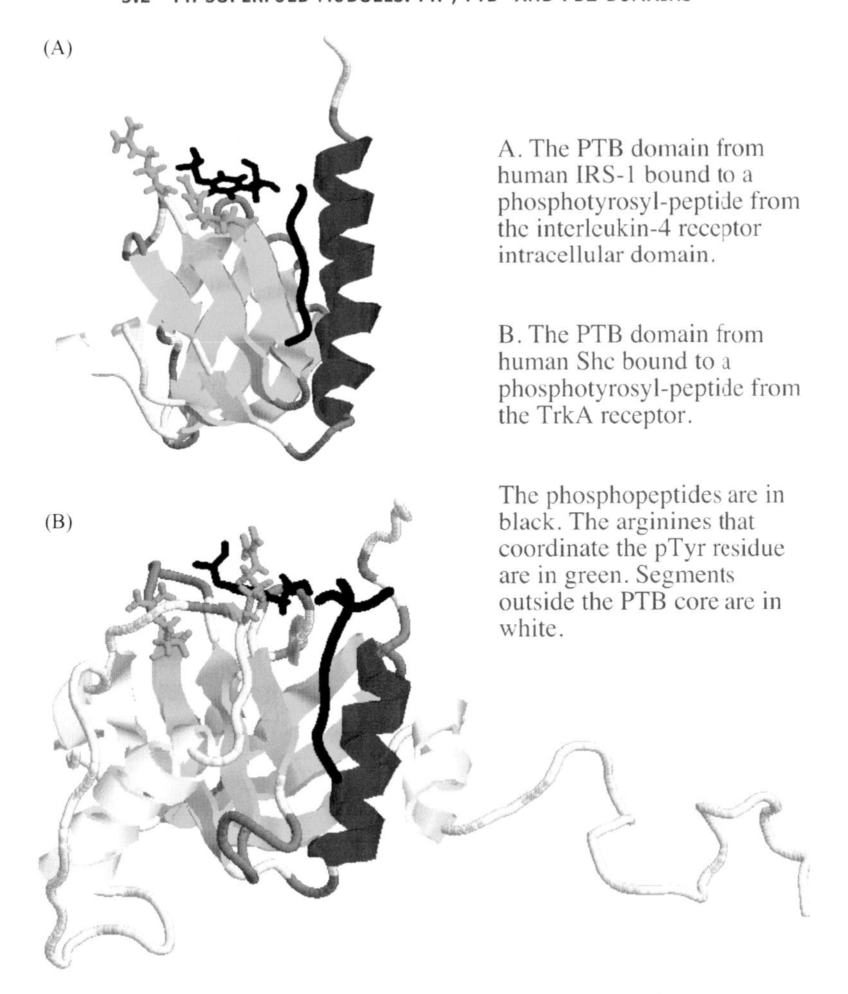

Figure 3.5 PTB domain structures (PDB file: 1SHC)

orientation also favoured by other PTB domains such as the insulin receptor substrate, IRS-1[44]. The β-turn is a result of the presence of a proline in the consensus ligand sequence Asn.Pro.Xxx.pTyr, and the result is an 'L-shaped' ligand[43] (Figure 3.5). The long arm of the 'L' forms an anti-parallel sheet with a β-strand from the PTB sandwich and the phosphotyrosine is typically coordinated by a pair of surface-exposed arginines[45]. In the IRS-1 PTB, the pair consists of Arg212 and Arg227 (these are numbered 57 and 72 in PDB file: 1IRS). In the Shc PTB domain, the arginine pair is made up of residues from completely different positions in the sequence: Arg67 and Arg175, human isoform $p52^{Shc}$ numbering (these are numbered Arg55 and Arg163 in PDB file: 1SHC).

A number of PTB modules bind ligands that do not contain a phosphotyrosine: either the ligand has the AsnProXxxTyr consensus (adopts a β-turn but is unphosphorylated), or the ligand lacks the consensus but instead adopts a helical turn[45].

3.2.3 PDZ domains – *C*-terminal (and *C*-terminal-like) peptide binding modules

PDZ domains were originally identified as intracellular binding partners for certain proteins containing sequences at their extreme *C*-termini that conform to a minimal four amino acid motif – crucially, this motif includes the terminating α-carboxylic acid. Although this motif is their predominant ligand preference, it has emerged that certain *internal* peptide motifs can also be recognised by certain PDZ proteins. The latter type of ligand uncovered a structural preference that resembles the PTB, in that these internal PDZ ligand motifs must be constrained with a β-turn that functionally mimics the carboxylate of the more conventional ligands[35,37]. Similar to the PTB ligand orientation, the PDZ ligand binds as an extra strand to the central β-sheet.

Many PDZ ligands are located in the *C*-terminal tails of integral membrane proteins, particularly ion channels, and because many PDZ-domain proteins contain multiple PDZ modules, their interaction can cause clustering of membrane proteins or provide a scaffold for recruitment and assembly of cytosolic proteins at the membrane[46]. PDZ proteins are particularly important in organising multi-protein signalling complexes in membranes of neurons and photoreceptor cells. They also play a key role in formation of cell-to-cell communication protein complexes in epithelial cells and contribute to cell polarity by acting as chaperones to deliver receptors to specific areas of the plasma membrane[37].

3.3 Bcr-homology (BcrH) domains

The Bcr gene was first discovered due to its involvement in chronic myelogenous leukaemia (CML). A majority of patients with CML show a shortened chromosome 22 with abnormal G-banding: the 'Philadelphia chromosome'. This proved to be due to a reciprocal translocation, t(9;22)-(q34;q11), from the long arm of chromosome 22, which recombines with part of the long arm of chromosome 9. The new chromosome carries part of the gene situated at the '***b***reakpoint ***c***luster ***r***egion' (Bcr) of chromosome 22 and this truncated gene's 5′ end is fused it with a 3′ fragment of a second truncated gene from chromosome 9, the '***Ab****e****l***son gene' (Abl)[47]. The normal cellular Abl gene product is a cytoplasmic tyrosine kinase, whereas the normal Bcr gene product is a serine/threonine kinase. However, the fused product of the Philadelphia translocation has lost important regulatory elements from both parent proteins and results in an oncoprotein with uncontrolled tyrosine kinase activity. It exists as three differently sized isoforms (with MW = 190, 210 or 230 kDa) that reflect three different breakpoints in the Bcr gene. The largest (p230) Bcr-Abl fusion protein retains a domain from Bcr that was found to have striking similarity with unrelated proteins such as RhoGAP and the p85 regulatory subunit of PI-3-kinase; this 'Bcr-homology' (BcrH) domain is missing from the smaller p210 and p190 Bcr-Abl chaemeric proteins[48].

Unfortunately, the abbreviation 'BH' was first proposed for the Bcr homology domain in order to avoid the unwieldy term, 'RhoGAP homology domain', as well as to account for the finding that not all of these modules were functional GTPase activators[49]; the p85α regulatory subunit of PI-3-kinase contains a non-functional BcrH.

I shall use 'BcrH' as the term for such disabled RhoGAP-like modules in order to distinguish them from Bcl-2 homology domains, which are now universally referred to as 'BH' domains (see Section 3.5).

RhoGAP domain is a conserved segment containing ~200 amino acids consisting of seven to nine α-helices connected by loops, with the core of homology being centred on a four helix bundle that contains an arginine-containing loop thought to represent the 'catalytic' residue needed to complete the active site of Rho-like GTPases and/or the closely related Rac proteins[48,49]. BcrH domains that lack this arginine (e.g., the BcrH module of the p85α subunit of PI-3-kinase) are unable to stimulate Rho/Rac-GTPase activity but can still bind to Rho/Rac-like proteins.

In functional RhoGAPs, the 'catalytic' arginine acts like the 'arginine finger' of the otherwise unrelated RasGAP protein (see Chapter 7) in that it helps to orientate the catalytic Gln residue of Rho (equivalent to Gln61 of Ras) that is responsible for coordinating a catalytic water molecule. The basic arginine sidechain at the same time neutralises emerging negative charges during GTP hydrolysis.

RhoGAP/BcrH-containing proteins may interact with just one or several of the Rac/Rho family. The Bcr protein, itself, acts as a GAP for Rac and Cdc42 but has no GTPase activating effect on Rho, Rap1A or Ras[50].

3.4 Dbl homology (DH) domains – partners of PH domains

Dbl was discovered because of its oncogenic role in ***d**iffuse **B**-cell **l**ymphoma*. The gene product was found to contain a unique central region that was subsequently detected in other proteins. These 'Dbl homology' (DH) domains appear to be invariably flanked by a *C*-terminal PH domain, and the presence of such a DH-PH pair in a protein is the signature of a guanine nucelotide exchange factor (GEF) for the Rho family of monomeric G proteins[51]. The PH domain is thought to direct intracellular localisation to membrane bound PIP3 and the DH domain is responsible for the GEF catalytic activity.

It is worth noting that DH modules are also found outside of the strictly defined RhoGEF family. For example, mSos has a paired DH-PH module in addition to its Ras-specific GEF module and can potentially catalyse guanine nucleotide exchange at either Ras or Rho, the latter being favoured by the appearance of the second messenger PIP3. Furthermore, the RhoGAP protein, Bcr, contains a DH-PH module in addition to its BcrH and thus can potentially activate or inactivate the Rho or Rac signal transduction cycle[51]. The DH-PH module of Bcr catalyses exchange of GDP for GTP in Rac, Rho and Cdc42 but has no such GEF activity towards Ras or Rap1A[50].

3.5 Bcl-2 homology (BH) domains

The gene for Bcl-2 was first discovered due to its involvement in ***B***-cell ***l***ymphomas. Like Bcr, the gene was located at a chromosomal breakpoint: in this case, t(14;18)-(q32;q21). In B-cell lymphomas, the Bcl-2 gene from the long arm of chromosome 18 is translocated and becomes fused with the immunoglobulin heavy chain joining region (IgJ_H) from the long arm of chromosome 14. The resulting chaemeric protein (Bcl-2-J_H) is over-expressed because it comes under the influence of the strong immunoglobulin promoter[52,53]. Both the normal and chaemeric Blc-2 proved to be apoptosis inhibitors (see Chapter 10); solving the Bcl-2 sequence and structure led to the identification of a family of related apoptosis regulators containing conserved 'Bcl-2 homology' (BH) domains. The growing Bcl-2 family includes Bax, Bad, Bcl-X_L, Bak and others (discussed in Chapter 10).

It is important to note that the term 'Bcl-2' simply indicates that it was the second of a group of number of breakpoint cluster oncogenes first found in B cell lymphomas. Other genes from this screen are completely unrelated to the Bcl-2 family. For example, Bcl-1 (also known as 'PRAD-1') is in fact an oncogenic form of cyclin D1[54], whereas Bcl-6 is a zinc finger protein[55].

The BH1-4 domains are not true modules. They are not transposable, but are restricted to the 'Bcl-2 family' of closely related proteins. The proteins in this family are similarly-sized and also share a common helix-only folding pattern (typified by Bcl-2 and Bcl-X_L), consisting of a core hydrophobic α-helix (α5) that is shielded from solvent by the hydrophobic surfaces of the remaining amphipathic helices (see Figure 3.6)[56]. Through this arrangement, a central hydrophobic groove is formed that can accommodate an additional amphipathic helix from another Bcl-2 family member. This addition is invariably due to the docking of a BH3 domain. In fact the presence of a single BH3 domain is the minimum attribute of a Bcl-2 protein (see Table 3.3). The binding site groove is made up of the hydrophobic surfaces of the BH1, BH2 and BH3 domains of the acceptor[57] and this binds an approximately 16 amino acid stretch from another BH3 domain, but only if this exogenous BH3 peptide can adopt a helical conformation[56,58].

The 'BH' domains are actually no more than conserved amino acid sequence motifs that are either present or absent from the discrete sequences of α-helices 1 (BH4), 2 (BH3), 4/5 (BH1) and 7/8 (BH2) of the various family members. In other words, the 3D structures look the same and only the primary amino acid sequence stretches can be used to discriminate between them. For example, BH1 domains (encompassing the *C*-terminal half of α4, the *N*-terminal half of α5 and their connecting loop) are characterised by the presence of a NWGR (Asn.Trp.Gly.Arg) motif at the start of α5 that is absent from the corresponding helix in non-BH1 members[58].

3.6 Ras binding domains

There are three main effectors for the monomeric G protein, Ras, namely: the serine/threonine kinase Raf; the guanine nucleotide exchange factor RalGDS; and

Table 3.3 BH-domain composition and properties of Bcl-2 family proteins

Protein	Domains (*N*-to-*C*)	Biological Effects
Death antagonists		
Bcl-2	BH4---BH3-BH1--BH2-TM	anti-apoptotic
BclXL	BH4---BH3-BH1--BH2-TM	anti-apoptotic
Mcl-1	BH4---BH3-BH1--BH2-TM	anti-apoptotic
Bcl-W	BH4-BH3-BH1-BH2-TM	anti-apoptotic
A-1	BH4-BH3-BH1-BH2-TM	anti-apoptotic
Multi-domain death agonists		
Bax	-BH3-BH1--BH2-TM	pro-apoptotic
Bak	-BH3-BH1--BH2-TM	pro-apoptotic
Mtd/Bok	-BH3-BH1--BH2-TM	pro-apoptotic
BH3-only death agonists		
Bik	---BH3---TM	pro-apoptotic
Bim	---BH3---TM	pro-apoptotic
Bad	---BH3---	pro-apoptotic
Bid	---BH3---	pro-apoptotic

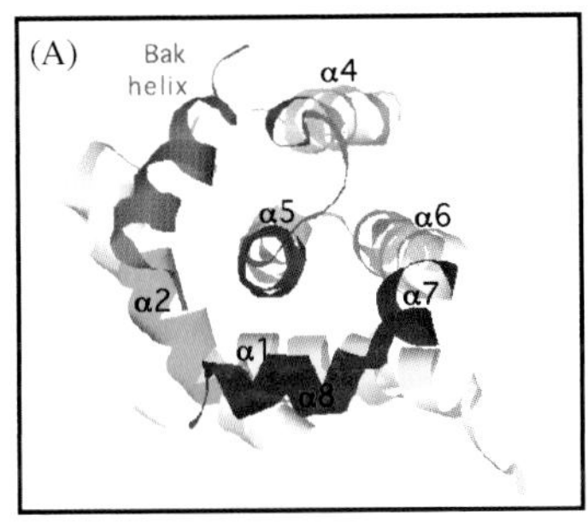

A. Helix nomenclature. Note the hydrophobic core helix-5

B. Spacefill rendering of the complex. Only BH1, BH2 and BH3 domains contact the peptide ligand.

C. Cartoon rendering of B, in the same orientation.

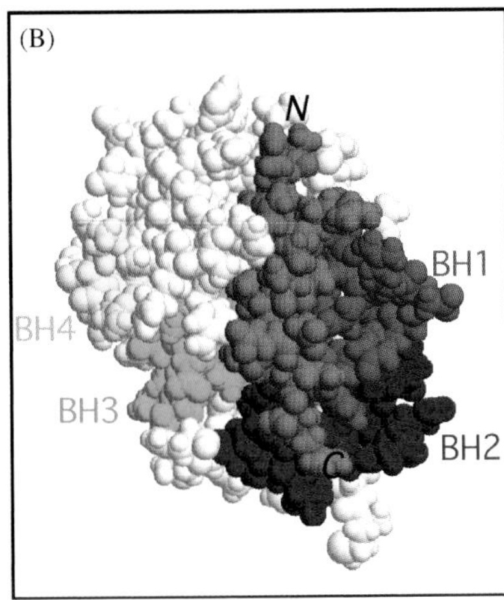

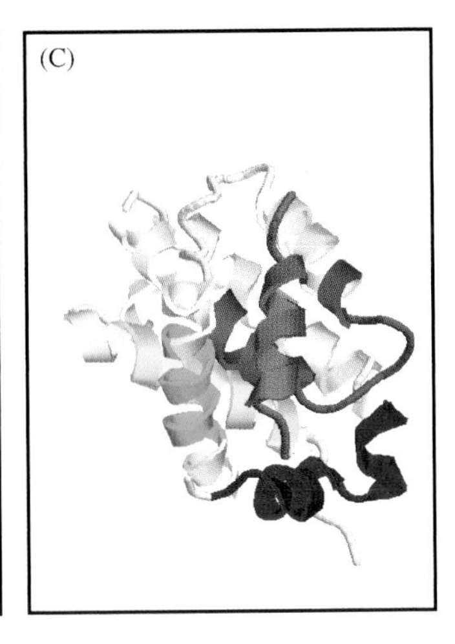

Figure 3.6 Bcl_{XL} complexed with a BH3 peptide

the class 1 forms of the lipid kinase, phosphoinositide 3-kinase (PI-3-K). The catalytic subunits of these heterodimers are divided into:

- Class 1A (p110α, β and δ catalytic subunits), which bind to SH2-containing p85α regulatory subunits and their splice variants (p85β and p55γ). The p85α regulatory subunits serve to recruit the catalytic subunit to phosphotyrosines on activated single pass receptors, thus activating its lipid kinase activity through simple translocation of the enzyme from the cytoplasm to the membrane where its substrate resides.

- A single Class 1B catalytic subunit (p110γ), which binds instead to p101 regulatory subunits that interact with Gβ/γ subunits downstream of 7-pass receptors. Again, the recruitment to the membrane allows the catalytic subunit to produce PIP3 from PIP2, explaining how certain 7-pass receptors can mimic the PI-3-K activation typical of single pass receptors.

An alternative (or perhaps reinforcing) activation occurs via a direct interaction of the class 1 catalytic subunits with membrane-associated, and activated, Ras<GTP> – an interaction recently elucidated in a crystal structure of PI-3-Kγ complexed with activated Ras[59]. In common with the Ras binding domain (RBD) of Raf and RalGDS, the RBD of the PI-3-K catalytic subunit contains a surface-exposed β-strand that forms an anti-parallel β-sheet with strand β2 of the effector region (switch I) of Ras. Like Raf, class 1 PI-3-K catalytic subunits only bind to GTP-occupied Ras and do not interact with GDP-occupied Ras. The overall RBD fold is similar in the three effectors and all three form the antiparallel β-sheet with the effector region β2 strand (see Chapter 7). However, in a divergence from Raf and RalGDS, the RBD of PI-3-K also makes extensive contacts with Ras switch II residues; no such contacts are made by the other two effectors.

3.7 Phosphoserine/phosphothreonine-binding domains

For many years, serine- and threonine-phosphorylation of target enzymes was only known to alter catalytic activity – either activating or inhibiting sensitive enzymes through changes in conformation brought about by the phosphorylation. However, proteins other than enzymes, such as transcription factors, were also found to be serine-phosphorylated and this had an activating effect that depended upon protein-to-protein binding and assembly. It has since become clear that serine- and threonine-phosphorylation not only changes enzymic activity but also has an alternative signalling role: the creation of protein-to-protein docking sites. Such phospho-Ser/Thr binding domains include examples of WW-, WD40-, leucine-rich (F-box), forkhead-associated-domains and 14-3-3 proteins. Space only permits discussion of the latter two.

As a very general rule, when presented in the correct sequence context, a serine *or* a threonine residue would be phosphorylated unselectively if the protein kinase recognises the surrounding amino acid motif. The same cannot be said for all the modules that subsequently bind to serines or threonines in their phosphorylated forms – 'forkhead-associated' modules, for example, only bind to phosphothreonine[60]. Like the SH2 and PTB phosphotyrosine-binding motifs, however, the selectivity and specificity of phosphoserine/theonine binding modules is provided by the amino acids directly flanking the target phospho-serine/threonine.

3.7.1 14-3-3 proteins

The first proteins definitively identified as having pSer/Thr binding domains were the 14-3-3 proteins – the name derives from their chromatographic fraction number when first purified from brain extracts using DEAE-cellulose columns[61]. Again, these domains are not transferable modules but are the signature of a group of nine related mammalian proteins: 14-3-3 types α, β, δ, γ, ε, η, σ, τ and ζ [62]. 14-3-3 types all bind to proteins that are Ser/Thr phosphorylated by PKA/PKC-related enzymes. Generally 14-3-3 types bind a single protein but there are instances where two proteins are bound simultaneously.

14-3-3 proteins serve a variety of functions. They often act to sequester or inhibit signalling molecules. For example, the tyrosine phosphatase Cdc25, which is responsible for the activation of the Cdk1/cyclinB complex by dephosphorylation, is excluded from the nucleus after DNA damage because it becomes Ser/Thr phosphorylated. This phosphorylated form of Cdc25 is now a ligand for 14-3-3 proteins and the 14-3-3/phosphoCdc25 complex is shuttled out of the nucleus due to exposure of a nuclear export signal (NES) that is found in the 14-3-3 structure (present on α-helix I, the most *C*-terminal of the nine helices). Nuclear exclusion prevents Cdc25 gaining access to its substrate (Cdk1) and mitosis is thereby blocked. Similarly, insulin-regulated forkhead transcription factors are excluded from the nucleus following phosphorylation by PKB and consequent complex formation with 14-3-3. Alternatively, 14-3-3 proteins inhibit function by taking the place of an alternative binding partner. For example, the BH3-only protein BAD is also Ser/Thr phosphorylated by PKB. Phospho-BAD binds 14-3-3 and is thus prevented from binding to Bcl-2. This effectively blocks the pro-apoptotic actions of BAD. On the other hand, 14-3-3 binding can enhance activation by scaffolding Raf and its activator PKC and, after activating phosphorylation has occurred, prolongs the activity of phospho-Raf by suppressing its dephosphorylation[60,61,62].

The proteins are helix-only structures, being made up of nine α-helices with the *N*-terminal four in a planar arrangement and the *C*-terminal five projecting downwards at 90° from the *N*-terminal helices, which present a dimerisation surface (Figure 3.7). The proteins are soluble and are found in either cytosol or nucleus, dependent upon activation/inactivation through ligand binding. All exist as homo- or heterodimers. Each monomer has an individual peptide-binding surface that forms a groove between the five *C*-terminal helices and contains a phosphate-binding pocket

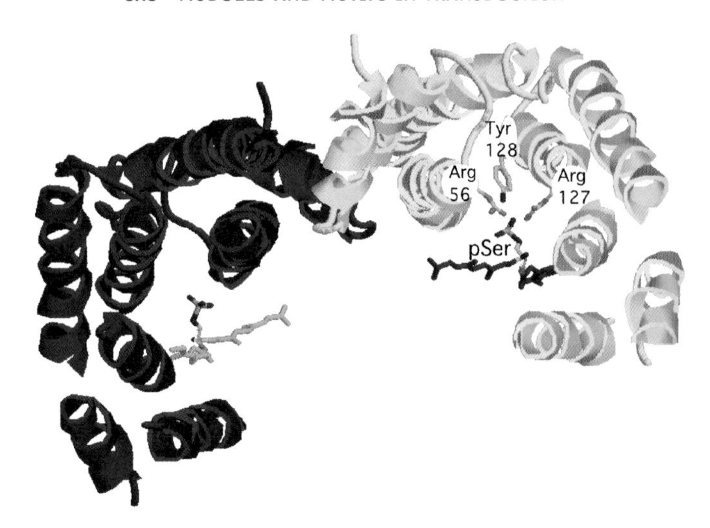

Figure 3.7 Two Raf-1 derived peptides bound to a 14-3-3-ζ dimer (PBD file: 1A37)

made up of regions of the *N*-terminal helices. Each dimer therefore has the *potential* to bind two phosphoserine/threonine residues – one at each of the half dimer sites. In the native dimeric form, 14-3-3 proteins bind pSer/Thr sequence motifs that often differ from those identified from *in vitro* studies scanning peptide libraries for optimal binding of their monomeric forms. Peptides that bind with high affinity to monomeric 14-3-3 are of two types: Arg.Ser.Xxx.pSer.Xxx.Pro (mode 1) or Arg. Xxx.Aro.Xxx.pSer.Xxx.Pro (mode 2, 'Aro' = aromatic). Such a 'mode 1' sequence is found in the protein kinase Raf, which can bind to both 14-3-3 monomers and dimers but only if mono-phosphorylated on the P^0 serine (Ser259 of cRaf-1) – phosphorylation at the P^{-2} serine destroys binding at P^0 pSer[61]. As with PTB ligands, the proline appears to provide a necessary 'kink' in the ligand that optimises its presentation to the binding cleft.

14-3-3 binding site of Raf-1 Arg.Ser.Thr.**pSer**.Thr.Pro.Asn

In some instances, 14-3-3 proteins can function as adaptors by virtue of their dual binding pockets being capable of binding two phosphoproteins simultaneously – of particular note is the finding that Raf and protein kinase Cζ are scaffolded by 14-3-3, thus facilitating an activating serine-phosphorylation of Raf by PKCζ[63].

Many other 14-3-3 ligand proteins do not contain mode 1/2 consensus sequences and yet bind with high affinity to dimers but only weakly to monomers. Such proteins include the protein phosphatase Wee1 and the insulin receptor substrate IRS-1, both of which require to be phosphorylated on two sites to achieve stable binding to 14-3-3 dimers.

The dimer structure of 14-3-3ζ shows a wide central channel made from the two grooves of the component monomers that accommodates two Raf1-derived phosphopeptides (Figure 3.7). The binding of the negatively charged pSer residue is accomplished by ionic bonds with a basic arginine pair (one each on helices E and C) and an H-bonding tyrosine (helix E). A conserved (helix C) lysine is also involved. All other interactions with the ligand residues adjacent to the phospho-amino acid are made by a face of helix I (the helix containing the NES) and it is this helix that determines the specificity of sequence context and switches between nuclear exclusion or entry dependent upon ligation or release of the target protein ligand[62].

3.7.2 Forkhead-associated domains

Forkhead-associated (FHA) domains are true transposable modules that are found in a wide variety of proteins and in species ranging from prokaryotes to eukaryotes[64]. They play important roles in protein trafficking (in the form of FHA-containing kinesins), in checkpoint control of mitosis (through FHA-containing checkpoint kinase Chk2), and in control of transcription (via FHA-containing forkhead transcription factors). The domain itself is an independently folded module that is predominantly β-sheet and terminates in *C*- and *N*-terminal ends that are close together. The module is specific for phosphothreonine-containing sequences – phosphoserine is not recognised.

A crystal structure of the FHA domain of Chk2 complexed with an optimal synthetic peptide ligand (His.Phe.Asp.**pThr**.Tyr.Leu.Ile.Arg) illustrates the general principles governing their mode of binding[65]. The phosphothreonine is bound in a manner different from the modes of binding in PTB, SH2 or 14-3-3 proteins, which all have at least one arginine residue that makes a salt-bridge with the phosphate. In FHA domains, the pThr is bound exclusively by H-bonds with residues found in loops between the beta-strands. In Chk2, H-bonds to the phosphoryl group form with Arg117 (β3/β4 loop), Lys141 (β4/β5 loop) and Ser140 (β6/β7 loop), and van der Waals binding occurs between the pThr hydrocarbon sidechain and conserved Asn166[64,65] (Figure 3.8). As with other phosphopeptide binding modules, specificity extends beyond the pThr due to specific recognition pockets that vary between different FHA domains. The P^{+3} position is particularly discriminatory – the Chk2 preference is for Ile, whereas the yeast ortholog of human Chk2, Rad53, is specific for Asp at this position.

The tumour-suppressing activity of Chk2 is discussed further in Chapter 10.

3.8 EF-hands – calcium-sensing modules

The ionic second messenger Ca^{2+} is recognised predominantly by the ubiquitous calcium-binding protein, calmodulin. Calmodulin is a dumbbell-shaped protein made up of a long central helix with globular ends that are the calcium binding sites. It has four Ca^{2+}-binding sites (two at each end) that are known as 'EF-hands'. The name refers to their helix-loop-helix motif that resembles the thumb and forefinger of a hand (helices 'E' and 'F') with the calcium ion bound in the loop, resembling the curled second finger.

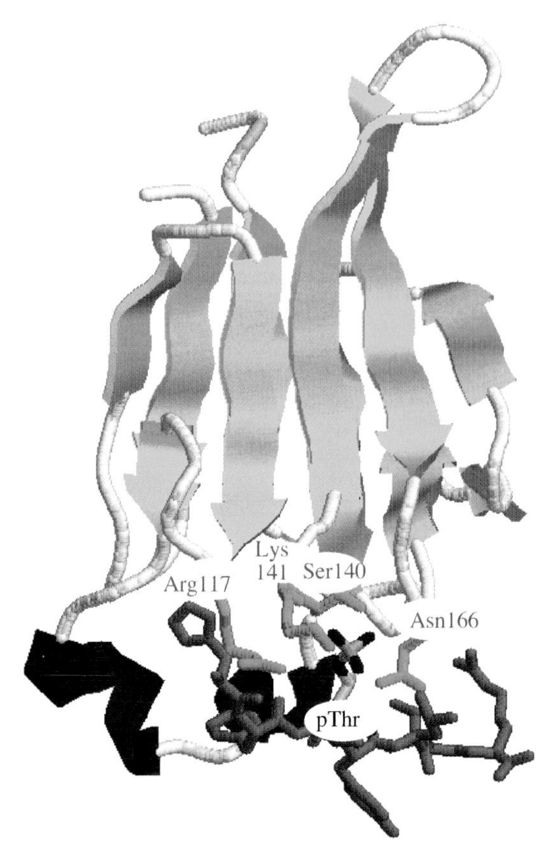

Figure 3.8 The FHA domain of Chk2 complexed with a phosphotheronine-containing peptide (grey sticks)

3.9 C1 and C2 domains – a Ca^{2+}-activated, lipid-binding, module

Proteins that reversibly associate/dissociate from lipid membranes are peripheral, being bound to the lipids of the inner leaflet of lipid bilayer. This is in contrast to integral membrane proteins that are irreversibly inserted into the membrane during translation via bilayer-spanning helices. The earliest example of such reversible recruitment to the membrane was encountered with 'conventional' forms of protein kinase C (cPKC) that are stimulated and recruited to the plasma membrane via binding the second messenger diacylglycerol (DAG) and the inner leaflet lipid, phosphatidyl serine. cPKCs undergo a calcium-activated and DAG-dependent translocation to the plasma membrane that is mediated by two domains named for their PKC-*c*onserved homology: C1 and C2[66] (see Chapter 5, Section 5.11.2).

References

1. Roskoski, R. (2004) Src protein-tyrosine kinase structure and regulation. *Biochem. Biophys. Res. Commun.*, **324**: 1155–1164.
2. Morra, M., Lu, J., Poy, F., Martin, M., Sayos, J., Calpe, S., Gullo, C., Howie, D., Rietdijk, S., Thompson, A., Coyle, A.J., Denny, C., Yaffe, M.B., Engel, P., Eck, M.J. and Terhorst, C. (2001) Structural basis for the interaction of the free SH2 domain EAT-2 with SLAM receptors in hematopoietic cells. *EMBO J.*, **20**: 5840–5852.
3. Waksman, G., Kominos, D., Robertson, S.C., Pant, N., Baltimore, D., Birge, R.B., Cowburn, D., Hanafusa, H., Mayer, B.J., Overduin, M., *et al.* (1992) Crystal structure of the phosphotyrosine domain SH2 of *v-src* complexed with tyrosine phosphorylated peptides. *Nature*, **358**: 646–653.
4. Waksman, G., Shoelson, S.E., Pant, N., Cowburn, D. and Kuriyan, J. (1993) Binding of a high affinity phosphotyrosyl peptide to the SrcSH2 domain: crystal structures of the complexed and peptide-free forms. *Cell*, **72**: 779–790.
5. Zhou, M-M., Meadows, R.P., Logan, T.M, Yoon, H.S., Wade, W.S., Ravichandran, K.S., Burakoff, S.J. and Fesik, S.W. (1995) Solution structure of the Shc SH2 domain complexed with a tyrosine-phosphorylated peptide from the T-cell receptor. *Proc. Natl. Acad. Sci. USA*, **92**: 7784–7788.
6. Songyang, Z., Shoelson, S.E., McGlade, J., Olivier, P., Pawson, T., Bustelo, X.R., Barbacid, M., Sabe, H., Hanafusa, H., Yi, T., Ren, R., Baltimore, D., Ratnofsky, S., Feldman, R.A. and Cantley, L.C. (1994) Specific motifs recognized by the SH2 domains of Csk, 3BP2, fps/fes, GRB-2 HCP, SHC, Syk and Vav. *Mol. Cell. Biol.*, **14**: 2777–2785.
7. Xu, W., Harrison, S.C. and Eck, M.J. (1997) Three dimensional structure of the tyrosine kinase c-Src. *Nature*, **385**: 595–602.
8. Musacchio, A., Noble, M., Pauptit, R., Wierenga, R. and Saraste, M. (1992) Crystal structure of a Src-homology 3 (SH3) domain. *Nature*, **359**: 851–855.
9. Pawson, T. (1997) New impressions of Src and Hck. *Nature*, **385**: 582–609.
10. Lim, W.A. and Richards, F.M. (1994) Critical residues in an SH3 domain from Sem-5 suggest a mechanism for proline-rich peptide recognition. *Structural Biology*, **1**: 221–225.
11. Rickles, R.J., Botfield, M.C., Zhou, X-M., Henry, P.A., Brugge, J.S. and Zoller, M.J. (1995) Phage display selection of ligand residues important for Src homology 3 domain binding specificity. *Proc. Natl. Acad. Sci.*, **92**: 10909–10913.
12. Sparks, A.B., Rider, J.E., Hoffman, N.G., Fowlkes, D.M., Quilliam, L.A. and Kay, B.K. (1996) Distinct ligand preferences of Src homology 3 domains from Src, Yes, Abl, cortactin, p53bp2, PLCγ, Crk and Grb2. *Proc. Natl. Acad. Sci.*, **93**: 1540–1544.
13. Pyper, J.M. and Bolen, J.B. (1990) Identification of a novel neuronal *C-SRC* exon expressed in human brain. *Molecular and Cellular Biology*, **10**: 2035–2040.
14. Pyper, J.M. and Bolen, J.B. (1989) Neuron-specific splicing of C-SRC RNA in human brain. *J. Neurosci. Res.*, **24**: 89–96.
15. Foster-Barber, A. and Bishop, J.M. (1998) Src interacts with dynamin and synapsin in neuronal cells. *Proc. Natl. Acad. Sci.*, **95**: 4673–4677.
16. Lim, W.A., Richards, F.M. and Fox, R.O. (1994) Structural determinants of peptide-binding orientation and of sequence specificity in SH3 domains. *Nature*, **372**: 375–379.
17. Musacchio, A., Cantley, L. and Harrison, S.C. (1996) Crystal structure of the breakpoint cluster region-homology domain from phosphoinositide 3-kinase p85a subunit. *Proc. Natl. Acad. Sci.*, **93**: 14373–14378.
18. Musacchio, A., Saraste, M and Wilmanns, M. (1994) High resolution crystal structures of tyrosine kinase SH3 domains complexed with proline-rich peptides. *Nature Structural Biology*, **1**: 546–551.

19. Boutin, J.A. (1997) Myristoylation. *Cell. Signal.*, **9**: 15–35.
20. Siverman, L. and Resh, M.D. (1992) Lysine residues form an integral component of a novel NH_2-terminal membrane targeting motif for myristoylated pp60$^{v\text{-}src}$. *J. Cell Biol.*, **119**: 415–425.
21. McLaughlin, S. and Aderm, A. (1995) The myristoyl-electrostatic switch: a modulator of reversible protein-membrane interactions. *TIBS*, **20**: 272–276.
22. Taniguchi, H. and Manenti, S. (1993) Interaction of myristoylated alanine-rich protein kinase C substrate (MARCKS) with membrane phospholipids. *J. Biol. Chem.*, **268**: 9960–9963.
23. Sheny-Scaria, A.M., Dietzen, D.J., Kwong, J., Link, D.C. and Lublin, D.M. (1994) Cysteine^3 of Src family protein kinases determines palmitoylation and localization in calveolae. *J. Cell Biol.*, **126**: 353–363.
24. Gallant, C., You, J.Y., Sasaki, Y., Grabarek, Z. and Morgan, K.G. (2005) MARCKS is a major PKC-dependent regulator of calmodulin targetting in smooth muscle. *J. Cell Sci.*, **118**: 3593–3605.
25. Carr, S.A., Biemann, K., Shoji, S., Parmelee, D.C. and Titani, K. (1982) n-Tetradecanoyl is the NH2-terminal blocking group of the catalytic subunit of cyclic AMP-dependent protein kinase from bovine cardiac muscle. *Proc. Natl. Acad. Sci. USA*, **79**: 6128–6131.
26. Pepperkok, R., Hotz-Wagenblatt, A., K—nig, N., Girod, A., Bossemeyer, D. and Kinzel, V. (2000) Intracellular distribution of mammalian protein kinase A catalytic subunit altered by conserved Asn2 deamination. *J. Cell Biol.*, **148**: 715–726.
27. Hantschel, O. and Superti-Furga, G. (2004) Regulation of the cAbl and Bcr-Abl tyrosine kinases. *Nat. Rev. Mol. Cell Biol.*, **5**: 33–44.
28. Schmitt, J.M. and Stork, P.J.S. (2002) PKA phosphorylation of Src mediates cAMP's inhibition of cell growth via Rap1. *Molecular Cell*, **9**: 85–94.
29. Schmitt, J.M. and Stork, P.J.S. (2002) $G\alpha$ and $G\beta\gamma$ require distinct Src-dependent pathways to activate Rap1 and Ras. *J. Biol. Chem.*, **277**: 43024–43032.
30. Walker, F., deBlaquire, J. and Burgess, A.W. (1993) Translocation of pp60c-src from the plasma membrane to the cytosol after stimulation by platelet-derived growth factor. *J. Biol. Chem.*, **268**: 19552–19558.
31. Gingrich, J.R., Pelkey, K.A., Fam, S.R., Huang, Y., Petralia, R.S., Wenthold, R.J. and Salter, M.W. (2004) Unique domain anchoring of Src to synaptic NMDA receptors via the mitochondrial protein NADH dehydrogenase subunit 2. *Proc. Natl. Acad. Sci. USA*, **101**: 6237–6242.
32. Vonakis, B.M., Chen, H., Haleem-Smith, H. and Metager, H. (1997) The unique domain as the site on Lyn kinase for its constitutive association with the high affinity receptor for IgE. *J. Biol. Chem.*, **272**: 24072–24080.
33. Gauen, L.K., Kong, A-N.T., Samelson, L.E. and Shaw, A.S. (1992) p59^{fyn} tyrosine kinase associated with multiple T-cell receptor subunits through its unique amino-terminal domain. *Mol. Cell. Biol.*, **12**: 5438–5446.
34. Xu, W., Doshi, A., Lei, M., Eck, M.J. and Harrison, S.C. (1999) Crystal structures of c-Src reveal features of its autoinhibitory mechanism. *Molecular Cell*, **3**: 629–638.
35. Cowburn, D.C. (1996) Adaptors and integrators. *Structure*, **4**: 1005–1008.
36. Blomberg, N. Baraldi, E., Nilges, M and Saraste, M. (1999) The PH superfold: A structural scaffold for multiple functions. *TIBS*, **24**: 441–445.
37. Harris, B.Z and Lim, W. (2001) Mechanism and role of PDZ domains in signaling complex assembly. *J. Cell Science*, **114**: 3219–3231.
38. Lemmon, M.A. (2003) Phosphoinositide recognition domains. *Traffic*, **4**: 201–213.
39. Srinivasan, N., Waterfield, M.D. and Blundell, T.L. (1996) Comparative analysis of the regions binding β/γ-subunits in Ga and PH domains. *Biochem. Biophys. Res. Commun.*, **220**: 697–702.

40. Falasca, M., Logan, S.K., Lehto, V.P., Baccante, G., Lemmon, M.A. and Schlessinger, J. (1998) Activation of phospholipase Cγ by PI 3-kinase-induced PH domain-mediated membrane targeting. *EMBO J.*, **17**: 414–422.
41. Singer, W.D., Brown, H.A. and Sternweis, P.C (1997) Regulation of eukaryotic phosphatidylinositol-specific phospholipase C and phospholipase D. *Annual Reviews in Biochemistry*, **66**: 475–509.
42. van der Geer, P. and Pawson, T. (1995) The PTB domain: a new protein module implicated in signal transduction. *TIBS*, **20**: 277–280.
43. Zhou, M-M., Ravichandran, K.S., Olejniczak, E.F., Petros, A.M., Meadows, R.P., Sattler, M., Harlan, J.E., Burakoff, S.J. and Fesik, S.W. (1995) Structure and ligand recognition of the phosphotyrosine binding domain of Shc. *Nature*, **378**: 584–592.
44. Eck, M.J., Dhe-Paganon, S., Trub, T., Nolte, R.T. and Shoelson, S.E. (1996) Structure of the IRS-1 PTB domain bound to the juxtamembrane region of the insulin receptor. *Cell*, **85**: 695–705.
45. Forman-Kay, J.D. and Pawson, T. (1999) Diversity in protein recognition by PTB domains. *Current Opinion in Structural Biology*, **9**: 690–695.
46. Pawson, T. and Scott, J.D. (1997) Signaling through scaffold, anchoring and adaptor proteins. *Science*, **278**: 2075–2080.
47. Faderl, S. Talpaz, M., Estrov, Z. O'Brien, S., Kurzrock, R. and Kantarjian, H.M. (1999) The biology of chronic myeloid leukemia. *New England Journal of Medicine*, **341**: 164–172.
48. Moon, S.Y. and Zheng, Y. (2003) Rho GTPase-activating proteins in cell regulation. *Trends in Cell Biology*, **13**: 13–22.
49. Musacchio, A., Cantley, L.C. and Harrison, S.C. (1996) Crystal structure of the breakpoint cluster region homology domain from phosphoinositide 3-kinase p85a subunit. *Proc. Natl. Acad. Sci.*, **93**: 14373–14378.
50. Chuang, T-H., Xu, X., Kaartinen, V., Heisterkam, N., Groffen, J. and Bokoch, G.M. (1995) Abr and Bcr are multifunctional regulators of the Rho GTP-binding protein family. *Proc. Natl. Acad. Sci.*, **92**: 10282–10286.
51. Zheng, Y. (2001) Dbl family guanine nucleotide exchange factors. *Trends in Biochemical Sciences*, **26**: 724–732.
52. Crescenzi, M., Seto, M., Herzig, G.P., Weiss, P.D., Griffith, R.C. and Korsmeyer, S.J (1988) Thermostable DNA polymerase chain amplification of t(14;18) chromosome breakpoints and detection of minimal residual disease. *Proc. Natl. Acad. Sci.*, **85**: 4869–4873.
53. Rabbitts, T.H. (1994) Chromosomal translocations in human cancer. *Nature*, **372**: 143–149.
54. Motokura, T., Bloom, T., Kim, H.G., Juppner, H., Ruderman, J.V., Kronenberg, H.M. and Arnold, A. (1991) A novel cyclin encoded by a bcl1-linked candidate oncogene. *Nature*, **350**: 512–515.
55. Ye, B.H., Lista, F., Lo Coco, F., Knowles, D.M., Offit, K., Chaganti, R.S.K. and Dallas-Favera, R. (1993) Alterations of a zinc finger-encoding gene, BCL-6, in diffuse large cell lymphoma. *Science*, **262**: 747–750.
56. Petros, D.M., Nettesheim, D.G., Wang, Y., Olejniczak, E.T., Meadows, R.P., Mack, J., Swift, K., Matayoshi, E.D., Zhang, H., Thompson, C.B. and Fesik, S.W. (2000) Rationale for Bcl-X_L/Bad peptide complex formation from structure, mutagenesis, and biophysical studies. *Protein Science*, **9**: 2528–2534.
57. Kelekar, A. and Thompson, C.B. (1998) Bcl-2-family proteins: the role of the BH3 domain in apoptosis. *Trends in Cell Biology*, **8**: 324–330.
58. Petros, A.M., Olejniczak, E.T. and Fesik, S.W. (2004) Structural biology of the Bcl-2 family of proteins. *Biochim. Biophys. Acta*, **1644**: 83–94.

59. Pacold, M.E., Suire, S., Perisic, O., Lara-Gonzalez, S., Davis, C.T., Walker, E.H., Hawkins, P.T., Stephens, L., Eccleston, J.F. and Williams, R.L. (2000) Crystal structure and functional analysis of Ras binding to its effector phosphoinoside 3-kinase γ. *Cell*, **103**: 931–943.
60. Yaffe, M.B. and Elia, A.E.H. (2001) Phosphoserine/threonine-binding domains. *Current Opinion in Cell Biology*, **13**: 131–138.
61. Tzivion, G. and Avruch, J. (2002) 14-3-3 proteins: active cofactors in cellular regulation by serine/threonine phosphorylation. *J. Biol. Chem.*, **277**: 3061–3064.
62. Rittinger, K., Budman, J., Xu, J., Volinia, S., Cantley, L.C., Smerdon, S.J., Gamblin, S.J. and Yaffe, M.B. (1999) Structural analysis of 14-3-3 phosphopeptide complexes identifies a dual role for the nuclear export signal of 14-3-3 in ligand binding. *Mol. Cell*, **4**: 153–166.
63. Van Der Hoeven, P.C., Van Der Wal, J.C., Ruurs, P., Van Dijk, M.C. and Bitterswijk, J. (2000) 14-3-3 isotypes facilitate coupling of protein kinase C-ζ to Raf-1: negative regulation by 14-3-3 phosphorylation. *EMBO J.*, **19**: 349–358.
64. Durocher, D. and Jackson, S.P. (2002) The FHA domain. *FEBS Letters*, **513**: 58–66.
65. Li, J., Williams, B.L., Haire, L.F., Goldberg, M., Wilker, E., Durocher, D., Yaffe, M.B., Jackson, S.P. and Smerdon, S.J. (2002) Structural and functional versatility of the FHA domain in DNA-damage signaling by the tumor suppressor kinase Chk2. *Mol. Cell*, **9**: 1045–1054.
66. Cho, W. (2001) Membrane targeting by C1 and C2 domains. *J. Biol. Chem.*, **276**: 32407–32410.

4

Protein kinase enzymes – activation and auto-inhibition

Protein kinase enzymes transfer a phosphoryl group from ATP to the alcohol or phenolic hyroxyl sidechains of either serine/threonine or tyrosine residues, respectively. The specificity of the peptide substrate for the kinase is largely determined by the amino acid sequences immediately adjacent to the target residue, particularly those on the *N*-terminal side. All protein kinase catalytic domains are globular bi-lobal structures with a small *N*-terminal lobe attached to a larger C-lobe, both being hinged like a pair of jaws by a linker peptide. The P-loop of the N-lobe binds ATP while the A-loop and C-loop of the C-lobe bind the peptide substrate. The two lobal faces of the cleft come together to construct the active site. Many, but by no means all, protein kinases have a primary activation switch that is triggered by phosphorylation of either a threonine or tyrosine(s) in the A-loop. After it is phosphorylated, the negatively charged phospho-sidechain shifts position to neutralise a positively charged 'basic cluster', which includes an invariant C-loop arginine adjacent to the catalytic aspartate (RD motif), plus a more distal basic residue (lysine or arginine) in the A-loop. In some RD kinases the A-loop movement is local. In others (like the IRK), A-loop phosphorylation alone causes a more widespread rearrangement that also re-positions the C-helix to interact with the P-loop, readying both lobes for substrate entry. Cyclin dependent kinases, in contrast, require cyclin binding to force C-helix movement. In PKA, a C-helix histidine joins the 'basic cluster' to interact with the phospho-threonine and 'close' the active site when both substrates are present. The A-loops in protein kinases display a spectrum of primary regulation ranging from none, through simple on/off, to more complex levels of control involving partner inhibitors and activator proteins.

Structure and Function in Cell Signalling John Nelson

4.1 The protein kinase fold

There are around 600 protein kinase enzymes in the human genome[1], of which three main families are involved in cell signalling:

- serine/threonine kinases – phosphorylate either serine or threonine (referred to as 'ST' kinases);
- tyrosine kinases – phosphorylate tyrosines (referred to as 'T' kinases);
- mixed function kinases – phosphorylate BOTH serine/threonine AND tyrosine simultaneously (referred to as 'STY' kinases).

Members of the tyrosine protein kinase family may be either receptor tyrosine kinases (RTKs) or non-receptor tyrosine kinases. A smaller proportion of ST kinases are also receptors (the TGFβ receptor is an example of a receptor ST kinase). ST-, T- and STY-kinases all share a homologous stretch of ≈ 300 amino acids that represents the core catalytic domain. *N*- and *C*-terminal extensions to this core are not conserved across the three families.

The common catalytic domain encompasses several invariant motifs and residues that together represent a kinase 'signature'. The domain may be divided into 12 subdomains, some of which fulfil structural roles with a smaller number involved in function[2] (see Table 4.1). A number of protein kinase structures have now been solved and all show a similar overall architecture of two lobes connected by a linker peptide 'hinge'[3].

All protein kinases follow the same folding pattern as c***A***MP-dependent ***p***rotein ***k***inase (PKA; also known as 'cAPK') – a serine/threonine protine kinase, and the first kinase to be crystalised. All protein kinase domains are made up of two lobes. The smaller *N*-terminal lobe largely consists of a five-stranded beta sheet and (in most protein kinases) contains a single α-helix: the 'C-helix'. This nomenclature arose because PKA (unusually) contains three helices in its N-lobe, named 'A-, B- and C-helices' – the C-helix is the only one conserved in other protein kinase structures (see Figure 4.1). The larger *C*-terminal lobe is predominantly helical. In broad terms, the nucleotide substrate (ATP) binding site is located in the N-lobe and the peptide substrate site in the C-lobe.

4.1.1 Invariant residues

The invariant protein kinase residues align with PKA-α amino acid positions: Gly52, Lys72, Glu91, Asp166, Asn171, Asp184, Gly186, Glu208, and Arg 280. Almost invariant are Gly50, Val57, Phe185, Asp220 and Gly225. Not surprisingly, many of these residues have been found to participate in the catalytic mechanism[2]. The key components of the active site are three loops: the *phosphate-binding loop* (P-loop), the *catalytic loop* (C-loop) and the *activation loop* (A-loop)[4]. The conserved Gly50 and 52

Table 4.1 Protein kinase catalytic sites – subdomain structure

	I	II	III
	43 64	65	84 98
PKA	FERKKTLGTGSFGRVMLVKHKA-----	TEQYYAMKILDKQKVVKLK	QIEHTLNEKRILQAV-
PKC	FNFLMVLGKGSFGKVMLSERKG-----	TDELYAVKILKKDVVIQDD	DVECTMVEKRVLALPG
Src	LRLEVKLGQGCFGEVWMGTWNG-----	-TTRVAIKTLKPGTM----	SPEAFLQEAQVMKKL-
IRK	ITLLRELGQGSFGMVYEGNARDIIKGE	AETRVAVKTVNESASLR--	ERIEFLNEASVMKGF-

	IV	V
	99	114 137
PKA	NFPFLVRLEYAFKDN	SNLYMVMEYVPGGEMFSHLRRIGR----------
PKC	KPPFLTQLHSCFQTM	DRLYFVMEYVNGGDLMYHIQQVGR----------
Src	RHEKLVQLYAVVSE-	EPIYIVTEYMSKGSLLDFLKGETGKY--------
IRK	TCHHVVRLLGVVSKG	QPTLVVMELMAHGDLKSYLRSLRPEAENNPGRPP

	VIa	VIb	VII	VIII-
	138	161	179 195	
PKA	FSEPHARFYAAQIVLTFEYLHSL	DLIYRDLKPENLLIDHQG	YIQVTDFGFAKRVKGRT---	--
PKC	FKEPHAVFYAAEIAIGLFFLQSK	GIIYRDLKLDNVMLDSEG	HIKIADFGMCKENIWDGVTT	--
Src	LRLPQLVDMAAQIASGMAYVERM	NYVHRDLRAANILVGENL	VCKVADFGLARLIEDNEYTA	R-
IR	PTLQEMIQMAAEIADGMAYLNAK	KFVHRDLAARNCMVAHDF	TVKIGDFGMTRDIYETDYYR	KG

	VIII	IX	X
	196	211	241 259
PKA	WTLCGTPEYLAPEII	LSKGYNKAVDWWALGVLIYEMAA-GYPPFFA	DQPIQIYEKIVSG-KVRFPS
PKC	KTFCGTPDYIAPEII	AYQPYGKSVDWWAFGVLLYEMLA-GQAPFEG	EDEDELFQSIMEH-NVAYPK
Src	QGAKFPIKWTAPEAA	LYGRFTIKSDVWSFGILLTELTTKGRVPYPG	MVNREVLDQVERGYRMPCPP
IRK	GKGLLPVRWMAPESL	KDGVFTTSSDMWSFGVVLWEITSLAEQPYQG	LSNEQVLKFVMDGGYLDQPD

	XI
	260 297
PKA	HFSSDLKD-LLRNLLQVDLTKRFGNLKNGVSDIKTHKWF
PKC	SMSKEAVA-ICKGLMTKHPGKRLGCGPEGERDIKEHAFF
Src	ECPESLHD-LMCQCWRKEPEERPTFEYL-------QAFL
IRK	NCPERVTD-LMRMCWQFNPKMRPTFLEIVNLL---KDDL

P-loop - green
C-helix - yellow
C-loop - red
A-loop - magenta

PKA= cAMP-dependent protein kinase β-type catalytic sub-unit (from amino acid 43)
PKC= Protein kinase C βI (from amino acid 339)
Src= Non-receptor protein tyrosine kinase (from amino acid 267)
IRK= Insulin receptor (from amino acid 996)

are present in the P-loop; invariant Asp166, Asn171 are in the C-loop; while Asp184, Gly186 and Glu208 are present in the A-loop (see Tables 4.1 and 4.2).

The PKA N-lobe extends from residues 40–126 and the C-lobe is from 127–280. An extended chain (hinge), residues 120–127, links the two lobes. The active site lies in the cleft between the two surfaces of the lobes and is composed of key residues from P-, C-, and A-loops (see Figure 4.2)[5]. The *C*-terminal residues of PKA, 281–350, are outside the conserved catalytic core; this 70 amino acid tail extends from the C-lobe to lie along one

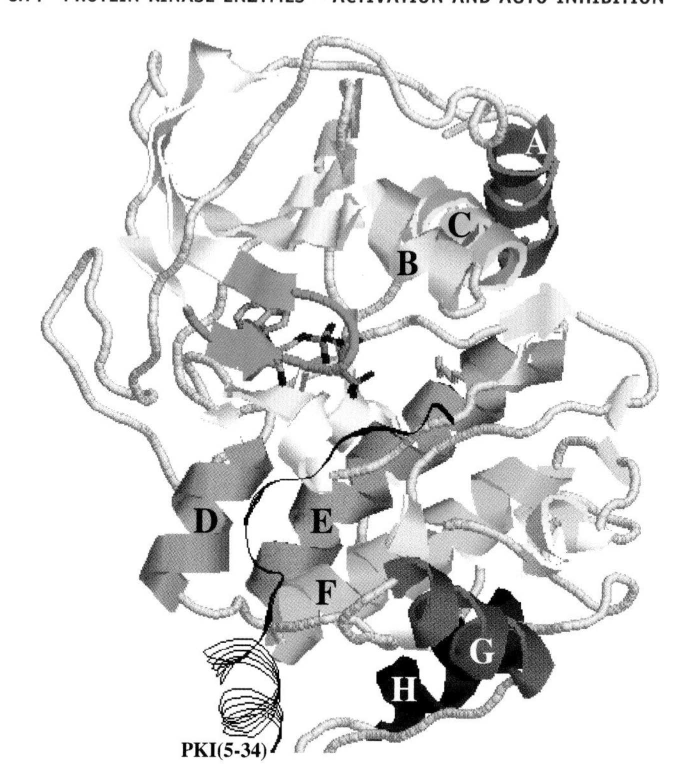

Figure 4.1 PKA in a closed conformation showing nomenclature for the main α-helices (PBD file: 1 CDK)

face of the N-lobe. The *N*-terminal 1–39 amino acids is usually disordered but can be seen as an ordered extension of the A-helix in some crystals[6].

4.1.2 The phosphate-binding loop or 'p-loop'

Almost all dinucleotide-binding proteins contain a common glycine-rich motif, first identified by Rossmann[7] – hence the name 'Rossmann motif'. A similar motif was later identified in mononucleotide-binding proteins, where such glycine-rich motifs are known as Walker A-box sequences or 'P-loops' (for ***p***hosphate-binding)[8]. Two consensus sequences have been found: G.X.G.X.X.G in dinucleotide-binding proteins, and G.X.X.G.X.G.K.[S/T] in mononucleotide-binding proteins. The mononucleotide motif is found in diverse proteins including adenylate kinase, p21Ras and protein kinases, but in the (mononucleotide-binding) protein kinases the structure differs.

Ras P-loop typifies mononucleotide-binding motifs that run between a beta strand (yellow) and a following alpha helix (magenta). In the Ras P-loop, the amide groups of the glycines form a 'giant anion hole' that holds the negatively charged phosphoryl groups of the nucleotide (GTP). The lysine16 sidechain interacts with the β- and γ-phosphates and helps stabilise the transition state by neutralising the negative charge of the γ-phosphate[9,10]. The lack of sidechains allows the 'anion hole' to accommodate

Table 4.2 Catalytic residue equivalence in PKA, Cdk2 and IRK

Kinase motif	Sub-domain	PKA	Cdk2	IRK
P-loop	I	L**G**T**G**SF**G**RV aas:— 49-57	I**G**E**G**TY**G**VV aas:— 10-18	L**G**Q**G**SF**G**MV aas:— 1002-1010
Invariant lysine — α/β phosphate-binding	II	Lys^{72}	Lys^{33}	Lys^{1030}
Invariant glutamate salt-bridged to Lys^{72}	III	Glu^{91}	Glu^{51}	Glu^{1047}
Basic residue interacting with γ-phosphate	VIb	Lys^{168}	Lys^{129}	Arg^{1136}
C-loop	VIb	aas:—164-171	aas:—125-132	aas:—1130-1137
A-loop	VII-VIII	aas:—184-208	aas:—145-172	aas:—1150-1179
catalytic Asp (RD motif)	VIb	Arg^{165}Asp^{166}	Arg^{126}Asp^{127}	Arg^{1131}Asp^{1132}
Mg^{2+}-coordinating residue	VII (DFG motif)	Asp^{184}	Asp^{145}	Asp^{1150}
Activating phosphorylation site	VII	Thr^{197}	Thr^{160}	Tyr$^{1162/1163}$
'Basic Cluster'	III VIb VII	C-helix-His^{87} C-loop-Arg^{165} A-loop-Lys^{189}	C-helix-Arg^{50} C-loop-Arg^{126} A-loop-Arg^{150}	C-loop-Arg^{1131} A-loop-Arg^{1155} A-loop-Arg^{1164}
'C'-helix	III	aas:—85-97	aas: — 45-55	aas:—1038-1051

PKA = cAMP-dependent protein kinase; Cdk2 = cyclin dependent kinase2; IRK = Insulin Receptor Kinase

the bulky phosphoryl groups. H-bonds are formed between the amide backbone of the glycines and the phosphoryl groups (see Figure 4.3).

Protein kinase P-loops, on the other hand, are made up of the first two strands in the anti-parallel beta sheet of the *N*-terminal lobe[9] – making a β-strand, β-turn, β-strand structure. The ATP is completely covered by this flap-like structure with the γ-phosphate positioned at the β-turn. In the PKA P-loop

(residues 49–57: Leu.**Gly**.Thr.**Gly**.Ser.Phe.**Gly**.Arg.Val)

the 'flap' is stabilised by backbone H-bonds between Gly50<–>Val57 and Gly52<–> Gly55[10]. The glycines perform a similar function of anchoring the ATP, but protein kinase P-loops (and those in dinucleotide binding proteins) do not include a lysine. Instead, the invariant lysine72 (in PKA subdomain II) binds the α and β-phosphates of ATP, neutralising their charges, while a second invariant lysine168 (in subdomain VIb, i.e., the 'C-loop' of the C-lobe) binds the γ-phosphate and probably stabilises the transition state[9] (see Figure 4.3).

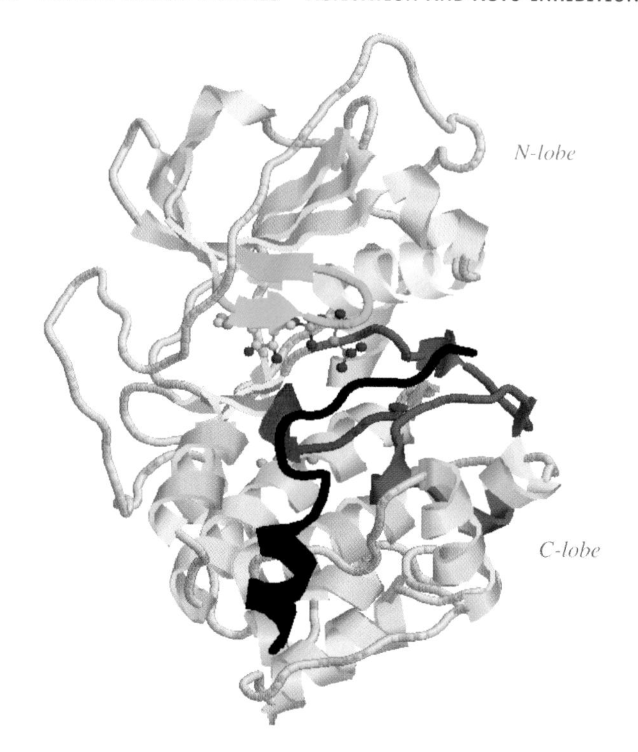

The P-loop is in green; C-loop in red; A-loop in purple; C-helix in yellow; and the inhibitor peptide is black

Figure 4.2 PKA in a complex with its inhibitor (PKI) and ATP analogue

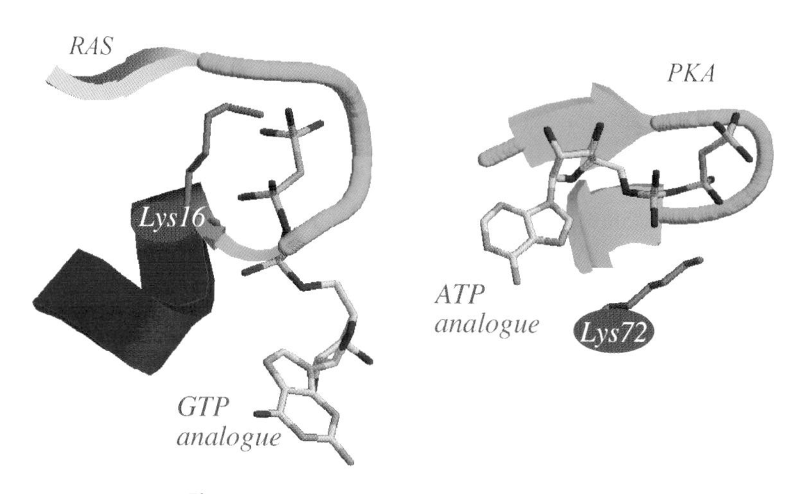

Figure 4.3 P-loop of Ras compared with PKA

Table 4.3 PKA substrate and inhibitor sequences

	P^{-4}	P^{-3}	P^{-2}	P^{-1}	P^{0}	P^{+1}
Inhibitor sequences...						
Heat stable PKA inhibitor sequence (PKI)	Gly	**Arg**	**Arg**	Asn	Ala	**Ile**
PKI Ala→Ser substitution (a substrate)	Gly	**Arg**	**Arg**	Asn	**Ser**	**Ile**
PKA RI subunit (a pseudosubstrate site)	Arg	**Arg**	**Arg**	Gly	Ala	**Ile**
PKA RII subunit (a phosphorylation site)	Asp	**Arg**	**Arg**	Val	**Ser**	Val
Substrate sequences...						
Kemptide		**Arg**	**Arg**	Ala	**Ser**	**Leu**
Phosphorylase kinase α-subunit	Phe	**Arg**	**Arg**	Leu	**Ser**	**Ile**
Phosphorylase kinase β-subunit	Lys	**Arg**	Ser	Asn	**Ser**	Val
Phosphofructokinase-2 (PFK-2)	Arg	**Arg**	**Arg**	Gly	**Ser**	Ser

In PKA crystal structures of activated (autophosphorylated) kinase complexed with either the heat stable protein kinase inhibitor (PKI, a pseudosubstrate) or the Ala → Ser substituted PKI substrates (see Table 4.3), the P-loop serine (Ser53) hydroxyl was found to be strongly H-bonded to the backbone of the P^0-site target Ser/Ala in the PKI pseudosubstrate sequence when crystalised with the stable ATP analogue, adenylyl imidophosphate (AMP-PNP)[9,10,11]. Since the backbone amide of Ser53 also H-bonds to the γ-phosphate of ATP, it was postulated that the transition state intermediate for phosphoryl transfer may involve Ser53 backbone H-bonding to ATP, and its hydroxyl group H-bonding to the peptide substrate[9]. However, subsequent site-directed mutagenesis and solution studies suggest that this is not a mechanism common to other ST-kinases.

Enzyme kinetics done on mutants in which Ser53 was substituted with glycine, threonine or proline, have shown that although the backbone amide of Ser53 may be required for efficient catalysis, the sidechain hydroxyl is not: proline substitution (lacking both the backbone amide and sidechain hydroxyl) was inefficient at binding ATP and phosphoryl transfer; glycine substitution (lacking hydroxyl only) was as efficient as wild type in phosphorylation of the substrate kemptide[12]. In comparison with wild-type PKA, it was found that the Ser → Gly mutant was more efficient than wild-type at phosphorylation of threonine residues suggesting that the Ser53 hydroxyl may play a part in substrate selectivity.

4.1.3 Critical differences between serine/threonine kinases and tyrosine kinases

In serine/threonine kinases, the recognition of the serine or threonine hydroxyl of the peptide substrate is dependent upon the A-loop residue corresponding to Thr201 of

PKA. This A-loop threonine is conserved in all ST-kinases but is replaced with a proline in tyrosine kinases. The second critical residue that selects for a serine/threonine hydroxyl over tyrosine is the C-loop lysine (Lys168 in PKA) that is conserved in all ST-kinases but is replaced with an arginine (Arg1136 in the insulin receptor kinase) in all tyrosine kinases[13]. The third difference lies in the activating phosphorylation site in the A-loop; in ST-kinases it is a threonine, in Tyr-kinases it is a tyrosine(s).

4.1.4 Closed and open conformations

The active apo-form of phosphorylated PKA crystalises in an 'open' conformation with the catalytic cleft ajar (see Figure 4.4). In contrast, both unphosphorylated apo-forms and tris-phosphorylated substrate-occupied forms of the insulin receptor kinase and the IGF-I receptor kinase adopt less open conformations, although the structures are 'more closed' in the presence of substrates[14] (see Figure 4.5). Autophosphorylated PKA can adopt open, intermediate, or closed conformations, the fully closed being the active conformation of the ternary complex of enzyme, ATP and substrate. This closed conformation is typified by the structure of activated PKA complexed with Mn^{2+}AMP-PNP and the PKA inhibitor polypeptide PKI (a pseudosubstrate, Table 4.3) (PDB file: 1CDK)[11]. In the closed conformation, the ATP analogue is now contacted, not only by N-lobe residues, but also by residues of the C-lobe, particularly via Lys168. The presence of the nucleotide's γ-phosphate is thus a trigger for adoption of the activated closed conformation. However, the fully-closed conformation of PKA is only possible

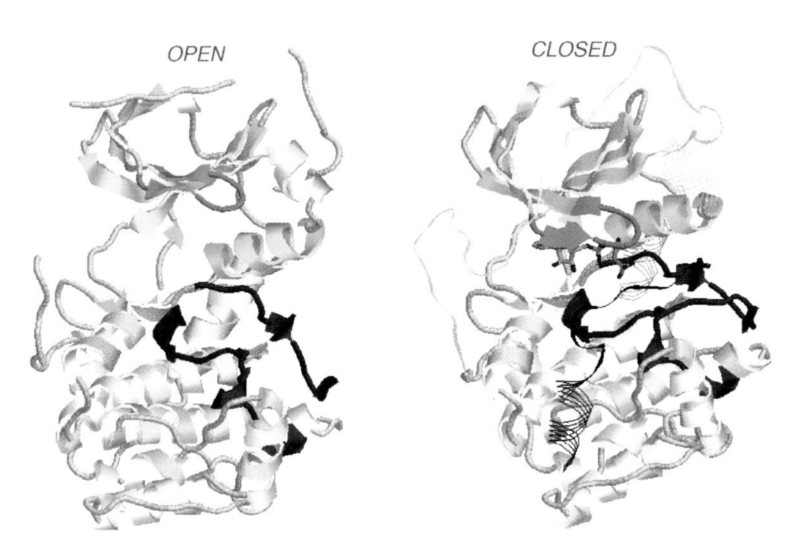

Figure 4.4 PKA catalytic subunit in open and closed conformations (PDB files: 1J3H and 1CDK, respectively)

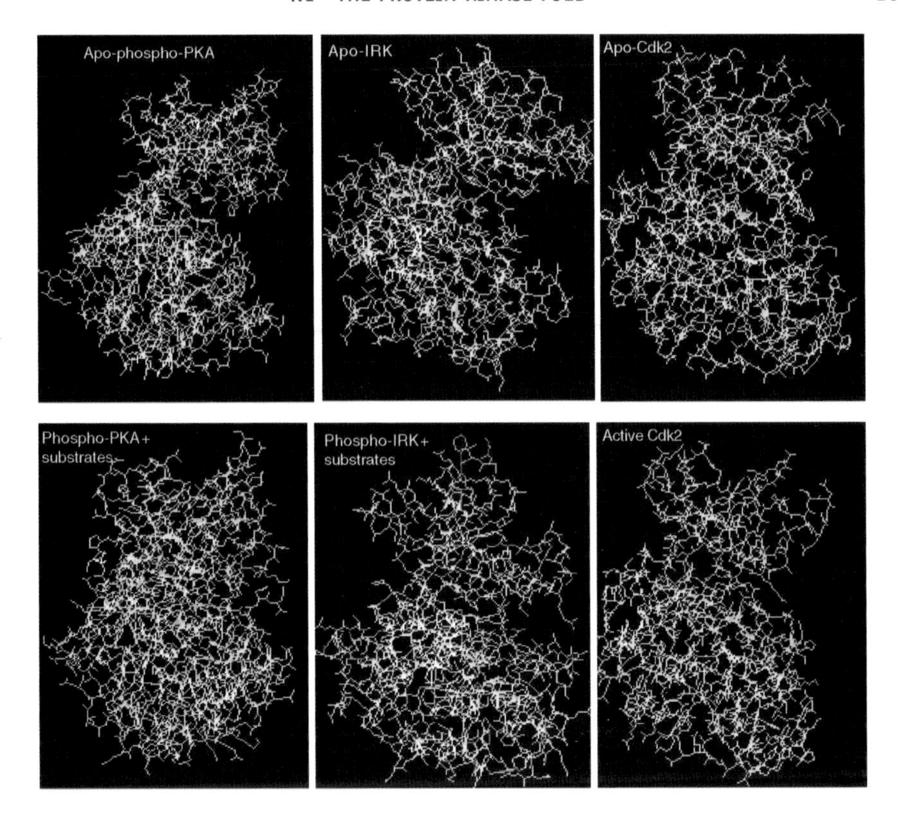

Figure 4.5 Gallery of open and closed kinase structures

when both ATP *and* substrate are present[15]. This is because more extensive connections between the two lobes are triggered by binding of the specificity-determining amino acids adjacent to the P^0 site, discussed in the next section.

The closed form of PKA is thought to exclude water that, if allowed in, would elicit wasteful ATPase activity (instead of phosphotransfer) by effectively substituting for the substrate hydroxyl[16]. The open form is needed to allow ATP and peptides access to the active site as well as to allow release of the phospho-peptide and ADP products[17].

Autophosphorylation of recombinant PKA occurs during synthesis in *E. coli* and is essential to its activity – mutants that cannot autophosphorylate are completely inactive unless *trans*-phosphorylated by co-incubation with wild-type PKA[12].

4.1.5 The catalytic loop or 'C-loop'

The catalytic loop of protein kinases is found in subdomain VIb (164–171 of PKA see Tables 4.1 & 4.2). Invariant residues contained in the C-loop of RD-protein kinases include an invariant Arg.Asp pair and an invariant Asn (Asp166, Asn171 of PKA). PKA catalytic subunits bind tightly to pseudosubstrate sequences in both their regulatory (R)

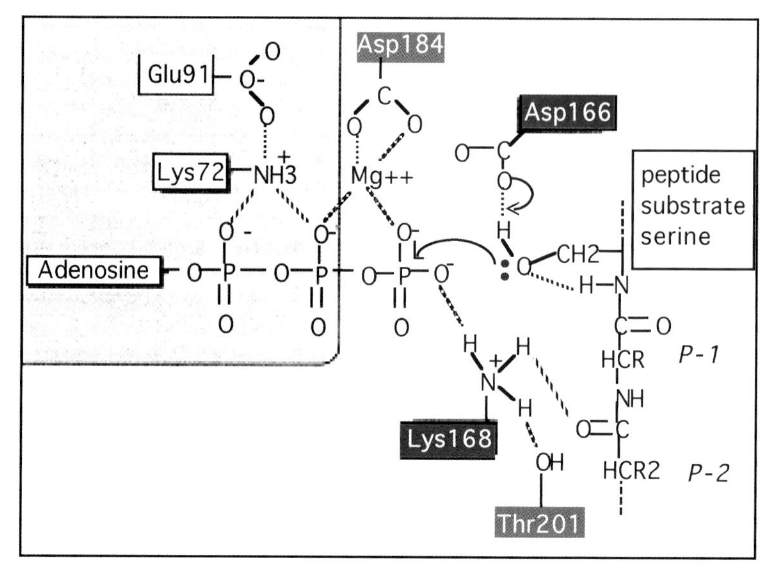


```
Typical substrate sequence         P-4 P-3 P-2  P-1  P0 P+1
Phosphofructokinase-2 (PFK-2):  Arg Arg Arg Gly Ser Ser
```

Figure 4.6 Phosphoryl transfer mechanism of PKA [reproduced from Madhusudan, et al. (1994)]

subunits and the naturally occurring heat stable protein kinase inhibitor protein (PKI). In crystal structures of PKA complexed with either substrate or pseudosubstrate peptide analogues of PKI, the carboxylate group of the invariant Asp166 is within 3 Å of the peptide's P^0-site residue (that is, the phosphorylatable serine of the substrate[13] or alanine in the pseudosubstrate[5]. This close proximity suggests an H-bond forms between Asp166 sidechain and the OH-group of Ser/Thr in the peptide substrate's P^0-site. Thus, the sidechain of Asp166 appears to act as a catalytic base to promote nucleophilic attack[13]. Substitution of PKA Asp166 residue with alanine leads a mutant with only 0.4% of the activity of wild-type enzyme[18]. Figure 4.6 illustrates the proposed phosphoryl transfer mechanism[13].

Another important interaction with PKA's peptide substrates is made by Lys168. This C-loop residue of PKA appears to fulfil the same function of the Lys16 residue incorporated into the P-loop motif of Ras, in that PKA Lys168 H-bonds to, and neutralises the negative charge of, the γ-phosphate of ATP and stabilises the transition state[10,13] but, importantly, Lys168 also H-bonds to the backbone P^{-2} α-carbonyl of the peptide substrate or pseudosubstrate. Thus, in the presence of both ATP and peptide substrate, Lys168 links N-lobe (via the P-loop-anchored ATP γ-phosphate) to the C-lobe (via the substrate peptide that is anchored to the A-loop), providing a trigger for closure of the active site cleft (see Figure 4.7).

This lysine is only invariant in serine/threonine kinases. Interestingly, tyrosine kinases lack an equivalent lysine (Table 4.2). Instead, tyrosine kinases contain an

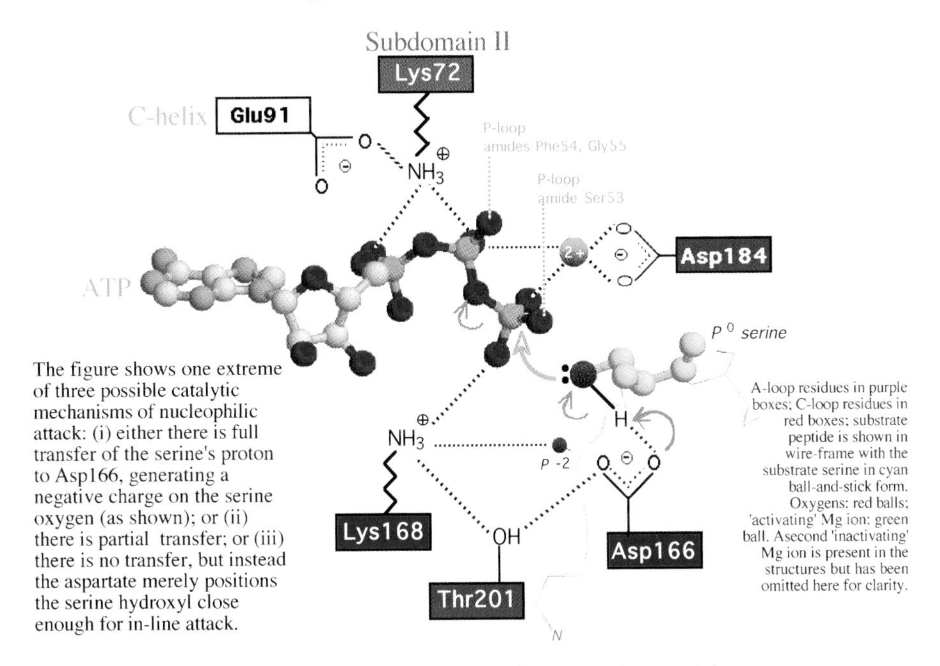

Figure 4.7 The PKA catalytic site on the way to the transition state

invariant arginine (Arg 1137 in IRK) in their C-loops (in the same position or two residues *C*-terminal) and this (positively charged) arginine residue is thought to perform the same charge-neutralising and transition state-stabilising effect as Lys168 interacting with the γ-phosphate of ATP[10].

When a peptide substrate or pseudosubstrate binds to PKA, Lys168 helps anchor the peptide to the surface of the large C-lobe by H-bonding to α-carbonyl group of the P^{-2} residue (see Figure 4.7). Replacement of this lysine with alanine yields a PKA mutant with 300-fold lower affinity for peptide substrate[13]. During actual catalysis, structural analysis suggests that Lys168 also stabilises the transition state by binding to both the γ-phosphate oxygen and, simultaneously, the target oxygen of the serine sidechain; it then continues to bind the transferred phosphate as it forms a covalent bond in the pSer reaction product (see Figure 4.8). In effect, Lys168 'shepherds' the phosphoryl moiety as it passes from ATP donor, through the transition state, to the final product[19]. A second key residue in recognition and active site assembly is the aforementioned A-loop residue Thr201, which makes a crucial H-bond with the catalytic Asp166 *only* when the peptide substrate (or pseudosubstrate) has docked with the protein. This interaction is thought to 'signal' to the enzyme that substrate is present and catalysis may be triggered[16].

The invariant Asn171 plays a mostly structural role: its sidechain amide H-bonds to the α-carbonyl group of Asp166 and this interaction stabilises the catalytic loop[13]. All tyrosine kinases and most Ser/Thr kinases have an arginine immediately *N*-terminal of

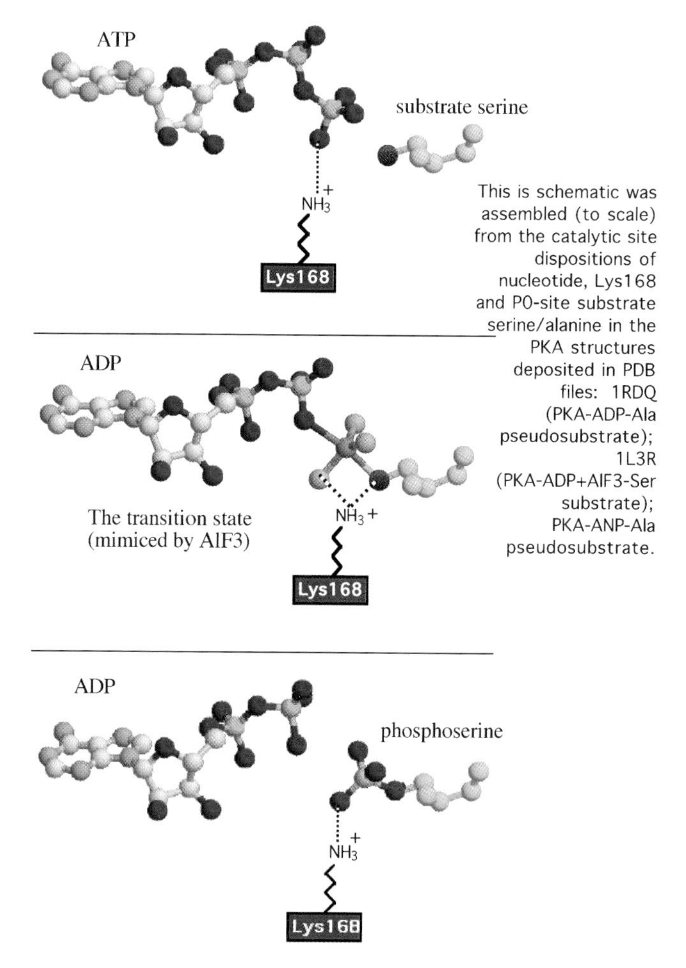

Figure 4.8 The role of Lys168 in phosphoryl transfer in the catalytic mechanism of PKA

the catalytic Asp. This mixed group of kinases are referred to as 'RD kinases' and includes all kinases that require A-loop phosphorylation for activity – the arginine of RD is part of the 'basic cluster', (see Section 4.2.1).

4.1.6 The activation segment/loop or 'A-loop'

A common theme in kinase activity regulation is the requirement for A-loop residues (subdomain VIII) to be phosphorylated to activate the enzyme. I shall use the term 'A-loop' for simplicity and consistency but the A-loop is also (more properly) termed the 'activation segment', because in certain kinase structures (often depending upon phosphorylation status) the loop is divided by a short helix or beta-strand.

In some protein kinases, phosphorylation and dephosphorylation of A-loop residues provide the sole mode of control over their activity. In others, the (A-loop-phosphorylated) mature active enzyme is subsequently regulated by a secondary mechanism under the control of second messengers: for example, the inhibition of PKA by pseudosubstrate sequences such as PKI or the cAMP-binding regulatory subunits RI and II. Still other kinases do not require A-loop phosphorylation. Unfortunately, no structures of nascent PKA (i.e., unphosphorylated) are available and so the following discussion relies upon inferences from the few inactive structures available at present.

In general, conversion of inactive to active kinase involves conformational changes in the A-loop that lead to:

- correct alignment of residues involved in catalysis and substrate binding;
- relief of steric blocking of active site allowing access of substrates (Mg2+, ATP, peptide);
- overall increase in the rate of phosphoryl transfer.

A-loops vary in sequence and length, but all are bounded by tripeptide sequences that are almost invariant: DFG (*N*-terminal end) and APE (*C*-terminal end). In PKA, the A-loop extends from Asp184–Glu208.

4.2 Protein kinases activated by A-loop phosphorylation

Examples requiring A-loop phosphorylation include members of the Ser/Thr-specific kinases, Tyr-specific kinases and Ser/Thr+Tyr-specific kinases. Kinases that *trans*-autophosphorylate must have A-loop sequences that conform to their own substrate specificity.

For example, PKA phosphorylates protein substrates with a consensus sequence containing a serine or threonine at the phosphorylation site (P^0); usually there is a large hydrophobic residue (Φ) at P^{+1}, and crucially, basic amino acids are present *N*-terminal of the P^0-site (Arg is usually found at P^{-2}; or at P^{-3} with additional *N*-terminal basic residues, usually at P^{-6}): **R.R**.X.**S/T**Φ. or **R**.X.X.**R**.X.X.**S/T**.Φ[13]. PKA's own A-loop sequence is: **R**.V.**K**.G.**R**.T.W.**T**.**L**, hence PKA can *trans*-autophosphorylate its A-loop Thr197 because it is presented in a substrate-like sequence context.

Other kinases cannot *trans*-autophosphorylate, and instead rely upon activation by a different upstream kinase. For example, MAPK (an ST-kinase) has a consensus substrate sequence **P**.X.**S/T**.P, but its own A-loop sequence is G.F.L.**T**^{183}.E.**Y**^{185}.V. Thus, MAPK is incapable of autophosphorylation because (i) its A-loop Thr is not in a substrate consensus and (ii) MAPK cannot phosphorylate Tyr residues. Instead, MAPK is activated by the STY-kinase MAPKK, which doubly phosphorylates the **T**.E.**Y** sequence, activating the enzyme[17].

4.2.1 Phosphorylation of A-loop residues and assembly of active site

The disposition of important residues in active kinase structures was first revealed in PKA crystal structures of the activated C subunit complexed with a pseudosubstrate peptide, Mg^{2+} and ATP. Its structure suggested that the need for Thr197 phosphorylation was related to the negative charge conferred by the phosphate. Thr197 phosphorylation increases catalytic activity by three orders of magnitude and mutation of Thr197 to alanine caused a reduction in affinity for ATP and reduced rates of phosphoryl transfer compared with wild-type[20]. All kinases that are activated by A-loop phosphorylation show increases in catalysis by two-to-five orders of magnitude after phosphorylation[21].

In the activated PKA ternary structures, the negatively charged phosphothreonine197 di-anion makes electrostatic interactions with (and stabilises) a positively charged cluster of amino acids: the C-helix residue His87, the A-loop residue Lys189, and the C-loop residue Arg165 (see Figure 4.9 and Table 4.2). Charge-stablised hydrogen bonds are made with the phosphate oxygens of pThr197 and the sidechains of Lys 189 and Arg165[10]. Interaction with Arg165 is thought to promote correct orientation and electrostatic environment of the catalytic base Asp166 (RD motif); interaction with Lys189 is thought to orient Asp184 (in the DFG motif) so that it can interact with the activating Mg^{2+} ion[4,17]. The positions of the basic cluster arginines from the C-lobe

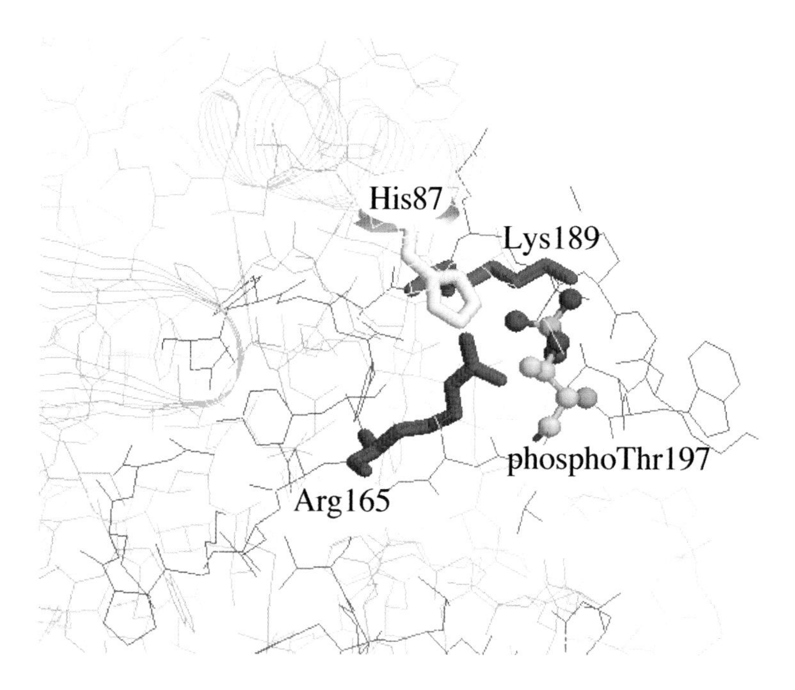

The P-loop: green strands; C-helix: yellow strands; A-loop: purple wireframe; C-loop: red wireframe.

Figure 4.9 The 'basic cluster' interacting with pThr197 of PKA

probably pre-exist, but the basic cluster histidine needs to be moved into position by substrate docking.

The pThr197 charge interaction with the C-helix residue His87 first of all requires nucleotide to dock so that the P-loop is stabilised and the Lys72 to C-helix-Glu91 salt-bridge is formed. Second, and crucially, the pThr197 ↔ His87 charge interaction is ultimately triggered by the docking of a substrate/pseudosubstrate peptide – the His87 sidechain is at least 9 Å away from the phosphate oxygen in the most open conformations of the apoenzyme[22,6] (PDB files: 1J3H; 1SYK) but is as close as 3 Å from pThr197 in the closed conformations[23] containing both nucleotide and peptidyl substrates[24,16] (PDB files: 1U7E and 1RDQ). This change in C-helix positioning may be brought about by the participation of the neighbouring Gln84 in the H-bonding network that docks the pseudosubstrate or substrate (as discussed in Section 4.5.1). The pThr197 ↔ His87 charge interaction is a second important trigger for PKA site closure linking N-lobe with C-lobe, but only when both nucleotide and substrate are present.

Kinases that do not require A-loop phosphorylation to be activated include the γ-subunit of phosphorylase kinase (PhK) (discussed in Section 5.8.2). This catalytic subunit resembles the common kinase structure overall but the equivalent residue to PKA's Thr197 is a glutamate (Glu182 in PhK). This anionic sidechain performs the same neutralising function upon its C-loop arginine (Arg148 in PhK) which, again, precedes its catalytic aspartate (Asp149 in PhK). In PhK, no lysine is present at the equivalent position to that of PKA and so a di-anion is not needed[17]. PhK activity is controlled by phosphorylation of its α and β regulatory subunits and calcium binding to its δ subunit (which is calmodulin).

A-loops typically terminate with a highly conserved APE motif. This is essential for kinase activity: alteration of any of the three residues results in inactive mutants of the Tyr-kinase, Src[2].

4.2.2 The A-loop and catalysis – transition state and site closure

Extensive solution and structural studies of PKA point to an in-line phosphoryl transfer mechanism that occurs through nucleophilic attack upon the γ-phosphorous of ATP by the sidechain oxygen of the substrate serine. The monomeric apo-enzyme exists in an 'open' conformation ready to receive substrates with its wedge-shaped catalytic cleft exposed to solvent. Binding of *both* Mg^{2+}ATP *and* peptide substrate (or pseudosubstrate) are required to cause the two lobes to close, thus excluding water as mentioned earlier. It appears important to exclude water because it could act as nucleophile instead of the P-site serine, and would thus cause unproductive ATPase activity rather than phosphoryl transfer. Equally, binding of regulatory pseudosubstrate inhibitors is ATP-dependent and also relies upon tight closure of the active site being maintained to protect the ATP from hydroysis[16,19].

4.2.3 ATP binding

The role of the P-loop is to provide H-bonds from its backbone amides to anchor the ATP phosphates: Phe54 and Gly55 amides bind to β and Ser53 amide to γ (as mentioned earlier, the hydroxyl of Ser53 also H-bonds to the substrate backbone carbonyl of the P^0-site serine). The negative charge at the α- and β-phosphates is neutralised by H-bonds from the sidechain of Lys72 (the residue stabilised by a salt-bridge from Glu91 of the C-helix) and by chelating bonds to the activating Mg^{2+} from β- and γ-phosphates.

The γ-phosphate is neutralised by both the metal and an H-bond with the C-loop Lys168. The net effect is to (i) withdraw electrons from the β–γ bond (weakening it) and (ii) to give the γ-phosphorous a partial positive charge, thus making it susceptible to nucleophilic attack.

4.2.4 A-loop and autoinhibition

The contribution of A-loop residues to kinase autoinhibition has provoked much discussion and the mechanism(s) are not entirely agreed. This situation can be expected to become clearer as more structures of phospho- and dephospho-kinase pairs are solved. Crystal structural data, mutation and solution studies have often appeared at odds but it is becoming clear that the auto-inhibitory mechanisms of A-loops vary from 'gated' to 'non-gated' extremes[21]. Nevertheless, a constant feature is a dramatic increase in catalytic rate following A-loop phosphorylation.

4.3 The insulin receptor kinase (IRK) – a 'gated' kinase

Tyrosine kinases like the insulin receptor and the Src kinase require A-loop tyrosine residues to be phosphorylated in order to activate. In Src, Tyr416 is phosphorylated; in the insulin receptor, three A-loop tyrosine (1158, 1162 and 1163) must be phosphorylated for full activation, Tyr1162 being phosphorylated first.

The truncated catalytic domain of the insulin receptor (beta-chain residues: Val978–Lys1283) was first crystalised in 1994 in its inactive, unphosphorylated form[25]; later, the activated tris-phosphoryated form was crystallised in complex with substrates[26]. The most striking difference between the two IRK structures is the position of the A-loop (see Figure 4.10). It is clear that phosphorylation of the three tyrosines in IRK's A-loop has caused the entire loop to be ejected from the active site cleft. By comparison, the P- and C-loops move relatively little.

In the original inactive IRK structure (unphosphorylated and without ATP), the A-loop obstructs the binding of both peptide and ATP substrates[25] (PDB file: 1IRK). The phenolic hydroxyl of Tyr1162 (equivalent to Thr197 of PKA) is hydrogen-bonded to the catalytic Asp1132 (equivalent to Asp166 of PKA), thus blocking binding of peptide substrates. The DFG motif is close to the IRK P-loop and the Phe1151 residue blocks the adenine recognition site for ATP. Apart from acting as an intrasteric inhibitor, Hubbard suggested that this conformation also placed Tyr1162 in the active site 'poised for *cis*-autophosphorylation'[25]. However, as explained below, this seems unlikely and it

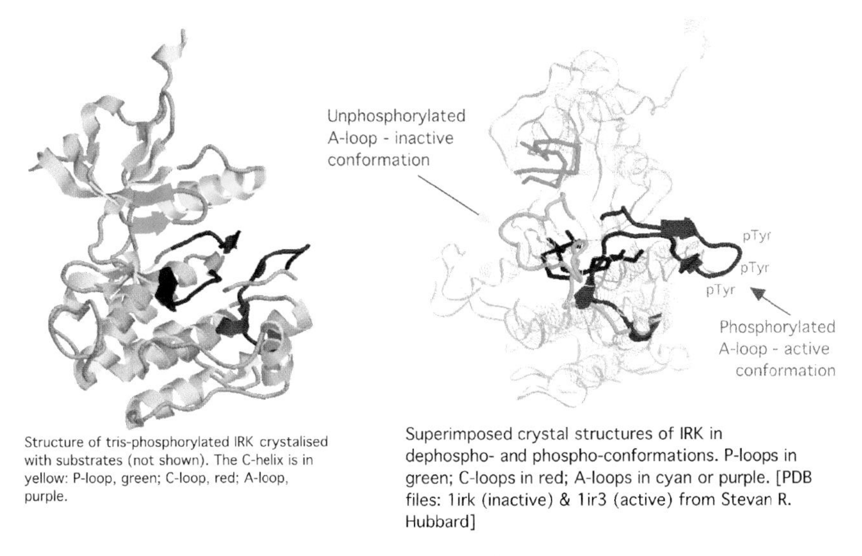

Structure of tris-phosphorylated IRK crystalised with substrates (not shown). The C-helix is in yellow: P-loop, green; C-loop, red; A-loop, purple.

Superimposed crystal structures of IRK in dephospho- and phospho-conformations. P-loops in green; C-loops in red; A-loops in cyan or purple. [PDB files: 1irk (inactive) & 1ir3 (active) from Stevan R. Hubbard]

Figure 4.10 Insulin receptor kinase structures

is generally accepted that the autophosphorylation is a *trans*-autophosphorylation – one IRK phosphorylates its twin and vice versa.

In the activated tris-phosphorylated form, the IRK A-loop has moved out of the active site cleft. In the region of the negatively charged triplet of phosphotyrosines, the A-loop now adopts a position similar to that surrounding phosphothreonine197 in active PKA[26] (PDB file: 1IR3). In active IRK one observes pTyr1162 and pTyr1163 neutralising a basic cluster of amino acids including the C-loop arginine1131 (of the RD motif), and two A-loop residues: arginine1155 (equivalent to PKA Lys189) and arginine1164. The phosphate of pTyr1162 forms charge-stabilised H-bonds with the sidechain of Arg1164; pTyr1163 is H-bonded to Arg1155; pTyr1158 is solvent-exposed and makes no contacts[26] (see Figure 4.11). So, large conformational change resulting from the ejection of the negatively charged phospho-A-loop of IRK away from the catalytic aspartate to a new interaction site of neutralising positively charged residues aids assembly and stabilisation of the active site in the same way as is thought to occur with that of PKA. This appears to be a general mechanism for phosphorylation-switched kinases.

More recently, it has been shown that the 'gate-closed' conformation of apo-IRK (crystallised in the absence of ATP) is unlikely to predominate *in vivo*. It was found that, in the presence of millimolar concentrations (1–10 mM) of a non-hydrolysable ATP analogue (AMP-PNP, aka 'ANP'), the unphosphorylated A-loop became exposed as judged by trypsin cleavage studies. Without ATP, the unphosphorylated A-loop is insensitive to trypsin (i.e., inactive 'gate-closed')[27]. As stated previously, the DFG motif occupies the adenine-binding site in unphosphorylated IRK. It appears that ATP can competitively displace this A-loop motif, causing the A-loop to be externally accessible (i.e., inactive 'gate open'). Since [ATP] = 2—8 mM in most cells, it suggests

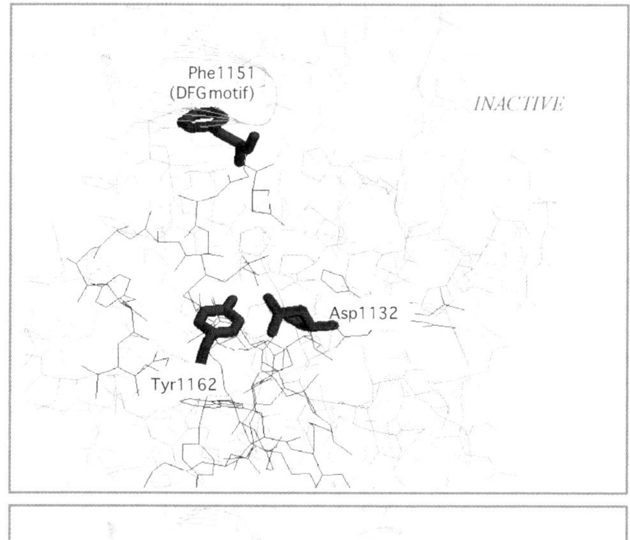

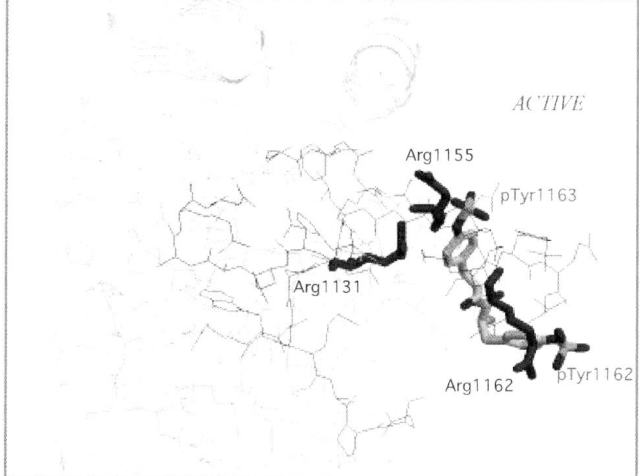

Figure 4.11 The activating movement of A-loop phosphotyrosines to the 'basic cluster' of IRK

that, in the absence of insulin, the A-loop is in equilibrium between 'gate-closed' and 'gate-open' forms. This intermediate 'gate-open', but inactive, form has been explored using a mutant IRK, thought to resemble the (high [ATP]) *in vivo* situation[28]. Crystal structures show ejection of DFG from the adenine-binding site is accompanied by ejection of Tyr1162 from the catalytic cleft. This confirms that IRK cannot phosphorylate its own A-loop (*cis*-) because its own A-loop and ATP cannot occupy the active site at the same time. Instead, one IRK (+ATP) relies upon another IRK (+ATP) to *trans*-phosphorylate it. It has been suggested that in the native insulin receptor, which is a di-sulphide-bonded dimer, the extracellular α-chains inhibit mutual transphosphorylation of the two IRK domains by simply keeping the kinase dimer partners apart in the absence of insulin.

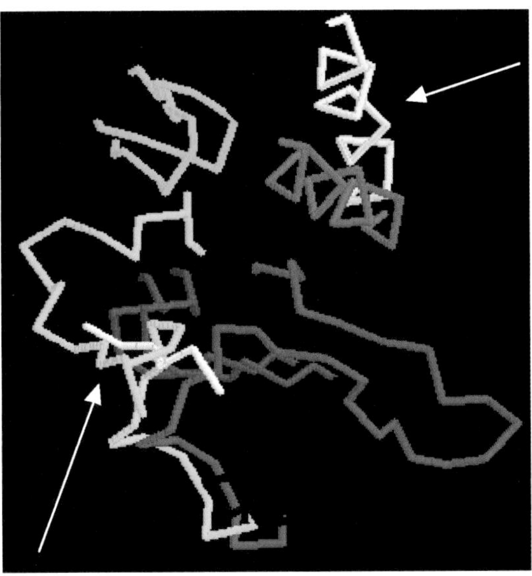

Figure 4.12 Movement of C-helix in IRK activation

Despite the effect of ATP upon the conformation of inactive IRK, the 'gate-open' kinase is not fully active until it is phosphorylated. The rate of phosphoryl transfer and the affinity for both ATP and peptide substrates are both low until all three A-loop tyrosines are phosphorylated. The interaction of phosphotyrosines with the basic cluster results in a activated catalytic site, along with an ordered peptide binding surface made up of A-loop residues 1169–1171, a region that is disordered in unphosphorylated 'gate open' form[28].

A further effect of A-loop phosphorylation is to cause movement of the C-helix of subdomain III. This is the only alpha helix in the N-lobe and is conserved throughout RD kinases. It contains a conserved glutamate. Movement of this alpha helix brings the conserved IRK glutamate1047 (Glu91 in PKA) into correct position to interact with the conserved lysine1030 of subdomain II (Lys72 in PKA) that is responsible for coordinating α and β phosphates of ATP. Thus the orientation of the C-helix in the activated tris-phosphorylated IRK structure closely resembles that of activated PKA (see Figure 4.12).

4.4 Cyclin dependent kinases

Cyclin dependent kinases (Cdks) are a family of ST-kinases that require the binding of another protein (a 'cyclin') to both activate and properly construct their active sites. Cdk

Table 4.4 Activation loop (T-loop) sequences in Cdk's

Type	ACCESSION	A-loop sequence	Thr no.
Cdk-1	NP_00177	**DFG**LARAFGIPIRVYTHEVVTLWYRS**PE**	161
Cdk-2	CAAA43985	**DFG**LARAFGVPVRTYTHEVVTLWYR**APE**	160
Cdk-4	AAC39521	**DFG**LARIYSYQM ALTPVVVTLWYR**APE**	172
Cdk-6	NP_001250	**DFG**LARIYSFQM ALTSVVVTLWYR**APE**	177
Cdk-7	NP_001790	**DFG**LAKSFGSPNRAYTHQVVTRWYR**APE**	176

enzymes follow the canonical kinase fold and like other ST-kinases they must be phosphorylated on a threonine residue in their activation loops. However, unlike other kinases, A-loop phosphorylation of the isolated catalytic subunit is insufficient to assemble the active conformation. In Cdk literature, the A-loop is frequently referred to as the 'T-loop' (short for 'Threonine-loop', see Table 4.4).

Like other protein kinases, activation by A-loop phosphorylation is accompanied by a re-arrangement of the catalytic site, including re-orientation of the C-helix. In Cdk literature, the C-helix is often referred to as the 'PSTAIRE-helix' because the PSTAIRE amino acid motif is semi-conserved among Cdks (the proline is the residue immediately preceding the helix, see Table 4.5).

Another feature that distinguishes Cdks from other ST-kinases is that Cdks can be inhibited by phosphorylation of their P-loops. In Cdk literature, the P-loop is sometimes referred to as the 'G-loop' ('glycine-rich loop', see Table 4.6).

The two positions for inhibitory phosphorylations are the conserved tyrosine and semi-conserved threonine residues of the P-loop of Cdk enzymes (see Table 4.6). As one might expect, phosphorylation of the P-loop disrupts the ATP binding site, rendering the kinase inactive.

4.4.1 Monomeric Cdk2 structures

In the absence of cyclin, A-loop phosphorylation is not enough to activate the kinase. Just like the insulin receptor kinase, the biggest conformational changes upon Cdk activation

Table 4.5 C-helix ('PSTAIRE-helix') sequences in Cdk's

Type	ACCESSION	C-helix sequence	Glu no.
Cdk-1	NP_00177	PSTAIREISLL	51
Cdk-2	CAAA43985	PSTAIREISLL	51
Cdk-4	AAC39521	PISTVREVALL	56
Cdk-6	NP_001250	PLSTIREVAVL	61
Cdk-7	NP_001790	NRTALREIKLL	62

Table 4.6 P-loop sequences in Cdk's

Type	ACCESSION	P-loop sequence	Thr/Tyr no.
Cdk-1	NP_00177	IGEG**TY**GVV	14/15
Cdk-2	CAAA43985	IGEG**TY**GVV	14/15
Cdk-4	AAC39521	IGVGA**Y**GTV	A/17
Cdk-6	NP_001250	IGEGA**Y**GKV	A/24

are centred on the A-loop and the C-helix. In the crystal structures of apo-Cdk2, the A-loop almost completely blocks the peptide substrate-binding cleft and the overall conformation prevents catalytic residues from gaining access to the γ-phosphate of the ATP co-substrate; although it is inactive, apo-Cdk2 can still bind ATP (see Figure 4.13)[29,30,31]. In this un-phosphorylated inactive structure, the A-loop contains a short helical segment (αL12) that displaces the C-helix by a 24° rotation (relative to its position in the activated PKA structure) such that its conserved glutamate51 (91 in PKA) is exposed to solvent and cannot form the activating salt bridge with the conserved lysine33 of subdomain II (72 in PKA). *This is the lysine that binds the* α and β-*phosphates of ATP.* Instead, lysine33

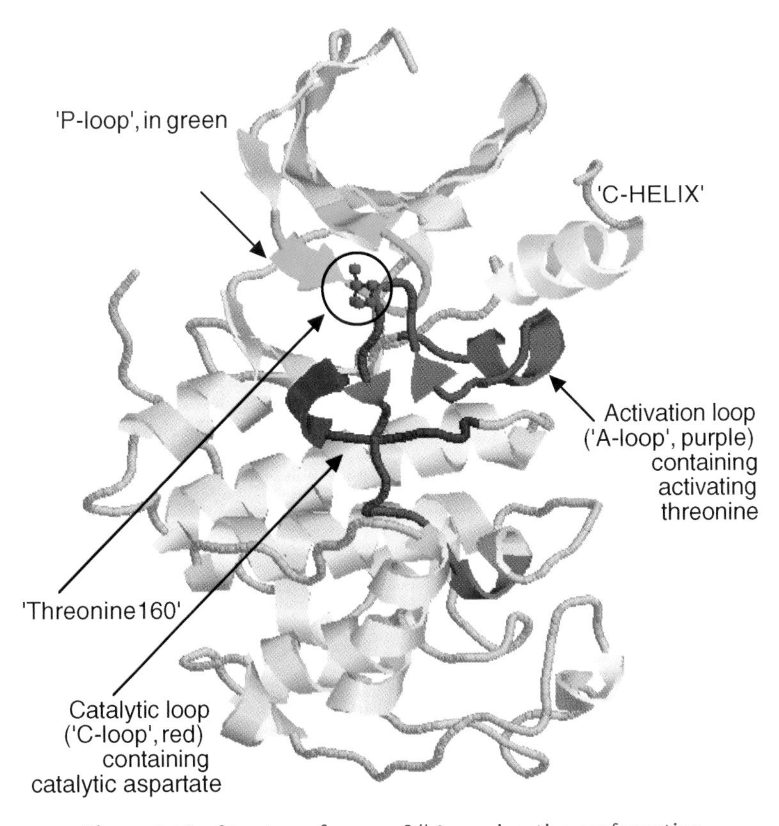

Figure 4.13 Structure of an apo-Cdk2 – an inactive conformation

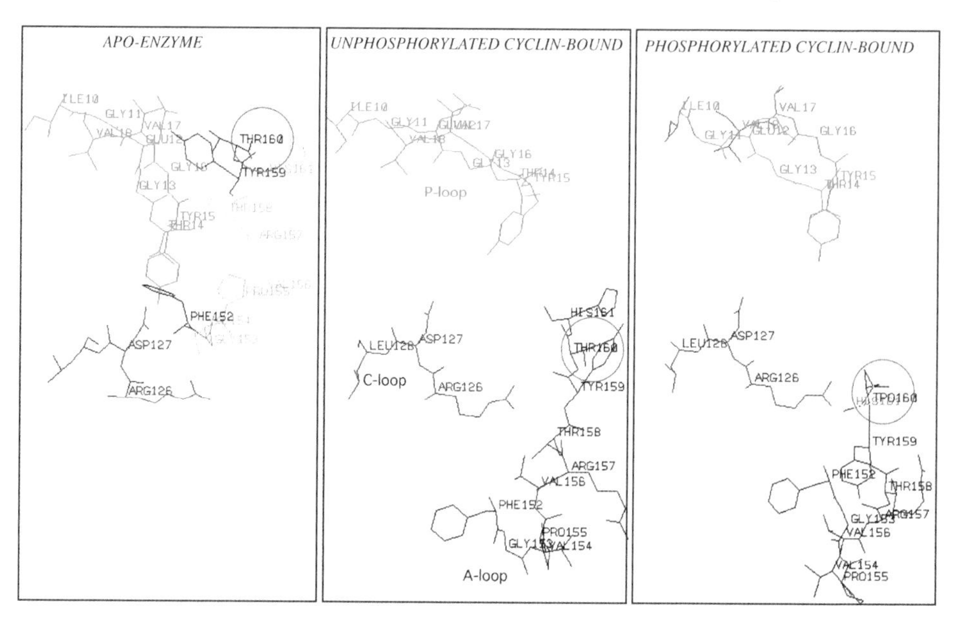

Figure 4.14 Three Cdk2 structures showing relative movement of the A-loop during staged activation

makes an unproductive salt-bridge with the A-loop aspartate145 (part of the DFG motif and equivalent to Asp184 of PKA)[32]. Aspartate145 is partially responsible for coordinating the Mg^{2+} ion in activated Cdk.

In isolated inactive Cdk2 (Figure 4.14), the activating threonine160 and surrounding A-loop residues are orientated towards, and very close to, the P-loop. In monomeric Cdk2 structures this A-loop region is relatively ill-defined (compared with the highly ordered structure it adopts when cyclin-bound) and atomic interactions with the P-loop cannot be assigned with certainty. However, two nucleotide-free apo-Cdk2 structures concur in placing the hydroxyl group of (A-loop) tyrosine159 less than 3 Å from the sidechain carbonyl of (P-loop) glutamate12 and the hydroxyl group of (P-loop) tyrosine15 within 3 Å of the backbone carbonyl of (A-loop) phenylalanine152 (PDB files: 1HCL and 1PW2). In ATP-bound monomeric Cdk2 structures, the Thr160 region is generally in the same position but is more disordered (PDB files: 1HCK and 1B38)[31] and becomes increasingly disordered when threonine160 is phosphorylated – in fact, in monomeric ATP-bound phosphoCk2 (PDB file: 1B39), the region is too disordered to resolve[30].

Although much of threonine160 is exposed in monomeric Cdk2 structures, its hydroxyl group is shielded from solvent[29,30] and this may explain why Cdks cannot be A-loop phosphorylated efficiently except when bound to a cyclin (Figures 4.13 and 4.14). It is obvious from the structures that A-loop phosphorylation alone cannot drive assembly of the active site. Unlike PKA and IRK, the massive structural re-arrangements that lead to active site assembly in Cdks is largely driven by docking with the cognate cyclin partner, with A-loop phosphorylation secondarily causing more modest adjustment (see also Chapter 10, Section 10.5).

4.4.2 Cyclin-bound unphosphorylated Cdk

Binding of cyclin is an essential prerequisite to the subsequent activation of Cdk by phosphorylation *in vivo*. Cyclin binding alone causes large conformational changes that result in:

- a complete re-orientation of both the A-loop and C-helix into conformations that begin to resemble activated PKA and IRK;
- correct positioning of threonine160 for phosphorylation by the Cdk-activating kinase, CAK.

The position of the C-helix in the cyclin A-bound complex corresponds with that of active PKA. Compared with free Cdk2, the C-helix of complexed Cdk2 swings by 90° into the active site upon cyclin docking. This movement is provoked by the formation of an extensive bonding network with the *N*-terminal cyclin box fold of cyclin A and is facilitated by the consequent 'melting' of the αL12 helix. The C-helix is 'clamped' by hydrophobic interactions with the 3rd and 5th helices of cyclin A, and residues at either end of the C-helix (Glu42, Val44 and Lys56) form hydrogen bonds with the cyclin box. The net result is that the conserved glutamate51is now positioned to make the activating salt-bridge with Lys33.

The other major conformational change is in the A-loop. In the structure of unphosphoryated Cdk2 in complex with cyclin A, the A-loop tyrosine159 has moved from under 3 Å to over 20 Å away from the sidechain of the P-loop glutamate12 (Figure 4.14). This movement is caused by the docking of the *N*-terminal domain of cyclin A with the N-lobe of Cdk2, resulting in van der Waals interactions between A-loop residues of Cdk2 (Ala151, Phe152 and Tyr159) and complimentary residues (Phe267, Ile182 and Ile270) in the first three helices of the first repeat of cyclin A, the 'cyclin box' fold[32].

4.4.3 Cyclin-bound phosphorylated Cdk

The conformational changes induced by cyclin binding result in exposure of threonine160, allowing it to be phosphorylated by ***c****yclin-dependent kinase-****a****ctivating* ***k****inase* (**CAK**). The creation of the phosphothreonine di-anion induces further conformational change, driven by the charge-neutralising interaction of negatively-charged phosphothreonine160 with a basic cluster of arginines: Arg50, Arg126 and Arg150 (from the C-helix, C-loop and A-loop, respectively)[33]. Similar charge-neutralising basic clusters maintain the active conformation in phosphorylated PKA and IRK (see Table 4.2). This movement is accompanied by additional van der Waals interactions being made between A-loop residues Val154 and Pro155 and residues of the cyclin A cyclin box (Leu320, His179, Ile182; and Thr316, Leu320, respectively). These contacts are not present in the unphosphorylated

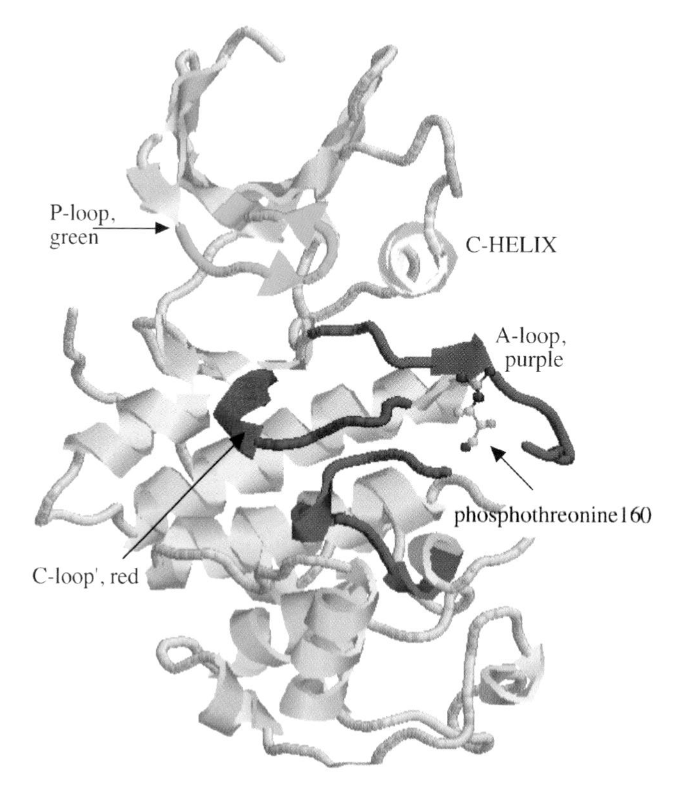

Figure 4.15 Structure of cyclin-bound-phospho-Cdk2 – an active conformation (cyclin erased for clarity)

complex. Superimposition of the active PKA-pseudosubstrate complex structure upon the Cdk2-cyclin A structures show that the unphosphorylated threonine160 partially occupies the presumed substrate-binding surface even in the cyclin-bound form. This final impediment to full activation is only relieved when the threonine160 residue is phosphorylated (see Figure 4.15)[33].

4.5 Secondary inhibition mechanisms – PKA

A structural study from Susan Taylor's group, captures PKA complexed with its regulatory RIα subunit and provides greater insight into the interaction of substrates and pseudosubstrates with the catalytic cleft of PKA[24]. Perhaps more importantly, it shows how the regulatory subunit changes conformation when it binds to the catalytic subunit and indicates a plausible mechanism for activation of PKA by cAMP.

Before looking at this in more detail, consider first the overall structure of inactive PKA as it is found in cells. *In vivo*, the PKA 'holoenzyme' is a tetramer founded upon a homodimer of regulatory subunits sequestering two trapped PKA catalytic subunits.

Table 4.7 cAMP-dependent protein kinase, regulatory subunit alpha 1 (Homo sapiens)

```
  1 MESGSTAASE EARSLRECEL YVQKHNIQAL LKDSIVQLCT ARPERPMAFL REYFERLEKE
 61 EAKQIQNLQK AGTRTDSRED EISPPPPNPV VKGRRRRGAI SAEVYTEEDA ASYVRKVIPK
121 DYKTMAALAK AIEKNVLFSH LDDNERSDIF DAMFSVSFIA GETVIQQGDE GDNFYVIDQG
181 ETDVYVNNEW ATSVGEGGSF GELALIYGTP RAATVKAKTN VKLWGIDRDS YRRILMGSTL
241 RKRKMYEEFL SKVSILESLD KWERLTVADA LEPVQFEDGQ KIVVQGEPGD EFFIILEGSA
301 AVLQRRSENE EFVEVGRLGP SDYFGEIALL MNRPRAATVV ARGPLKCVKL DRPRFERVLG
361 PCSDILKRNI QQYNSFVSLS V
```

```
  1    SGSTAASE EARSLRECEL YVQKHNIQAL LKDSIVQLCT ARPERPMAFL REYFERLEKE
                                             94    98
 59 EAKQIQNLQK AGTRTDSRED EISPPPPNPV VKGRRRRGAI SAEVYTEEDA ASYVRKVIPK
                               αA:A        β1:A       β2:A         β3:A
119 DYKTMAALAK AIEKNVLFSH LDDNERSDIF DAMFSVSFIA GETVIQQGDE GDNFYVIDQG
       β4:A       β5:A      β6:A  αB:A 205      β7:A       β8:A     αB:A  234 αC:A
179 ETDVYVNNEW ATSVGEGGSF GELALIYGTP RAATVKAKTN VKLWGIDRDS YRRILMGSTL
       αC:A                260   αA:B      β1:B       β2:B          β3:B
239 RKRKMYEEFL SKVSILESLD KWERLTVADA LEPVQFEDGQ KIVVQGEPGD EFFIILEGSA
    β4:B         β5:B       β6:B  αB':B 329     β7:B       β8:B        αB:B
299 AVLQRRSENE EFVEVGRLGP SDYFGEIALL MNRPRAATVV ARGPLKCVKL DRPRFERVLG
       αC:B         371
359 PCSDILKRNI QQYNSFVSLS V
```

Tyr205 & Met329 are at centres of PBC hydrophobic pocket in CBD-A, -B, respectively
Trp260 & Tyr 371 are affinity-labelled by cAMP analogues

β-STRAND HELIX PBC LINKER PSEUDOSUBSTRATE LINKER

Each of these (potentially active) PKA catalytic subunits is stuck in an unproductive transition state with a substrate-like loop that the individual regulatory subunits present.

There are two types of regulatory subunit (R^{SUB}), based upon the type of loop they present. RI-types present a 'pseudosubstrate', whereas RII-types present an 'autophosphorylation site'. RI-RI dimers and RII-RII dimers form due to interactions between their unique *N*-terminal dimerisation domains – heterodimers do not form. The dimerisation domains have a second function in 'scaffolding', in that they bind to so-called 'A-kinase anchoring proteins' (AKAPs see Table 5.5). The RI-or-II-dimerisation domain is a four-helix bundle connected to the rest of the protein by the 'linker' peptide.

The RIα linker (or 'hinge') contains a consensus substrate sequence (see Table 4.3) that has either an alanine at P^0 (= pseudosubstrate) or a serine at P^0 (= autophosphorylation site). The linker connects the dimerisation domain to a *C*-terminal pair of cAMP-binding domains (CBD). The CBD module is derived from an ancient protein that acts as a cAMP receptor in bacteria through to humans. The R^{SUB} shares structural similarity with the ***c****atabolite* ***a****ctivator* ***p****rotein* (CAP) from *E. coli* and R^{SUB} studies tend to use nomenclature (see Table 4.7) based upon the CAP folding pattern (the actual sequence homology is very low)[34]. An individual domain consists of an 8-stranded β-barrel or 'basket' followed by a *C*-terminal α-helical segment. In both CBDs, the cAMP binding site is centred upon the 'phosphate-binding cassette' that is located in each of the tandem β-barrels.

In the structure in question (see Figure 4.16), a truncated RIα subunit (minus the dimerisation domain and the second CBD) has been utilised, and it is complexed with a Thr197-phosphorylated PKA catalytic subunit ('PKA^C').

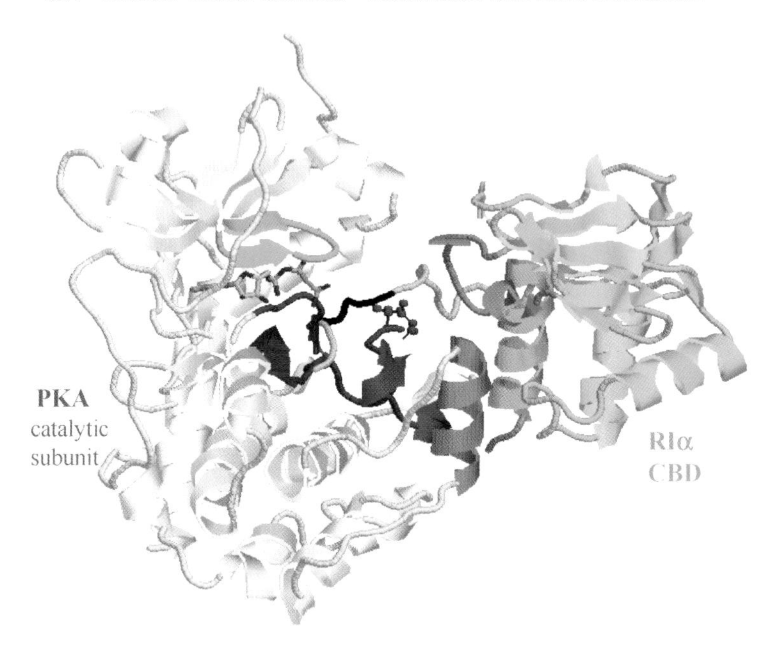

Figure 4.16 The complex between PKA^C and RIα

4.5.1 Substrate or pseudosubstrate binding to the catalytic cleft of PKA

The substrate preference of PKA for arginine residues *N*-terminal of the P^0 site is explained by their ion-pair interactions with three glutamate residues of PKA^C. Glu127, Glu170 (C-loop) and Glu230 make charge interactions with the first two arginines (at P^{-3} and P^{-2}) of the RIα pseudosubstrate sequence. This is true of all PKA substrates and pseudosubstrates (see Figure 4.17).

A second set of pseudosubstrate/substrate interactions are made with the A-loop, the P-loop and the *N*-terminus of the C-helix. P-loop residues and the C-helix terminus join an H-bonded network that locates the pseudosubstrate $P^0 \rightarrow P^{+3}$ residues in an antiparallel β-sheet with a portion of A-loop between Gly200 ← Lys198 (Table 4.8 and Figure 4.18). The result is that the linker adopts an ordered structure. Interestingly, the IRK also binds substrates in a similar manner, with P^0 and the *C*-terminal residues docking in an antiparallel strand addition to the IRK A-loop[26]. Such 'β-strand addition' is a relatively common way for two proteins to interact.

4.5.2 The extended binding surface of R_{SUB}

The major natural inhibitors of PKA^C, the PKI polypeptides and R-type proteins, share a similar primary docking site, in the form of consensus pseudosubstrate sequences. These hinge-like linkers are disordered when free (invisible in X-ray or NMR), but become highly ordered after binding to the PKA^C catalytic cleft by β-strand addition.

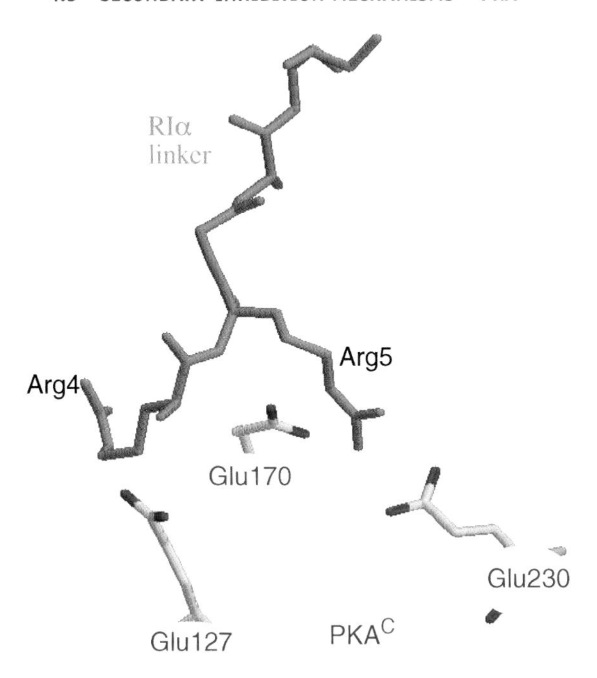

Figure 4.17 The acidic nature of the PKA active site cleft complements the basic residues of the consensus sequence in RIα

The two inhibitors differ greatly, however, in the extended binding surface that is used to give the protein a higher affinity (above that of the pseudosubstrate alone). On the one hand, PKI has a binding surface that extends *N*-terminally of the pseudosubstrate sequence, in the form of an amphipathic helix that docks hydrophobically with PKA^C (see Figure 4.2).

In contrast, the regulatory subunit of PKA, typified by RIα, has a much larger extended binding surface that is *C*-terminal of the pseudosubstrate. The corresponding,

Table 4.8 The H-bonds linking PKA^C with RIα linker region

PKA^C		RIα
Thr51 carbonyl	+	Arg94 sidechain
Ser53 sidechain	+	Ala97 amide
Gly200 carbonyl	+	Ile98 amide
Gly200 amide	+	Ile98 carbonyl
Leu198 carbonyl	+	Ala100 amide
Gln84 sidechain	+	Ser99 sidechain

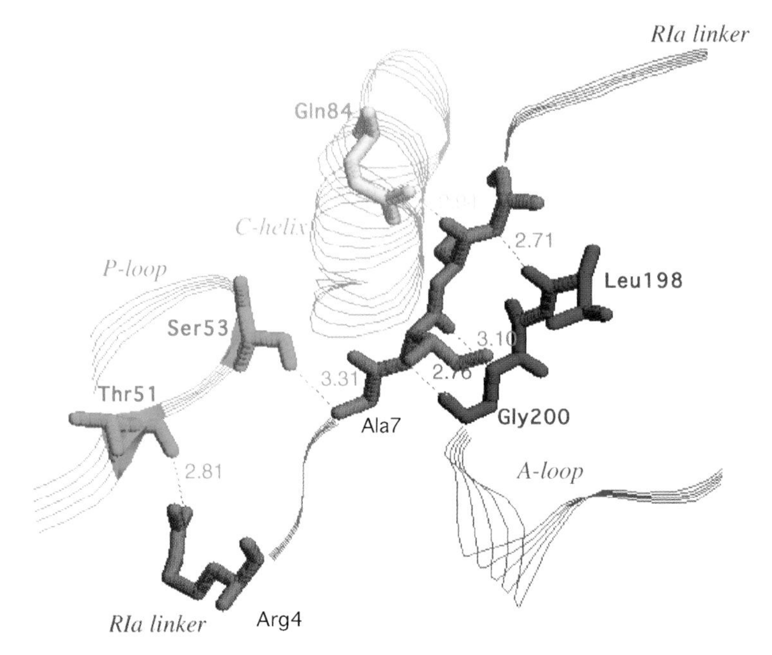

Figure 4.18 RIα pseudosubstrate binds by β-strand addition to A-loop

and complimentary, binding surface of PKA^C is similarly extensive and is centrally organised by the phosphorylation of PKA^C's primary activation switch, that is, Thr197. Only active Thr197-phosphoryated PKA^C monomers will bind to R-subunits – ironically, PKA^C becomes re-inhibited almost immediately after being 'born', just after its primary autoinhibition has been relieved during processing.

The mature catalytic subunit changes little when it binds to RIα, but the regulatory subunit undergoes dramatic re-arrangement when it docks. Apart from the ordering of its linker peptide, two major changes occur in the folded domains. First, the phosphate-binding cassette of RIα becomes distorted when it binds hydrophobically with a complimentary face of PKA^C's 'G-helix' (located in the C-lobe). Second, PKA^C's A-loop provides a docking site for two helices of RIα's helical segment. These latter helices are contiguous (but at an angle to one another) in free RIα[35] (see Figure 4.19). However, binding to PKA^C causes them to merge into a single continuous helix: the B/C helix[24] (see Figure 4.16).

As inferred from Cdk and IRK structures, Thr197-phosphorylation of PKA^C causes its A-loop to swing out to the edge of the catalytic cleft. This movement of pThr197 (towards 'basic cluster' residues Arg165 and Lys189) pulls its neighbour, Trp196, into an exposed position that allows it to interact with RIα. The positioning of pThr197, and the docking and ordering of the pseudosubstrate, collaborate to assemble the RIα binding surfaces for the docking of the PKA^C A-loop and G-helix.

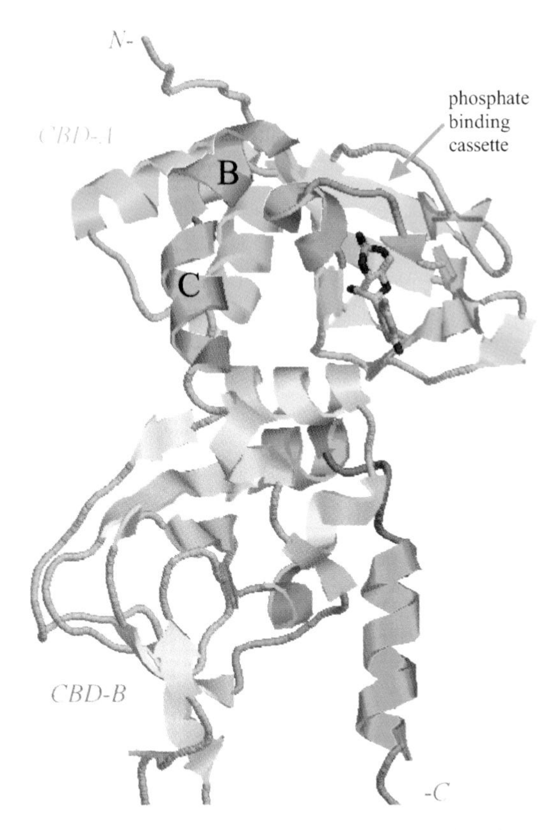

Figure 4.19 Structure of tandem RIα CBD domains

When Thr197 is phosphorylated, its neighbouring A-loop residue Trp196 is available for direct interaction with the Met234 residue of the RIα B/C-helix. Trp196 is the centre for a packing of the RIα B/C helix to the A-loop. Trp196 also interacts directly with Tyr103 of the RIα linker segment. This pulls the B/C helix and PBC near to the interaction centre of the PKA^{C}-G helix at Tyr247. The Tyr247 residue in the G-helix provides a direct interacting centre with Tyr205 of the RIα PBC (see Figure 4.20). Tyr247 (like Trp196) also interacts with Ile98 pf the RIa linker/pseudosubstrate and it is suggested that the ordered linker serves to 'zip' the complex together. Many other contacts are made, too numerous to mention here[24].

4.5.3 Effects on cAMP binding at CBD-A

The docking of PKA^{C} causes 'stretching' of the PBC and so it loses its capacity to bind cAMP (Figures 4.19 and 4.20). A structure of a truncated RIα containing a tandem CBD-A and CBD-B pair has cGMP (mimicking cAMP) bound to CBD-A, whereas CBD-B is

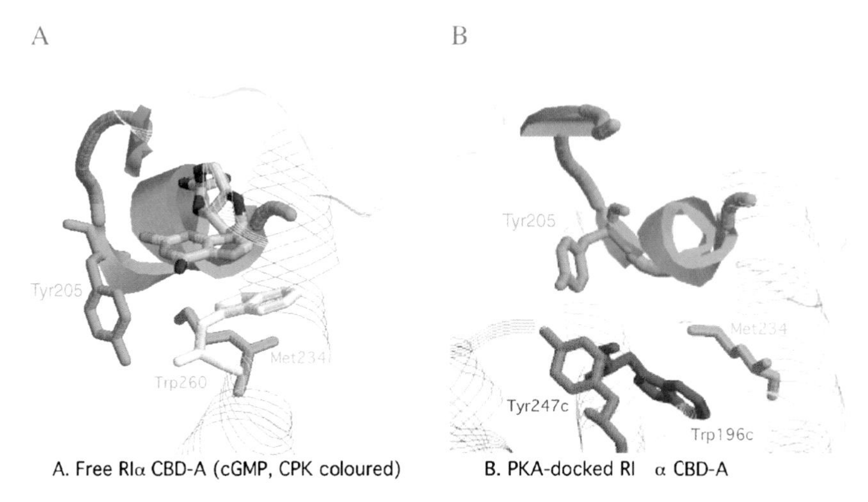

Figure 4.20 The phosphate-binding cassette of the RIα cAMP-binding domain A

empty[35]. The bound cyclic nucleotide lies in the β-basket where a highly conserved arginine (Arg209 in CBD-A; Arg233 in CBD-B) interacts with the cyclic phosphate. In each PBC a highly conserved glutamate (Glu200 in CBD-A; Glu233 in CBD-B) interacts with the 2′OH of the ribose.

Each cAMP is capped by a distal aromatic residue stacking against its adenine ring – in CBD-A this is Trp260, in CBD-B it is Tyr371 [34]. Both of these residues (from the A-helix and *C*-terminal tail of CBD-B, respectively) had earlier been found to be photoaffinity-labelled by reactive cAMP mimetics[35]. The residues perform a gate-like function, locking the nucleotide away from phosphodiesterases, and then opening to release it when PKA^C docks. It is thought that activation of the holoenzyme occurs by cAMP binding to CBD-B first and that delivers conformational change to CBD-A, allowing it to open sufficiently for the second cAMP to out-compete the PKA^C binding conformation.

It seems likely that the order is reversed when cAMP levels fall. The C-lobe of PKA^C is thought to bind primarily (if not exclusively) to the first CBD, where it dislodges cAMP first. PKA^C residue Tyr247 'competes' for the cAMP binding site by forming an H-bond to Tyr205 of the PBC (CBD-A). This forced re-orientation of Tyr205 distorts the PBC (Figure 4.20) and cAMP is ejected. This presumably leads to disruption of the CBD-B site, which loses cAMP next. The holoenzyme is then fully inhibited.

The wider consequences of these submicroscopic movements are dealt with in Chapter 5.

References

1. Lander, E.S., *et al.* (2001) Initial sequencing and analysis of the human genome. *Nature*, **409**: 860–921.
2. Hanks, S.K., Quinn, A.M. and Hunter, T. (1988) The protein kinase family: conserved features and deduced phylogeny of the catalytic domains. *Science*, **241**: 42–51.
3. Engh, R.A. and Bossemeyer, D. (2002) Structural aspects of protein kinase control – role of conformational flexibility. *Pharmacology & Therapeutics*, **93** 99–111.
4. Bossemeyer, D. (1995) Protein kinases – structure and function. *FEBS Letters*, **369**: 57–61.
5. Knighton, D.R. *et al.* (1991) Crystal structure of the catalytic subunit of cyclic adenosine monophosphate-dependent kinase. *Science*, **253**: 407–414.
6. Wu, J., Yang, J., Kannan, N., Madhusudan, Xuong, N-H., Ten Eyck, L.F. and Taylor, S.S. (2006) Crystal structure of the E230Q mutant of cAMP-dependent protein kinase reveals an unexpected apoenzyme conformation and an extended A helix. *Protein Science*, **14**: 2871–2879.
7. Rossmann, M.G., Moras, D. and Olsen, K.W. (1974) Chemical and biological evolution of a nucleotide-binding protein. *Nature*, **250**: 194–199.
8. Walker, J.E., Saraste, M., Runswick, M.J. and Gay, N.J. (1982). Distantly related sequences in the alpha- and beta-subunits of ATP synthase, myosin, kinases and other ATP-requiring enzymes and a common nucleotide binding fold. *EMBO J.*, **1**: 945–951.
9. Bossemeyer, D. (1994) The glycine-rich sequence of protein kinases: a multi-functional element. *TIBS*, **19**: 201–205.
10. Bossemeyer, D., Engh, R.A., Kinzel, V., Ponstingl, H. and Huber, R. (1993) Phosphotransferase and substrate binding mechanism of the cAMP-dependent protein kinase catalytic subunit from procine heart as deduced from the 2.0 Å structure of the complex with Mn^{2+} adenylyl imidophosphate and inhibitor peptide PKI(5-24) *EMBO J.*, **12**: 849–859
11. Narayana, N., Cox, S., Shaltiel, S., Taylor, S.S. and Zuong, N-H. (1997) Crystal structure of a polyhistidine-tagged recombinant catalytic subunit if cAMP-dependent protein kinase complexed with the peptide inhibitor PKI95-24 and adenosine. *Biochemistry*, **36**: 4438–4448.
12. Aimes, R.T., Hemmer, W. and Taylor, S.S. (2000) Serine-53 at the tip of the glycine-rich loop of the cAMP-dependent protein kinase: role in catalysis, P-site specificity, and interaction with inhibitors. *Biochemistry*, **39**: 8325–8332.
13. Madhusudan, Trafny, E.A., *et al.* (1994) cAMP-dependent protein kinase: crystallographic insights into substrate recognition and phosphotransfer. *Protein Science*, **3**: 176–187.
14. Munshi, S.,Kornienko, M., Hall, D.L., Reid, J.C. Waxman, L., Stirdivant, S.M., Darke, P.L. and Kuo, L.C. (2002) Crystal structure of the apo, unactivated insulin-like growth factor-1 receptor kinase. *J. Biol. Chem.*, **277**: 38797–38802.
15. Zheng, J., Knighton, D.R., Xuong, N-H., Taylor, S.S., Sowadski, J.M. and Ten Eyck, L.F. (1993) Crystal structure of the myristylated catalytic subunit of cAMP-dependent protein kinase reveal open and closed conformations. *Protein Science*, **2**: 1559–1573.
16. Yang, J., Ten Eyck, L.F., Xuong, N-H. and Taylor, S.S. (2004) Crystal structure of a cAMP-dependent protein kinase mutant at 1.26 Å: new insights into the catalytic mechanism. *J. Mol. Biol.*, **336**: 473–487.
17. Johnson, L.N., Noble, M.E.M. and Owen, D.J. (1996) Active and inactive protein kinases: structural basis for regulation. *Cell*, **85**: 149–158.
18. Gibbs, C.S. and Zoller, M.J. (1991) *Biochemistry*, **30**: 5329–5334.
19. Madhusudan, Akamine, P., Xuong, N-H. and Taylor, S. (2002) *Nature Structural Biology*, **9**: 273–277.
20. Adams, J.A., McGlone, M.L., Gibson, R. and Taylor, S.S. (1995) Phosphorylation modulates catalytic function and regulation in the cAMP-dependent protein kinase. *Biochemistry*, **34**: 2447–2454.

21. Adams, J.A. (2003) Activation loop phosphorylation and catalysis in protein kinases: is there functional evidence for the autoinhibitor model? *Biochemistry*, **42**: 601–607.
22. Akamine, P., Madhusudan, Wu, J., Xuong, N.H., Ten Eyck, L.F. and Taylor, S.S. (2003) Dynamic features of cAMP-dependent protein kinase revealed by apoenzyme crystal structure. *J. Mol. Biol.*, **327**: 159–171.
23. Johnson, D.A., Akamine, P., Radzio-Andzelm, E., Madhusudan and Taylor, S.S. (2001) Dynamics of cAMP-dependent protein kinase. *Chemical Reviews*, **101**: 2243–2270.
24. Kim, C., Xuong, N-H. and Taylor, S.S. (2005) Crystal structure of a complex between the catalytic and regulatory (RIα) subunits of PKA. *Science*, **307**: 690–696.
25. Hubbard, S.R., Wei, L., Ellis, L. and Hendrickson, W.A. (1994) Crystal structure of the tyrosine kinase domain of the human insulin receptor. *Nature*, **372**: 746–754.
26. Hubbard, S.R. (1997) Crystal structure of the activated insulin receptor tyrosine kinase in complex with peptide substrate and ATP analog. *EMBO J.*, **16**: 5572–5581.
27. Frankel, M., Bishop, S.M., Ablooglu, A.J., Han, Y-P. and Kohanski, R.A. (1999) Conformational changes in the activation loop of the insulin receptor's kinase domain. *Protein Science*, **8**: 2158–2165.
28. Till, J. H., Ablooglu, A.J., Frankel, M., Bishop, S.M., Kahanski, R.A. and Hubbard, S.R. (2001) Crystallographic and solution studies of an activation loop mutant of the insulin receptor kinase. *Journal of Biological Chemistry*, **276**: 10049–10055.
29. DeBont, H.L., Rosenblatt, J., Jancarik, J., Jones, H.D., Morgan, D.O. and Kim, S-H. (1993) Crystal structure of cyclin-dependent kinase 2. *Nature.*, **363**: 595–602.
30. Brown, N.R., Noble, M.E.M., Lawrie, A.M., Morris, M.C., Tunnah, P., Divita, G., Johnson, L.N. and Endicott, J.A. (1999) Effects of phosphorylation of threonine 160 on cyclin-dependent kinase 2 structure and activity. *J. Biol. Chem.*, **274**: 8746–8756.
31. Wu, S.Y., McNae, I., Kontopidis, G., McClue, S.J., McInnes, C., Stewart, K.J., Wang, S., Zheleva, D.I., Marriage, H., Lane, D.P., Taylor, P., Fischer, P.M. and Walkinshaw, M.D. (2003) Discovery of a novel family of CDK inhibitors with the program LIDAEUS: structural basis for ligand-induced disordering of the activation loop. *Structure*, **11**: 399–410.
32. Jeffrey, P.D., Russo, A.A., Polyak, K., Gibbs, E., Hurwitz, J., Massague, J. and Pavletich, N. (1995) Mechanism of CDK activation revealed by the structure of a cyclinA-CDK2 complex. *Nature*, **376**: 313–320.
33. Russo, A.A., Jeffrey, P.D. and Pavletich, N.P. (1996) Structural basis of cyclin-dependent kinase activation by phosphorylation. *Nature Structural Biology*, **3**: 696–700.
34. Su., Y., Dostmann, W.R.G., Herberg, F.W., Durick, K., Xuong, N-H., Ten Eyck, L.F., Taylor, S.S. (1995) Regulatory subunit of protein kinase A: structure of deletion mutant with cAMP binding domains. *Science*, **269**: 807–813.
35. Wu, J., Brown, S., Xuong, N-H. and Taylor, S.S. (2004) RIa subunit of PKA: a cAMP-free structure reveals a hydrophobic capping mechanism for docking cAMP into site B. *Structure*, **12**: 1057–1065.

5

7-pass receptors and the catabolic response

Catabolic hormones that bind to 7-pass receptors play a dominant role in the activation of glycogen breakdown in liver and skeletal muscle, as well as stimulating lipolysis in adipose tissue. These hormones shut down glycogen synthesis. The enzymes that synthesise or degrade glycogen (glycogen synthase and glycogen phosphorylase, respectively) as well as enzymes that control their activities (phosphorylase kinase and protein phosphatase-1) are closely associated with glycogen granules that are themselves bound to the smooth endoplasmic reticulum. The initial activating steps occur at the plasma membrane where the activated 7-pass receptors couple with heterotrimeric G proteins that in turn couple with effector enzymes that produce soluble second messengers (cAMP, or IP3 $\rightarrow$ Ca^{2+}) capable of disseminating the signal through the cytosol. cAMP activates the cAMP-dependent protein kinase PKA, which activates phosphorylase kinase through serine phosphorylation. Alternatively, increases in cytosolic calcium can also activate phosphorylase kinase and in most circumstances both signals are needed to fully (co-)activate phosphorylase kinase. Whereas phosphorylase kinase activates glycogen phosphorylase through serine phosphorylation, PKA and phosphorylase kinase also phosphorylate glycogen synthase, inhibiting it and thus preventing a 'futile cycle'.

Muscle glycogen phosphorylase is highly sensitive to metabolic status of the tissue. Independently of hormonal input, muscle glycogen phosphorylase is activated by increases in [AMP] (indicative of low energy charge) but inhibited by increases in [ATP] or [glucose-6-phosphate] (indicative of high energy charge). Liver glycogen phosphorylase is relatively insensitive to AMP or glucose-6-phosphate (G6P). Both tissue-specific isoforms are activated by a single phosphorylation at serine-14. The liver enzyme, in particular can be inhibited by high glucose (even if phosphorylated) – liver acutely 'senses' and responds to the levels of circulating glucose.

Structure and Function in Cell Signalling John Nelson

5.1 7-pass receptor phylogeny

Classification systems for 7-pass receptors based upon sequence homologies and evolutionary relationships divided the receptors into six clans (or classes) (named A–F) that were subdivided into subclans (or families), but some more recently identified mammalian receptor families do not readily fit into any of the original clans[1]. Note also, that some families in the six clans do not exist in mammals. For example, clans D and E represent fungal pheromone receptors, and slime mould cAMP receptors, respectively whereas clan F contains archaebacterial opsins.

- Clan A: Rhodopsin-related receptors
- Clan B: Secretin-related receptors
- Clan C: Glutamate-related receptors
- Clan D: Fungal pheromone receptors
- Clan E: cAMP receptors
- Clan F: *Archaea* opsin receptors

Two other mammalian 7-pass receptor families not included in the 'clan' system are

- The adhesion receptor family – related to, but distinct from Clan B
- The 'frizzled'/taste receptor family – related to but distinct from Clan C

By far the largest mammalian group of 7-pass receptors is Clan A, which includes the olfactory receptors and visual rhodopsin as well as receptors for a diverse set of medically important ligands such as prostaglandins, biogenic amines (like adrenaline, serotonin, dopamine), melatonins, tachykinins, chemokines and nucleotides. The secretin receptor class (Clan B) includes physiologically important receptors for secretin, glucagon, growth hormone releasing hormone and vasoactive intestinal peptide. Developmental signalling is dominated by the adhesion- and frizzled-families (the latter group includes the receptors for Wnt ligands, discussed in Chapter 10).

5.2 Functional mechanisms of 7-pass receptors

The alternative name for 7-pass receptors is 'G protein-coupled receptors' or GPCRs. This term is used because all 7-pass receptors utilise heterotrimeric G proteins as their primary means of signal transduction. Broadly, GPCR-types can be subdivided into mechanistically functional groups that share specificity for one class of G protein. These preferences are not reflected in shared homologies within each group of receptors, rather the shared G protein specificity was arrived at by convergence.

5.2.1 Gαs-coupling receptors – glucagon- and β-adrenergic receptors – stimulation of cAMP production

7-pass receptors that use adenylate cyclase activation (via Gαs) as their predominant signal transduction mechanism include disparate receptors for a variety of ligands, including catecholamines such as adrenaline, and polypeptides such as glucagon. Despite belonging to different 'clans', such receptors are best considered together. Central to 7-pass receptor signalling by this functional group of receptors is the classical (cholera toxin-sensitive) pathway leading from stimulation of cAMP production to PKA activation. As we have seen in Chapter 1, PKA functions to cause activation (or inactivation) of target enzymes by serine phosphorylation.

In the hepatocytes of most animals, stimulation by adrenaline or glucagon causes PKA activation and this results in the phosphorylation and activation of phosphorylase kinase (PhK), the only enzyme capable of activating glycogen phosphorylase (GP). PhK phosphorylates GP on a single *N*-terminal serine. Simultaneously, both PKA and PhK inactivate glycogen synthase (GS) by serine phosphorylation. In adipocytes, the same βAR- or glucagon-receptor pathway results in activation of PKA that culminates in activation of 'hormone-sensitive lipase' (HSL), again by PKA-catalysed serine phosphorylation.

5.2.2 Gαq-coupling receptors – bombesin- and α1-adrenergic receptors – stimulation of calcium release from the endoplasmic reticulum

A second functional group uses the enzyme phospholipase Cβ (PLCβ) as prime effector. Its activation (via the G protein, Gq) leads to calcium release. Again, a disparate set of receptors from different clans and subclans share specificity for Gαq through evolutionary convergence. Such receptors include those for polypeptide ligands (such as gastrin-releasing hormone (GRH), also known as 'bombesin') and catecholmines (via the α1-type adrenergic receptors). In hepatocytes, Gq-coupled receptors activate PLCβ and this 'signal-activated phospholipase' cleaves the membrane lipid PIP2 to yield two second messengers: soluble IP3 and membrane-bound diacylglycerol (see Chapter 1, Section 1.1.9). IP3 diffuses to the endoplasmic reticulum, where membrane-bound IP3 receptors bind the second messenger. The IP3 receptor is a calcium-selective ion channel that opens upon IP3 binding, and releases calcium into the cytosol. Calcium is an activator of PhK. Calcium alone can activate PhK, but in most animal livers it is thought that Gs-coupled and Gq-coupled pathways collaborate in co-activating PhK through the combined effects of PKA catalysed phosphorylation of PhK's regulatory α- and β-subunits (downstream of Gs) and Gq-induced calcium release that results in activation through binding to PhK's δ-subunit (which is identical to the ubiquitous calcium-binding protein, calmodulin (see Figure 5.1, discussed in Section 5.8).

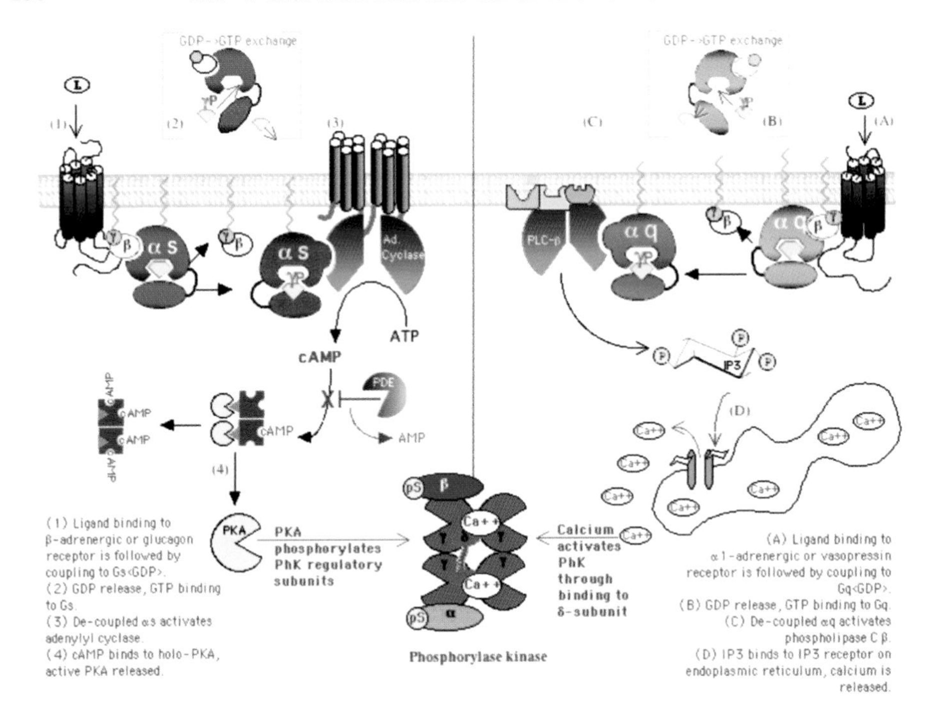

Figure 5.1 Two routes to activation of phosphorylase kinase

5.2.3 Gαi-coupling receptors – somatostatin and α2-adrenergic receptors – inhibition of adenylyl cyclase, activation of K^+ ion channels, inhibition of Ca^{2+} channels, and activation of phospholipase Cβ2

The third functional group of 7-pass receptors couple to the Gαi/o-type proteins. Again, receptors that couple to Gαi-type subunits include members from unrelated clans. These 7-pass receptors have cognate ligands ranging from small neurotransmitters (muscarinic acetyl choline receptors) through catecholamines (α2-adrenergic receptor) to polypeptides such as somatostatin and Met-enkephalin[2]. Gi α-subunits inhibit adenylyl cyclase (antagonising Gαs effects at certain adenylyl isoforms), activate K^+ ion channels, inhibit plasma membrane Ca^{2+} channels, while their β/γ subunits stimulate calcium release from intracellular stores through pertussis toxin-sensitive pathways. This alternative route to calcium release is due to β/γ coupling to PLCβ2 isoform, which is uniquely sensitive to β/γ-stimulation and has a resultant effect (PIP2 cleavage, IP3 release, IP3-receptor opening) indistinguishable from Gq activation, except for pertussis sensitivity. However, this pathway does not appear to operate in hepatocytes, where PhK activation is always pertussis-insensitive. Gi-type proteins are particularly important signalling in neuronal tissue and immune cells.

The visual receptor (rhodopsin) system deploys a Gi-homologous G protein called 'transducin' that is pertussis sensitive but in all other regards its transduction mechanism is quite different from those above. Transducin uniquely activates a cGMP-specific phosphodiesterase and differs from Gi in being both pertussis toxin- *and* cholera toxin-sensitive.

5.2.4 Glucagon- and β-adrenergic-receptors – the catabolic cAMP-dependent protein kinase (PKA) pathway leading to glycogenolysis

There are two inherent problems in sending signals from a discrete site of hormone secretion to a distant target organ:

- The signals (adrenaline or glucagon) are not cell permeant and so need to be captured by cell surface receptors that must be able to transduce this information to the interior of the cell.

- The signal arrives at its target after travelling through the circulation, and is highly diluted. The dilute hormone can only bind one-to-one with receptors, but its intracellular targets are at high concentrations.

We will use the Gαs → adenylyl cyclase → cAMP → PKA pathway to explore these challenges and how the cell deals with them.

The classic PKA 'kinase cascade' was the first 'linear' signalling pathway to be characterised (Figure 5.1). However, the concept of linear transduction was soon superseded by the realisation that even such a seemingly simple pathway is subject to interventions from other pathways, most notably the calcium-mobilising α-adrenergic receptors. The catabolic pathway leading from 7-pass receptors (for glucagon and adrenaline) is typical of what came to be known as a protein kinase 'cascade'. The original concept was a tiered cascade of soluble enzymes, with each step leading to an amplification over the previous step, this being a characteristic expected of the catalytic components.

Caveat The 'cascade' model really only holds true for the PKA pathway, and even here is subject to reinterpretation (being partly scaffolded, as discussed later in Section 5.7). Far from being the norm, true kinase 'cascades' are the exception to the rule. The majority, such as the various MAPK pathways, are not diffusible cascades but exist as scaffolded complexes of two or more distinct kinases.

5.3 Amplification

The first stage in signal amplification is generated from the activating interaction of receptor with G protein; and, during the subsequent G protein couplings with the effector enzyme, adenylyl cyclase. Relative values of cAMP concentration (before and after stimulation) are available but no data sets from homogenous conditions exist to make definitive statements about the exact 'gain' in the system. Nevertheless, a hypothetical calculation can be made with what is at hand.

A 'thought experiment' that helps one think about cellular scale...

- An activated human Gαs subunit has an estimated catalytic rate (k_{cat}) of GTP hydrolysis of around 4 min^{-1}.
- Meaning an individual activated Gαs<GTP> lasts around 15 seconds[3].
- The turnover rate of type I mammalian adenylyl cyclase is 890 min^{-1}, or around 15 sec^{-1} [4,5].
- So, during the 15 seconds it is coupled to Gαs<GTP>, a potential 225 cAMP molecules could be produced during a single interaction, before Gαs hydrolyses GTP to GDP and the complex dissociates.

What would this mean in terms of cellular concentration?

- A cultured human cell line (HeLa) has a volume of 1,200 μm^3 [6] or 1.2 picolitres.
- A cell is around 70% aqueous[7].
- So the volume of a cell's 'cytosol' is = 0.84 picolitres, or 0.84×10^{-12} litres.

What is the cAMP concentration produced from one coupling event?

- Avogadro's number = 6×10^{23} (i.e. the number of molecules per mole).
- So the number of moles in 225 cAMP molecules = $(225) \div (6 \times 10^{23}) = 3.75 \times 10^{-22}$ moles (in a volume of 0.84 picolitres).
- Therefore the molarity of cAMP produced (by one receptor inducing a single Gαs-to-adenylyl cyclase interaction) is $= (3.75 \times 10^{-22} \div 0.84) \times (1 \times 10^{12}) = 4.46 \times 10^{-10}$ M.
- If 1000 receptors successfully couple and activate a single Gαs each, the amount of cAMP produced (in 15 seconds) would be 4.46×10^{-7} M or 0.446 μM.
- If each occupied receptor coupled ten times with Gs proteins then 4.46 μM cAMP could be produced.

Caveat

The compliment of Gs-coupling receptors in different cell types varies widely — hepatocytes $\approx 1 \times 10^5$ glucagon receptors per cell; erythrocytes and S49 $\approx 1 \times 10^3$ β-adrenergic receptors per cell; neutrophils have only 500 β-adrenergic receptors per cell[8,9]. The number of receptor activations and couplings is likely to vary between extremes — an estimate in S49 cells, where the ratio of β-adrenergic receptors to Gs is 1:100, is that each receptor couples with 5 to 10 Gs molecules[10].

So how do the above hypothetical calculations compare with reality?

In a concerted assault on this problem, an estimate of $\approx 10{,}000$-fold signal amplification was calculated for cAMP production arising from stimulation of a single β-adrenergic receptor (based on 500 β-adrenergic receptors coupling with 5×10^4 Gs molecules to give an output of 5×10^6 cAMP molecules in neutrophils)[9]. Neutrophils, however, have untypically low levels of β-adrenergic receptor – many cell types have 'spare receptors' – and levels of amplification are likely to vary widely.

5.3.1 Stimulated changes in cytosolic cAMP concentration

The levels of cAMP before and after stimulation are notoriously difficult to assay, because cAMP is constantly being destroyed by the action of ***p**hospho**d**iest**e**rases* (PDEs). However, measurements in rat atrium cells suggest resting levels of free cAMP at 0·32 μM, rising to 8·77 μM (using the authors' conversion factor) after stimulation with β-adrenergic agonist and in the presence of PDE inhibitor – an increase of ≈27-fold[7]. Others find a stimulated rise in [cAMP] of ≈15-fold in myocyte cells in response to β-adrenergic stimulation[11].

5.3.2 'Spare receptors' – maximum signal transduction from partial occupancy of receptors

In hepatocytes, stimulation of glucagon receptors leads to an instantaneous rise in [cAMP] that peaks at 20 seconds[12]. However, binding of glucagon at the receptor continues to rise, peaking at 60 seconds, and this suggests the receptor population is not fully occupied during the initial burst of cAMP production. It is estimated that only 10% receptor occupancy is sufficient for maximal cAMP production. This would mean only around 2,000 out of the total of 20,000 high affinity sites are actually required to give a full effector response. Although the notion of spare receptors has been controversial, and may only apply where receptors are at such high levels, the many early biochemical studies that support the hypothesis have been reinforced by recent real-time FRET measurements of fluorescently labelled receptors and G proteins also showing maximal response from partial occupancy[13].

5.3.3 Collision coupling *versus* pre-coupling

The original model of receptor interactions with G protein and the subsequent G protein interaction with effector was one of 'collision-coupling'[14]. In this model, individual receptors, G proteins and effector enzymes diffuse freely in the 2-dimensional space of the plasma membrane. G protein collisions with receptors are unfruitful if the receptor is unoccupied; adenylyl cyclase collisions with Gαs are unproductive if the Gα subunit is bound to its β/γ dimer. Kinetic analysis of most biochemical data was best modelled as a 'collision-coupling' but there remained controversy over the possibility that some unoccupied 7-pass receptor types might exist in 'pre-coupled' complexes with Gα/β/γ heterotrimers. However, very recent evidence using fluorescently tagged receptors and G proteins strongly supports the early model and shows that receptors and G proteins are individually highly mobile in the plane of the membrane and only 'collision couple' after stimulation[13].

5.3.4 How much cAMP is needed?

In vivo, the hyper-stimulated levels measured *in vitro* (with PDE activity blocked) are unlikely to occur under normal conditions, where cAMP rises are transient due to PDE actions. It is important to note that the estimated half-maximal stimulation of PKA occurs at 1 μM cAMP, and because activation is governed by the *positively cooperative* binding of cAMP to the regulatory subunits of PKA, only 2- to 3-fold increases in [cAMP] are necessary for full stimulation of PKA[15]. PKA is on a knife edge as far as activation potential is concerned, being switched on or off by relatively small changes in cAMP concentration. Nevertheless, the potential exists for very high chronic levels in disease states and such high levels of cAMP are potentially toxic. Cholera can cause chronically high levels of cAMP to build up in infected cells because its toxin locks Gαs into a permanently GTP-occupied state.

5.3.5 The PKA 'cascade'

The next levels of amplification are due to the kinase cascade itself. cAMP binding releases the PKA catalytic subunits (PKA^C) from the inactive holoenzyme. Each PKA^C can then phosphorylate several molecules of its substrate, phosphorylase kinase (PhK), and each newly activated PhK can, in turn, phosphorylate several molecules of glycogen phosphorylase (GP). The amplifying steps are reflected in the increasing cellular concentrations of enzyme at each tier, going downstream. The ratio of PKA:PhK:GP is estimated to be 1:10:240, giving a further amplification of $\approx 10,000$[16].

5.4 Adenylyl cyclase – signal limitation

Adenylyl cyclase (formerly called 'adenylate cyclase' or 'adenyl cyclase') is the predominant effector enzyme for 7-pass receptors that couple to the Gs protein (such G proteins are designated 's' because they ***s***timulate adenylyl cyclase activity).

In the case of S49 murine lymphoma cells, the parameter that limits the amount of cAMP produced appears to be the level of adenylyl cyclase. In these cells the ratio of Gs: AC is 100:3. A similar adenylyl cyclase 'bottleneck' has been described in cardiac cells. Overexpressing the upstream receptor has little effect upon cAMP output, but over-expressing the cardiac adenylyl isoform does cause an increase in cAMP output that is proportional to the level of AC expression[17].

5.4.1 Adenylyl cyclase – signal termination and PDE isoforms

The cAMP signal is also limited (and ultimately terminated) by ***p***hospho***die***sterases (PDEs, discussed further in Section 5.1.2.9) that hydrolyse cAMP to AMP. In cells where Gs-activating receptors are overexpressed, constitutive signalling from unoccupied receptors may reach high enough levels to cause AC activation in the absence of ligand. Such 'basal' AC signalling is also a feature of certain AC isoforms. AC2, for example, has a 15-fold higher level of basal signalling than AC6[18]. In tissues with a high basal activity of adenylyl cyclase, such as adipocytes, cAMP levels are actively kept in check by the action of PDE. In such cases, PDE inhibitors alone can cause pathway activation. In hepatocytes, where basal AC activity is low, PDE inhibitors alone have little effect and the pathway must be activated through ligand-induced AC activation[19].

5.4.2 Crosstalk and negative feed back

There are four major types of PDE distinguished by different regulatory characteristics. PDE1 enzymes (PDE1A, B, C and splice variants thereof) are calmodulin-binding proteins and all activated by increases in calcium (Table 5.1). In the heart, the α1-AR pathway causes Ca^{2+} release and this attenuates the simultaneous β-AR-induced cAMP signal resulting from adrenaline stimulation – an example of crosstalk. PDE3B isoforms (and the PDE4D3 isoform) are activated by phosphorylation by the cAMP-dependent protein kinase A. In this case, PKA attenuates its own signalling through activation of an enzyme that destroys its second messenger – an example of a negative feedback loop that is important in liver, heart and fat cells[19]. PKA can also phosphorylate and inactivate the AC5 and AC6 isoforms of adenylyl cyclase, again negatively regulating it own continued activation[20]. Opposing the action of adrenaline or glucagon is insulin, which causes activation of PKB. PKB is a serine threonine kinase that also phosphorylates (and activates) the same serine residue of PDE3B in adipocytes and thus insulin reverses the catabolic effects of adrenaline/glucagons by removing cAMP. This crosstalk has the same effect as the negative feedback loop from PKA[21,22] (see Figure 5.2 and Chapter 9).

5.5 Adenylyl cyclase isoforms

There are nine types of membrane-bound, signal-regulated adenylyl cyclase (AC) in the human genome. A tenth adenylyl cyclase is also present but is only expressed in sperm: it is a soluble enzyme that only responds to changes in bicarbonate, is not regulated by extracellular signals and will therefore not be considered here[23,24]. The nine membrane-bound isoforms share a similar topology resembling a tandem repeat, with a 6-pass transmembrane domain followed by a cytoplasmic domain (C1) connected to a second 6-pass transmembrane domain and a second linked cytoplasmic domain (C2). The C1 and C2 domains represent a 'split' catalytic site. Although the C2 portion has weak activity if expressed alone, both C1 and C2 are required for full activity. These catalytic domains are homologous between the different isoforms and retain high similarity with prokaryote forms. However, the transmembrane sequences differ between the family members[19].

5.5.1 Transmembrane isoforms of adenylyl cyclase

The nine transmembrane adenylyl cyclase isoforms form four families that share common functions and homologies[23]:

- Ca^{2+}/calmodulin-stimulated ACs;
- ACs that are stimulated by the β/γ-subunits released by G proteins (this β/γ-stimulation is only possible if αs is co-stimulating, see Chapter 8, Section 8.6);
- Ca^{2+}- and Gαi-inhibited ACs;

Table 5.1 Phosphodiesterase enzymes

Substrate	Family (*genes*)	Tissue distribution	Subcellular location	Stimulators	Inhibitors
cAMP	PDE4 (*4*)	Brain, lung, liver, kidney, Sertoli cells, immunocytes	All bind activated βAR via β-arrestin. 'Short' isoform — plasma membrane/Golgi-associated. 'Long' isofrom — cytosolic (or cytoskeletal: PDE4D & PDE4D variants bind to SH3 proteins like Src). • PDE4D5 interacts with RACK-1, a PKC scaffold • PDE4D3 binds AKAP450 and mAKAP, a PKA scaffold • PDE4A binds AKAP95 [nucleus], AKAP149 [mitochondria] or the AKAP 'MTG' [Golgi])	PKA (PDE4D3)	
	PDE7 (2)	Skeletal muscle, heart, kidney, brain, pancreas, T-lymphocytes	PDE7A (in T lymphocytes) localised to the Golgi apparatus via MTG		
	PDE8 (2)	Testes, eye, skeletal muscle, heart, kidney, ovary, brain, T-lymphocytes			
cAMP ***or*** cGMP	PDE1 (*3*)	Heart, brain, lung, smooth muscle	Cytoplasmic in contractile smooth muscle; nuclear in proliferating smooth muscle	Ca^{2+}/calmodulin	
	PDE2 (*1*)	Adrenal, heart, lung, liver, platelets		cGMP	
	PDE3 (2)	Heart, lung, liver, platelets, adipose, immunocytes	Endoplasmic reticulum membrane	PKA PKB (PDE3B)	cGMP
	PDE10 (*1*)	Testes, brain			
	PDE11 (*1*)	Skeletal muscle, prostate, kidney, liver, pituitary, testes, salivary glands			
cGMP	PDE5 (*1*)	Lung, platelets, smooth muscle			*Viagra
	PDE6 (*3*)	Photoreceptor cells			
	PDE9 (*1*)	Kidney, liver, lung, brain			

Information sources for Table: -

Asirvathan, A.L., Galligan, S.G., Schillace, R.V., Davey, M.P., Vasta, V., Beavo, J.A. and Carr, D.W. (2004) A-kinase anchoring proteins interact with phosphodiesterases in T lymphocyte Cell lines. *J. Immunol.*, **173**: 4806-4814.

Nagel, D.J., Aizawa, T., Jeon, K-I., Liu, W., Mohan, A., Wei, H., Miano, J.M., Florio, V.A., Gao, P., Korshunov, V.A., Berk, B.C. and Yan, C. (2006) Role of nuclear Ca2+/calmodulin-stimulated phosphodiesterase 1A in vascular smooth muscle cell growth and survival. *Circulation Research*, **98**: 777-784.

Houslay, M.D. and Milligan (1997) Tailoring cAMP-signalling responses through isoform multiplicity. *TIBS*, **22**: 217-224.

Essayan, D.M. (2001) Cyclic nucleotide phosphodiesterases. *J. Allergy Clin. Immunol.*, **108**: 671-680.

*Mehats, C., Andersen, C.B., Filopanti, M., Jin, S-L.C. and Conti, M. (2002) Cyclic nucleotide phopshodiesterases and their role in endocrine cell signalling. *TRENDS in Endocrinology & Metabolism*, **13**:29-35.

Tasken, K and Aandahl, E.M. (2003) Localized effects of cAMP mediated by distinct routes of protein kinase A. *Physiol. Rev.*, **84**: 137-167.

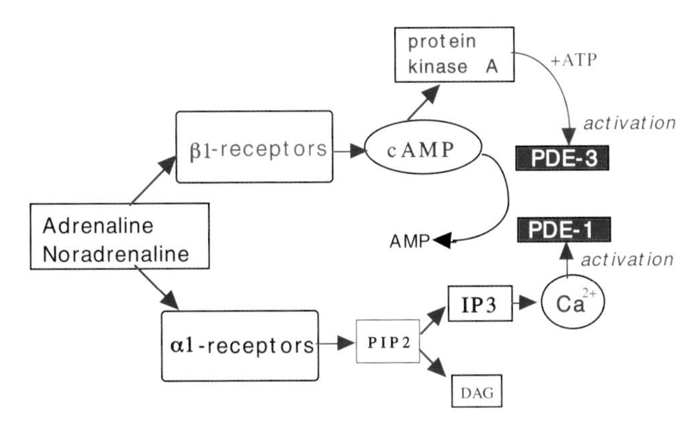

Figure 5.2 Negative feedback and crosstalk in cardiomyocytes

- Forskolin-insensitive AC. Forskolin is a diterpine extracted from the roots of the *Coleus forskohlii* plant (a member of the mint family). Forskolin directly activates most forms of AC, independently of receptors, and is a commonly used reagent in G protein research (see also Chapter 8, Section 8.4.5).

Note that *all forms* of AC are *stimulated* by the alpha subunit of Gs, and *all except AC9* are *stimulated by forskolin*. A further important point is that the Ca^{2+}-sensitive forms of AC are unaffected by Ca^{2+}-released from intracellular IP3-sensitive stores (downstream of Gαq activation). Instead, AC1, 3, 8 are activated, and AC5, 6 are inhibited, by extracellular Ca^{2+}-influx through capacitative entry or via voltage-gated channels in the plasma membrane; the brain isoforms are functionally compartmentalised with voltage-gated Ca^{2+}-channels[23]. The properties and tissue distribution of the signal-activated AC isoforms is summarised in Table 5.2.

5.6 G proteins and the adenylyl cyclase effector isoforms

The heterotrimeric G proteins are obligate mediators of 7-pass receptor signals, being responsible for 'transducing' the signal from the receptor to the effector (Table 5.3). Receptors for adrenaline appear to be potentially antagonistic towards each other (Table 5.4), and considering the adenylyl cyclase effector first, it is clear from Tables 5.2, 5.3 and 5.4 that its control is far from simple. First, forms of the effector enzyme may be inhibited or stimulated by G protein alpha subunits. Second, some isoforms recapitulate the slime moulds in being activated by β/γ subunits whereas others are by contrast inhibited by the β/γ heterodimer. Calcium can stimulate or inhibit certain AC isoforms. Finally, some isoforms are also subject to inhibiting or activating phosphorylations by PKA or PKC.

Table 5.2 Adenylyl cyclase: membrane-spanning isoforms

FAMILY	Isoform	Tissue distribution	G protein effects	PKA	PKC/TPA	Ca^{2+}
Ca^{2+}/calmodulin-stimulated	AC1	Brain & adrenal medulla only	$\alpha_s\uparrow$ $\alpha_o\downarrow$ $\beta/\gamma\downarrow$		$\uparrow$	$\uparrow$
	AC3	Brain, brown adipose, testis (less in heart & liver)	$\alpha_s\uparrow$		$\uparrow$	$\uparrow$
	AC8	Brain, lung (less in testis, heart, adrenal)	$\alpha_s\uparrow$			$\uparrow$
Gβγ-stimulated	AC2	Brain, skeletal muscle, lung (less in heart)	$\alpha_s\uparrow$ $\beta/\gamma\uparrow$		$\uparrow$	
	AC4	Liver, lung, heart, brown adipose (less in brain)	$\alpha_s\uparrow$ $\beta/\gamma\uparrow$		$\downarrow$	
	AC7	UBIQUITOUS (but absent in adrenal)	$\alpha_s\uparrow$ $\beta/\gamma\uparrow$			
Ca^{2+}- and Gαi-inhibited	AC5	UBIQUITOUS (but absent in skeletal muscle)	$\alpha_s\uparrow$ $\alpha_i\downarrow$ $\beta/\gamma\downarrow$	$\downarrow$	$\uparrow$	$\downarrow$
	AC6	UBIQUITOUS	$\alpha_s\uparrow$ $\alpha_i\downarrow$ $\beta/\gamma\downarrow$	$\downarrow$	$\downarrow$	$\downarrow$
Forskolin-insensitive	AC9	UBIQUITOUS (but absent in testis)	$\alpha_s\uparrow$			

Information sources for this table: Defer, N., *et al.* (2000); Houslay, M.D. & Milligan, G. (1997); Sunahara, R.K. & Taussig, R. (2002)
$\uparrow$ = 'stimulates'; $\downarrow$ = 'inhibits'

Table 5.3 Heterotrimeric G protein isoforms

FAMILY	Subunit isoform	Tissue distribution	Effectors	Disease
Gs	α_s (S,L,XL)	UBIQUITOUS)	↑ All AC isoforms	Cholera (ADP-ribosylation of Switch I Arg201 of α_s). Pituitary & thyroid cancer (adenomas caused by somatic point mutation of Switch I Arg201 or Switch II Gln227 of α_s). McCune-Albright syndrome (mosaic of symptoms caused by embryonic point mutation of Switch I Arg201 of α_s).
	α_{olf}	Brain/olfactory neurons	↑ All AC isoforms	
Gi	α_i (1-3)	Wide distribution including neurons	↓ AC5 & AC6 ↓ Ca^{2+} channels ↑ K^+ channels	Whooping cough (ADP-ribosylation of Cys352 at the *C*-terminus of α_{i2} by pertussis toxin). Adrenal & ovarian adenomas (point mutation of Switch I Arg179 of α_{i2}).
	α_o (1/2)	Brain, neurons, heart	↓ AC1 ↓ Ca^{2+} channels ↑ K^+ channels	
	α_z	Platelets, neurons, adrenal chromaffin cells	↓ Ca^{2+} channels ↑ K^+ channels	
	α_t (1/2)	Retina	↑ cGMP phosphodiesterase	Night blindness (point mutation)
Gq	α_q	UBIQUITOUS	↑ phospholipase Cβ1,3,4	Melanocytosis
	α_{11}	UBIQUITOUS	↑ p63-RhoGEF*	
	α_{14}	Stromal/epithelial cells	↑ Bruton's tyrosine kinase*	
	α_{15}	Myeloid cells	↑ K^+ channels*	
	α_{16}	Myeloid cells		
G12	α_{12}	UBIQUITOUS	↑ Phospholipase D	
	α_{13}	UBIQUITOUS	↑ Phospholipase Cε	
Gβ/γ			↑ phospholipase Cβ2 ↑ AC2,4,7; ↓ AC1,5,6	

*Also stimulated by α_q

Table 5.4 Human adrenergic receptors

Family/ isoform		Tissue distribution‡						Agonists [isoform selective]	Antagonists [isoform selective]	G protein	Effectors
		Heart	Lung	Liver	Kidney	Muscle	Adipose				
β-AR §	β_1	+ +	+ +	+	+	0/+	+ + +	Isoproterenol>>NA=A† [Xamoterol]	Bupranolol, Propanolol [Atenolol, Bisoprolol]	Gs	
§	β_2	+ +	+ + +	+ +	+ +	+/ + + +	+ + +	Isoproterenol≥A>>NA [Clenbuterol]	Bupranolol, Propanolol [ICI-118,551]		
◊	β_3	0	0	0	0	0	+ + +	Isoproterenol>NA>>A [BRL 37344]	Bupranolol [SR59230A]		
α_1-AR ∂	α_{1A}*	+ + +		+ + + +	+/-	✓^{f} muscle arteries		A/NA>Phenylephrine* [Oxymetazoline, A61603]	Prazosin, Corynanthine [(*S*)-Niguldipine]	Gq	
∂	α_{1B}	+		+	+ +	0^f muscle arteries		A/NA> Phenylephrine*	Prazosin, Corynanthine		
∂	α_{1D}*	+		+/-	+/-	0^f muscle arteries		A>>NA> Phenylephrine	Prazosin, Corynanthine [BMY 7378]		
α_2-AR ¢	α_{2A}	✓	✓	✓	✓	✓	0 (sub-cutaneous)	(NA=A¶) Clonidine [Oxymetazoline, Guanfacine]	Yohimbine>> Prozosin [BRL 44408]	Gi	
¢	α_{2B}	✓	✓	✓	✓	✓	✓	(NA=A¶) Clonidine	Yohimbine>> Prozosin [Imiloxan]		
¢	α_{2C}	✓	✓	0	✓	✓	0	(NA=A¶) Clonidine	Yohimbine>> Prozosin [MK 912]		

‡ Entries in blue: detection of the protein by radioligand displacement/bioassays ("+++" = 70-150; "++" = 20-40; "+" = 7-13 fmol/mg protein). Entries in red: mRNA detection (authors' indication of relative levels). ✓ = present

† "NA" = noradrenaline; "A" = adrenaline

* α_{1A} was formerly known as "α_{1C}"; α_{1D} was formerly known as "$\alpha_{1A/D}$". *Adrenaline and noradrenaline have lower potency toward α1A and α1B than α1D. [see Schwinn, D.A. (1994) & Hieble, J.P., Bondinell, W.E. and Ruffolo, R.R.Jr. (1995)]

¶ Hieble, J.P., Bondinell, W.E. and Ruffolo, R.R.Jr. (1995) α- and β-Adrenoceptors: from the gene to the clinic. 1. Molecular biology and adrenoceptor subclassification. *J. Med. Chem.*, **38**: 3415-3444.

Radioligand studies: -

§Sano, M., Yoshima, T., Yagura, T. and Yamamoto, I. (1993) Non-homogenous distribution of β1- and β2-adrenoceptors in various human tissues. *Life Sciences*, **52**: 1063-1070.

§Liggett, S.B., Shah, S.D. and Cryer, P.E. (1988) Characterization of β-adrenergic receptors of human skeletal muscleobtained by needle biopsy. *Am. J. Physiol.*, **254**: E795-798.

*f*Jarajapu, Y.P.R., Coats, P., McGrath, J.C., Hillier, C. and MacDonald, A. (2001) Functional characterization iof α1-adrenoceptor subtypes in human skeletal muscle resistance arteries. *Br. J. Pharmacol.*, **133**: 679-686.

mRNA studies: -

∂Price, D.T., Lefkowitz, R.J., Caron, M.G., Berkowtiz, D. and Schwinn, D.A. (1993) Localization of mRNA for three distinct α1-adrenergic receptor subtypes in human tissues: implications for human a-adrenergic physiology. *Molecular Pharmacology*, **45**: 171-175.

◊Krief, S., Lonnqvist, F., Raimbault, S., Baude, B., Van Spronsen, A., Arner, P., Strosberg, A.D., Ricquier, D. and Emorine, L.J. (1993) Tissue distribution of b3-adrenergic receptor mRNA in man. *Advances in Pharmacology*, **31**: 333-341.

Schwinn, D.A. (1994) Adrenergic receptors: unique localization in human tissues. *Advances in Pharmacology*, **31**: 333-341.

¢Eason, M.G. and Liggett, S.B. (1993) Human a2-adrenergic receptor subtype dsitribution: widespread and subtype-selective expression of α2C10, α2C4, and α2C2 mRNA in multiple tissues. *Mol. Pharmacol.*, **44**: 70-75. [α2C10 is the gene for α2A; α2C2 = α2B; α2C4 = α2C

5.6.1 Gs-coupling catabolic receptors

The two main catabolic hormones, glucagon and adrenaline, are agonists for Gs-coupling 7-pass receptors. The **β-*a*drenergic *r*eceptors** (**$\beta_{1,2,3}$AR**) and the glucagon receptor both stimulate GDP- and β/γ-dissociation from the αs subunit – GTP binding causes the alpha subunit to lose its β/γ- and receptor-binding interfaces, which are replaced with an adenylyl cyclase-binding surface. AC is thereby activated by αs (certain forms are also co-activated by the liberated β/γ subunits).

5.6.2 β–Adrenergic/glucagon-receptor stimulation of glycogenolysis

In dog liver, glycogenolysis is upregulated predominantly through β2-adrenergic receptors, which couple with Gs. The αs subunit (freed via receptor-induced GDP → GTP exchange and β/γ dissociation) transduces the signal through coupling and activation of liver AC isoforms. The cAMP that is consequently produced can diffuse through the cytosol, activating protein kinase A. PKA catalytic subunits are released from their regulatory subunits and PKA can then serine-phosphorylate its substrates, in particular phosphorylase kinase (PhK). Phosphorylation of the α- and β-regulatory subunits of PhK relieves its autoinhibition (probably by conformational change resulting in ejection of pseudosubstrate sequence loop(s)). PhK now activates glycogen phosphorylase through serine phosphorylation.

Skeletal muscle has glycogen stores that are degraded in response to adrenaline but are unaffected by glucagon because muscle lacks glucagon receptors.

5.6.3 β–Adrenergic/glucagon-receptor inhibition of glycogen synthesis

In order to avoid a ‘futile cycle’ of simultaneous glycogen degradation and synthesis, glycogen synthase (GS) is simultaneously inhibited by serine-phosphorylations carried out by both PhK and PKA (among other enzymes, as discussed later).

5.6.4 α1-adrenergic receptor stimulation of glycogenolysis

Soon after the discovery of PhK and its activation by cAMP-dependent PKA, it was realised that PhK is also calcium-dependent. In young male rats, β2-adrenergic receptor agonists such as isoproterenol have no effect on glycogenolysis in the liver. Instead, PhK is activated by α1-adrenergic receptors, which couple to the Gq-type G protein. Release of its αq subunit causes activation of phospholipase Cβ, a phosphoinositide-specific lipase that cleaves PIP2 to release IP3. IP3 diffuses to the endoplasmic reticulum where it binds to IP3 receptors that are ligand-gated Ca^{2+}-channels. The channels open to release calcium into the cytosol. Calcium binds to the δ-regulatory subunit of PhK, causing activation (via release of a glycogen phosphorylase-like pseudosubstrate sequence loop in the δ-subunit).

5.6.5 Diffusible cascade or scaffolded pathway

In resting cells, the low basal amounts of cAMP are present in both the soluble and the particulate (membranes, organelles) fractions. Basal cAMP is bound (at suboptimal levels) to cAMP-binding proteins, in particular the RI and RII regulatory subunits of PKA. Although thought originally to be exclusively cytosolic, it later transpired that inactive PKA holoenzymes differ in subcellular localisation. RI-containing PKA is largely (but not exclusively) cytosolic whereas a large proportion of RII-containing PKA is localised to membranes and organelles. The particulate localisation of RII holoenzymes proved to be due to their binding to AKAP proteins (named for c***A***MP-dependent protein ***k***inase ***a***nchoring ***p***roteins) that are themselves anchored to membranes, microtubules and other subcellular structures[25].

In the same vein, the phosphodiesterase (PDE) enzymes (responsible for cAMP degradation) are likewise split between cytosolic and particulate fractions of the cell. Certain forms of PDE are co-localised to AKAP proteins along with RII PKA.

Finally, the downstream targets of a key branch of the pathway (PhK, GP, GS) are largely found bound to glycogen particles associated with the smooth endoplasmic reticulum. This localisation is due (in part at least) to glycogen-targetting proteins that act as scaffolds for protein serine-phosphatases and their substrates (PhK, GP, GS).

5.7 Regulatory subunits of PKA and A-Kinase Anchoring Proteins

RI-type PKA signals cytosolically, but RII-type PKA signals in more discrete locations where its release of catalytic subunits is presumably more effective because of the close proximity of substrates. Why would this be? One obvious answer is that it allows PKA catalytic subunits to be specially targetted to discrete locations where particular substrates are sequestered. A second advantage of this arrangement is that the signal can be fine-tuned by AKAPs that not only bind PKA, but also PDE isoforms potentially capable of rapid local elimination of cAMP.

But why would RII, in particular, be preferentially scaffolded? RII-PKA has a lower affinity for cAMP than RI-PKA. RI-PKA activates at cAMP levels that are typically 2–8-fold lower than those required for RII[25]. RI-PKA activation can also be accelerated by the presence of its substrates. The pseudosubstrate linker in RI (peptide strand linking dimerisation domain to the tandem cAMP binding modules, which contains alanine instead of serine in the P^0 position) occupies and blocks the active site of the catalytic subunit. However, this inhibitor sequence of RI-PKA, but not that of RII-PKA, can be displaced by substrate loops from its target proteins, with the result that RI-PKA can activate at lower cAMP levels if it encounters substrate[26]. Perhaps the higher [cAMP] activation threshold RII-PKA necessitates scaffolding because only highly localised substrates can be reached by diffusion before the signal is turned off.

5.7.1 RII regulatory subunits – reversible phosphorylation and scaffolding

Unlike RI, RII regulatory subunits contain autophosphorylatable 'pseudosubstrate' linker sequences between their dimerisation domains and cAMP modules. In unstimulated interphase cells, the pseuodsubstrate serine90 (RIIα) is autophosphorylated by the bound PKA catalytic subunits and remains bound in a stable complex. In resting cells the phospho-holoenzyme is mainly particulate. During mitosis, RII subunits are phosphorylated on a second site (*N*-terminal threonine54 of RIIa) by the cyclin B-dependent kinase, Cdk-1[27], and this has the effect of redistributing RII-PKA to chromatin via binding to chromatin-anchored AKAP95[28].

After activation by cAMP, the autophosphorylated RII will not rebind PKA unless it is first de-phosphorylated[29] – phospho-RII has a 10-fold lower affinity for the catalytic subunits than dephospho-RII[30]. This is in direct contrast to RI-PKA, which is autoinhibited by the true pseudosubstrate hinge found in RI subunits. RI cannot be phosphorylated, but rebinds as soon as cAMP levels fall; rebinding of RII is retarded. Perhaps the potentially long-lasting RII-PKA output is attenuated by protein phosphatases, which in some cases are also bound to the same AKAP scaffold (see Table 5.5).

The discovery of scaffolding proteins has begun to transform understanding of signalling. There is a bewildering array of classes, families and isoforms, of proteins that have been shown, in protein interaction screens (such as yeast 2-hybrid), to bind certain signal components. However, the wider implications of many of these in vitro interactions is still unexplored and exciting research remains to be done to unravel their potential multiple effects on pathways. For instance, it is still too early to make definitive statements about their role in glycogen metabolism. Nevertheless, important recent findings are helping explain previous anomalies. One very important discovery was that the protein phosphatases, which reverse serine kinase signals, are mostly rather non-specific and only become regulated and specific when bound to a scaffolding or regulatory co-factor protein. Proteins that target phosphatase to glycogen (glycogen-targetting proteins) modify specificity (by localisation) and activity – when scaffolded, the phosphatase is generally active. AKAP proteins similarly increase specificity by localisation – while scaffolded to the AKAP, the phosphatase is generally inhibited.

Although most recent research has naturally focused on how the scaffolded phosphatase counteracts the effects of locally released PKA (by de-phosphorylating exogenous substrates originally phosphorylated by the kinase), it is interesting to note that the serine protein phosphatases PP1, PP2A and PP2B ('calcineurin') are also capable of dephosphorylating the RII pseudosubstrate pSer90[27]. The calcium/ calmodulin-dependent phosphatase, calcineurin, was originally purified from rabbit skeletal muscle but is found in many other tissues including brain, where it is found tightly associated with RII-PKA. For a protein phosphatase, the enzyme has an unusually narrow substrate specificity. It is particularly active against the phosphorylated pseudosubstrate sequence of RII and shows a preference for minimal peptide sequence:

DLDVPIPGRFDRRV**pS**VCAE

Table 5.5 Selected AKAP Proteins and their functions

AKAP gene	AKAP proteins	Tissue distribution	Subcellular compartment	Proteins bound
AKAP1	AKAP121		Mitochondria, mitotic spindle (tubulin-binding AKAP). 3'UTR of mRNA[*]	RII–PKA, PTP-D1 (protein tyrosine phosphatase, activator of Src)
	S-AKAP84	Heart, lung, liver, skeletal muscle, kidney, testis, thyroid	Endoplasmic reticulum	RII–PKA, PTP-D1
	D-AKAP1		'N0' variant : localisation similar to AKAP121 'N1' variant: sarcoplasmic reticulum[†]	*Dual specific*, binds either RII-PKA or RI-PKA
	AKAP149		Mitochondria , 3'UTR of mRNA* Nuclear lamins of the nuclear envelope‡	RII–PKA, PKC, PP-1[‡], PDE4A[§]
AKAP5	AKAP79 (human ortholog of bovine AKAP75)	Brain: hypocampus postsynaptic neurons	Plasma membrane	RII-PKA, PKC, PP-2B ('calcineurin')[¶]
AKAP8	AKAP95	Heart, liver, skeletal muscle, kidney, pancreas	Nuclear matrix (interphase) Chromatin (mitosis)	RII-PKA, AMY-1 (a Myc binder), MCM2 (minichromosome maintenance protein)[¥]
AKAP10	D-AKAP2	Liver, lung, spleen, brain	Mitochodria	*Dual specific*, binds either RII-PKA or RI-PKA. It may act as a RGS protein.
AKAP11	AKAP220	Testis, brain	Peroxisomes, vesicles, centrosomes	RII-PKA, (RI-PKA?[$]), PP-1[¢], GSK3[f]
MTG		Lymphocytes	Golgi	RII-PKA, PDE4A, PDE7A§

Table 5.5 (*Continued*)

Information sources for Table...

Tasken, K. and Aandahl, E.M. (2003) Localized effects of cAMP mediated by distinct routes of protein kinase A. *Physio. Rev.*, **84**; 137-167.

† see: Feliciello, A., Gottesman, M.E. and Avvedimento, E.V. (2001) The biological functions of A-kinase anchor proteins. *J. Mol. Biol.*, **308**: 99-114.

* Specifically binds mRNA of certain mitochodrial proteins [see: Feliciello, A., Gottesman, M.E. and Avvedimento, E.V. (2005) cAMP-PKA signaling to the mitochondria: protein scaffolds, mRNA and phosphatases. *Cellular Signalling*, **17**: 279-287.

‡Kuntziger, T., Rognac, M., Folstad, R.L.S. and Collas, P. (2006) Association of PP1 with its regulatory subunit AKAP149 is regulated by serine phosphorylation flanking the RVXF motif of AKAP149. *Biochemistry*, **45**: 5868-5877.

§Asirvatham, A.L. Galligan, S.G., Schillace, R.V., Davey, M.P., Vasta, V., Beavo, J.A. and Carr, D.W. (2004) A-kinase anchoring proteins interact with phosphodiesterases in T lymphocyte cell lines. *J. Immunol.*, **173**: 4806-4814.

¢Schillace, R.V. and Scott, J.D. (1999) Association of the type 1 protein phosphatase PP1 with the A-kinase anchoring protein AKAP220.

¶ Faux, M.C., ROllins, E.M. Edwards, A.S., Langeberg, L.K. and Newton, A.C. (1999) Mechanism of A-kinase-anchring protein 79 (AKAP79) and protein kinase C interaction. *Biochem. J.*, **343**: 443-452.

*f*Tanji, C., Yamamoto, H., Yorioka, N., Kohno, N., Kiruchi, K and Kikuchi, A. (2002) A-kinase anchoring protein AKAP220 binds glycogen synthase kinase-3β (GSK-3β) and mediates protein linase A-dependent inhibition of GSK-3β. *J. Biol. Chem.*, **277**: 36955-36961.

¥Eide, T., Tasken, K.A., Carlson, C., Williams, G., Jahnsen, T., Tasken, K. and Collas, P. (2003) Protein kinase A-anchoring protein AKAP95 interacts with MCM2, a regulator of DNA replication.

$ Reinton, N., Collas, P., Hangen, T.B., Skalhegg, B.S., Hansson, V., Jahnsen, T. and Tasken, K. (2000) Localization of a novel human A-kinase-anchoring protein, hAKAP220, during spermatogenesis. *Dev. Biol.*, **223**: 194-204.

This is a peptide capable of forming an amphipathic β-sheet, which is similar to the structural determinant of substrate/pseudosubstrate-binding to PKA's catalytic subunit (see Chapter 4)[31].

PKA substrate and pseudosubstrate sequences

Phosphorylase kinase α-subunit	Phe	Arg	Arg	Leu	**Ser**	Ile
Phosphorylase kinase β-subunit	Lys	Arg	Ser	Asn	**Ser**	Val
PKA RI subunit pseudosubstrate site	Arg	Arg	Arg	Gly	Ala	Ile
PKA RII subunit autophosphorylation site	Asp	Arg	Arg	Val	**Ser**	Val
6-phosphofructo-2-kinase	Arg	Arg	Arg	Gly	**Ser**	Ser

Consensus sequence for PKA substrates= . . .**Arg.Arg.X.Ser**. . . (Hubbard, M.J. and Cohen, P. (1993) *TIBS*, **18**: 172–177)

5.7.2 Segregation of pathways

Like examples of the AC isoforms, many G protein α-subunits are virtually ubiquitous (see Table 5.2). A few G protein α-subunits have restricted expression. Brain and neuronal tissue uniquely express Ca^{2+}/calmodulin-stimulated adenylyl cyclases in combination with αolf- and αo-containing G proteins ('olf' for 'olfactory; 'o' for 'other'); retinae are the only organs expressing αt-containing G protein ('t' for 'transducin').

Although the ratios of various AC isoforms vary between cell types, it is nevertheless expected that most cells will have multiple types of adenylyl cyclase, G proteins, and (potentially competing) upstream receptors. Such complexity is inferred from disparate studies but is hypothetical in the sense that no single study has yet monitored such multiplex systems in real time in a single cell. Furthermore, even though a cognate receptor-G protein pair may be expressed in a tissue, that does not necessarily mean that the expected signal outcome will be present. Competing signal components may be at too low a level, or may be separated by compartmentalisation in membrane micro-domains or discretely scaffolded structures. In addition, second messengers (although diffusible) are often restricted to distinct subcellular pools because the enzymes that destroy the signal are often localised to the same compartment as the effector[17,19].

A prominent example is the control of glycogen breakdown in rat livers (alluded to in Chapter 1). Male rat liver contains the G proteins Gs, Gi and Gq as well as β-, α1- and α2-adrenergic receptors. In juveniles and post-mature phosphorylase kinase is predominantly activated through the β-AR → Gs → AC↑ axis; by contrast in mature animals it is instead activated via the α1-AR → Gq → PLC-β↑ axis. Despite the presence of potential α2-α2AR → GαI → AC ↓ or α2AR → Giβ/γ → PLC-β2↑ axes, there is little evidence that adrenaline causes a futile cycle of AC activation (via β-AR and Gαs) with simultaneous inhibition (via α2-AR and Gαi). Nor does it seem that the Ca^{2+}-activation of phosphorylase kinase by α1-AR and Gαq is in any way supplemented by Gi-released β/γ subunits (downstream of α2-AR), because the β-*AR-independent* activation of phosphorylase kinase (see Section 5.8) by calcium release is pertussis-insensitive (pertussis specifically prevents Gi-coupling to activated α2-AR)[32,33].

A second example of signal segregation can be found in cardiomyocytes, which contain potentially competitive pathways. Cardiomyocytes contain both AC-stimulatory (Gs) and AC-inhibitory G proteins (Gi2, Gi3 and Go). They express both β1- and β2-adrenergic receptors, along with adenylyl cyclases, AC5 and AC6, and both type I and type II PKA. Also present are M2-muscarinic acetyl choline receptors (m2-mAChR) that couple with Gi. However, m2-mAChR only suppresses the β1-AR activation of AC5/6, and is without effect on β2-AR signals. Although the β2-AR is at lower levels than β1-AR, its activation of AC is disproportionately stronger, because the two adrenergic pathways are partitioned from one another. β2-AR, AC5/6, and type II PKA are confined to specialised membrane lipid microdomains called calveolae that are functionally separate from the rest of the plasma membrane that contains β1-AR and m2-mAChR[34].

With this in mind, the present discussion is limited to a few prominent effects and the subject of crosstalk is, of necessity, restricted.

5.7.3 PKA and its inhibitors

Before looking at the pathway, let us first review how PKA exists in a resting cell. The prime regulators of PKA are the so-called regulatory subunits (R^{SUB}) that keep the constitutively active catalytic subunits (PKA^{C}) in an inhibited form; either in the cytosol or scaffolded to specific subcellular compartments. The inactive holoenzyme is specifically excluded from the nucleus by the presence of the R subunits – PKA^{C} can only enter the nucleus when it is released from R^{SUB}.

An R^{SUB} monomer is a multi-domain protein containing a single dimerisation domain at the *N*-terminus, connected to a *C*-terminal tandem pair of cAMP binding modules via a flexible linker or 'hinge'. The cAMP-binding domain is a subclass of ***c***yclic ***n***ucleotide-***b***inding (CNB) module – CNB modules are also found in guanine nucleotide exchange factors, phosphodiesterases and ion channels[35].

The linker peptide contains a pseudosubstrate (RI types) or a substrate consensus sequence (RII types). When PKA^{C} is released and active, the R^{SUB} exists as a cAMP-occupied homodimer: RI only dimerises with RI; RII only dimerises with RII. cAMP is so tightly bound in the homodimer that it cannot be removed, even with extensive dialysis, a finding explained by the way cAMP is buried away from solvent in free R^{SUB}[36]. However, aside from keeping the dimer permanently attached, these *N*-terminal homo-dimerisation domains have a second important function, as docking sites for the PKA scaffolding proteins: ***A***-***k***inase ***a***nchoring ***p***roteins (AKAP). AKAPs play an important role in sequestering PKA^{C} subunits to specific localities in the cell, such as the plasma membrane, nuclear matrix[37] or the mitochondrial outer membrane[38]. Such AKAP binding regions of R^{SUB} ensure that PKA^{C} is released close to its intended substrates, providing economy of signal integration (see Table 5.5).

The *C*-terminal portion of R^{SUB} contains a pair of modules known as ***c***AMP-***b***inding ***d***omains (CBD-A and CBD-B) that are related to the bacterial CAP protein. The first of these R^{SUB} modules, CBD-A, cannot bind cAMP and PKA^{C} at the same time because of their competition for the same site on R^{SUB}. PKA^{C} docking causes ejection of cAMP from R^{SUB} and a stable complex is formed (its formation is ATP-dependent). The catalytic site is now closed and shielded from solvent.

The heat stable PKI-type polypeptide inhibitors also have a signal integration motif in the form of a nuclear export signal sequence in their *C*-terminal region. This is essential to remove free PKA catalytic subunits from the nucleus after the upstream signals terminate[35].

The structural details of PKA inhibition are discussed in Chapter 4.

5.8 Phosphorylase kinase

Phosphorylase kinase (PhK) is the prime activator of glycogen phosphorylase (GP) and thereby controls the production of glucose-6-phosphate from the breakdown of glycogen in storage organs such as liver and skeletal muscle. Although liver GP is nominally an allosteric enzyme, its activation by the catabolite AMP is only achieved *in vitro* by concentrations unlikely to occur *in vivo*. The main way in which liver GP is activated, therefore, is by serine phosphorylation by PhK.

5.8.1 PhK structure

The Phk holoenzyme is thought to exist principally as a tetramer of tetramers with an overall molecular mass of 1,300 kDa. The mass of the holoenzyme varies due to the existence of varying subunit isoforms. Each tetramer component of the holoenzyme consists of the following:

- **α–subunit** – MW ≈ 140 kDa. A regulatory protein that de-supresses PhK activity when serine-phosphorylated.
- **β–subunit** – MW ≈ 130 kDa. A regulatory protein that de-supresses PhK activity when serine-phosphorylated. Highly homologous with the α-subunit, and probably arising from the same ancestral gene.
- **γ–subunit** – MW ≈ 45 kDa. The catalytic subunit that is constitutively active when dissociated; a Ser/Thr-kinase with very narrow substrate specificity.
- **δ–subunit** – MW ≈ 17 kDa. A regulatory protein that de-supresses PhK activity when cytosolic Ca^{2+} levels rise. It is identical to calmodulin (a ubiquitous Ca^{2+}-binding protein). Unusually, the δ-subunit/calmodulin remains bound to the holoenzyme under all conditions – most calmodulin-activated enzymes only associate with calmodulin while Ca^{2+} levels are high, with de-binding and de-activation occurring simultaneously when Ca^{2+} drops to resting levels.

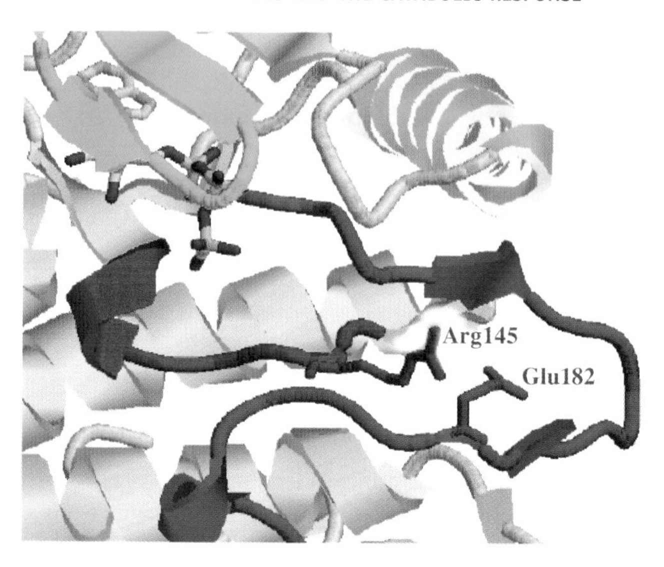

Figure 5.3 The catalytic γ-subunit of phosphorylase kinase showing the interaction of A-loop glutamate with the RD motif of the C-loop

5.8.2 The catalytic γ-subunit of PhK

The PhK γ-subunit exhibits the usual bi-lobal protein kinase fold and conserved Ser/Thr-kinase signature residues, but shows sequence divergence beyond the kinase core in a unique *C*-terminal extension that makes up about one third of the molecular mass of the protein.

Two crystal structures of residues 1–298 of the Phk γ-subunit from rabbit muscle show the catalytic core of the enzyme complexed with either Mn^{2+} and non-hydrolysable ATP analogue (AMPPNP) or Mg^{2+} and ADP[39]. Both exhibit 'closed' conformations similar to activated PKA ternary complexes. The monomeric γ-subunit is a constitutively active enzyme that does not require A-loop phosphorylation. The reason behind this lack of regulation by the A-loop is seen to be a substitution of a negatively-charged glutamate for the usual activating A-loop threonine residue. Figure 5.3 shows that the anionic A-loop glutamate (Glu182) of PhK interacts with the cationic arginine (Arg145) of the RD motif in the same way that negatively-charged phosphate of pThr197 of PKA does with its own RD motif. In contrast to pThr 197 of PKA, however, the PhK Glu182 residue makes no interdomain contact with the C-helix of the N-lobe.

5.8.3 Regulatory subunits

A region of the catalytic γ-subunit (γ^{SUB} residues 343–386) acts as a binding surface for the regulatory α-subunit (α^{SUB} residues 1060–1237 of α); this same region of γ^{SUB} is also one of two high affinity calmodulin (δ^{SUB}) binding sites that exist in the catalytic subunit's unique *C*-terminal regulatory domain[40]. Of the two nanomolar affinity binding sites for calmodulin within the catalytic subunit – residues: 301–327 and residues 332–371 – the latter motif has the highest affinity for calmodulin. It has been suggested that the tandem calmodulin binding sites may explain how δ^{SUB} can bind to γ^{SUB} in either the presence or absence of calcium[41]. α- and β-subunits also have calmodulin-binding sites: 1070–1094 of α^{SUB}, 5–28 and/or 770–794 of β^{SUB}.

This overlap of regulatory binding sites on γ^{SUB} explains the observed interdependence of the two PhK-activating signals: serine-phosphorylation of the α^{SUB} and β^{SUB} by PKA and Ca^{2+} activation of the δ^{SUB}. Activating phosphorylation of α^{SUB} and β^{SUB} (by PKA) has the effect of increasing the affinity of δ^{SUB} for Ca^{2+}; equally, binding of calcium to the δ^{SUB} causes perturbance in α^{SUB}'s binding surface for γ^{SUB}. This offers a partial explanation as to how calcium signals are communicated to the δ-subunit, which is thought to act as the primary coordinator of regulatory signals in the PhK holoenzyme.

5.8.4 PhK substrates and autophosphorylation sites

Thus far, only a very limited number of *in vivo* PhK substrates are known. These include glycogen phosphorylase, glycogen synthase and its own regulatoryα^{SUB} and β^{SUB} (*in vitro* autophosphorylation sites). From these proteins, a consensus sequence can be identified, as shown in the following table:

P^{-4}	P^{-3}	P^{-2}	P^{-1}	P^{0}	P^{+1}	P^{+2}
Lys/Arg	Lys/Arg	Gln	Xxx	**Ser**	Phe/Φ	Arg/Φ

Undoubtedly, the most important physiological PhK substrate is glycogen phosphorylase, which is phosphorylated by PhK on a single residue, Ser14, and is thus activated. Indeed, phosphorylase kinase is the only enzyme that can perform this activation step. Glycogen synthase is also serine-phosphorylated by PhK at the same time and this inhibits the synthase, thus preventing a potential futile cycle where glycogen would be broken down and at the same time synthesised in the same cell. However, the inactivating phosphorylation of glycogen synthase can be carried out by other enzymes, including PKA and GSK-3.

The presence of an arginine at P^{+2} in the Ser14-containing PhK substrate sequence of glycogen phosphorylase explains why PKA cannot phosphorylate this protein. The large sidechain of the P^{+2} arginine can be accommodated in the PhK structure and may be neutralised by nearby glutamates, but in PKA the P^{+2} substrate site is much smaller and an arginine could not fit[39].

5.8.5 Possible mechanisms of activation and holoenzyme conformation

Although the catalytic subunit has been crystallised in a truncated form (minus the *C*-terminal regulatory domain), there is scant structural information on the α- and β-subunits. Much of the knowledge available on the holoezyme's behaviour comes from solution studies looking at calcium- and phosphorylation-provoked changes to the holoenzyme conformation that are reflected in accessibility to antibodies, proteases or crosslinkers. Such information has been mapped onto recent high-resolution cryo-electron microscopic structures to build a slowly emerging view on how the enzyme is controlled[42]. Most studies have used isolated PhK from fast-twitch glycolytic rabbit skeletal muscle, where the protein is found at staggeringly high levels – up to 1% of the soluble protein.

- The *C*-terminal third of γ^{SUB} contains a regulatory domain with two calmodulin (CalM) binding sites (aas: 301–327 (C1) and 332–371 (C2)) bracketing a proposed 'autoinhibitory domain' that resembles the substrate, glycogen phosphorylase.
- The CalM-identical δ^{SUB} binds to γ^{SUB} in isolation, an interaction that is greatly strengthened by the addition of α^{SUB} and calcium.
- Both C1 and C2 of γ^{SUB} can bind to δ^{SUB} simultaneously, but the binding of C2 is disturbed by calcium[40].
- The holoenzyme can be activated by adding excess (extrinsic) CalM, which binds to surfaces of α^{SUB} and β^{SUB} (aas; 1070–1093; 5–28 and 770–794, respectively.
- Both α^{SUB} and β^{SUB} CalM-binding sites are close to regulatory phosphorylation sites (both PKA-targetted and potential autophosphorylation) – their interaction with CalM is disturbed by PKA phosphorylation.
- In the absence of calcium, δ^{SUB} only binds to γ^{SUB} at the C1 site and the other end of δ^{SUB} is thought to interact with CalM-binding sites on α^{SUB} and/or β^{SUB}.
- In the activated holoenzyme, the phosphorylation of α^{SUB} and/or β^{SUB} by PKA disrupts their δ^{SUB} binding ability and an increase in calcium activates a two-pronged binding by δ^{SUB} to both C1 and C2 of γ^{SUB} – this removes the glycogen phosphorylase-like inhibitor from the active site.
- The holoenzyme is held together by β^{SUB} interactions. The β^{SUB} epitopes map to the centre of the hexadecamer. These can be seen in the form of four bridges in the most recent cryo-electron microscopy structures[43].

The above is summarised in a speculative scheme of activation (Figure 5.4). It must be appreciated that this is probably very far from the likely complexity of how this enzyme is controlled. For example, both α^{SUB} and β^{SUB} have sequences that are PhK substrates. Such autophosphorylation-type pseudosubstrates are found in the RII-type PKA regulatory subunits where they act to inhibit PKA by blocking the active site cleft.

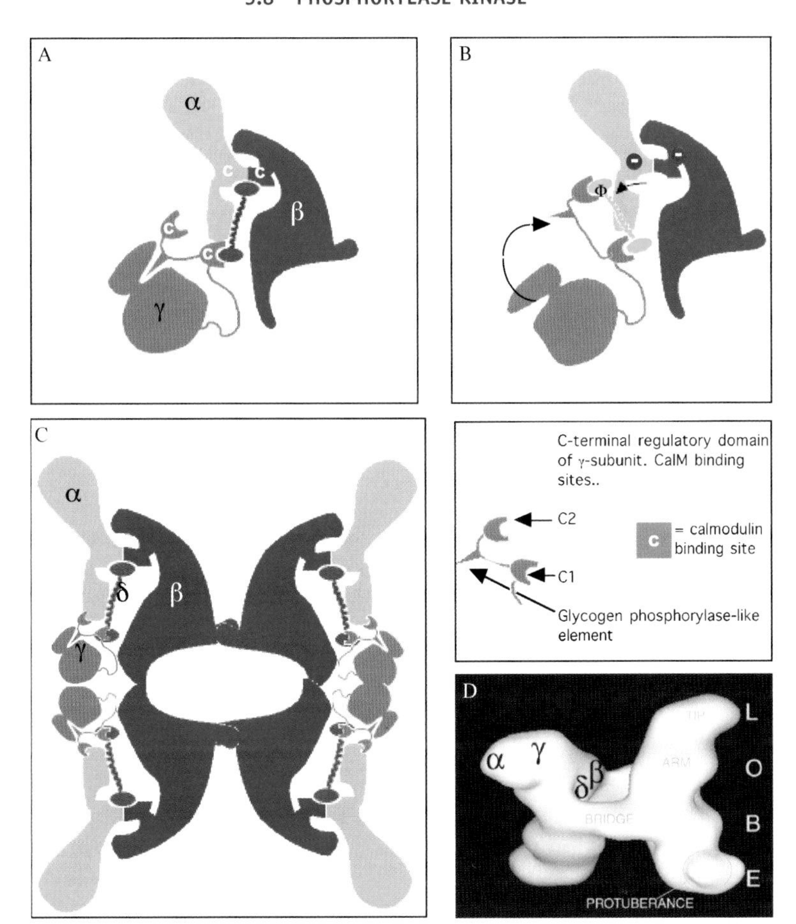

Figure 5.4 Showing possible activation mechanism, schematic of holoenzyme composition and actual orientation of subunits superimposed on cryo-EM structure (as immuno-detected). **A.** δ-SUB is bound to γ-SUB (C1) and to the CalM binding site(s) of the regulatory subunits. The active site is blocked by C-tail GP-like domain of δ-SUB. **B.** Calcium-binding and/or serine-phosphorylation disturbs regulatory subunit binding. δ-SUB now binds both C1 and C2. GP-like domain is thus removed from active site. **C.** Schematic holoenzyme composition. **D.** CryoEm structure indicating epitope mapping. Reproduced from *Structure* **10**: 23–32 with permission from Elsevier (© Copyright 2002, Elsevier)

PhK's substrate, glycogen phosphorylase, is a dimer. Therefore, it was speculated that the PhK holoenzymes might be capable of phosphorylating the dimer in one step, because two of its γ^{SUB}s could theoretically engage both constituent monomers at the same time. However, a PhK-glycogen phosphorylase complex has been imaged[44] and this indicates that each γ^{SUB} binds one end of a glycogen phosphorylase individually (see Figure 5.5).

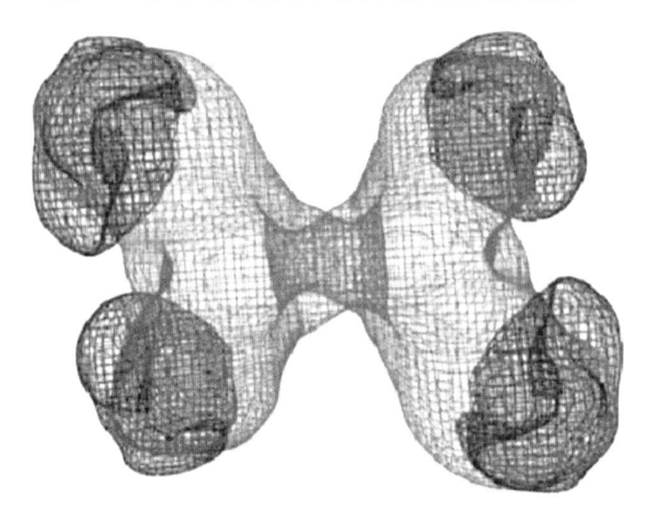

Figure 5.5 Cryo-EM structure of PhK-glycogen phosphorylase complex (PhK in purple, glycogen phosphorylase in green). *Reproduced from: Venien-Bryan, C., Lowe, E.M., Boisset, N., Traxler, K.W., Johnson, L.N. and Carlson, G.M. (2002) Three-dimensional structure of phosphrylase kinase at 22Å resolution and its complex with glycogen phosphorylase b*. Reproduced from *Structure* **10**: 33–41 with permission from Elsevier (© Copyright 2002, Elsevier)

5.9 Glycogen phosphorylase

Glycogen phosphorylase (GP) is a homodimeric protein that was the subject of perhaps the earliest structure/function analysis of how a single phosphorylation site can reversibly alter the conformation of an entire protein. The enzyme is so-named because it causes *phosphorolysis* of α1-4 linked glucose residues from the ends of glycogen chains (Figure 5.6). It has two substrates: (i) the non-reducing end of a strand of glycogen and (ii) inorganic phosphate. It also requires the co-factor pyridoxal phosphate, which is covalently bonded adjacent to the active site. Glucose can also bind to the active site but acts as an inhibitor.

5.9.1 Glycogen phosphorylase isoforms

There are three types of GP named after the tissues in which they predominate: brain, muscle and liver. Of these, muscle and liver GP isozymes are the best understood. All GP enzymes are functional dimers, with a tendency to tetramer formation when activated by phosphorylation (this is assumed to occur *in vitro* only)[45]. The tetramer only forms in the absence of glycogen because a crucial part of this tetramer-forming surface is the so-called 'glycogen storage' site. This glycogen store-binding site is over 30 Å distant from the catalytic site[46]. Although tetramer formation is a marker of activation, the 'activated' tetramers do not turnover substrate unless dissociated into dimers by the addition of glycogen. The activating effect of glycogen binding on phospho-GPa can be mimicked

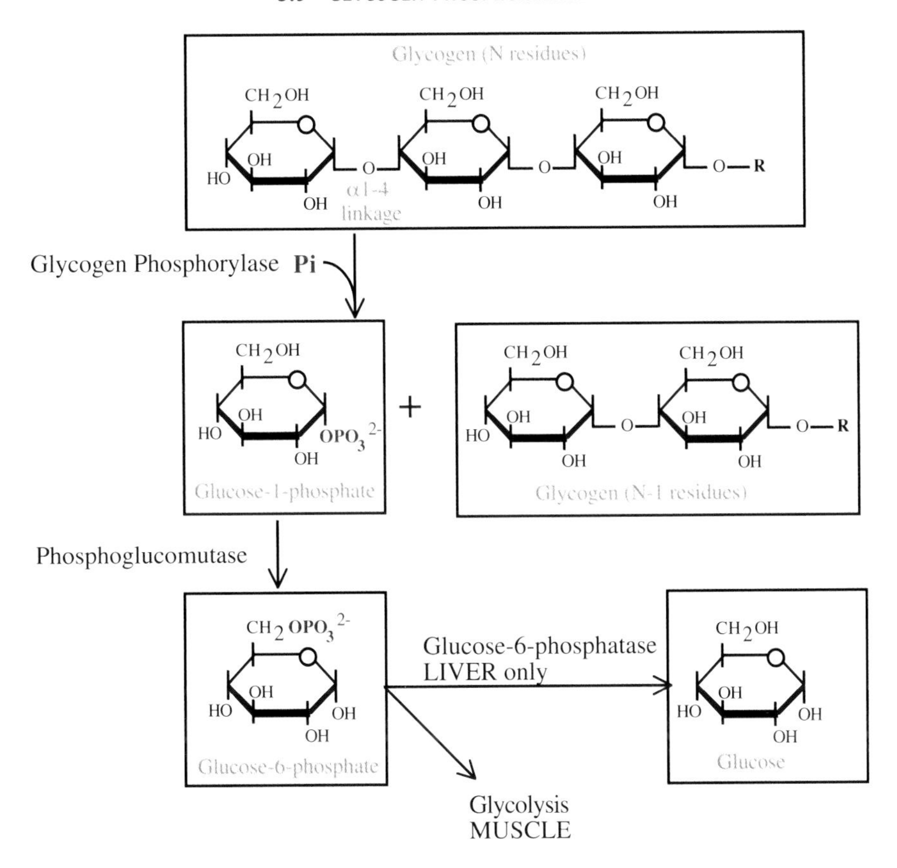

Figure 5.6 Glycogenolysis and glucose production versus glycolysis

by covalent attachment of oligosaccharide, which renders the phospho-enzyme maximally active and dimeric even in the absence of glycogen. Aside from the activating allosteric effect, glycogen (in association with glycogen-targetting proteins) also acts to scaffold GP and the other enzymes involved in glycogen metabolism, including glycogen synthase, protein phosphatase-1 and phosphorylase kinase. GP is bound to glycogen granules in the cytosol and on the endoplamic reticulum membrane via this (non-catalytic) glycogen-binding site.

5.9.2 Glycogen phosphorylase allosteric sites

GP has two physiologically relevant allosteric sites far removed from the catalytic site:

- The first site binds AMP. AMP acts as a positive regulator of the activity of the enzyme. The same site can also bind ATP, or the phosphorylated metabolite G6P, both of which antagonise the effects of AMP.

- The second allosteric site binds to glycogen particles and acts as a store-sensing and glycogen-localising regulator – hence the name 'glycogen-storage' site.

A third site known as the 'purine site' binds caffeine and this causes inhibition of the enzyme. The site's physiological relevance is unknown but its existence has led to speculation that there may be some undiscovered endogenous ligand for this site[46].

Dephospho-GP can also be activated by ammonium sulphate – sulphate ions mimic the effects of phosphorylation by substituting for the missing phosphate at the serine14 hydroxyl group – in the presence of sulphate, dephospho-GP becomes activated and adopts an 'R-state' conformation that includes an ordered *N*-terminal tail[46].

5.9.3 Control by hormones or metabolite effectors – functional differences between muscle and liver isoforms

The dimeric nature of phosphorylase allows its allosteric regulators and substrates to display cooperative effects. In the holoenzyme, both catalytic sites face into the interface between the dimers and activation is accompanied by rotation of the dimers by about 10° with respect to one another, the axis of rotation running approximately through the α2 helix. The dimerisation interface arrangement is controlled by changes in the conformation of two subdomains known as the 'tower' and the 'cap' (see Figure 5.7). Although the liver and muscle enzymes are very similar in their monomeric

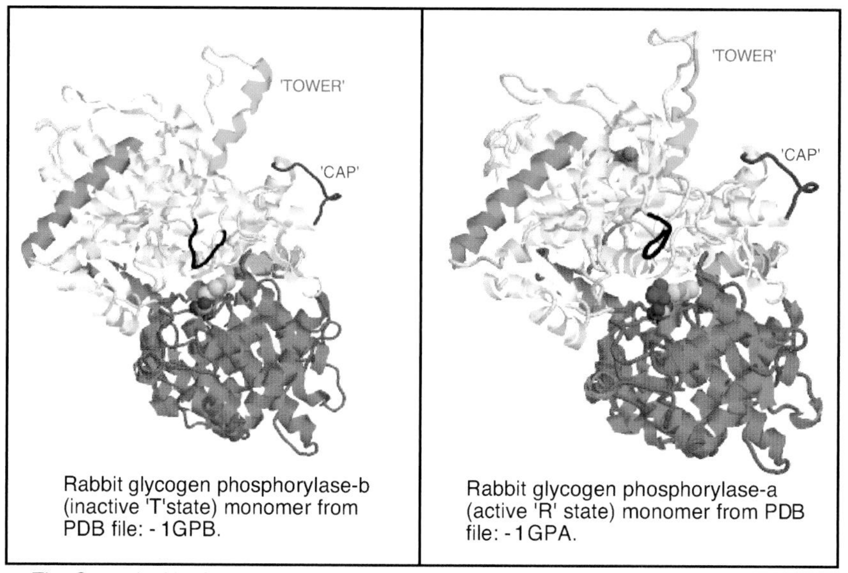

Figure 5.7 Active and inactive conformations of a glycogen phosphorylase protomer (PDB files: -1GPB & -1GPA)

tertiary structure, they differ in their quaternary structures and this appears to be the source of their differing response to regulators[47]. The following list summarises their respective physiological properties:

- GP can exist as a (PhK-)phosphorylated enzyme (classically referred to as 'GPa') and a (PP1-)dephosphorylated enzyme (classically referred to as 'GPb').
- Conditionally speaking, all GP isozymes are active when phosphorylated on a single serine (Ser14) at the *N*-terminus.
- Both phosphorylated and non-phosphorylated muscle and liver GP can adopt active (classically referred to as 'R-state', for 'relaxed') or inactive conformations (classically referred to as 'T-state', for 'tense').
- Both muscle GP (GP_M) and liver GP (GP_L) are inhibited by excess free glucose, which binds to, and blocks, their active sites.
- Dephospho-GP_M is potently activated (80%) by AMP (an indicator of low energy charge in the muscle). Even the activated *phospho*-GP_M can be further activated by AMP (10–20%).
- The positive regulator of GP_M, AMP, is antagonised by ATP and/or glucose-6-phosphate (indicators of high energy charge), which both bind at the same GP_M site as AMP but instead cause inhibition of catalytic activity.
- Dephospho-GP_L is almost completely inactive, insensitive to inhibition by ATP or G6P and barely activated by AMP.
- Phospho-GP_L is virtually maximally active and is only susceptible to inhibition by glucose.

5.9.4 How do these properties of GP isoforms fit with metabolic necessities?

Striated skeletal muscle Skeletal muscle needs to be able to respond to conscious commands regardless of the overall energy status of the body – insensitivity to starvation, for example, is ensured because striated muscle contains no glucagon receptors. Muscle GP must have forms of control that act independently of hormonal factors so that glycogen levels can be depleted and refilled during work without reference to adrenaline or insulin.

When muscle runs out of glucose, AMP levels rise and this alone is capable of inducing glycogenolysis by activating dephospho-GP_M (this is *not* conditional upon adrenaline stimulation). Conversely, if ATP or G6P levels are high (an

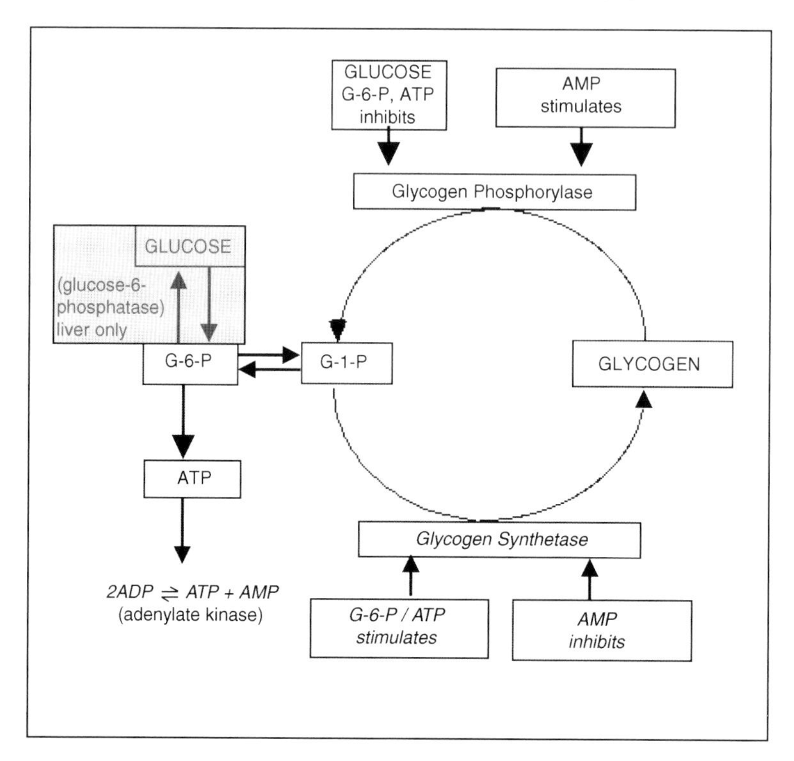

Figure 5.8 Summary of allosteric control of MUSCLE glycogen phosphorylase

indication of high energy status) GP_M is inhibited (independently of insulin) (Figure 5.8).

During stressless exercise, another non-hormonal control is thought to be exerted by the levels of glycogen storage in the muscle. *In vitro*, binding of oligosaccharides to its 'glycogen storage site' dramatically stimulates the catalytic activity of phospho-GP_M by an order of magnitude over its activity alone[48]. *In vivo*, this complex multi-factorial control is not entirely understood but it appears that GP_M activity is critically dependent upon the levels of muscle glycogen. GP_M is activated by binding to replete glycogen storage granules and inhibited when glycogen stores become depleted[49]. An opposing control is exerted over glycogen synthase (GS), which is also (indirectly) bound to glycogen granules. GS is inhibited by replete glycogen store and stimulated by low glycogen storage. It appears that these opposing enzymes may work rhythmically with contraction and rest to deplete, and then replete, the muscle stores[50,51].

Liver Liver is the body's major exporter of glucose – it is the only organ expressing significant amounts of glucose-6-phosphatase (remember: unlike glucose, phosphorylated sugars are 'trapped' inside the cell because they are charged). Being an energy-expensive commodity, glucose must only be exported by the liver when necessary. Crucially, it must be exported when demanded, no matter what the energy status of the liver happens to be. So, liver GP 'ignores' indicators of its own energy charge (AMP, ATP, G6P) and instead only 'senses' circulating glucose levels and only responds to that metabolic cue and to hormonally-induced phosphorylation/dephosphorylation. Glucagon and adrenaline lead to activation of liver GP by serine14 phosphorylation. However, an excess of circulating glucose can over-ride that signal by binding to the enzyme's active site.

Apart from blocking substrate entry to the active site, glucose binding also changes the conformation of the *N*-terminal peptide containing pSer14 such that the phosphorylated residue becomes exposed to solvent. Thus, glucose makes pSer14 available for dephosphorylation by protein phosphatase-1 (PP-1) and this aids the dephosphorylation of GP when glucose levels are high enough.

5.9.5 Structural changes induced by GP activating signals

T-state The dimeric inactive holoenzyme is held together by two pairs of complementary subdomains that interact with each other. The first contact is between the so-called 'tower' helices (α7), which noticeably diverge from the main protein fold (Figure 5.7). The tower helix of one protomer extends across the domain interface and makes an antiparallel association with the tower′ of the other protomer′ (the superscript′ indicates the second protomer throughout) (Figure 5.9).

The second dimer-forming site is termed the cap/α2′ subdomain, which is situated at the *N*-terminus immediately following the phosphorylation peptide. It consists of helices α1 and α4, the *N*-terminal portions of α2 and α5, and is centred on the loop (termed the 'cap') between the first and second α-helices. The cap of one protomer associates with the α2′ helix and β7′ strand of the other protomer′. A reciprocal, symmetrical cap′ to α2/β7 association is also made (Figure 5.9).

The cap/α2 subdomain also corresponds with the main allosteric site (for AMP or G6P). This dimerisation site is modified by the binding of allosteric modulators and/or by the phosphorylation of the *N*-terminal peptide tail.

5.9.6 T-state to R-state transition

In inactive T-state GP dimers, the *N*-terminal phosphorylation peptide (aas 1–20) is not visible, being disordered. The non-phosphorylated serine14 of one protomer makes no contact with the other protomer'. Instead Ser14 is bound to acidic residues of its own chain (Glu501 and Glu105) and Ser14′ of the other protomer′ is bound to the corresponding glutamates′ of *its own* chain[52,53] [see for example PDB file 7GBP].

5.9.7 Activation by phosphorylation

The *hormonally*-driven T- to R-state transition is induced by the action of PhK, which phosphorylates Ser14 and Ser14′. The negative charge of the added phosphate

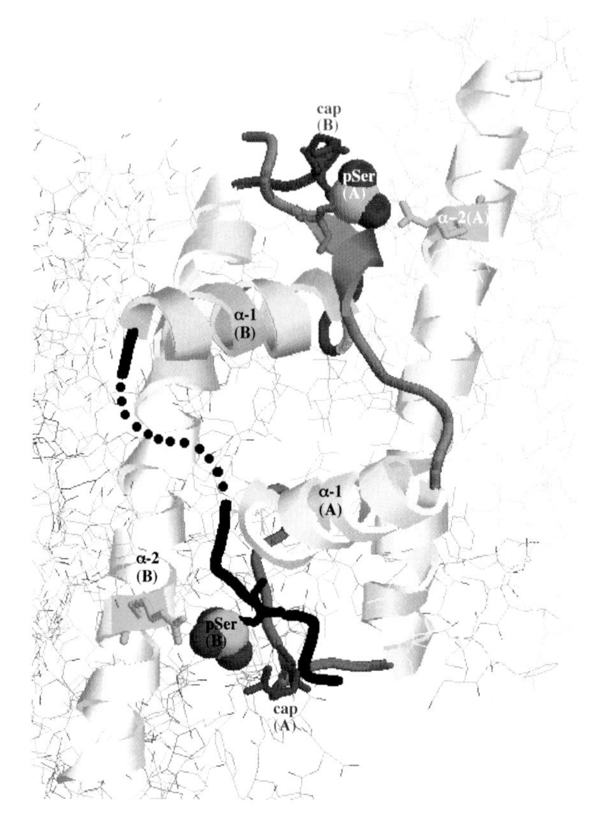

Figure 5.9 The 'cap' to α-2 helix intersubunit binding surfaces centred upon the phosphorylated serine-14 residues [PDB file: 1GPA]

eliminates *intra*chain contacts (through electrostatic repulsion) and results in an ordered 3_{10} helix conformation of the *N*-terminal tails, which now make *inter*chain contacts by binding across the top of the cap of the apposing protomer. The negatively-charged pSer14 sidechain makes salt-bridges with two basic residues: Arg69 of its own α2 helix and Arg43′ from the cap′ of the partner protomer; similarly pSer14′ ion pairs with Arg69′ of α2′ and the cap Arg43 (Figure 5.9). This draws the cap/α2′ and cap′/α2 contacts together at one end of the dimer interface and pulls apart the tower helices' contacts at the other end. The result is a rotation of protomer′ by about 10° with respect to the other (see Figure 5.10), accompanied by displacement and disordering of the *C*-terminal tail.

The tower helices (aas 262–276) are connected directly to the 280s loop (aas 280–289), which acts as a 'gate' to the active site. In the T-state, this 'gate' loop of dephospho-GP packs against the 380s loop (aas 376–386), inserts an aspartate (Asp283) into the active site and effectively blocks substrate entry[45]. In R-state, the rotation of the dimers causes tilting of the tower helices, which both unwind by

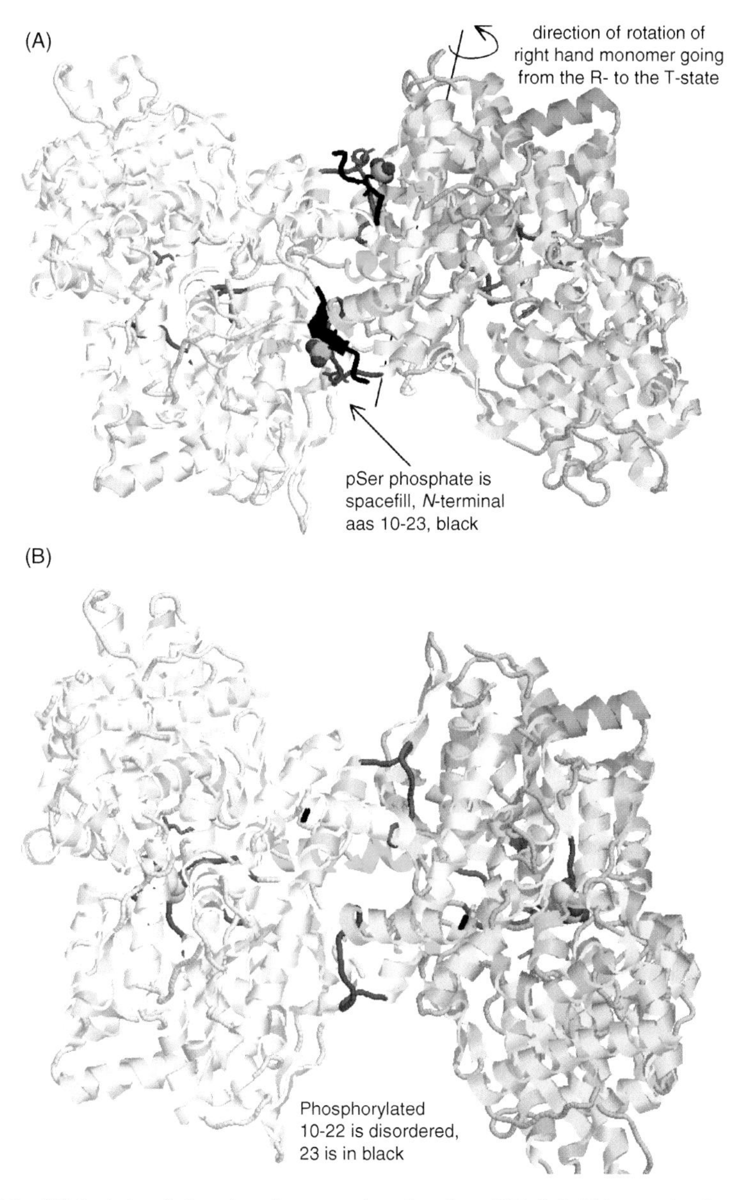

Figure 5.10 (A) R-state of phospho-glycogen phosphorylase ('GPa') Rabbit muscle glycogen phosphorylase: R-state, Ser14 phosphorylated. (B) T-states of phospho-glycogen phosphorylase ('GPa'). Human liver glycogen phosphorylase: T-state, Ser14 phosphorylated, inhibited by glucose analogue (PDB files:- 1GPA & 1FC0)

two turns, breaking interchain H-bonds between the towers and pulling the gate loop open (see Figure 5.11)[47]. The movement of the gate removes the Asp283 from the active site, to be replaced with an arginine (Arg569) – this basic residue creates a high affinity phosphate-binding site adjacent to the catalytic co-factor, pyridoxal

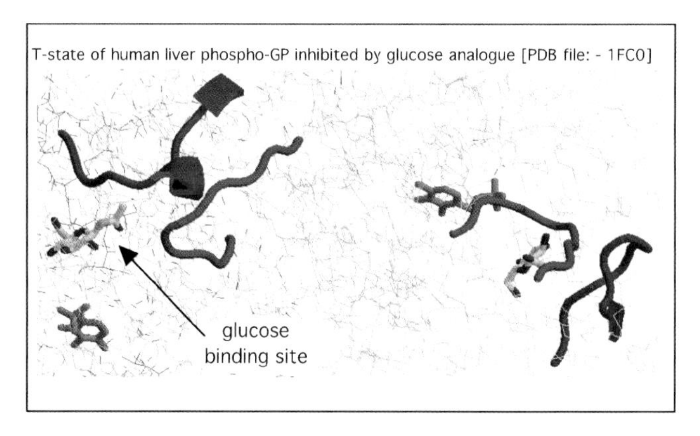

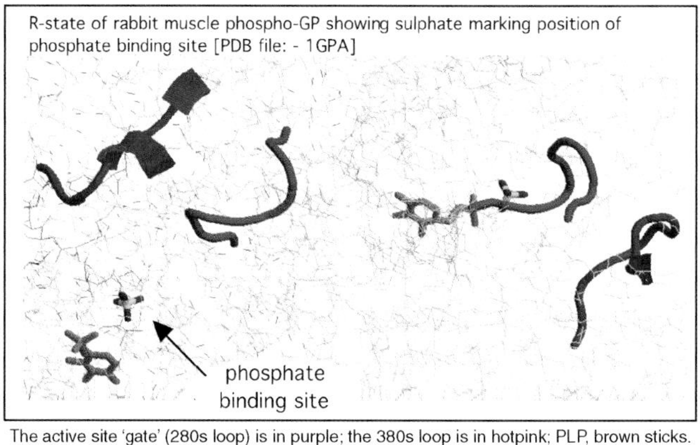

Figure 5.11 Conformations of the GP active site in the T-state versus the R-state (PDB files:- 1GPA & 1FCO)

5′-phosphate. The disordering/opening of the 280s gate combined with the protomer rotation, opens the catalytic tunnel, completes the active site and allows access to the oligosaccharide substrate.

5.9.8 Activation by 5′-AMP

Muscle-type dephospho-GP can be alternatively activated by elevated [AMP], which also binds to the cap/$\alpha2$ subdomain. Like the pSer peptide, AMP also spans the dimer interface but binds below the cap. One AMP binds cap′ and $\alpha2$, the other binds cap and $\alpha2'$ and the result is a stabilised and tightened cap/$\alpha2$ interface resembling the phospho-GP_M. This AMP-activated dephospho-GP_M structure also results in an ordering of the *N*-terminal peptide, which now adopts a dimer-spanning orientation that resembles the PhK-phosphorylated structure (see for example PDB file 7GBP, where the pSer phosphate is mimicked by a sulphate ion)[53].

Phosphorylation of the muscle enzyme results in a higher affinity for the activator AMP and a lowered affinity for the inhibitor G6P[52]. AMP can bind to the phospho-GP_M

and co-occupies the allosteric site – pSer binds above the cap, AMP below: this can further stimulate the muscle phospho-enzyme's activity.

The muscle enzyme binds AMP more tightly than the relatively insensitive liver enzyme, because the 'adenine loop' (aas315–325) of GP_M makes additional H-bonds with AMP that are not possible in the liver enzyme due to its 'adenine loop' having a different conformation (despite sequence homology). The differing conformation of the liver adenine loop is theorised to be due to differences in the configuration of the dimer interface[47].

5.9.9 Inhibition by glucose

Glucose binding to the catalytic site causes closure of the active site gate loop, thus blocking substrate access, and leads to adoption of the T-state even when the enzyme is phosphorylated. Indeed, when the glucose analogue *N*-acetylglucosamine is bound to phospho-GP_L, the *N*-terminal peptide becomes disordered even though it is phosphorylated (Figure 5.10). This glucose-induced T-state conformation ensures that the liver enzyme will cease glycogenolysis when glucose levels rise, no matter what its phosphorylation status. It is also likely that the glucose-induced disordering of the *N*-terminal tail exposes the pSer to solvent, making it more easily de-phosphorylated by PP-1. The muscle enzyme is also glucose-inhibited, but in GP_M the effects are dependent upon AMP levels[54].

5.10 Glycogen synthase

Glycogen synthase (GS) is somewhat less well understood than GP, and the only structures available at the time of writing are from bacteria[55]. The bacterial GS enzymes do not share primary sequence homology with mammalian GS[56]. However, a number of key residues appear in near identical positions in both glycogen synthases and glycogen phosphorylase and they share similar folds.

The control of GS activity by phosphorylation is more complex than GP, which has only a single phosphorylation site. GS, by contrast, is subject to multiple serine phosphorylations that are concentrated at *N*-terminal and *C*-terminal regions. These phosphorlyations contribute to its inhibition *in vitro* and *in vivo*. Seven serine residues are phosphorylated by nine different protein kinases *in vitro*, at least six of which do so *in vivo*. In the early literature, the kinases were preliminarily named 'glycogen synthase kinase-1, -2, -3' etc., but a few were later identified with previously known protein kinases. For example, the originally-named 'glycogen synthase kinase-1' (GSK-1) is identical to casein kinase-I[57]; 'GSK-2' is identical to phosphorylase kinase[58]. GSK-3, on the other hand, is unique.

The phosphorylation sites on GS are traditionally named after their locations in the rabbit skeletal muscle form of the enzyme (Figure 5.12), as originally identified after controlled proteolytic cleavage of the polypeptide chain[59]. Adrenaline stimulation of

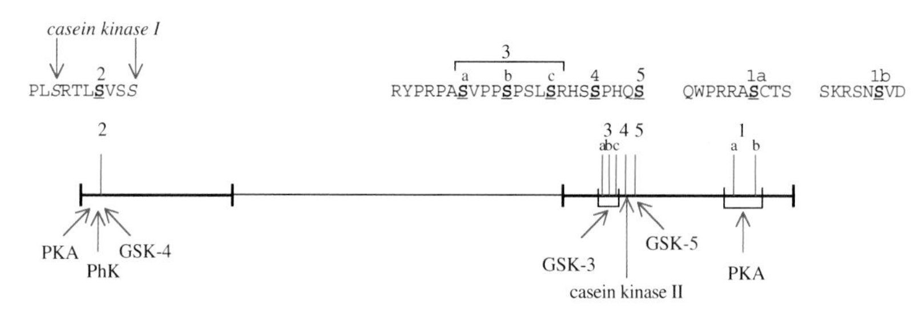

Information from: Fiol, C.J. et al. (1990) J. Biol. Chem., 265: 6061-6065
Kuret, J. et al. (1985) Eur. J. Biochem., 151: 39-48.
Parker, P.J. et al. (1982) Eur. J. Biochem., 124: 47-55.

Figure 5.12 Schematic diagram of glycogen synthase serine phosphorylation sites

muscle results in inhibitory serine phosphorylation at the *N*-terminal 'site 2' (Ser7) by PKA and PhK – this negative feedback loop ensures that GS is inhibited during the time that GP is activated. The *C*-terminal 'sites 3a, b and c' are specific for GSK-3, an enzyme that is unusual in being constitutively active in resting cells[60]. GSK-3 is theorised to maintain a background level of GS phosphorylation by default. So, when adrenaline/glucagon stimulation ceases and is replaced with insulin stimulation (in response to a meal), the process of activating glycogen synthesis involves not only the termination of the cAMP-induced PKA → PhK → phosphorylation of GS, but also necessitates the inhibition of GSK-3 to allow unimpeded dephosphorylation and activation of GS by protein phosphatase.

There is *in vitro* evidence that GSK-3 may be scaffolded by AKAP220 that also binds PKA and PP1. When thus scaffolded to its two controlling enzymes, PKA phosphorylates and inactivates GSK-3 more efficiently, and PP-1 dephosphorylates and activates with a similar increase in efficiency[61].

5.10.1 GSK-3 – a multi-tasking enzyme

GSK-3 turns out to be a very unusual enzyme – one that is involved in a myriad of tasks besides its original role in glycogen metabolism. In particular, it has a key influence on the cell cycle and gene transcription because it is also controlled by insulin receptor signal pathways (see Chapter 9, Section 9.8) and Mitogenic signals (see Chapter 10, Sections 10.8.7 and 10.17.2).

GSK-3 is also unusual in having a substrate preference for target serine residues that are *N*-terminal of a (pre-existing) phosphoserine. In other words, GSK-3 substrates need to be 'primed' by the action of another kinase in order for efficient phosphorylation by GSK-3 itself. GS, for example, is primed by casein kinase II, which phosphorylates 'site 4' (see Figure 5.12). This then allows GSK-3 to phosphorylate 'sites 3a–c', which may be processive as each phosphorylation (going in the *N*-terminal direction from site 4, to site 3c, to site 3b, etc.) creates a new GSK-3 consensus sequence, viz: **S**.X.X.X.pS – the target serine (bold) is always separated by three residues from the priming phosphoserine (underlined)[62].

In the same vein, the cAMP response element binding protein (CREB, a transcription factor that is activated by nuclear-translocated PKA) is also a target of GSK-3 provided it is 'primed' by prior phosphorylation by PKA[63].

5.11 Remaining questions – scaffolds and alternate second messenger 'receptors'

We are moving into an era where the significance of protein-to-protein interactions, identified in *in vitro* screens such as yeast 2-hybrid, will be explored in a cell physiological setting. At the time of writing, frustratingly little is known about how scaffolding proteins such as AKAPs contribute to the spatio-temporal control of glycogen metabolism by PKA. It seems likely, for example, that AKAPs that target PKA (and PP-1) to the liver endoplasmic reticulum (or muscle sarcoplasmic reticulum) would increase signal fidelity by placing the signal node in the immediate vicinity of the mature glycogen granules that are anchored to the endoplasmic reticulum membrane.

5.11.1 Protein kinase C

Protein kinase C (PKC) is a serine/threonine kinase discovered in 1971, and was so-named for its activation by calcium. It later transpired that PKC also needed to dock with the membrane-bound second messenger, diacylgycerol (DAG), and with inner leaflet phosphatidyl serine. As we have seen, both DAG and calcium may be produced by either Gαq-coupled α1AR or β/γ release from Gαi-coupled α2AR.

This 'conventional' PKC is the founder member of a large family of mammalian PKC isozymes that are all members of the AGC group of kinases (the group that includes PKA, see Appendix 1). PKCs can be divided into 'conventional' (cPKC), 'atypical' (aPKC) and 'novel' (nPKC) (Table 5.6); additional PKC-related kinases have also been discovered[64]. All PKCs are self-inhibited by an *N*-terminal pseudosubstrate domain that performs the same function as that of the PKA pseudosubstrate loop of R^{SUB}.

Table 5.6 Protein kinase C classes

Class	Members	*N*-terminal Regulatory Domain	Ca^{2+}	DAG/ Phorbol Ester	Other stimuli
cPKC	PKCα PCKβI/II PKCγ	pseudosubstrate-[CR|CR]-C2-	activates	both activate	phosphatidyl serine
nPKC	PKCδ PKCε PKCη PKCθ	C2-pseudosubstrate-[CR|CR]-	no effect	both activate	phosphatidyl serine
aPKC	PKCζ PKCι PKCλ	pseudosubstrate-[CR]-	no effect	no effect	PIP3

CR: 'cysteine-rich' zinc finger domain (a classic PKC C1 domain has a tandem pair of CR's)
PIP3: phosphatidylinositol-3,4,5-trisphosphate

5.11.2 Lipid activation of PKC – DAG-binding isoforms are also activated by phorbol esters

'Conventional' PKC types are constitutively A-loop phosphorylated and only need the pseudosubstrate to debind in order to become active. The resting cytosolic cPKC is activated by calcium (which binds to its C2 domain and increases its affinity for phosphatidyl serine), and DAG, which binds to its C1 domain. PKC activation results from occupancy of these two regulatory elements by DAG, and Ca^{2+}/phosphatidyl serine. This causes both tight binding to the membrane and ejection of the pseudosubstrate from the active site[65].

'Atypical' PKC types such as PKCζ need to be phosphorylated by PIP3-dependent kinase (PDK-1) before they can be activated by lipid-binding. PKCζ lacks a C2 domain and is insensitive to calcium. PKCζ also lacks a complete C1 domain (because it contains only one of the cysteine-rich repeats) and is insensitive to DAG. Instead, it is activated by PIP3 (produced by *PI*-3-kinase).

5.11.3 Alternative DAG/phorbol ester receptors

Phorbol ester (***P***horbol ***M***yristoyl ***A***cetate or PMA, also known as ***T***etradecanoyl ***P***horbol ***A***cetate or TPA) is a plant-derived polycyclic alcohol derivative (Figure 5.13) that mimics DAG but is long lasting and produces a chronic DAG-like signal. For this reason, it has been relied upon as a useful tool in PKC-related research – and for many years, PKC was considered to be the only phorbol ester-activatable protein and therefore the only DAG target. For example, the growth-promoting and cancer-promoting effects of phorbol esters were thought to be solely mediated by PKC. Recently, however, novel and unsuspected TPA/DAG 'receptors' have been found that may be solely responsible for effects previously considered to be PKC mediated, or may act in a parallel, overlapping manner with PKC.

Whereas PKC does indeed stimulate or co-stimulate the MAPK pathway (Chapter 7, Section 7.3.9), some of the mitogenic effects of TPA may be mediated by the alternative phorbol ester 'receptors'– a type of Ras-GEF, known as 'RasGRP' (for Ras ***g***uanyl-***r***eleasing ***p***rotein)[66]. TPA stimulates RasGRP translocation to the membrane, which allows it to activate Ras and thus the MAPK pathway. In thymocytes, at least, the proliferative effects of TPA are PKC-independent and are entirely due to RasGRP, as are

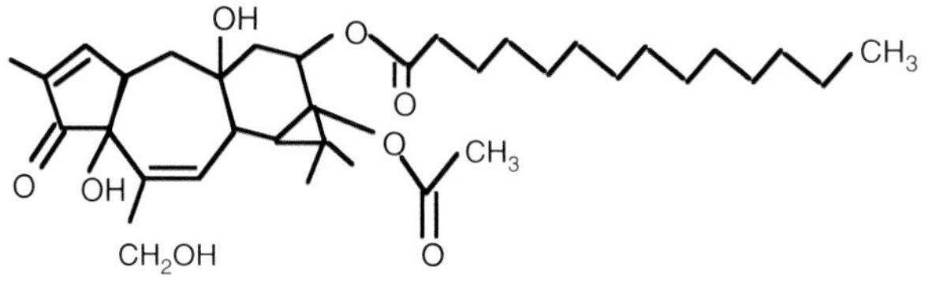

Figure 5.13 Commonly used Phorbol Ester:- Phorbol Myristoyl Acetate (PMA) - also known as Tetradecanoyl Phorbol Acetate (TPA)

the growth-promoting effects of activation of the T cell receptor, to which RasGRP is naturally linked in these cells. Other novel DAG/TPA receptors include the α- and β-chimaerins (that display RacGAP activity), Unc proteins (involved with synaptic core complex) and DAG kinase, the enzyme that 'deactivates' the DAG signal[67].

5.11.4 PKC scaffolds

Although much is known about PKC activation, its actual physiological function is in many cases enigmatic. Indeed, despite the great differences in activation strategies between isoforms, their substrate preferences are remarkably similar. It seems likely that newly discovered PKC scaffold proteins may be the way the cell confers more specificity – by simply localising the kinase with a discrete substrate. PKC scaffolding proteins include those that only bind activated PKC: '***r***eceptors for ***a***ctivated ***C*** ***k***inase' (RACKs); '***s***ubstrates that ***i***nteract with ***C*** ***k***inases' (STICKs); and others that bind inactive PKC, including members of the AKAP family (AKAP 79 and gravin) and 14-3-3 proteins[66]. RACK proteins cause translocation and activation of PKC at the membrane and increase the rate of PKC substrate phosphorylation. AKAP79 scaffolds PKA, the phosphatase PP-2B and PKC, the latter being inhibited by the interaction. Calcium/DAG releases and activates PKC, which phosphorylates AKAP79, changing its compartmentalisation. STICK protein subcellular location is also controlled by PKC phosphorylation. STICKs include the ubiquitous '***m***yristoylated ***a***lanine-***r***ich ***C***-***k***inase ***s***ubstrate' (MARCKS) (see Chapter 3, Section 3.14).

5.11.5 What does PKC actually do?

With respect to our discussion of glycogenolysis, α1AR stimulation of liver does result in activation of PKC[68], but there is little or no evidence that this has any direct effect on glycogenolysis. Instead, activation of PKC stimulates the Na^+/H^+ exchanger (causing alkalinisation of cytosol) and increases glucose export in perfused rat liver, but has no effect on gluconeogenesis. Most PKC metabolic effects fall into the categories of crosstalk and feedback, PKC causing phosphorylation and desensitisation of receptors such as βAR, the glucagon receptor, α1AR, the EGF receptor and the insulin receptor[69].

Studies of knockouts of PKC avoid involvement of non-PKC targets, but are sometimes confounded by the redundancy of multiple PKC isoforms and resultant mild phenotypes. However, clear roles were identified in some knockouts. For example, PKCβ-knockout leads to immunodeficiency and lack of PKCζ leads to impaired B-cell function. Both knockouts display impaired NFκB signalling.

PKC has two prominent cytoskeletal substrates in the form of MARCKS and pleckstrin. Both substrates are involved in actin remodelling[64]. MARCKS phosphorylation by PKC causes subcellular translocation and calmodulin release. MARCKS, like pleckstrin, also binds and crosslinks actin filaments of the cytoskeleton and both are involved in actin polymerisation and depolymerisation during cell shape changing activities, such as exocytosis. Both MARCKS and pleckstrin actin-binding is inhibited by PKC phosphorylation. In platelets, MARCKS phosphorylation appears to be the predominant trigger for serotonin release – phorbol ester-stimulation of PKC leads to phosphorylation of the MARCKS central basic/effector domain, followed by actin disassembly and secretion[70].

5.11.6 Alternate cAMP receptors

For a long time, PKA was considered to be the only target of cAMP, and PDE enzymes were thought to be the only other proteins that bind (then destroy) the second messenger. However, a new family of cAMP-binding proteins have been identified as being the key to the control of insulin secretion from pancreatic β-cells. Specifically, these proteins – collectively known as Epac proteins ('***e***xchange ***p***rotein ***a***ctivated by ***c***AMP) – explain the complex signalling of the 7-pass receptor for glucagon-like peptide-1 (GLP-1), a receptor coupled to Gs and thus expected to produce simple PKA-mediated responses. Epac-1 was originally characterised as a protein that (when cAMP-occupied) stimulated GDP-for-GTP exchange at the Rap1/2 monomeric G protein – effectively, Epac-1 is a cAMP-activated RapGEF. Epac-2 interacts indirectly with the monomeric G protein Rab via an insulin granule-associated protein, Rim1. Epac-1 has a single cAMP-binding domain with homology to R^{SUB-I}, and Epac-2 has two cAMP-binding modules. The engagement of Epacs by the GLP-1 receptor allows it to add signals to its basal PKA response, including Ca^{2+} mobilisation from internal stores, 'priming' of insulin granules, and exocytosis of insulin. The complexity of this system has been recently reviewed[71].

5.12 G protein coupled receptor kinases – downregulators, signal integrators

G-protein coupled ***r***eceptor ***k***inase-1 (GRK-1) was originally known as rhodopsin kinase, a serine/theronine kinase that desensitises rhodopsin in association with arrestin, causing decoupling from transducin. Lefkowitz discovered a non-retinal homologue of rhodopsin kinase in the form of 'β-***a***drenergic ***r***eceptor ***k***inase' (βARK). βARK is commonly referred to as 'GRK-2'. When chronically stimulated, βAR (and other 7-pass receptors with correct substrate sequences) becomes multiply serine-phosphorylated in the third intracellular loop and C-tail. This phosphorylation of G protein coupling sites renders βAR (and others) incapable of G-coupling and thus blocks signalling. This non-specific heterologous negative feedback response is caused by PKA and PKC, both being unselective over whether a receptor is activated or not.

GRK-2 is more selective and only phosphorylates activated, ligand occupied receptors leading to homologous desensitisation. Chronic stimulation of βAR causes a surplus of free β/γ subunits to build up in the membrane. GRK-2 contains a *C*-terminal β/γ-binding form of PH domain that causes its translocation from the cytosol to the membrane where it serine-phosphorylates the C-tail of agonist-occupied 7-pass receptors through docking with the activated conformation of the third intracellular loop. The phosphorylated C-tail of the receptor acts as a docking site for β-arrestin, which translocates from the cytosol, binds the receptor and prevents further G-coupling. The disabled receptor/arrestin complex becomes sequestered in clathrin-coated vesicles (arrestin is a clathrin binder) and is internalised. Interestingly, in the case of the 7-pass formyl peptide receptor, the agonist-receptor-arrestin complex locks the receptor into a high-affinity agonist binding state, much like the ternary complex with 'empty pocket G proteins (except, of course, that the arrestin-stablised complex is resistant to the affinity-lowering effects of GTPγS[72]).

GRK proteins are subject to multiple regulating signals including modulation by phosphorylation by PKA, Erk1/2, Src and a sequestering effect of $Ca2^{+}$/calmodulin-binding. GRK, like arrestin, has a calveolin-binding site and both GRK and arrestin can serve as scaffolds. The complex nature of this protein's many interactions has been recently reviewed[73], and is discussed again in Chapter 8.

References

1. Fredriksson, R., Lagerstrom, M.C., Lundin, L-G. and Schioth, H.B. (2003) The G-protein-coupled receptors in the human genome from five main families: phylogenetic analysis, paralogon groups, and fingerprints. *Mol. Pharmacol.*, **63**: 1256–1272.
2. Hepler, J.R. and Gilman, A.G. (1992) G-proteins. *TIBS*, **17**: 383–387.
3. Landis, C.A., Masters, S.B., Spada, A., Pace, A.M., Bourne, H.R. and Vallar, L. (1989) GTP'ase inhibiting mutations activate the α chain of Gs and stimulate adenylyl cyclase in human pituitary tumours. *Nature*, **340**: 692–696.
4. Dessauer, C.W. and Gilman, A.G. (1996) Purification and characterization of a soluble form of mammalian adenylyl cyclase. *J. Biol. Chem.*, **271**: 16967–16974.
5. Taussig, R., Quarmby, L.M. and Gilman, A.G. (1993) Regulation of purified type I and type II adenylyl cyclases by G-protein βγ subunits. *J. Biol. Chem.*, **268**: 9–12.
6. Freshney, R. (2000) *Culture of Animal Cells: A Manual of Basic Techniques*; Wiley-Liss: New York.
7. Terasaki, W.L. and Brooker, G. (1977) Cardiac adenosine 3′:5′-monophosphate. *J. Biol. Chem.*, **252**: 1041–1050.
8. Kahn, C.R. (1976) Membrane receptors for hormones and neurotransmitters. *J. Cell Biol.*, **70**: 261–286.
9. Meuller, H., Weingarten, R., Ransnas, L.A., Bokoch, G.M. and Sklar, L.A. (1991) Differential amplification of antagonistic receptor pathways in neutrophils. *J. Biol. Chem.*, **266**: 12939–12943.
10. Ransnas, L.A. and Insel, P.A. (1988) Quantitation of the guanine nucleotide binding regulatory protein Gs in S49 cell membranes using antipeptide antibodies to as. *J. Biol. Chem.*, **263**: 9482–9485.
11. Laflamme, M. and Becker, P.L (1998) Do β_2-adrenergic receptors modulate Ca^{2+} in adult rat ventricular myocytes? *Am. J. Physiol. Heart Circ.*, **274:** H1308–H1314.
12. Sonne, O., Berg, T. and Christoffersen, T. (1978) Binding of ^{125}I-labelled glucagons and glucagons-stimulated accumulation of adenosine 3′:5′-monophosphate in isolated intact rat hepatocytes. *J. Biol. Chem.*, **253:** 3202–3210.
13. Johnston, C.A. and Siderovski, D.P. (2006) Resolving G-protein-coupled receptor signaling mechanics *in vivo* using fluorescent biosensors. *Cell Science*, **2**: 16–24.
14. Tolkovovsky, A.M. and Levitzki, A. (1978) Mode of coupling between the β-adrenergic receptor and adenylate cyclase in turkey erythrocytes. *Biochemistry*, **17**: 3795–3810.
15. Heldin, C-H. and Purton, M. (1996) *Signal Transduction*. Chapman & Hall.
16. Lodish, H., Baltimore, D., Berk, A., Zipursky, S.L., Matsudaira, P. and Darnell, J. (1995) *Molecular Cell Biology*, 3rd Edition. Scientific American Books, p. 885.

17. Ostrom, R.S., Post, S.R. and Insel, P.A. (2000) Stoichiometry and compartmentalisation in G-protein-coupled receptor signalling: implications for therapeutic intervention involving G_S. *The Journal of Pharmacology and Experimental Therapeutics*, **294**: 407–412.
18. Pieroni, J.P., Harry, A., Chen, J., Jacobowitz, O., Magnusson, R.P. and Iyengar, R. (1995) Distinct characteristics of the basal activities of adenylyl cyclases 2 and 6. *J. Biol. Chem.*, **270**: 21368–21373.
19. Houslay, M.D. and Milligan, G. (1997) Tailoring cAMP-signalling responses through isoform multiplicity. *TIBS*, **22**: 217–224.
20. Iwami, G., Kawabe, J., Ebina, T., Cannon, P.J. Homcy, C.J. and Ishiwara, Y. (1995) Regulation of adenylyl cyclase by protein kinase A. *J. Biol. Chem.*, **270**: 12481–12484.
21. Sette, C and Conti, M. (1996) Phosphorylation and activation of a cAMP-specific phosphodiesterase by the cAMP-dependent protein kinase. *J. Biol. Chem.*, **271**: 16526–16534.
22. Degerman, E., Landstrom, T.R. Wijkander, J., Holst, L.S., Ahmad, F., Belfrage, P. and Manganiello, V. (1998) Phosphorylation and activation of hormone-sensitive adipocyte phsophodiesterase type 3B. *Methods*, **14**: 43–53.
23. Sunahara, R.K. and Taussig, R. (2002) Isoforms of mammalian adenylyl cyclase: multiplicities of signaling. *Molecular Interventions*, **2**: 168–184.
24. Defer, N., Best-Belpomme, M and Hanoune, J. (2000) Tissue specificity and physiological relevance of various isoforms of adenylyl cyclase. *Am. J. Physiol. Renal Physiol.*, **269**: F400–F416.
25. Tasken, K. and Aandahl, E.M. (2003) Localized effects of cAMP mediated by distinct routes of protein kinase A. *Physio. Rev.*, **84**; 137–167.
26. Viste, K., Kopperud, R.K., Christensen, A.E. and Doskeland, S.O. (2005) Substrate enhances the sensitivity of type I protein kinase A to cAMP. *J. Biol. Chem.*, **280**: 13279–13284.
27. Keryer, G., Yassenko, M., Labbe, J-C., Castro, A., Lohmann, S.M., Evian-Brion, D. and Tasken, K. (1998) Mitosis-specific phosphorylation and subcellular redistribution of the RIIa regulatory subunit of cAMP-dependent protein kinase. *J. Biol. Chem.*, **273**: 34594–34602.
28. Landsverk, H.B., Carlson, C.R., Steen, R.L., Vossebein, L., Herberg, F.W., Tasken, K. and Collas, P. (2001) Regulation of anchoring of the RIIα regulatory subunit of PKA to AKAP95 by threonine phosphorylation of RIIα: implications for chromosome dynamics at mitosis. *J. Cell Science*, **114**: 3255–3264.
29. Rangel-Aldao, R. and Rosen, O.M. (1977) Effect of cAMP and ATP on the reassociation of phosphorylated and nonphosphorylated subunits of the cAMP-dependent protein kinase from bovine cardiac muscle. *J. Biol. Chem.*, **252**: 7140–7145.
30. Scott, C.W. and Mumby, M.C. (1985) Phosphorylation of type II regulatory subunit of cAMP-dependent protein kinase in intact smooth muscle. *J. Biol. Chem.*, **260**: 2274–2280.
31. Blumenthal, D.K., Takio, K., Hansen, S. and Krebs, E.G. (1986) Dephosphorylation of cAMP-dependent protein kinase regulatory subunit (type II) by calmodulin-dependent protein phosphatase. *J. Biol. Chem.*, **261**: 8140–8145.
32. Morgan, N.G., Blackmore, P.F. and Exton, J.H. (1983) Age-related changes in the control of hepatic cyclic cAMP levels by α1- and β2-adrenergic receptors in male rats. *J. Biol. Chem.*, **258**: 5103–5109.
33. Blank, J.L., Ross, A.H. and Exton, J.H. (1991) Purification and characterization of two G-proteins that activate the β1 isoform of phosphoinositide-specific phospholipase C. *J. Biol. Chem.*, **266**: 18206–18216.
34. Rybin, V.O., Xu, X., Lisanti, M.P. and Steinber, S.F. (2000) Differential targeting of the β-adrenergic receptor subtypes and adenylyl cyclase to cardiomyocyte caveolae. *J. Biol. Chem.*, **275**: 41447–41457.
35. Johnson, D.A., Akamine, P., Radzio-Andzelm, E., Madhusudan and Taylor, S.S. (2001) Dynamics of cAMP-dependent protein kinase. *Chemical Reviews*, **101**: 2243–2270.
36. Wu, J., Brown, S., Xuong, N-H. and Taylor, S.S. (2004) RIα subunit of PKA: cAMP-free structure reveals a hydrophobic capping mechanism for docking cAMP into site B. *Structure*, **12**: 1057–1065.

37. Akileswaran, L., Taraska, J.W., Sayer, J.A., Gettemy, J.M. and Coghlan, V.M. (2001) A-kinase-anchoring protein AKAP95 is targeted to the nuclear matrix and associates with p68 RNA helicase. *J. Biol. Chem.*, **276**: 17448–17454.
38. Feliciello, A., Gottesman, M.E. and Avvedimento, E.V. (2005) cAMP-PKA signaling to the mitochondria: protein scaffolds, mRNA and phosphatases. *Cellular Signalling*, **17**: 279–287.
39. Owen, D.J., Noble, M.E., Garman, E.F., Papageorgiou, A.C. and Johnson L.N. (1995) Two structures of the catalytic domain of phosphorylase kinase: an active protein kinase complexed with substrate analogue and product. *Structure*, **3**: 467–82.
40. Rice, N.A., Nadeau, O.W., Yang, Q. and Carlson, G.M. (2002) The calmodulin-binding domain of the catalytic γ subunit of phosphorylase kinase interacts with its inhibitory α subunit. *J. Biol. Chem.*, **277**: 14681–14687.
41. Brushia, R.J. and Walsh, D.A. (1999) Phosphorylase kinase: the complexity of its regulation is reflected in the complexity of its structure. *Frontiers in Bioscience*, **4**: 618–641.
42. Nadeau, O.W., Carlson, G.M. and Gogol, E.P. (2002) A Ca^{2+}-dependent global conformational change in the 3D structure of phosphorylase kinase obtained from electron microscopy. *Structure*, **10**: 23–32.
43. Nadeau, O.W., Gogol, E.P. and Carlson, G.M. (2005) Cryoelectron microscopy reveals new features in the three-dimensional structure of phosphorylase kinase. *Protein Science*, **14**: 914–920.
44. Venien-Bryan, C., Lowe, E.M., Boisset, N., Traxler, K.W., Johnson, L.N. and Carlson, G.M. (2002) Three-dimensional structure of phosphrylase kinase at 22 Å resolution and its complex with glycogen phosphorylase b. *Structure*, **10**: 33–41.
45. Barford, D. and Johnson, L.N. (1992) The molecular mechanism for the tetrameric association of glycogen phosphorylase promoted by protein phosphorylation. *Protein Science*, **1**: 472–493.
46. Johnson, L.N. (1992) Glycogen phosphorylase: control by phosphorylation and allosteric effects. *FASEB J.*, **6**: 2274–2282.
47. Rath, V.L., Ammirati, M., LeMotte, P.K., Fennell, K.F., Mansour, M.N., Danley, D.E., Hynes, T. R., Schulte, G.K., Wasilko, D.J. and Pandit, J. (2000) Activation of human liver glycogen phosphorylase by the alteration of the secondary structure and packing of the catalytic core. *Molecular Cell*, **6**: 139–148.
48. Kasvinsky, P.J., Madsen, N.B., Fletterick, R.J. and Sygusch, J. (1978) X-ray crystallographic and kinetic studies of oligosaccharide binding to phosphorylase. *J. Biol. Chem.*, **253**: 1290–1296.
49. Vandenberghe, K., Richter, E.A. and Hespel, P. (1999) Regulation of glycogen breakdown by glycogen level in contracting rat muscle. *Acta Physiol. Scand.*, **165**: 307–314.
50. Nielsen, J.N. and Richter, E.A. (2003) Regulation of glycogen synthase in skeletal muscle during exercise. *Acta Physiol. Scand.*, **178**: 309–319.
51. Franch, J., Aslesen, R. and Jensen, J. (1999) Regulation of glycogen synthesis in rat skeletal muscle after glycogen-depleting contractile activity: effects of adrenaline on glycogen synthesis and activation of glycogen synthase and glycogen phosphorylase. *Biochem. J.*, **344**: 231–235.
52. Sprang, S.R., Acharya, K.R., Goldsmith, E.J., Stuart, D.I., Varvill, K., Fletterick, R.J., Madsen, N.B. and Johnson, L.N. (1988) Structural changes in glycogen phosphorylase induced by phosphorylation. *Nature*, **336**: 215–221.
53. Barford, D., Hu, S-H. and Johnson, L.N. (1991) Structural mechanism for glycogen phosphoylase control by phosphorylation and AMP. *J. Mol. Biol.*, **218**: 233–260.
54. Lukacs, C.M., Oikonomakos, N.G., Crowther, R.L., Hong, L-N., Kammlott, R.U., Levin, W., Li, S., Liu, C-M., Lucas-McGady, D., Pietranico, S. and Reik, L. (2006) *PROTEINS: Structure, Function and Bioinformatics*, **63**: 1123–1126.
55. Buschiazzo, A., Ugalde, J.E., Guerin, M.E., Shepard, W., Ugalde, R.A. and Alzari, P.M. (2004) Crystal structure of glycogen synthase: homologous enzymes catalyse glycogen synthesis and degradation. *EMBO J.*, **23**: 3196–3205.

56. Browner, M.P., Nakano, K., Bagn, A.G. and Fletterick, R.J. (1989) Human muscle glycogen synthase cDNA sequence: a negatively charged protein with an asymmetric charge distribution. *Proc. Natl. Acad. Sci.*, **86**: 1443–1447.
57. Kuret, J., Woodgett, J.R. and Cohen, P. (1985) Multisite phosphorylation of glycogen synthase from rabbit skeletal muscle. *Eur. J. Biochem.*, **151**: 39–48.
58. Embi, N., Rylatt, D.B. and Cohen, P. (1979) Glycogne-synthase kinase-2 and phosphorylase kinase are the same enzyme. *Eur. J. Biochem.*, **100**: 339–347.
59. Parker, P.J. Embi, N., Cauldwell, B. and Cohen, P. (1982) Glycogen synthase from rabbit skeletal muscle. State of phosphorylation of the seven phosphoserine residues *in vivo* in the presence and absence of adrenaline. *Eur. J. Biochem.*, **124**: 47–55.
60. Doble, B.W. and Woodgett, J.R. (2003) GSK-3: tricks of the trade for a multi-tasking kinase. *J. Cell Sci.*, **116**: 1175–1186.
61. Tanji, C., Yamamoto, H., Yorioka, N., Kohno, N., Kikuchi, K. and Kikuchi, A. (2002) A-kinase anchoring protein AKAP220 binds to glycogen synthase kinase-3β (GSK-3β) and mediates protein kinase A-dependent inhibition of GSK-3β. *J. Biol. Chem.*, **277**: 36955–36961.
62. Fiol, C.J., Wang, A., Roeske, R.W. and Roach, B.J. (1990) Ordered multisite protein phosphorylation. Analysis of glycogen synthase kinase 3 action using model peptide substrates. *J. Biol. Chem.*, **265**: 6061–6055.
63. Fiol, C.J., Williams, J.S., Chou, C-H., Wang, M., Raoch, P.J. and Andrisani. O.M. (1994) A secondary phosphorylation of CREB341 at Ser129 is required for the cAMP control of gene expression. A role for glycogen synthase kinase-3 in the control of gene expression. *J. Biol. Chem.*, **269**: 32187–32193.
64. Toker, A. (1998) Signaling through protein kinase C. *Frontiers in Bioscience*, **3**: 1134–1147.
65. Newton, A.C. (1997) Regulation of protein kinase C. *Current Opinion in Cell Biology*, **9**: 161–167.
66. Ron, D. and Kazanietz, M.G. (1999) New insights into the regulation of protein kinase C and novel phorbol ester receptors. *FASEB J.*, **13**: 1658–1676.
67. Brose, N. and Rosenmund, C. (2002) Move over protein kinase C, you've got company: alternative cellular effectors of diacylglycerol and phorbol esters. *J. Cell Science*, **115**: 4399–4411.
68. Urecelay, E., Butta, N., Manchon, C.G., Cipres, G., Reguero, A.M., Ayuso, M.S. and Parrilla, R. (1993) Role of protein kinase C in the a1-adrenoceptor-mediated response of perfused rat liver. *Endocrinology*, **133**: 2105–2115.
69. Houslay, M.D. (1991) 'Crosstalk': pivotal role for protein kinase C in modulating relationships between signal pathways. *Eur. J. Biochem.*, **195**: 9–27.
70. Elzagallaai, A., Rose, S.D. and Trifaro, J-M. (2000) Platelet secretion induced by phorbol esters stimulation is mediated through phosphorylation of MARCKS: a MARCKS-derived peptide blocks MARCKS phosphorylation and serotonin release without affectin pleckstrin phosphorylation. *Blood*, **95**: 894–902.
71. Holz, G.G. (2004) Epac: a new cAMP-binding protein in support of glucagon-like peptide-1 receptor-mediated signal transduction in the pancreatic b-cell. *Diabetes*, **53**: 5–13.
72. Key, T.A., Bennett, T.A., Foutz, T.D., Gurevich, V.V., Sklar, L.A. and Prossnitz, E.R. (2001) Regulation of formyl peptide receptor agonist affinity by reconstitution with arrestins and heterotrimeric G-proteins. *J. Biol. Chem.*, **276**: 49204–49212.
73. Theilade, J., Haunso, S. and Sheikh, S.P. (2001) G-protein-coupled receptor kinase 2 – a feed back regulator of Gq pathway signalling. *Current Drug Targets – Immune, Endocrine & Metabolic Disorders*, **1**: 139–151.

6

Single pass growth factor receptors

Single pass receptors may be catalytic or non-catalytic. The majority of the catalytic types are members of the receptor tyrosine kinase (RTK) superfamily and members of this group are the primary receptors for the polypeptide 'growth factors' that stimulate cell division. Many of these follow the activation mechanism of the insulin receptor kinase and are controlled by reversible tyrosine phosphorylations of their A-loops. Others, such as the EGF- and PDGF-receptor kinases, exhibit a more complex activation that involves the region immediately adjacent to the membrane anchor (the 'juxtamembrane' region). The RTK signalling pathway centres upon the formation of a multi-protein 'signal transduction particle' that is assembled via phosphorylated tyrosines on the receptor's intracellular domain (these are mostly a result of autophosphorylation). SH2 and PTB domain proteins (Grb2, mSos, Shc, PI-3-kinase, RasGAP, SHP-2 and PLCγ) are then recruited from the cytosol (where they normally reside in unstimulated cells) and membrane-bound monomeric G proteins (Ras, Rac Rho) subsequently collision-couple with GEFs or GAPs that are scaffolded to the RTK. RTK signalling is extremely complex but robust, with a high degree of in-built redundancy. RTKs are often able to activate their downstream targets by use of alternative pathways and they also crosstalk extensively with other receptor systems. Adding to the diversity of signalling is their ability to heterodimerise within their own family so that two distinct sets of signalling molecules may be assembled by one ligand.

6.1 Receptor tyrosine kinases – ligands and signal transduction

Receptor tyrosine kinase (RTK) intracellular signals are a great deal more diverse than 7-pass receptors and are not solely reliant upon generation of soluble second

Structure and Function in Cell Signalling John Nelson

messengers. Their dominant mode of signal transduction is the formation of oligomeric 'signal transduction particles' centred on a phosphorylated RTK dimer. The make-up and temporal membership of these multi-protein complexes varies with each receptor and the cell type. In addition, each assembly is highly responsive to the cell's dynamics. The human spectrum of diverse scaffolded structures, such as these, has been titled the 'interactome'.

The vast majority of RTKs contain between one and three tyrosine residues in their A-loops in positions roughly equivalent to the activating threonine of PKA or the activating tyrosine triad of the insulin receptor kinase (IRK). Many RTKs follow the example of IRK, their kinase activities being simply controlled by the phosphorylations/dephosphorylations of A-loop tyrosine(s), which cause conformational changes culminating in reversible re-configuration of the active site and/or unblocking of substrate access. RTKs that are simply controlled by A-loop phosphorylation status include the insulin receptor, the IGF-I receptor, the FGF receptor, the hepatocyte growth factor receptor (MET), nerve growth factor receptor (TRKA), brain-derived neurotrophic factor receptor (TRKB), among others[1]. This mechanism was discussed in Chapter 4 in relation to the IRK and will not be discussed further. Two exceptions, however, serve to illustrate more subtle conditional A-loop control (as exemplified by the PDGF receptor) and its abandonment (as found with ErbB-type receptors).

6.1.1 RTK ligands and receptors

Broadly speaking, the tyrosine kinase-type growth factor receptors and their ligands appear to have evolved together as systems. For example, the PDGFR family ligands (such as PDGF and VEGF) are dimeric, homo-bivalent, disulphide-bonded polypeptides and their receptors (PDGFR and the VEGF receptors like KDR) all contain a characteristic 'insert' in their kinase domains and share an ectodomain architecture made up of multiple immunoglobulin-like repeats. EGFR-type ligands (EGF, TGF-α, heregulin, etc.) are all monomeric, monovalent, triple-looped and disulphide-constrained peptides, and their receptors (EGFR, erbB2-4) share an ectodomain architecture of tandem cysteine-rich ('CR') repeats separating paired ligand-binding ('L') modules each with a 'β-solenoid' fold. Insulin, IGF-I and IGF-II are homologous hetero-bivalent ligands and the insulin receptor and IGF-I receptors are both disulphide-bonded and intrinsically dimerised proteins made up from two identical monomers that are themselves disulphide-bonded proteins – originally single gene-products with a missing section lost during processing of the receptor (Figure 6.1).

A major exception is the high affinity insulin-like growth factor-II receptor, which is totally unrelated to RTKs. The IGF-II receptor is identical to the CI-type mannose-6-phosphate receptor, a protein that appears to have developed an endocytic role in IGF-II clearance by convergent evolution while retaining its original role as a lysosomal trafficking mannose-6-phosphate receptor.

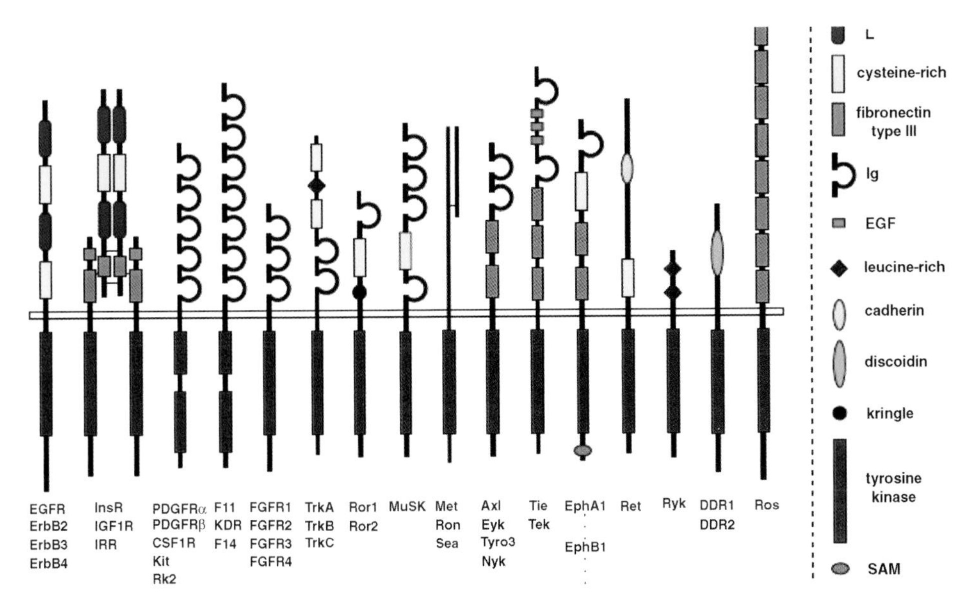

Figure 6.1 Domain organisation for a variety of RTKs. The extracellular portion of the receptors is on top and the cytoplasmic portion is on bottom. Some RTKs (e.g., PDGF receptors) contain a large insert in the tyrosine kinase domain, which is represented as a break in the rectangular symbol. The lengths of the receptors as shown are only approximately to scale

6.2 The PDGFR family – signal transduction

PDGF receptors and VEGF receptors belong to a family of single pass receptor tyrosine kinases (see Appendix 1), which are characterised by ectodomains containing immunoglobulin repeat modules and a 'split' catalytic site containing a long insert not found in other RTKs. Although PDGFR requires A-loop phosphorylation on a tyrosine for full activity, its control is rather more complicated and, at least in some members, appears conditional upon phosphorylation of tyrosine(s) outside of the kinase domain, in the juxtamembrane region of the protein.

There are two PDGFR genes, PDGFR-α (yielding a mature glycosylated protein of MW = 170 kDa) and PDGFR-β (mature MW = 190 kDa), and there are two genes for the ligand PDGF-A and PDGF-B. The ligands are disulphide-bonded dimers (see Figure 6.2) that are either homodimers (PDGF-BB and PDGF-AA) or heterodimeric PDGF-AB. The receptors have differing affinities for the three possible ligands. PDGF-BB can bind to both PDGFR-α and PDGFR-β; PDGF-AA effectively only binds to PDGFR-α. The ligands are bivalent so one ligand can bind two receptors at once, providing a simple means to cause receptor dimerisation. PDGF-AA can only bridge between two PDGFR-α monomers to give an α–α homodimer; PDGF-AB induces α–α and α–β dimers; PDGF-BB induces both α–α and β–β homodimers as well as the α–β heterodimer[2]. The existence of different ligands and the distinct receptor dimers that they induce leads to subtle differences in signalling – PDGFR-β–β and PDGFR-α–α

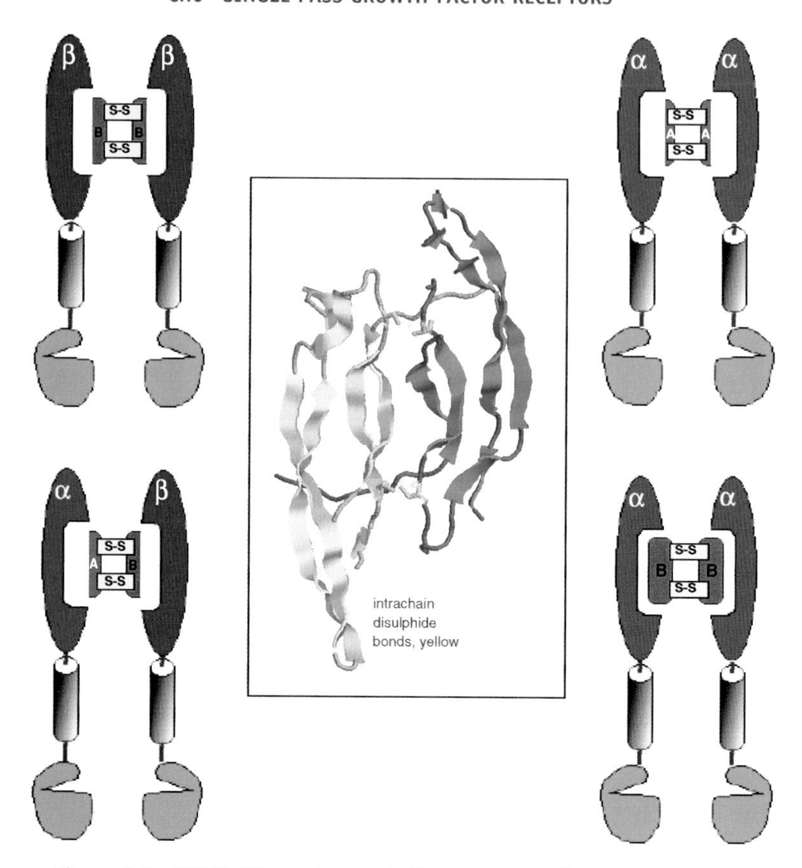

Figure 6.2 PDGFR-BB structure and allowed receptor-ligand combinations

differ in signal molecule recruitment patterns, with PDGFR-α–β presumably exhibiting both. These subtle differences due to ligand/receptor combinations are much expanded in the EGFR family, which exhibits multiple ligand cross-reactivities and dramatically different signalling due to heterodimerisation.

6.2.1 PDGFR signal transduction particle

Before discussing the activation of the receptor's kinase domain, let us first look at the composition of its oligomeric signal transduction particle (after full autophosphorylation) and its signalling output. *Numbering refers to human PDGFR-β unless otherwise stated.*

During activation, the ligand-occupied and dimerised β-type PDGFR creates docking sites for phosphotyrosine-binding proteins through autophosphorylation of intracellular tyrosine residues, including: Tyr579 and Tyr581 in the juxtamembrane region; Tyr716, Tyr740, Tyr751 and Tyr771 in the kinase 'insert' segment; and Tyr1009 and Tyr1021 in the *C*-terminal tail[3,4,5,6] (Table 6.1, Figure 6.3).

Table 6.1 Sequence numbering of three PDGFR family members

Precursor of human platelet-derived growth factor receptor-β

```
   1 MRLPGAMPAL ALKGELLLLS LLLLLEPQIS QGLVVTPPGP ELVLNVSSTF VLTCSGSAPV
  61 VWERMSQEPP QEMAKAQDGT FSSVLTLTNL TGLDTGEYFC THNDSRGLET DERKRLYIFV
 121 PDPTVGFLPN DAEELFIFLT EITEITIPCR VTDPQLVVTL HEKKGDVALP VPYDHQRGFS
 181 GIFEDRSYIC KTTIGDREVD SDAYYVYRLQ VSSINVSVNA VQTVVRQGEN ITLMCIVIGN
 241 DVVNFEWTYP RKESGRLVEP VTDFLLDMPY HIRSILHIPS AELEDSGTYT CNVTESVNDH
 301 QDEKAINITV VESGYVRLLG EVGTLQFAEL HRSRTLQVVF EAYPPPTVLW FKDNRTLGDS
 361 SAGEIALSTR NVSETRYVSE LTLVRVKVAE AGHYTMRAFH EDAEVQLSFQ LQINVPVRVL
 421 ELSESHPDSG EQTVRCRGRG MPQPNIIWSA CRDLKRCPRE LPPTLLGNSS EEESQLETNV
 481 TYWEEEQEFE VVSTLRLQHV DRPLSVRCTL RNAVGQDTQE VIVVPHSLPF KVVVISAILA
 541 LVVLTIISLI ILIMLWQKKP RYEIRWKVIE SVSSDGHE I VDPMQLPYD STWELPRDQL
 601 VLGRTLGSGA FGQVVEATAH GLSHSQATMK VAVKMLKSTA RSSEKQALMS ELKIMSHLGP
 661 HLNVVNLLGA CTKGGPIYII TEYCRYGDLV DYLHRNKHTF LQHHSDKRRP PSAELYSNAL
 721 PVGLPLPSHV SLTGESDGGY MDMSKDESVD YVPMLDMKGD VKYADIESSN YMAPYDNYVP
 781 SAPERTCRAT LINESPVLSY MDLVGFSYQV ANGMEFLASK NCVHRDLAAR NVLICEGKLV
 841 KICDFGLARD IMRDSNYISK GSTFLPLKWM APESIFNSLY TTLSDVWSFG ILLWEIFTLG
 901 GTPYPELPMN EQFYNAIKRG YRMAQPAHAS DEIYEIMQKC WEEKFEIRPP FSQLVLLLER
 961 LLGEGYKKKY QQVDEEFLRS DHPAILRSQA RLPGFHGLRS PLDTSSVLYT AVQPNEGDND
1021 YIIPLPDPKP EVADEGPLEG SPSLASSTLN EVNTSSTISC DSPLEPQDEP EPEPQLELQV
1081 EPEPELEQLP DSGCPAPRAE AEDSFL
```

Precursor of human platelet-derived growth factor receptor-α

```
   1 MGTSHPAFLV LGCLLTGLSL ILCQLSLPSI LPNENEKVVQ LNSSFSLRCF GESEVSWQYP
  61 MSEEESSDVE IRNEENNSGL FVTVLEVSSA SAAHTGLYTC YYNHTQTEEN ELEGRHIYIY
 121 VPDPDVAFVP LGMTDYLVIV EDDDSAIIPC RTTDPETPVT LHNSEGVVPA SYDSRQGFNG
 181 TFTVGPYICE ATVKGKKFQT IPFNVYALKA TSELDLEMEA LKTVYKSGET IVVTCAVFNN
 241 EVVDLQWTYP GEVKGKGITM LEEIKVPSIK LVYTLTVPEA TVKDSGDYEC AARQATREVK
 301 EMKKVTISVH EKGFIEIKPT FSQLEAVNLH EVKHFVVEVR AYPPPRISWL KNNLTLIENL
 361 TEITTDVEKI QEIRYRSKLK LIRAKEEDSG HYTIVAQNED AVKSYTFELL TQVPSSILDL
 421 VDDHHGSTGG QTVRCTAEGT PLPDIEWMIC KDIKKCNNET SWTILANNVS NIITEIHSRD
 481 RSTVEGRVTF AKVEETIAVR CLAKNLLGAE NRELKLVAPT LRSELTVAAA VLVLLVIVII
 541 SLIVLVVIWK QKPRYEIRWR VIESISPDGH E I VDPMQL PYDSRWEFPR DGLVLGRVLG
 601 SGAFGKVVEG TAYGLSRSQP VMKVAVKMLK PTARSSEKQA LMSELKIMTH LGPHLNIVNL
 661 LGACTKSGPI YIITEYCFYG DLVNYLHKNR DSFLSHHPEK PKKELDIFGL NPADESTRSY
 721 VILSFENNGD YMDMKQADTT QYVPMLERKE VSKYSDIQRS LYDRPASYKK KSMLDSEVKN
 781 LLSDDNSEGL TLLDLLSFTY QVARGMEFLA SKNCVHRDLA ARNVLLAQGK IVKICDFGLA
 841 RDIMHDSNYV SKGSTFLPVK WMAPESIFDN LYTTLSDVWS YGILLWEIFS LGGTPYPGMM
 901 VDSTFYNKIK SGYRMAKPDH ATSEVYEIMV KCWNSEPEKR PSFYHLSEIV ENLLPGQYKK
 961 SYEKIHLDFL KSDHPAVARM RVDSDNAYIG VTYKNEEDKL KDWEGGLDEQ RLSADSGYII
1021 PLPDIDPVPE EEDLGKRNRH SSQTSEESAI ETGSSSSTFI KREDETIEDI DMMDDIGIDS
1081 SDLVEDSFL
```

FLT3 Precursor Sequence

```
1   MPALARDAGT VPLLVVFSAM IFGTITNQDL PVIKCVLINH KNNDSSVGKS SSYPMVSESP
61  EDLGCALRPQ SSGTVYEAAA VEVDVSASIT LQVLVDAPGN ISCLWVFKHS SLNCQPHFDL
121 QNRGVVSMVI LKMTETQAGE YLLFIQSEAT NYTILFTVSI RNTLLYTLRR PYFRKMENQD
181 ALVCISESVP EPIVEWVLCD SQGESCKEES PAVVKKEEKV LHELFGTDIR CCARNELGRE
241 CTRLFTIDLN QTPQTTLPQL FLKVGEPLWI RCKAVHVNHG FGLTWELENK ALEEGNYFEM
301 STYSTNRTMI RILFAFVSSV ARNDTGYYTC SSSKHPSQSA LVTIVGKGFI NATNSSEDYE
361 IDQYEEFCFS VRFKAYPQIR CTWTFSRKSF PCEQKGLDNG YSISKFCNHK HQPGEYIFHA
421 ENDDAQFTKM FTLNIRRKPQ VLAEASASQA SCFSDGYPLP SWTWKKCSDK SPNCTEEITE
481 GVWNRKANRK VFGQWVSSST LNMSEAIKGF LVKCCAYNSL GTSCETILLN SPGPFPFIQD
541 NISFYATIGV CLLFIVVLTL LICHKYKKQF RYESQLQMVQ VTGSSDNE F VDFREYEYD
601 LKWEFPRENL EFGKVLGSGA FGKVMNATAY GISKTGVSIQ VAVKMLKEKA DSSEREALMS
661 ELKMMTQLGS HENIVNLLGA CTLSGPIYLI FEYCCYGDLL NYLRSKREKF HRTWTEIFKE
721 HNFSFYPTFQ SHPNSSMPGS REVQIHPDSD QISGLHGNSF HSEDEIEYEN QKRLEEEEDL
781 NVLTFEDLLC FAYQVAKGME FLEFKSCVHR DLAARNVLVT HGKVVKICDF GLARDIMSDS
841 N VVRGNARL PVKWMAPESL FEGIYTIKSD VWSYGILLWE IFSLGVNPYP GIPVDANFYK
901 LIQNGFKMDQ PFYATEEIYI IMQSCWAFDS RKRPSFPNLT SFLGCQLADA EEAMYQNVDG
961 RVSECPHTYQ NRRPFSREMD LGLLSPQAQV EDS
```

Juxtamembrane- E= end of β-strand in structure; JMBJMSJMZ

Table 6.1 (*Continued*)

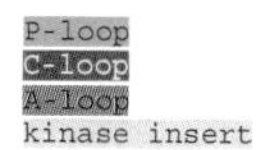

§The 'kinase insert' of PDGFR family members is situated between the P-loop and C-loop of the linear sequence, inserted between α-helices E and F. In β-type PDGFR this is from aas 697-797; in the a-type it is from 670-789 (boxed in cyan). [*According to:-* Bohmer, F.D., Karagyozov, L., Uecker, A., Botzi, A., Mahboobi, S. and Dove, S. (2003) A single amino acid exchange inverts susceptibility of related receptor tyrosine kinases for the ATP site inhibitor STI-571. *J. Biol. Chem.*, **278**: 5148-5155. See 'Figure 1' in this paper, but note there is a typographical error: Flt3 sequence should start at 596, not '598'].

‡The juxtamembrane domains of PDGFR have been aligned with that of the FLT3 structure — boxed in blue (white lettering) with the linker joining to the transmembrane domain boxed in grey. [*According to:-* Griffith, J., Black, J., Faerman, C., Swenson, L., Wynn, M., Lu, F., Lippke, J. and Kumkum, S. (2004) The structural basis for autoinhibition of FLT3 by the juxtamembrane domain. *Molecular Cell*, **13**: 169-178. Note: in this paper there are several errors in numbering stemming from a mistake in Figure 2 — the numbering in the top line skips from 590 to 610 (missing ten residue numbers). I have corrected these].

Also autophosphorylated is the Tyr857 residue in the A-loop. As we shall see later, the A-loop and juxtamembrane tyrosine-phosphorylations collaborate to trigger reversal of the autoinhibition of the catalytic site.

6.2.2 MAP kinases and MAPK kinases

A characteristic of growth factor signalling pathways is the downstream activation of dual-specificity STY-kinases, collectively referred to as 'MAP kinase kinases' (MAPKK). Their targets are the 'MAP kinases', which are cytoplasmic serine/threonine protein kinases that translocate to the nucleus when activated, where they activate transcription factors via serine phosphorylation. MAP kinase pathways were first identified as transducers of mitogenic signals and are activated by the coupling of GTP-bound Ras with its effector enzyme, Raf. Ras-activated Raf then serine phosphorylates MAPK kinase, which in turn doubly-phosphorylates MAPK. However, other monomeric G proteins such as Rac and Rap can also influence MAP kinase pathways. MAP kinase was originally termed microtubule-associated-protein2 kinase for its first known substrate, but is now more aptly named: ***m***itogen-***a***ctivated ***p***rotein ***k***inase, or MAPK. The classic idea of a MAPK 'cascade' of soluble enzymes interacting with each other by diffusion and collision has been superseded by the finding that MAPK pathway enzymes are present as two- or three-membered cassettes that enable fast relay of information from the effector enzyme (a MAPKK-kinase, such as Raf) through MAPKK to MAPK.

There are five families of MAPK enzymes, all of which require dual A-loop phosphorylation of a Thr.Xxx.Tyr motif for activity[7]:

- ***E***xtracellular signal-***r***egulated ***k***inase (members: Erk1 [p44] and Erk2 [p42])

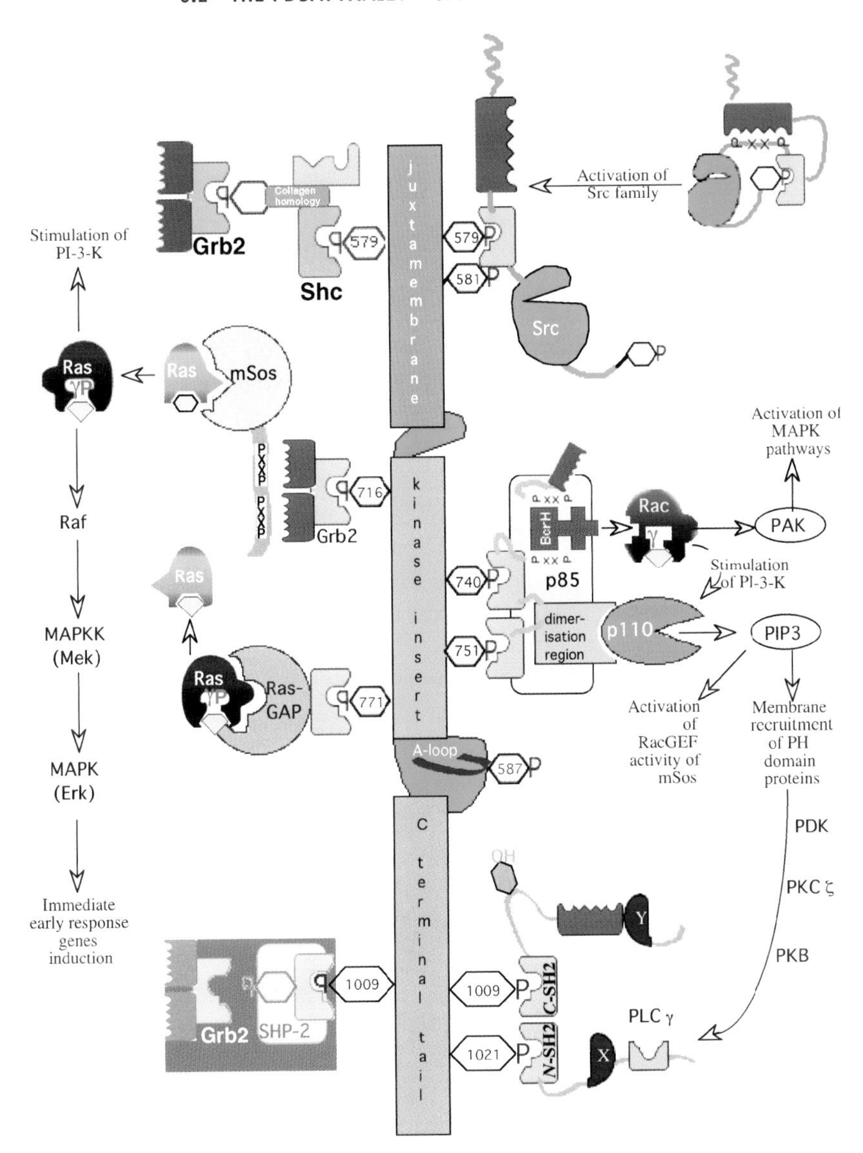

Figure 6.3 PDGFR-β signal transduction particle

- ***S****tress-****a****ctivated* ***p****rotein* ***k****inase (SAPK), also known as c-****J****un* ***N****-terminal* ***k****inase (members: Jnk1, Jnk2, Jnk3)

- p38 kinase homologous (p38α, p38β, p38γ, p38δ)

- Erk3/4

- Erk5

Associated with Erk1/2-, SAPK- and p38-pathways are distinct and specific upstream MAPKKs that are generally specific to each. In addition, the members of each MAPK pathway are oligomerised in the form of a linear cassette by distinct scaffold proteins.

6.2.3 PDGFR kinase insert tyrosines – PI-3-kinase, and Ras *versus* Rac

The insert tyrosines of PDGFR-α (731 and 742; equivalent to PDGFR-β Tyr740, Tyr751) can be mutated to phenylalanine with no effect on ligand induced autophosphorylation or the mutants' abilities to phosphorylate exogenous substrate, but their point mutation does impair the receptor's ability to recruit and stimulate (Class 1A) PI-3-kinase[8]. Without these phosphotyrosines, the p85 regulatory subunit of PI-3-kinase is not recruited from cytosol to the membrane and PIP3 is not produced. This tandem pTyr-binding site for PI-3-kinase p85 subunit ($p85^{PI3K}$) is particularly specific and has high-affinity because it docks with a tandem SH2 pair on this PI-3-kinase adaptor/regulator subunit[6]. The minimal consensus sequence for these SH2 domains is: pTyr.Xxx.Xxx. Met, a sequence found at the tandem insert target tyrosines in both PDGFR-α and PDGFR-β, but not in Flt-3 (see Table 6.1).

The $p85^{PI3K}$ subunit is an adaptor protein consisting of an *N*-terminal SH3 domain, followed by a BcrH domain that is bracketed between a pair of proline-rich regions (see Figure 6.4). The *C*-terminal half of the molecule consists of a pair

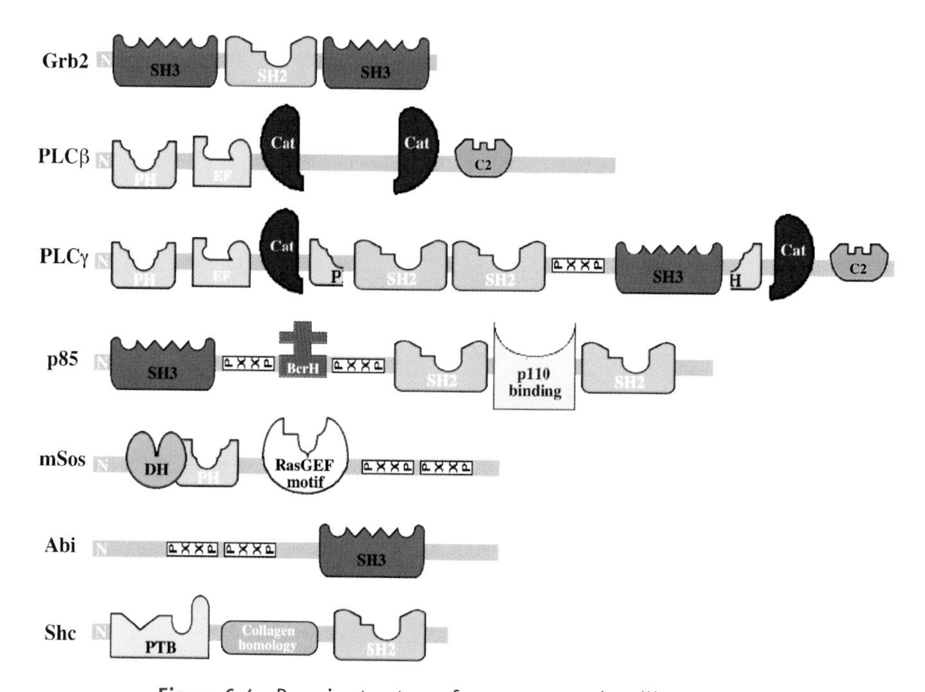

Figure 6.4 Domain structure of non-receptor signalling proteins

of SH2 domains, on either side of a unique domain that is responsible for dimerisation with the p110 catalytic subunit ($p110^{PI3K}$)[9]. Recruitment of PI-3-kinase from the cytosol to the plasma membrane (by SH2 docking) is sufficient for activation. In other words, the p110 catalytic subunit is actually constitutively active and only needs to be brought close to its substrate PIP2 (membrane-bound to the inner leaflet). The consequent production of PIP3 causes the translocation and activation of a number of PH-domain serine/threonine protein kinases, including ***P***IP3-***d***ependent ***k***inase (PDK) and atypical protein kinase C-ζ (PKCζ). This is discussed further in Chapter 9 (Section 9.9). The appearance of PIP3 also causes PLCγ to translocate from cytosol to membrane, as discussed below.

6.2.4 PDGFR, PI-3-kinase, Ras and mitosis

Aside from its well-established PIP3 signal, PI-3-kinase is also inextricably linked with regulation of monomeric G protein signalling. For example, GTP-occupied Ras binds to the p110 PI-3-kinase catalytic subunit and stimulates its phosphoinositide 3-kinase activity, downstream[10]. There appears to be a reciprocal arrangement that also places PI-3-kinase upstream of Ras. At low levels of growth factor stimulation, for example, the activation of the Ras/MAPK pathway is conditional upon basal levels of PI-3-kinase activity[11]. At least in some circumstances, PI-3-kinase can help activate the growth-promoting Ras/MAPK pathway.

Peaks of both PI-3-kinase activity and Ras<GTP> accumulation occur in two distinct phases after mitogenic stimulation of quiescent cells – both exhibit an immediate transient peak followed by another peak later in the G1 phase of the cell cycle.

The early rise in Ras levels (see below) disappears a few minutes after growth factor stimulation and remains low only to peak again after 4 hours[12]. The very early activation of the Ras/MAPK pathway is transient and is followed by a delayed accumulation of Ras during mid-G1, beginning at 2 hours and persisting until 6 hours. The later activation phase results in up to half the cell's compliment of Ras accumulating as Ras<GTP>. The mid-G1 Ras activation phase appears to be independent of Shc-Grb2 complex formation and is uncoupled from MAPK activation.

PI-3-kinase activation follows a similar pattern. In the continued presence of PDGF, PIP3 levels peak after only 5 minutes and then decline after 30 minutes; a longer-lasting accumulation of PIP3 appears between 3 and 7 hours[13].

6.2.5 PDGFR, PI-3-kinase, Rac and motility

The monomeric G protein, Rac, is absolutely essential for cell migration; inhibition of Rac prevents the movement of macrophages, epithelial cells and fibroblasts[14]. Activated Rac stimulates actin polymerisation in focal complexes at membrane protrusions (lamellipodia) that form at the leading edge of migrating cells; lamellipodia that fail to adhere to substratum are swept backwards in the form of membrane 'ruffles'. Rac is also a downstream target of PI-3-kinase recruitment to the membrane.

It is known that the BcrH domain of $p85^{PI3K}$ binds Rac<GTP>. The BcrH domain is a disabled Rho/Rac GTP'ase activating module (RhoGAP-like) – although it binds directly to GTP-activated Rac, it has no down-regulatory activity and cannot stimulate the Rac GTP'ase. It appears essential for the full range of $p85^{PI3K}$ interactions and, significantly, the short isoforms of $p85^{PI3K}$ (p50 and p55) that are missing this domain display different signalling behaviour[15]. Perhaps the BcrH domain serves to concentrate Rac<GTP> at the RTK signal particle, or to prolong Rac activity by antagonising binding of functional RhoGAPs. Certainly, the *N*-terminal half of $p85^{PI3K}$ (containing the BcrH domain and polyproline regions) is essential for cytoskeletal signalling, although not for mitogenesis[9]. It has been shown, for example, that the *N*-terminal section of $p85^{PI3K}$ is essential for its relocation to specialised areas of the plasma membrane where Rac activation is concentrated during cell migration. GFP-tagged $p85^{PI3K}$ co-migrates with activated EGFR/erbB3 dimers, moving from the base of microvilli to membrane ruffles then to surface patches[16].

The p85 PI-3-kinase subunit has an alternative, regulatory, way of interacting with Rac via activation of RacGEF activity in Sos, as discussed below.

Interestingly, the interaction of Rac with p85/p110 dimers appears to increase their phosphoinositide kinase activity, just as Ras does[15].

6.2.6 PDGFR insert phosphotyrosines and Ras regulators

A second kinase insert position, pTyr771 of PDGFR-β, acts as a docking site for the SH2 domain of RasGAP, the downregulator of Ras signalling[17].

A third insert site, pTyr716, acts as direct docking site for Grb2. Grb2 is an adaptor that consists of a single SH2 domain sandwiched between two SH3 domains (Figure 6.4). It forms a cytoplasmic constitutive complex with the guanine nucleotide exchange factor, mSos, by binding to the proline-rich *C*-terminal tail of mSos via SH3 interaction. Recruitment of the Grb2/mSos complex to the pTyr-docking site allows mSos to activate Ras, and thereby activate the MAPK cascade[3]. It must be emphasised that, although Grb2 is the primary adaptor involved in mSos activation of the Ras/MAPK pathway, MAPK activation can be brought about, or at least strengthened, by other PDGFR pathways, as discussed above[18].

Ras<GTP> selectively activates the Erk1/2-type MAPK cascade by recruitment of Raf to the plasma membrane where Raf is further activated by phosphorylation. In contrast, Rac<GTP> primarily activates the stress-activated protein kinase-(SAPK)-type MAPK cascade, with a similar but lesser effect on the p38-type MAPK pathway[19]. However, it has been found that Rac also activates the Erk-type MAPK pathway in a PI-3-kinase-dependent manner. Either the physical interaction of $p85^{PI3K}$ with Rac<GTP>, and/or the increased production of PIP3 this induces, stimulates RacGEF activity, upregulating high enough levels of Rac<GTP> to activate p21-activated protein kinase (PAK, a serine/threonine kinase). PAK has been shown to collaborate with Raf in activating the Erk1/2 pathway. PAK3 serine phosphorylates Raf-1 in an activating step; PAK1 phosphorylates Mek1 (but not Mek2), creating an important docking site for Erk[20]. Interestingly, this activating effect of Rac-PAK is stimulated by cell adherence (cells in suspension show basal activity) and is suggested to be caused due to integrin activation[21]. So, although Ras is primarily thought of as mitogenic and Rac as cytokinetic, both pathways potentially overlap.

Despite the potential overlap revealed in the above studies (which were mostly done using constitutively active or dominant-negative mutants of Ras, Rac, and downstream elements), complete pathways exhibit more conditional controls. For example, both PDGFR-α and -β receptors activate PI-3-kinase and both might therefore be expected to activate the same Rac-PAK pathways discussed above. Instead, it was found that only PDGFR-α was able to activate the SAPK/Jnk1-type MAPK pathway, although both PDGFR-α and PDGFR-β could activate the Erk2-type MAPK pathway[22]. This difference in PDGFR signalling results in PDGFR-α antagonising PDGFR-β signalling, in that the former receptor's unique activation of Jnk-1 suppresses the 'transforming' activity of Erk1/2 by the latter. In cells expressing only the PDGFR-β receptor, Erk1/2 activation is unopposed – PDGFR-β–β 'transforms' fibroblasts by allowing them to grow in an anchorage-independent manner, like cancer cells.

In fibroblasts, the activation of Ras by growth factors precedes that of Rac. An immediate short-lived activation of Ras (3 minutes) is followed by a more sustained activation of Rac (3–15 minutes)[23].

6.2.7 Sos-1 – a bi-functional *g*uanine nucleotide *e*xchange *f*actor (GEF)

Son-of-sevenless (Sos) was first discovered as a mutated protein present in the R7 photoreceptor pathway of fruit flies. Mammalian mSos was first characterised as a RasGEF (Chapter 7, Section 7.7.2) but recently its alternative RacGEF activity has been recognised as being important[23]. mSos's RasGEF activity is mediated via its Cdc25-like domain, which represents its RasGEF motif, whereas its RacGEF activity is due to its *N*-terminal tandem Dbl-homology/plextrin-homology region. mSos *C*-terminal tail is proline-rich and is a target for SH3 domain proteins such as Grb2 (Figure 6.4). mSos can act as either a RasGEF or RacGEF and, in each guise, mSos is complexed with a different, distinct adaptor.

In vivo, mSos is found in the cytosol of unstimulated cells; in activated cells, a proportion is associated transiently with the activated RTK signal transduction particle. The cellular compliment of resting Sos is divided between a large pool of Grb2-Sos complex and a 10-fold smaller pool of a trimeric complex containing Sos complexed with Eps8 (an RTK substrate) and Abi-1 (Abelson-interactor-1; or 'E3b1'). The former dimer acts to mediate RTK activation of Ras, and the latter trimeric complex mediates Rac activation.

6.2.8 Sos – the switch from RasGEF to RacGEF

Abi-1 is an SH3-containing adaptor (Figure 6.4) that competes with Grb2-SH3 for binding to the proline-rich *C*-terminal tail of Sos (VPVPPPVPPRRR). Abi-1 is strictly required for RTK stimulation of Rac and PAK. A unique proline-rich stretch in Abi-1 (PPPPPV**DY**TEDEE, critical residues are underlined) provides a binding site for the SH3 domain of Eps8. Thus, Abi-1 acts to scaffold these two proteins. Additionally, Abi-1 targets its complexes to lamellipodia and filopodia, sites where Rac is involved in actin polymerisation regulation[24]. Abi-1 contains a central tyrosine that is constitutively phosphorylated and this can bind to the more *N*-terminal of the two SH2 domains of p85 (the regulatory subunit of PI3k)[25].

The switch in Grb2-Sos to Abi-1-Eps8-Sos is driven by disruption of the Grb2-containing complex after the initial phase of Ras signalling is over. Hyperphosphorylation of Sos is suggested to be responsible[23]. However, formation of the trimeric complex is not enough, it needs the participation of activated PI-3-kinase before it can interact with Rac. Receptor-docked PI-3-kinase produces a local high concentration of PIP3 that serves to activate RacGEF activity of the trimeric complex (a tandem PIP3-dependent DH/PH corresponding to a Rho/RacGEF domain is found in Sos).

The type of phosphoinositide bound to the DH/PH domain of mSos controls whether RasGEF or RacGEF activity is displayed (Figure 6.5). The PH module can bind either PIP2 (the substrate of both PLCγ and PI-3-kinase) or PIP3 (the product of PI-3-kinase). mSos binding to PIP2 induces tight association between the DH and PH modules; the domain is thus disabled and cannot bind to Rac. In this conformation, mSos acts as a RasGEF when it is recruited by receptor-bound Grb2. When PI-3-kinase is activated, mSos binds PIP3; this loosens the inhibitory interaction between DH and PH modules, and the protein can now bind Rac and manifest RacGEF activity[26].

Production of PIP3 and the physical association of PI-3-kinase with the trimeric complex are essential to forming functional RacGEF activity. The *N*-terminal of PI-3-kinase's two SH2 (*N*-SH2) domains binds to the activated mSos-Abi-1-Eps8 complex via the pTyr of Abi-1. Interestingly, the *C*-terminal of $p85^{PI3K}$'s pair of SH2 domains (*C*-SH2) has a high affinity for PIP3, which competes for binding with the receptor's pTyr motif[27], leaving open the possibility that PI-3-kinase, bound to the Sos-containing RacGEF complex via its *N*-SH2, may remain simultaneously receptor-bound or even undock from the activated receptor if high levels of PIP3 displace binding of *C*-SH2 to PDGFR pTyr. The latter is a distinct possibility because, although PI-3-kinase is associated with the Sos-centred RacGEF complex, mSos is not immunoprecipitated with PDGFR during the time when RacGEF activity is expressed (this is after the initial phase of Ras activation is over)[23,25].

Cytoplasmic PI-3-kinase is recruited to the plasma membrane by binding to phosphotyrosines on the erbB3 component of activated EGFR/erbB3 receptor dimers, an interaction that depends crucially upon the presence of both SH2 domains of $p85^{PI3K}$ and, to a lesser extent, on retention of its *N*-terminal region (containing the SH3, Bcr and proline-rich domains)[16]. The *N*-terminal SH2 of $p85^{PI3K}$ also binds PIP3 and this would

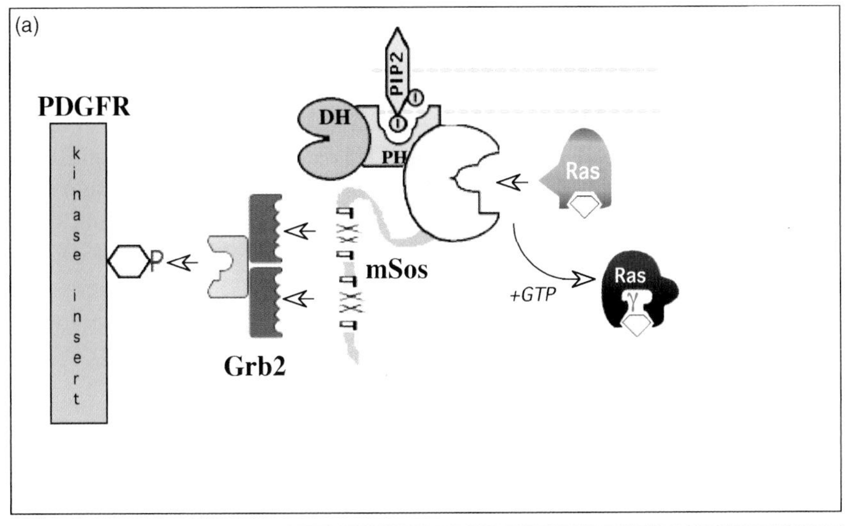

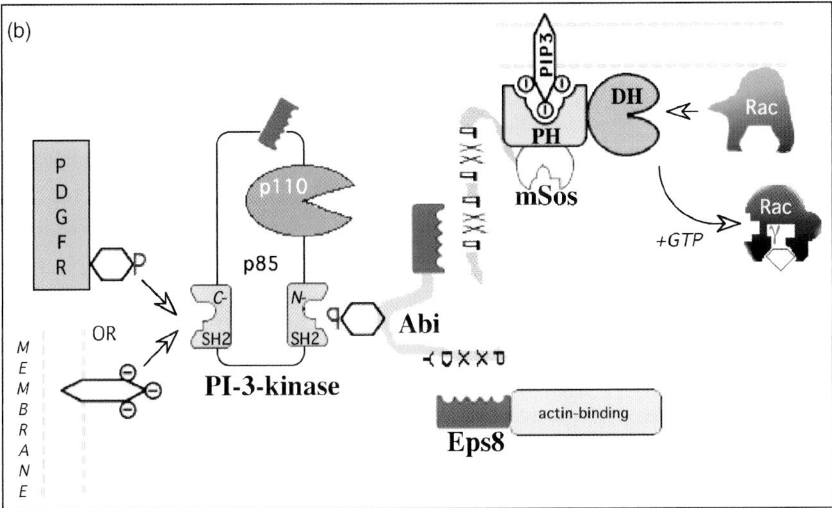

Figure 6.5 mSos can act as RasGEF or RacGEF

suggest that PI-3-kinase sets up its own negative feedback loop by binding stepwise to its own product, eventually dissociating itself from activated receptors when both SH2 domains swap from pTyr- to PIP3-binding. Indeed, this was how the mechanism was discovered – through the observation that the amount of $p85^{PI3K}$ associating with pTyr residues on the insulin receptor substrate was doubled when wortmannin (a PI-3-kinase active site inhibitor) was added along with the activating insulin[27]. The *C*-terminal SH2 of PLC-γ also displays dual pTyr-*versus*-PIP3-binding, and undocking by PIP3 competition has been suggested as a way to redistribute activated PLC-γ too[27,28].

The PIP3 signal from PI-3-kinase also activates a number of other RacGEFs via DH/PH domain binding, including Vav and α-PIX[29].

6.2.9 PDGFR *C*-terminal tail tyrosines

The *C*-terminal tail pTyr1009 and pTyr1021 of PDGFR-β act as a dual binding site for PLCγ[6]. Like $p85^{PI3K}$, PLCγ contains a pair of SH2 domains so the binding is high affinity. Moreover, like $p85^{PI3K}$, the *C*-SH2 of PLCγ can also bind PIP3 (as an alternative ligand to pTyr) while its *N*-SH2 plays the predominant role in pTyr-docking with PDGFR[30]. Disabling the *C*-SH2 of PLCγ1 only slightly reduces the mutant's binding to phosphorylated PDGFR (with similarly slight lowering of phospholipase activation), but disabling the *N*-SH2 completely prevents binding to PDGFR and blocks activation of the enzyme. The phosphotyrosine that predominates in mediating the docking of PLCγ is PDGFR-β pTyr1021.

6.2.10 Alternative Grb-2 docking sites: SHP-2 and Shc

There is evidence that other SH2 domain proteins compete for the minor PLCγ-binding site (i.e., pTyr1009). Here we see another difference in the two receptors. In PDGFR-β, pTyr1009 acts as a docking site for the SH2-containing protein tyrosine phosphatase SHP-2 (also known as Syp, SH-PTP-2 or PTP-1D). SHP-2 becomes tyrosine phosphorylated upon interaction with the receptor and thus provides a docking site for Grb2-SH2, and there is evidence that this results in mSos recruitment and Ras activation, at least in some cell types[2,3]. In PDGFR-α on the other hand, SHP-2 instead binds to pTyr720 in the insert segment with the same consequence of phosphorylation and Grb2 recruitment but in this case, there is no activation of Ras[31].

6.2.11 Shc

Shc is a PTB-SH2-domain adaptor (with no enzymic activity) that becomes tyrosine phosphorylated on interaction with RTKs and can act as a bridging ligand between the receptor and Grb2/mSos complex. The *N*-terminal PTB domain of Shc is followed by a collagen-like GlyPro-rich linker that connects to the *C*-terminal SH2 domain (Figure 6.4). It is well established that Shc can provide an alternative way for Grb2/mSos dimers to interact with RTKs, and this is discussed below in relation to PTB-binding to the EGF receptor, and in Chapter 9 in relation to it bridging between Grb2 and the insulin receptor substrate. Shc can bind to phosphotyrosines in two ways: either via its PTB domain, which has a specificity for Asn.Pro.Xxx.pTyr[32], or via its SH2 domain, which has specificity for pTyr.Glu/Φ.Xxx.Met/Φ. A phosphorylatable tyrosine in the central collagen homology region provides a docking site for Grb2-SH2 and this link with the RTK allows recruitment of mSos that then stimulates GTP-for GDP exchange at Ras.

Whereas Shc associates with the activated PDGFR, becomes tyrosine phosphorylated and allows Grb2 to dock, the exact docking site is not well-characterised. In porcine aortic endothelial cells Shc associates with PDGFR-β via its SH2 domain, becomes tyrosine-phosphorylated and induces Grb2 recruitment to the receptor. There is no evidence that it interacts via its PTB domain[33]. Studies with synthetic pTyr peptides from the PDGFR-β intracellular domain indicated binding to pTyr-579, -740, -751, and - 771; the non-phosphorylated peptides or the remaining phosphotyrosine motifs pTyr-581, -857, -1009, and -1021, showed little or no binding to Shc SH2[5]. Tyrosine579 seemed to be a preferred site. In contrast, in vascular smooth muscle cells, Shc associated with PDGFR even when the receptor was unactivated; Grb2, however, was only recruited to the receptor (via Shc) upon PDGF stimulation[34]. This may suggest that Shc SH2 can associate with unphosphorylated sequences.

6.2.12 PLCγ

PLCγ is homologous with PLCβ but contains extra inserted modules that allow it to interact with RTKs and other signalling molecules that are outside the scope of PLCβ's signalling routes (see Figure 6.4). PLCγ is cytosolic in resting cells, but is recruited to the membrane within seconds of activation of wild-type PDGFR – IP3 begins to be produced within 15 seconds, peaking at 30 seconds; calcium release peaks later at around 75 seconds[27]. Eliminating the primary PLCγ binding site on PDGFR (Phe-for-Tyr1021 substitution) reduces and delays the calcium release but does not block it completely. Significantly, eliminating the tandem PI-3-kinase site (Tyr740 and Tyr751) on PDGFR also reduces and delays the calcium response; substituting both *C*-terminal and kinase insert sites completely blocks PLCγ activation by growth factor. So, the fast IP3/Ca^{2+}-response is a result of synergising recruitment signals: the appearance of the correct pTyr on the receptor and the generation of PIP3 by PDGFR-recruited PI-3-kinase.

Interaction with the receptor allows the kinase to tyrosine-phosphorylate PLCγ; this covalent modification step is necessary for full activity because it is theorised to drive a conformational re-arrangement of the enzyme such that the 'split catalytic site' is re-united[30]. Docking with PDGFR brings PLCγ to the membrane via SH2 interaction; a translocation that is strengthened by the activation of PI-3-kinase docked to the same receptor. The local production of PIP3 provides an alternative membrane-anchoring point via PLCγ's PIP3-specific PH domain. The activated receptor phosphorylates PLCγ on its regulatory tyrosine (Tyr783 of PLCγ1) and this is thought to relieve some conformational restraint that allows (or is assisted by) the *C*-SH2 domain to re-direct its binding from PDGFR-pTyr1009 to PIP3. The active site is thus assembled and positioned at the source of PIP2 substrate, the plasma membrane inner leaflet (Figure 6.6).

6.2.13 PDGFR Juxtamembrane tyrosines

The juxtamembrane pTyr residue pair of PDGFR (Tyr579 and Tyr581) acts as a docking site for Src family members, resulting in Src activation via displacement of the inhibitory intrachain SH2-pTyr bond of inactive Src and its replacement with

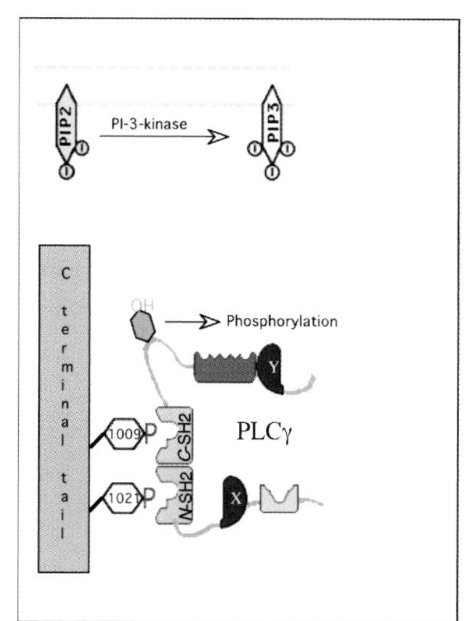

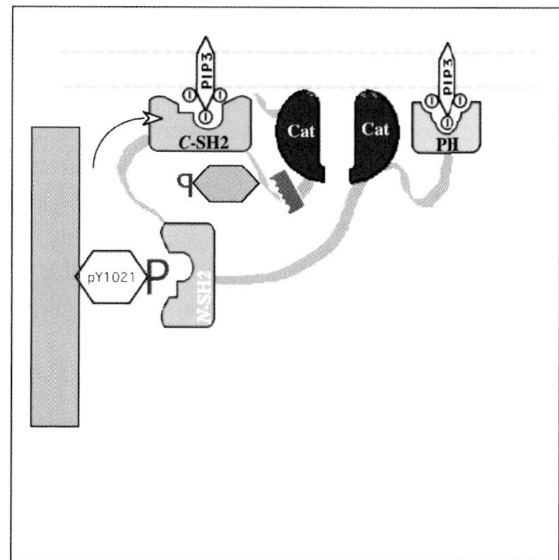

Figure 6.6 Possible PLCγ activation steps

SrcSH2-PDGFR-pTyr579/581 binding. (See Chapter 3, Section 3.1.5); also discussed below in relation to the effect on kinase activation.) Note that Src does not appear to be able to phosphorylate and activate PDGFR as it can the insulin receptor[35]. pTyr 579 is also capable of binding Shc, and PDGFR-β phosphorylates Shc, thereby providing an alternative binding site for Grb2. However, this is unlikely to activate MAPK and mitogenesis, as point mutation of Tyr579 has no effect on PDGF-stimulated DNA synthesis[4]. pTyr716 appears to be the primary site for MAPK activation, because its elimination reduces the number of immediate early response genes induced in response to ligand[18].

6.3 PDGFR family autoinhibition: juxtamembrane and A-loop tyrosines

In the PDGFR/VEGFR family (the PTK-XV class of kinases) a single tyrosine in the A-loop needs to be autophosphorylated for the full range of PDFGR-like signalling and a pair of juxtamembrane tyrosine residues are also involved in kinase activation. The way these two regulatory elements interact has been delineated in solution studies of PDGFR/VEGFR point and deletion mutants.

6.3.1 PDGFR juxtamembrane and A-loop tyrosines → phenylalanines

Tyrosine to phenylalanine is a fairly conservative substitution – phenylalanine is aromatic and of a similar size, and only lacks the H-bonding potential of the phenolic hydroxyl. Most relevant to the present discussion is the fact that it cannot undergo phosphorylation.

Point mutation of the juxtamembrane Tyr579 to a Phe (Y579F) produced a PDGFR with a severely reduced ability to bind Src (via Src's SH2 domain), although it still responded to PDGF stimulation with autophosphorylation; a Phe substitution of the juxtamembrane Tyr581 (Y581F) produced similar results: retention of ligand response but impaired Src binding[4]. Binding studies with synthetic peptides encompassing this region (+/– tyrosine phosphorylation) confirmed that, despite only having one SH2 domain, Src bound cooperatively to the doubly-phosphorylated juxtamembrane peptide (pTyr579+pTyr581) in preference to singly-phosphorylated peptides (either Tyr579 *or* pTyr581), possibly because of the Src family preference for negatively charged acidic residues *C*-terminal from the pTyr. The PDGFR pTyr579 is thought to be the prime site (P^0) because a related receptor (the CSF-1 receptor) retains this tyrosine but has a phenylalanine at the 581 position.

6.3.2 PDGFR juxtamembrane (Y-Y → A-A) mutant unresponsive to ligand

Unexpectedly, the double mutant PDGFR protein lacking both phosphorylatable tyrosines (Y579/581**F**) was activation-defective, inasmuch as stimulation with PDGF did not induce any autophosphorylation. This finding was explored further, comparing the double juxtamembrane mutant with the single A-loop mutant.

6.3.3 PDGFR Y579/581F is stuck in an autoinhibited state

The basal kinase activity of the Y579/581**F** double mutant is actually comparable with unstimulated wild-type PDGFR so it is not 'kinase dead'. The juxtamembrane double mutant Y579/581**F** can be artificially stimulated with ATP-loading to the same extent as wild type but PDGF cannot upregulate its autophosphorylation activity[35]. Consequently, the Y579/581**F** protein is unable to bind any of its SH2-containing docking partners: binding of Src, RasGAP, PLCγ and the p85 subunit of PI3K is undetectable.

6.3.4 PDGFR A-loop (Y → F) mutant cannot bind exogenous substrate polypeptides

In contrast to the tandem Phe juxtamembrane mutation, the single A-loop mutant (Phe substituted for Tyr857, 'Y857**F**') is fully competent to autophosphorylate when stimulated with PDGF and can dock with SH2-containing proteins when so activated. However, this mutant cannot be activated by ATP-loading and is unable to phosphorylate other protein substrates, including an ideal consensus peptide. This suggests that substitution of the A-loop tyrosine does not produce a 'kinase dead' receptor, but rather that its active site is not optimally assembled for substrate recognition.

Finally, it is noteworthy that both Phe-substituted juxtamembrane and A-loop PDGFR-β mutants dimerise normally in response to PDGF.

6.3.5 PDGFR juxtamembrane and A-loop tyrosines → alanines

Substitution of tyrosine with an alanine residue is a much more drastic mutation and makes a significant change at that position – alanine is much smaller, unphosphorylatable and non-aromatic. Lack of the benzene ring means alanine cannot π-bond with other aromatics.

6.3.6 PDGFR Y579/581A is constitutively active

Interestingly, juxtamembrane alanine substitutions tend to produce constitutively active forms of PDGFR-β (Table 6.2). Tandem substitution of the central juxtamembrane tyrosine pair with alanines (human PDGFR-β: Y579/581**A**) produced a constitutively activated receptor[36] – *note this study was on mouse PDGFR and so numbering in the paper differs from the human numbering I have substituted for clarity.* Indeed, scattered single alanine substitutions throughout the juxtamembrane region (corresponding to human PDGFR-β positions: 562, 566, 569, 587, 589) had similar effects. Notably, these sensitive residues have either bulky hydrophobic or aromatic sidechains.

A similar study on the closely related Kit-type PDGFR family member confirmed the importance of the juxtamembrane in autoinhibition and further showed that mutating its C-loop catalytic aspartate residue (equivalent to Asp826 in human PDGFR-β) to an alanine did not produce a consititutively active kinase (as juxtamembrane deletions did), but one that was unable to phosphorylate substrate[37]. The same study conclusively showed the importance of the juxtamembrane domain in autoinhibition with synthetic peptides.

Deletion of the juxtamembrane region from Kit yields a constitutively active kinase that can be inhibited by adding a synthetic peptide encompassing the missing stretch of

Table 6.2 Alanine-scanning of murine PDGFR-β juxtamembrane region

541 LVVLTIISLI ILIMLWQKKP R EIR KV E SVSSDGHE I VDPMQ P D STWELPRDQL

541 LVVLTIISLI ILIMLWQKKP RAEIRAKVAE SVSSDGHEAI AVDPMQAPAD STWELPRDQL

541 LVVLTIISLI ILIMLWQKKP RYEIRWKVIE SVSSDGHE I VDPMQLPYD STWELPRDQL

JMB JMS JMZ

Above shows the positions of activating alanine substitutions that caused constitutive activation in mouse PDGFR-β. Positions are mapped onto the human sequence for comparison. Upper panel, native; middle, alanine substitutions in yellow; lower, juxtamembrane sections indicated (as revealed in Flt-3). Note, the numbering of the murine receptor differs, but the juxtamembrane sequence is identical to human save for a valine-for-methionine substitution at human position 585 (italics).

Irusta, P.M., Luo, Y., Bakht, O., Lai, C-C., Smith, S.O. and DiMaio, D. (2002) Definition of an inhibitory juxtamembrane WW-like domain in the platelet-derived growth factor β receptor. *J. Biol. Chem.*, **277**: 38627-38634.

sequence. However, the doubly-phosphorylated juxtamembrane peptide (with tandem pTyr's equivalent to PDGFR-β Y579/581) has no inhibitory effect. The wild-type juxtamembrane peptide was found to have an ordered structure that was disrupted by the dual tyrosine-phosphorylation. Finally, the constitutive activity of juxtamembrane-deletion mutant Kit was inhibited *in vivo* by co-expression of wild-type juxtamembrane domain[37].

6.4 Crystal structure of kinase domain of PDGFR family-A member: Flt-3

The unique kinase insert has proved a great hindrance to crystallisation and PDGFR family members have only been crystallised with this region deleted. A recent crystal structure of the PDGFR-related Flt-3 receptor (missing the kinase insert) complements studies on the biochemistry of PDGFR activation and largely confirms predictions from the solution studies[38].

Flt-3 ('FMS-like kinase 3') is a close relative of PDGFR, both being in the PTK-XV (A) family of protein kinases (see Appendix 1). The receptor is also known as FLK-2 ('foetal liver kinase 2') or STK-1 (stem cell kinase 1) and its ligand FL ('Flt-3 ligand') acts as a growth and differentiation factor for haemopoetic stem cells[39]. Flt-3 mutations are associated with acute myeloid leukaemias and the oncoproteins are seen as a promising therapeutic target[40]. These Flt-3 kinase-activating mutations point to regulatory elements, particularly the juxtamembrane region, which are probably utilised by the other PTK-XV(A) family members (like PDGFR).

A recent Flt-3 crystal structure (PDB file: 1RJB) is unusual in containing the full-length juxtamembrane domain (these are often missing or fragmentary in RTK crystals) and its interaction with the kinase core suggests how PDGFR family members are autoinhibited[38]. The structure is of a cytoplasmic domain-only protein (aas: His564–Ser993) engineered to remove the kinase insert segment (His711–His761), which would otherwise interfere with crystallisation (see cyan box Table 6.1). Also missing from the coordinate file are two disordered segments (see grey boxes in Table 6.1). A RasMol-rendered view of Flt-3 is shown in Figure 6.7, using the usual catalytic site colouring and with the juxtamembrane coloured in shades of blue; compare this with PKA and the active and inactive insulin receptor kinase (IRK) structures in Figure 4.10. It is immediately apparent that the A-loop of autoinhibited Flt-3 is in a conformation similar to that of inactive unphosphorylated IRK, while the C-helix is different from either active or inactive IRK positions. The key feature, however, is that the juxtamembrane domain has invaded the active site and thus blocks access to substrates as well as disabling catalysis.

6.4.1 Flt-3 juxtamembrane interactions and autoinhibition/activation

The juxtamembrane domain is in three parts: the 'binding' section (JMB), the 'switch' section (JMS) and the 'zipper' section (JMZ)[38](see Figure 6.8).

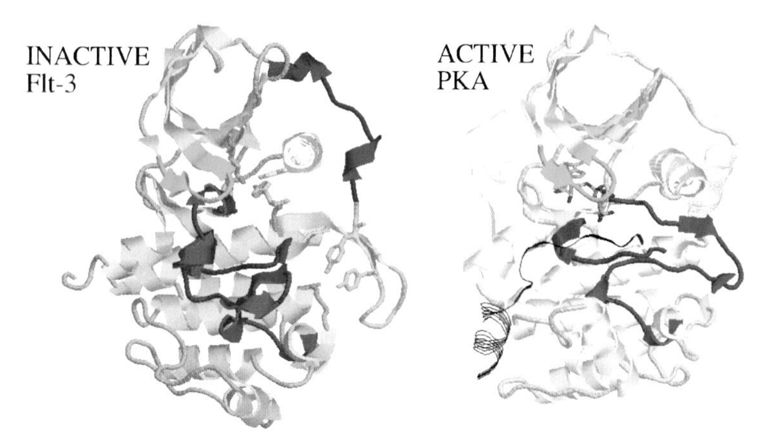

Figure 6.7 Showing autoinhibited Flt-3 compared with the active PKA

The most *N*-terminal JMB residue, Tyr572 is buried in a hydrophobic pocket in the N-lobe and its phenolic group hydrogen bonds with the invariant Glu661 – this is the conserved C-helix glutamate that salt-bridges with the subdomain II lysine, which coordinates the α/β phosphates of ATP (see Table 4.1, Chapter 4). The anchoring

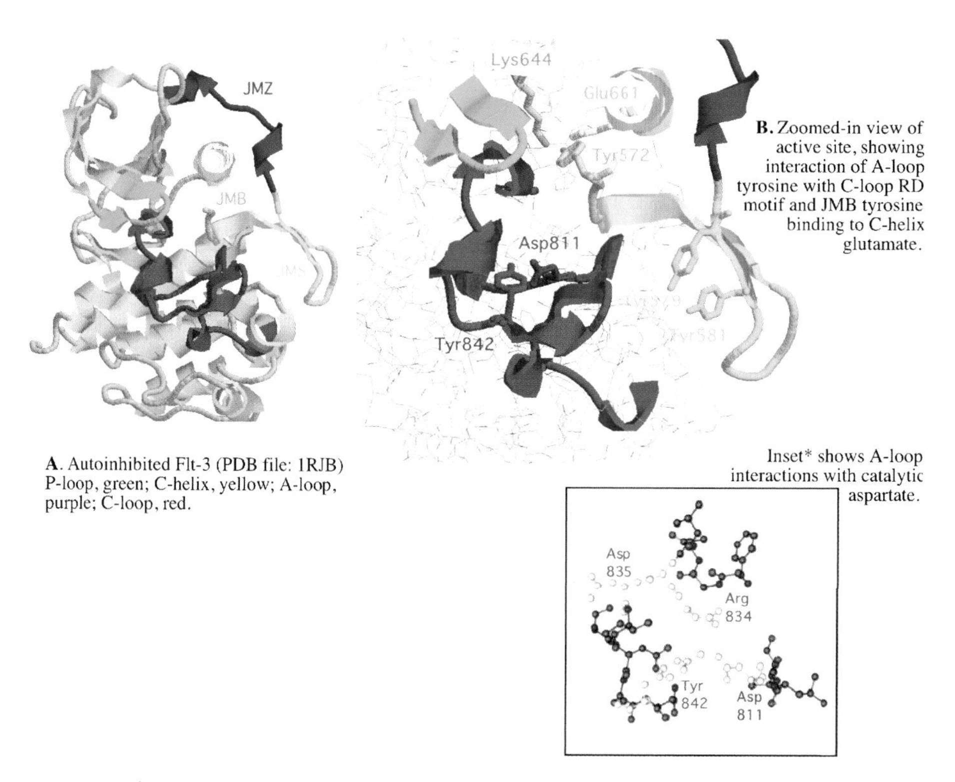

Figure 6.8 Auto-inhibited Flt-3 – juxtamembrane and A-loop interactions

JMB section is followed by JMS, which contains a loop with the two phosphorylatable tyrosines Tyr589 and Tyr591 (equivalent to PDGFR-β Tyr579 and Tyr581). The tyrosine pair project towards the C-lobe and this 'switch' hangs like a latch over the lip of the catalytic cleft. The JMZ section provides a second anchor point for the switch. In particular, a conserved tyrosine (Tyr599) is buried in a hydrophobic pocket of the N-lobe and H-bonds with a conserved glutamate (Glu604).

The single A-loop tyrosine of Flt3 (Tyr842) is hydrogen bonded to the catalytic aspartate of the C-loop 'RD' motif (Asp811, equivalent to PKA Asp166 or PDGFR-β Asp826) in exactly the same manner as in the disabled active site of autoinhibited insulin receptor kinase[41] (see Figure 6.8 and Chapter 4, Section 4.3). Furthermore, the catalytic aspartate is ion-paired with Arg834 of the A-loop (equivalent to PKA-Lys189; Cdk-2-Arg150; IRK-Arg1155 – the residue forming the 'charged cluster' when A-loop is phosphorylated); significantly, the latter's residue neighbour is Asp835, a major oncogenic point mutation in human acute myeloid leukaemia. This point mutation of Asp835 (predominantly tyrosine substitution) leads to an oncoprotein that is constitutively activated, being autophosphorylated and mitogenic without ligand stimulation[42].

Superimposing the structures of inactive Flt3 and the activated insulin receptor kinase reveals that the central section of the juxtamembrane region occupies exactly the same space that a section of the activated A-loop of IRK would occupy (see Figure 6.9). As well as hindering the activating movement of the A-loop, the juxtamembrane interaction has clearly pulled the C-helix out of position. A final similarity with the insulin receptor kinase is that the phenylalanine of the A-loop DFG motif of Flt3 is blocking the adenosine-binding site in exactly the same way that the DFG-phenylalanine of the inactive insulin receptor interferes with ATP access.

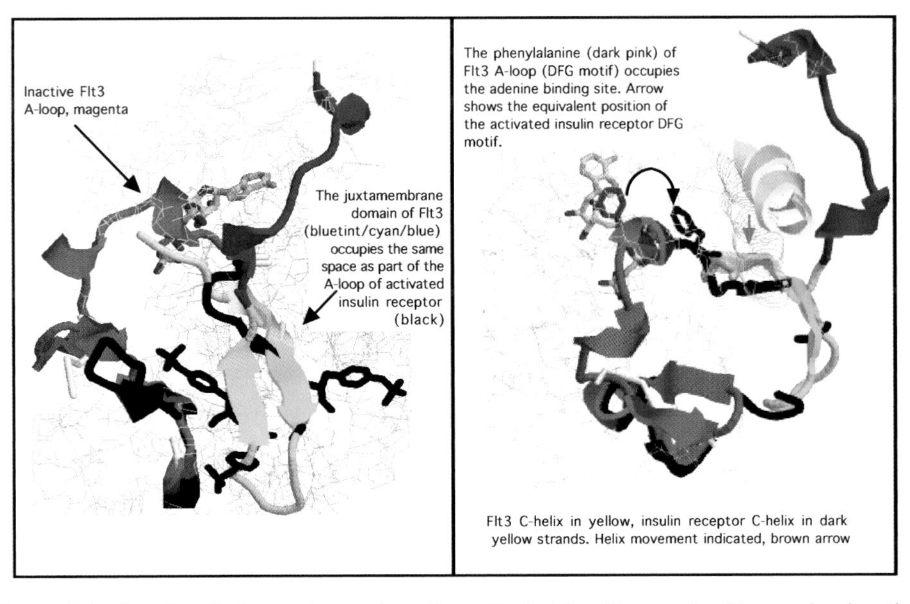

Figure 6.9 Inactive Flt-3 superimposed on the activated insulin receptor kinase, showing the inhibiting positions of Flt-3's juxtamembrane and DFG motifs

The authors of the research[38] suggest that ligand-binding to Flt3 renders the JMS switch sufficiently mobile and exposed enough to undergo phosphorylation easily. Such changes in its electronegativity would be likely to disturb the entire juxtamembrane domain and, after it detached, the A-loop could become exposed and phosphorylated so enabling the correct assembly of ATP- and peptide substrate-binding sites. The importance of the anchoring JMR residue Tyr572 is seen in PDGFR point mutations of its equivalent Tyr.

In summary, the PDGFR family activation probably resembles that described for Flt3 where the juxtamembrane region and the A-loop mount a joint regime of auto-inhibition that disables the catalytic base, blocks ATP binding and prevents the correct positioning of the A-loop for peptide substrate binding.

6.5 The ErbB family

The founder member of this clan of receptors is the epidermal growth factor (EGF) receptor. Its existence was first suspected after the independent discoveries of 'EGF by Stanley Cohen and Harry Gregory. EGF is secreted in an active form in human urine and was detected as early as 1942 as a gastric acid release-inhibiting factor called 'urogastrone'[43]. Gregory isolated this factor from urine and noted its similarity to the alternative isolate from Cohen's lab. Cohen had just purified EGF from male mouse submaxillary glands, so named because it stimulated the proliferation of epidermal epithelial cells. Cohen shared the Nobel Prize for this discovery with Rita Levi-Montalcini (the discoverer of nerve growth factor). Cohen's lab later identified the EGF receptor as a 170 kDa membrane protein that became phosphorylated in response to EGF. This has recently been reviewed[44].

It later transpired that a truncated version of the chicken EGF receptor (EGFR) was present as the second of two oncogenes incorporated into the genome of the avian ***e***rythro***b***lastosis retro***v***irus (AEV). Hence the alternative name ErbB (the first AEV oncogene, named ErbA, turned out to be a thyroid hormone receptor). The 'ErbB' classification became adopted as more receptors were revealed to be homologous.

The mature cellular form of EGFR is a 1186-residue transmembrane protein that is derived from a 1210 precursor through cleavage of its *N*-terminal signal sequence. Its (*N*-terminal) extracellular portion is heavily *N*-glycosylated (sugars making up to 20% of its molecular mass). The extracellular part of the protein represents the ligand-binding domain and consists of residues: 1-621. It is separated from the intracellular portion by a transmembrane helix approximately encompassing residues 626–647[45]. The transmembrane helix merges with the 'juxtamembrane' domain (644–687), followed by the catalytic core (688–955). A unique *C*-terminal tail (956–1186) completes the molecule.

6.5.1 EGFR family members and ligands

There are now four members of the human EGFR family: ErbB1 (the EGF-receptor, also binds TGF-α), ErbB2 (an 'orphan' receptor with no known high affinity ligand), ErbB3

Table 6.3 High affinity ligand binding specificities of ErbB-type receptors

	ErbB1	ErbB2	ErbB3	ErbB4
Epidermal growth factor (EGF)	+	–	–	–
Transforming growth factor-α (TGF-α)	+	–	–	–
Amphiregulin	+	–	–	–
Epiregulin	+	–	–	+
Heparin-binding EGF (HB-EGF)	+	–	–	+
Betacellulin	+	–	–	+
Neuregulin-1 (NRG-1, also known as heregulin or neu differentiation factor)	–	–	+	+
NRG-2	–	–	+	+
NRG-3	–	–	–	+
NRG-4	–	–	–	+

(a receptor for heregulin with an inactive kinase domain) and ErbB4 (receptor for betacellulin and heregulin). Since the original assignments of ligand specificity it has emerged that the receptors are more promiscuous in ligand binding than previously thought[46] (see Table 6.3). All of the ligands contain the EGF-fold of three disulphide-constrained loops, and are generally thought to be monovalent. That is to say, a single ErbB ligand cannot span between two members of an ErbB dimer.

Aberrant expression of EGF receptor family members, ErbB1 and ErbB2, is common in breast cancer, and is associated with poor prognosis[47,45]. ErbB1 and ErbB2 are often co-expressed and collaborate in signalling[48]. Co-expression of the receptor and its cognate ligand (for example, ErbB1 and TGF-α) in the same cell contributes to cancer growth by autocrine stimulation of cell proliferation.

Although homologous and having similar overall architecture, the ErbB-type receptors differ markedly from one another in function. The differences within the ErbB family are much greater than the differences, say, between PDGFR-α versus –β.

For example, ErB2 is unique in having no high affinity extracellular ligand, or first messenger. For this reason, it is usually thought of as an 'orphan' receptor. Nonetheless, a membrane-associated activator of ErbB2 does exist in the form of a transmembrane sialomucin, MUC-4, which can induce ErbB2 phosphorylation[49]. However, since MUC-4 is endogenous and, furthermore, does not bind to the defective ligand-binding domain of ErbB2, it is convenient to think of ErbB2 as purely a co-receptor. Having no extracellular ligand of its own, ErbB2 relies upon dimerisation with another activated ErbB monomer (ErbB1, ErbB3 or ErbB4) to activate its kinase – ErbB2/ErbB2 homodimers are inactive.

ErbB3, by contrast, binds extracellular ligands but has an inactive catalytic site. Therefore, ErbB3/ErbB3 homodimers are also inactive. ErbB3, however, dimerises with ErbB2 to assemble a powerful signalling entity from otherwise defective partners – signalling from ErbB2/ErbB3 heterodimers is thought to play an important role in breast cancer progression[46].

The formation of the various ErbB heterodimers is obviously driven by ligand: NRG-1 and betacellulin inducing ErbB4/ErbB4 or ErbB2/ErbB4 formation, and NRG-1 inducing ErbB2/ErbB3. Although crystal structures of ErbB1 ectodomain (see below) show EGF binding in a monovalent manner, there is evidence that EGF-like ligands can either act bivalently or transactivate unliganded partners. For example, EGF alone is thought to activate cell motility through formation of ErbB1/ErbB3 dimers that then recruit PI-3-kinase in wild-type A431 cells[16].

6.6 ErbB-type receptor signal transduction particles

The principles of particle assembly discussed above in relation to the PDGFR are equally applicable to EGFR (see Figure 6.10) and its relatives except that with ErbB-type receptors, heterodimerisation plays an essentially important part of their signalling flexibility. Since no individual family member possesses the full range of pTyr docking sites displayed by PDGFR, ErbB receptors are forced to collaborate (see Table 6.4)[46]. The receptors vary greatly in their intracellular tyrosine phosphorylation sites and by selectively heterodimerising, they can assemble all or just a few of the motifs offered by PDFGR. We will concentrate on EGFR as an example of ErbB signal particles (Figure 6.10).

As alluded to above, EGFR lacks the appropriate pTyr motifs for recruitment of $p85^{PI3K}$ and would be motility-impaired without erbB3, which has an abundance of $p85^{PI3K}$-specific pTyr motifs (Table 6.4). All ErbB-types, however, are able to recruit Shc and thus all can activate Ras. EGFR is unusual in being very rapidly internalised after ligand activation, a finding in part explained by the unique presence of a direct Cbl binding site. Cbl is a multiple domain adaptor/downregulator, which binds to EGFR pTyr1045 via an *N*-terminal 'TKB' region (for 'tyrosine kinase binding) that contains an SH2 domain. Next to this is a RING finger domain that is a ubiquitin-ligase, which ubiquitinates EGFR, marking it for internalisation and degradation when the EGF signal needs to be terminated[50]. Then, Cbl has a central proline-rich region followed by a *C*-terminal tail with multiple phosphorylation sites, one of which is a docking site for $p85^{PI3K}$. So, Cbl not only acts to downregulate EGFR

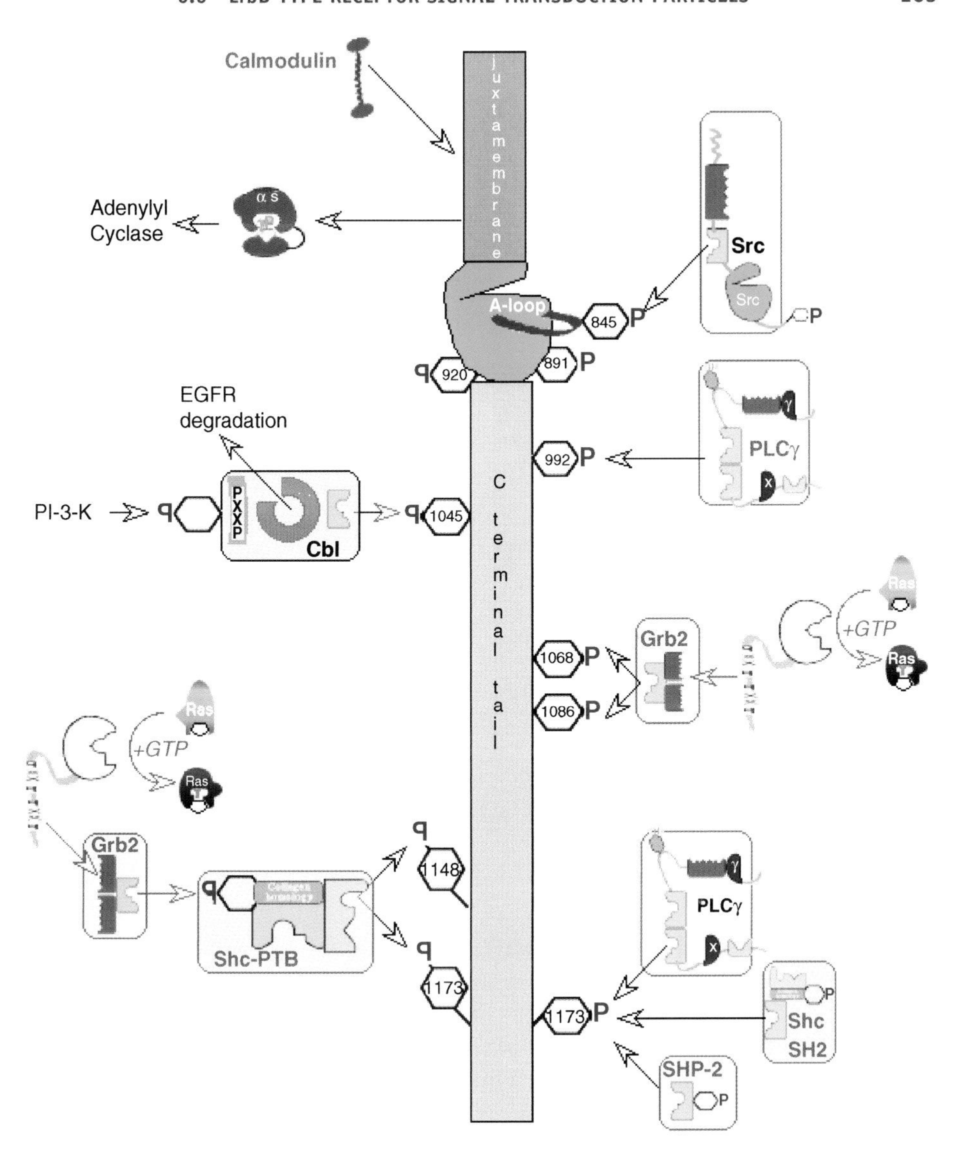

Figure 6.10 The EGF receptor signal transduction particle

signalling, but in some cells at least, provides an alternative way of recruiting PI-3-kinase to stimulate motility[51], although it has also been reported to regulate actin assembly negatively[52].

ErbB-type receptors have multiple Shc binding sites and Shc can bind to ErbB receptors via its SH2 domain or its PTB domain. For example, pTyr1148 is a high affinity binding motif for Shc-PTB[33], whereas pTyr1173 acts as a binding site for either the SH2 or PTB domain of Shc.

Table 6.4 Phosphotyrosines in ErbB receptor intercellular domains and the proteins that dock with each motif

ErbB1		ErbB2		ErbB3		ErbB4	
845	Src			1035	$p85^{PI3K}$		
992	PLCγ			1178	$p85^{PI3K}$		
1045	Cbl	1023		1180	Grb7	1056	$p85^{PI3K}$
1068	Grb2	1139	Grb2	1203	$p85^{PI3K}$	1188	Shc
1086	Grb2	1196	Shc	1205	$p85^{PI3K}$	1242	Shc
1148	Shc	1221	Shc	1241	$p85^{PI3K}$		
1173	Shc	1222	Shc	1243	Grb7		
	PLCγ	1248	Shc	1257	$p85^{PI3K}$		
	SHP-2		Chk	1270	$p85^{PI3K}$		
				1309	Shc		

6.6.1 The epidermal growth factor receptor kinase – a pre-assembled active site

The EGFR family is a major exception to the standard model of protein kinase primary autoinhibition. The tyrosine kinase domain of EGFR (as well as those of ErbB2 and ErbB4) does not follow the general rule that A-loop phosphorylation is required for the activation and assembly of the RTK catalytic site. EGFR does contain an A-loop tyrosine (Tyr845) in the same 'activating' position as is present in IRK (see Chapter 4) but in EGFR, Tyr845 has no regulatory role. It can be substituted with a phenylalanine without effect on activity. At present, it is known that

- EGFR is catalytically inactive (and mostly monomeric) in the absence of ligand. Native EGFR dimers can form in the absence of ligand, but are inactive.

- Ligand-binding induces activation via dimerisation, followed by *trans*-autophosphorylation of tyrosines (primarily) in the *C*-terminal tail.

- *v*ErbB (the viral oncogene) is constitutively active (presumably) because it is truncated and is missing both the extracellular domain and parts of the *C*-terminal tail.

- Even in the absence of EGF, the normal unphosphorylated EGFR kinase appears to be in an active conformation with catalytic cleft poised for substrate loading and (potentially) competent to phosphorylate.

- Mutants missing only the extracellular domain (the 'ectodomain') exist as membrane-bound, constitutively active dimers[45].

Like the insulin receptor, dimerisation in the absence of ligand is not in itself enough to activate the kinase. Ligand binding induces conformational changes in the ectodomain that spreads right through the membrane to the intracellular domain, directly or indirectly activating the kinase without A-loop phosphorylation. Clearly, its autoinhibition mechanism is a good deal more complex than other protein kinases.

If we compare the apo-EGFR kinase structure with that of phosphorylase kinase (another protein kinase that does not require A-loop phosphorylation, see Chapter 5), a number of similarities seem evident but the disposition of key residues is not identical (see below).

As mentioned above, Tyr845 of EGFR is in a position equivalent to pTyr1163 of IRK and also interacts (albeit non-electrostatically) with two residues that match those of IRK's 'basic cluster'[53]. It binds, by van der Waals interaction, to the aliphatic portion of Lys836 (equivalent to Arg1115 of IRK). It also H-bonds to Arg812 in the RD motif (equivalent to Arg1131 of IRK). The missing phosphoryl dianion is mimicked by a triad of glutamate residues of EGFR's A-loop: Glu842; Glu844; Glu848 (see Figure 6.11). The result of these deviations from the RD kinase core is that the A-loop of unphosphorylated apo-EGFR resembles the conformation of active IRK.

Another catalytically important component of the active site is also pre-assembled in apo-EGFR; namely, the salt-bridge between the invariant glutamate of the C-helix and the lysine of subdomain II that binds α- and β-phosphates of ATP.

6.7 Autoinhibition of EGFR and activation

At the time of writing, there is no firm mechanism of autoinhibition for unliganded EGFR. Current models, unsurprisingly, focus on the demonstrable importance of the *C*-terminal tail (a regulatory element missing in *v*ErbB) and the juxtamembrane region (the site of a down-regulating phosphorylation by protein kinase C as well as a target for calmodulin binding). Two models have been proposed; both emphasise an inactive receptor geometry that keeps the autophosphorylation substrate sequences away from the active site cleft, and both depend upon reversible electrostatic attraction/repulsion cycles to explain inhibition and activation.

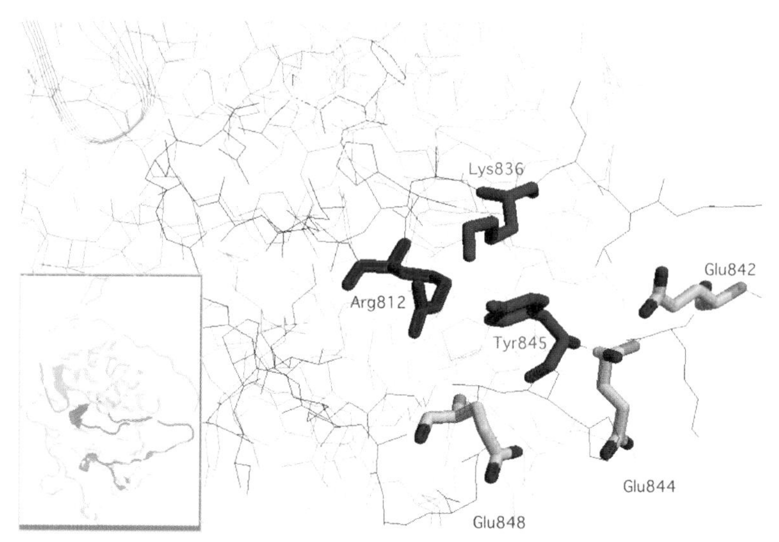

C-helix is in yellow; P-loop, green; A-loop purple; C-loop red. The A-loop triad of glutamates are cpk-coloured.
Inset shows the orientation of the zoomed-in structure

Figure 6.11 The EGFR tyrosine kinase 'basic cluster' equivalent

One model is dependent upon there being pre-existing, inactive EGFR dimers that become re-arranged by ligand binding. The suggestion is that the unusually ordered, and negatively charged, *C*-terminal extended tail of one monomer interacts with the positively charged back face of the catalytic domain, and that this interaction inhibits *trans*-autophosphoryation by sequestering the *C*-terminal substrate sequences away from the active site cleft[54]. Alternatively, the basic juxtamembrane domain interacts with the acidic *C*-tail region to produce an inactive dimer[55].

A second model posits a mechanism whereby the inactive monomeric EGFR is autoinhibited by electrostatic binding of both the basic juxtamembrane region and the back face of the catalytic domain to acidic lipids of the inner leaflet[56]. The evidence for this model is discussed below. However, it is perfectly possible that both models are correct and, individually, represent two stages in the full activation of the kinase.

6.7.1 Ligand binding, dimerisation and activation

EGFR monomers bind a single EGF ligand each and this promotes dimerisation. Unlike the PDGFR and the insulin receptor, where a single ligand bridges and ligates two receptor monomers, the EGFR-EGF dimer has a 2:2 composition rather than 2:1. Clearly, dimerisation of ErbB receptors cannot be simply explained with a simple bridging action of a bivalent ligand.

N-terminally truncated EGFR mutants (encompassing transmembrane and intracellular domains only) are constitutively active and dimerised, without ligand. *C*-terminally truncated receptors (containing the soluble ectodomain only) can also dimerise, but in this case the dimerisation is largely ligand-dependent. Thus, there are both intracellular and extracellular dimerisation domains. In the ligand-occupied ectodomain, a long 'dimerisation loop' (Asp238–Lys260) of the CR1 domain projects from one receptor and binds to a complementary pocket at the base of the CR1 loop of the dimer partner, while a second dimerisation region is situated within the intracellular kinase domain between residues 835–918 [45].

The ligand binding 'ectodomain' has been crystallised in association with its ligand, transforming growth factor-α (TGF-α, homologous with EGF and an equipotent agonist)[57]. The ectodomain folds in four separate modules: two 'ligand binding domains' (L1: aas 1–165; L2: aas 310–481) and two cysteine-rich domains (CR1: aas 166–309: CR2: aas 482–618). In the activated bound structure, the TGF-α ligand is held clamped between the L1 and L2 domains that are hinged together by CR1 (PDB file: 1MOX). Most of the CR2 domain is missing from the structure and in Figure 6.12, a pastiche of the activated receptor has been constructed using a copy of CR1 and a transmembrane helix to fill the gap between the ectodomain structure and the separate crystal structure of apo-EGFR kinase (PDB file: 1M14). Remember, this is only a construct – if the complete receptor structure is ever solved, it may not look like this.

Another crystal structure has been solved for the EGFR ectodomain, this time in an inactive (but bound) form[58] (PDB file: 1NQL). In this case EGF is bound, but at a low pH. Such low pH binding is well known to be low affinity and incapable of stimulating EGFR activity via dimerisation, so the structure is thought to resemble inactive unoccupied receptor. The structure is quite different. First, the ligand is not clamped, but instead only binds to the L1 domain. More importantly, the dimerisation loop is not exposed but is buried in an intrachain interaction with its CR2 domain. This neatly explains why the unliganded receptor rarely dimerises. Figure 6.13 shows an idealised construct of inactive EGFR with this inactive ectodomain joined to the apo-EGFR kinase. Again, this is merely a pastiche of structures and may not reflect the real position.

When ligand binds, it is obvious that a great degree of conformational change is produced in the ectodomain, particularly with regard to the reorganisation that forces the dimerisation loop to swing out ready to engage a second activated monomer. The conformational changes induced first by ligand, and then by dimerisation, are transmitted through the membrane, possibly by a reorientation of the transmembrane helix, and cause unknown changes to the intracellular domain.

6.7.2 The EGFR juxtamembrane domain – a nexus for crosstalk

The juxtamembrane region does not have a defined structure in aqueous solution and is mostly missing from the kinase crystal structures, but in a lipid micelle it folds into three amphipathic helices tightly bound to the micelle surface; this strongly suggests that it lies along the plane of the inner leaflet of the plasma membrane *in vivo*[59].

The juxtamembrane region contains a PXXP-type basolateral sorting signal, a lysosomal degradation motif, and a nuclear localisation signal along with the binding

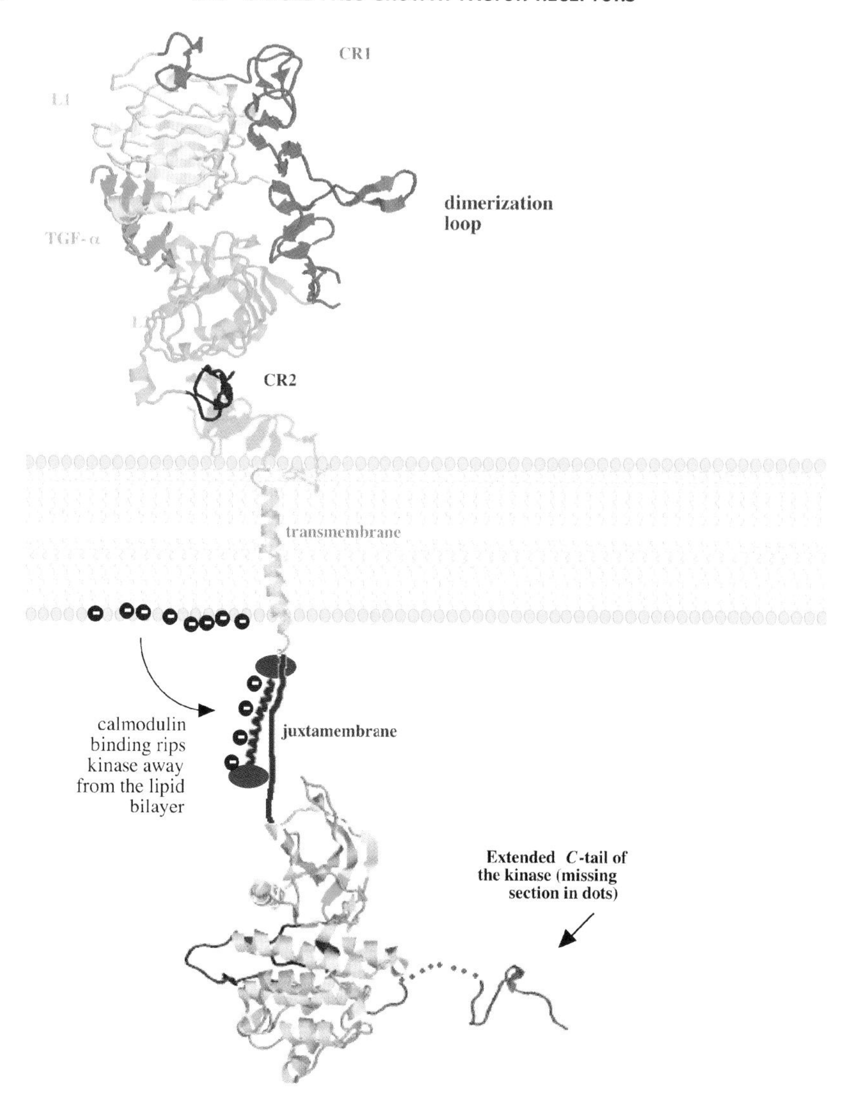

Figure 6.12 EGFR pastiche illustrating possible activation mechanism

site for calmodulin. The juxtamembrane region (particularly amino acid stretch 645–657) of one EGFR monomer may interact in an inhibitory manner with the polypeptide backbone of its partner EGFR in inactive dimers, as judged by studies using synthetic peptides derived from juxtamembrane sequences[60]. In addition, the juxtamembrane region of EGFR can directly bind to, and activate, the heterotrimeric G protein Gαs, allowing EGF to activate adenylyl cyclase. Thus the juxtamembrane domain enables EGFR to crosstalk with signalling pathways normally activated by 7-pass receptor ligands[61].

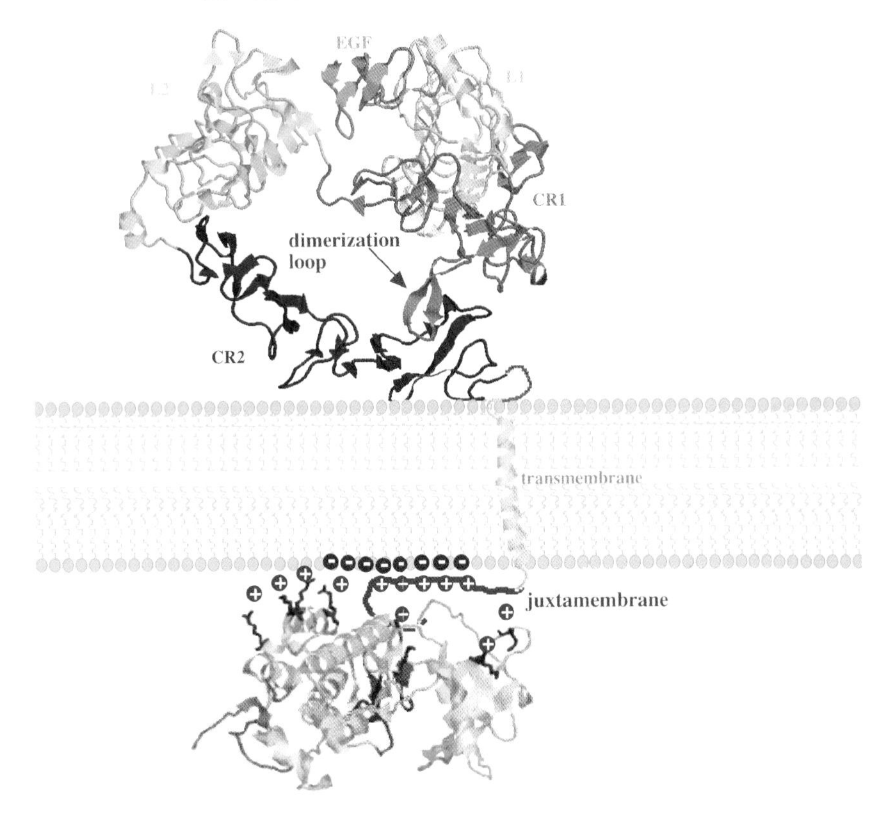

Figure 6.13 EGFR pastiche illustrating possible autoinhibited conformation

Finally, this region of EGFR is subject to inhibitory phosphorylations by protein kinase C (at Thr654) and CaM-kinase (at Thr669)[62]. These latter effects are calcium-dependent and act as negative feedback loops, turning off the receptor by initiating its endocytosis after EGF-stimulated calcium release becomes high or persistent enough. In particular, it should be noted that phosphorylation of Thr654 by PKC prevents calmodulin binding and this may explain the inhibitory effect of this covalent modification (see below).

In summary, the juxtamembrane domain represents a concentrated focus for competing signals that may be revealed or hidden by reversible lipid binding, ion fluxes, phosphorylation, or occupancy by other proteins. Its functional significance is still not completely understood and is obviously an area that will remain the subject of intense investigation.

6.7.3 EGFR activation and calcium

One of the earliest signals produced after EGF binding is a rapid and transient rise in cytosolic calcium and this appears to be an absolute prerequisite for full activation of the receptor. Indeed, calcium alone can activate EGFR in the complete absence of EGF. Crosstalk from the calcium-mobilising bradykinin receptor, membrane-depolarisation by KCl, or treatment with calcium ionophore can all activate unliganded EGFR, the effects being, variously, calmodulin-dependent or calmodulin-independent[63].

The immediate rapid rise in calcium caused by EGF is not from the normal endoplasmic reticulum stores[64], as the effect is not dependent upon IP3 production. Although, phospholipase C is involved this is not the classical PLC cascade (see Chapter 5). It appears that the initial transient EGF-stimulated spike in [Ca^{2+}] is due to influx of extracellular calcium through 'store-operated Ca^{2+}-channels' in the plasma membrane and it was suggested that the channel involved might be one of the ***t***ransient ***r***eceptor ***p***otential (TRP) family[64]. This tentative prediction was recently proved correct. Knockdown of TRP family member, TRPC4 was found to eliminate EGF-stimulated calcium flux while at the same time suppressing the mitogenic effects of EGF in corneal epithelial cells[65].

A classical PLC-IP3 calcium signal *is* activated downstream of EGFR (Figure 6.10) but this is not associated with relief of autoinhibition of the receptor kinase.

The EGFR juxtamembrane domain – a calcium-activated switch? The juxtamembrane domains of both EGFR and ErbB2 (as well as the juxtamembrane region of the insulin receptor) are well established as being calmodulin-binding modules. Although calmodulin-EGFR binding was at first shown to be inhibitory, this early study used solubilised EGFR that was subsequently EGF-stimulated in solution[66]. *In vivo*, where EGFR is membrane-bound, EGF-activation is dependent upon both calcium mobilisation and the consequent activation of calmodulin[67]. Calcium-activated calmodulin binds to the portion of receptor closest to the membrane – the 'juxtamembrane' domain. The same mode of kinase activation applies to ErbB2, another ErbB member with a calmodulin-binding juxtamembrane domain[68].

The juxtamembrane domain contains a preponderance of hydrophobic and basic residues that can interact with the acidic lipid headgroups of the plasma membrane bilayer, particularly phosphoinositides. In the absence of ligand, the juxtamembrane domain is theorised to anchor the bulk of the intracellular domain to the inner leaflet, aided by the concentration of positively charged residues at the back face of the kinase domain.

As mentioned earlier, one of the first consequences of EGF binding to its receptor, after dimerisation has occurred, is a transient rise in cytosolic calcium that leads to the activation of calmodulin. Ca^{2+}-calmodulin can bind to the juxtamembrane domain and this recruitment of (highly negatively charged) calmodulin effectively reverses the charge on that part of the receptor from positive to negative, 'ripping' both juxtamembrane and kinase domains off the inner leaflet by electrostatic repulsion (see Figures 6.12 and 6.13)[56].

In summary, EGFR's catalytic site is kept inactive in its unliganded state by being anchored to the negatively charged membrane via charge interactions with the positively charged juxtamembrane and back face of the kinase domains. This keeps its catalytic cleft in a position such that it cannot dock with its substrate (the *C*-terminal tail). The likelihood of substrate docking is then increased by the proximity effect of dimerisation (provoked by the ligand-stimulated exposure of ectodomain dimerisation loops). The immediate calcium flux produced after dimerisation recruits acidic calmodulin to the juxtamembrane domain, reversing its polarity and rotating the kinase domain away from the membrane and into a new position that allows free access to its partner's *C*-tail. Thus, the activation acts by an electrostatic switch mechanism.

References

1. Hubbard, S.R. and Till, J.H. (2000) Protein tyrosine kinase structure and function. *Annu. Rev. Biochem.*, **69**: 373–398.
2. Claesson-Welsh, L. (1994) Platelet-derived growth factor receptor signals. *J. Biol. Chem.*, **269**: 32023–32026.
3. Arvidsson, A-K., Rupp, E., Nanberg, E., Downward, J., Ronnstrand, L., Wennstrom, S., Schlessinger, J., Heldin, C-H. and Claesson-Welsh, L. (1994) Tyr-716 in the platelet derived growth factor β-kinase insert is involved in GRB2 binding and Ras activation. *Mol. Cell. Biol.*, **14**: 6715–6726.
4. Mori, S., Ronnstrand, L., Yokote, K., Engstrom, A., Courtneidge, S.A., Claesson-Welsh, L.C. and Heldin, C-H. (1993) Identification of two juxtamembrane autophosphorylation sites in the PDGF β-receptor; involvement in the interaction with the Src family tyrosine kinases. *EMBO J.*, **12**: 2257–2264.
5. Yokote, K., Mori, S., Hansen, K., McGlade, J., Pawson, T., Heldin, C-H. and Claesson-Welsh, L. (1994) Direct interaction between Shc and the platelet-derived growth factor β-receptor. *J. Biol. Chem.*, **269**: 15337–15343.
6. Nollau, P. and Mayer, B.J. (2001) Profiling the global tyrosine phosphorylation state by Src homology 2 domain binding. *PNAS*, **98**: 13531–13536.
7. Qi, M. and Elion, E.A. (2006) MAP kinase pathways. *J. Cell Science*, **118**: 3569–3572.
8. Yu, J-H., Heidaran, M.A., Pierce, J.H., Gutkind, S., Lombardi, D., Ruggiero, M. and Aaronson, S.A. (1991) Tyrosine mutations within a platelet-derived growth factor receptor kinase domain abrogate receptor-associated phosphatidylinositol-3 kinase activity without affecting mitogenic or chemotactic signal transduction. *Molecular and Cellular Biology*, **11**: 3780–3785.
9. Hill, K.M., Huang, Y., Yip, S-C., Yu, J., Segall, J.E. and Backer, J.M. (2001) N-terminal domains of the Class IA phosphoinositide 3-kinase regulatory subunit play a role in cytoskeletal but not mitogenic signalling. *J. Biol. Chem.*, **276**: 16374–16378.
10. Pacold, M.E., Suire, S., Persic, O., Lara-Gonzalez, S., Davis, C.T., Walker, E.H., Hawkins, P.T., Stephens, L., Eccleston, J.F. and Williams, R.L. (2000) Crystal structure and functional analysis of Ras binding to its effector phosphoinositide 3-kinase. *Cell*, **103**(6): 931–943.
11. Wennstrom, S. and Downward, J. (1999) Role of phosphoinositide 3-kinase in activation of Ras and mitogen-activated protein kinase by epidermal growth factor. *Mol. Cell. Biol.*, **19**: 4279–4288.
12. Taylor, S.J. and Shalloway, D. (1996) Cell cycle-dependent activation of Ras. *Current Biology*, **6**: 1621–1627.
13. Jones, S.M., Klinghoffer, R., Prestwich, G.D., Toker, A. and Kazlauskas, A. (1999) PDGF induces an early and a late wave of PI 3-kinase activity, and only the late wave is required for progression through G1. *Current Biology*, **9**: 512–521.
14. Bar-Sagi, D. and Hall, A. (2000) Ras and Rho GTPases: a family reunion. *Cell*, **103**: 227–238.
15. Beeton, C.A., Das, P., Waterfield, M.D. and Shepherd, P.R. (1999) The SH3 and BH domains of the p85α adaptor subunit play a critical role in regulating Class IA phosphoinositide 3-kinase function. *Molecular Cell Biology Research Communications*, **1**: 153–157.

16. Gillham, H., Golding, M.C.H.M., Pepperkok, R. and Gullick, W.J. (1999) Intracellular movement of green fluorescent protein-tagged phosphatidyl 3-kinase in response to growth factor receptor signalling. *J. Cell Biol.*, **146**: 869–880.
17. Kazlauskas, A., Kashishian, A., Cooper, J.A. and Valius, M. (1992) GTPase-activating protein and phosphatidylinositol 3-kinase bind to distinct regions of the platelet-derived growth factor receptor 1β subunit. *Molecular and Cellular Biology*, **12**: 2534–2544.
18. Pawson, T. and Saxton, T.M. (1999) Signalling networks – do all roads lead to the same genes? *Cell*, **97**: 675–678.
19. Frost, J.A., Xu, S., Hutchison, M.R., Marcue, S. and Cobb, M.H. (1996) Actions of Rho family small G proteins and p21-activated protein kinases on mitogen-activated protein kinase family members. *Mol. Cell. Biol.*, **16**: 3707–3713.
20. Frost, J.A., Steen, H., Shapiro, P., Lewis, T., Ahn, N., Shaw, P.E. and Cobb, M.H. (1997) Cross-cascade activation of ERKs and ternary complex factors by Rho family members. *EMBO J.*, **16**: 6426–6438.
21. Eblen, S.T., Slack, J.K., Weber, M.J. and Catling, A.D. (2002) Rac-PAK signaling stimulates extracellular signal-regulated kinase (ERK) activation by regulating formation of MEK1-ERK complexes. *Mol. Cell. Biol.*, **22**: 6023–6033.
22. Yu, J., Deuel, T.F. and Kim, H-R.C. (2000) Platelet-derived growth factor (PDGF) receptor-a activates c-Jun NH2-terminal kinase-1 and antagonizes PDGF receptor-β-induced phenotypic transformation. *J. Biol. Chem.*, **275**: 19076–19082.
23. Innocenti, M., Tenca, P., Frittoli, E., Faretta, M., Tocchetti, A., DiFiore, P.P. and Scita, G. (2002) Mechanisms through which Sos-1 coordinates the activation of Ras and Rac. *J. Cell Biol.*, **156**: 125–136.
24. Stradal, T., Courtney, K.D., Rottner, K., Hahne, P., Small, J.V. and Pendergast, A.M. (2001) The Abl interactor proteins localize to sites of actin polymerisation at the tips of lamellipodia and filopodia. *Current Biology*, **11**: 891–895.
25. Innocenti, M., Frittoli, E., Ponzanelli, I., Falck, J.R., Brachmann, S.M., Di Fiore, P.P. and Scita, G. (2003) Phosphinositide 3-kinase activates Rac by entering in complex with Eps8, Abi1, and Sos-1. *J. Cell Biol.*, **160**: 17–23.
26. Das, B., Shu, X., Day, G-J., Han, J., Krishna, M., Falck, J.R. and Broek, D. (2000) Control of intramolecular interactions between the pleckstrin homology and Dbl homology domains of Van and Sos1 regulates Rac binding. *J. Biol. Chem.*, **275**: 15074–15081.
27. Rameh, L.E., Rhee, S.G., Spokes, K., Kazlauskas, A., Cantley, L.C. and Cantley, L.G. (1998) Phosphoinositide 3-kinase regulates phospholipase Cγ-mediated calcium signalling. *J. Biol. Chem.*, **273**: 23750–23757.
28. Singer, W.D., Brown, H.A. and Sternweiss, P.C. (1997) Regulation of eukaryotic phosphatidylinositol-specific phospholipase C and phospholipase D. *Annual Reviews of Biochemistry*, **66**: 475–509.
29. Welch, H.C.E., Coadwell, W.J., Stephens, L.R. and Hawkins, P.T. (2003) Phosphoinositide 3-kinase-dependent activation of Rac. *FEBS Letters*, **546**: 93–97.
30. Rhee, S.G. (2001) Regulation of phosphoinositide-specific phospholipase C. *Annual Reviews of Biochemistry*, **70**: 281–312.
31. Bazenet, C.E., Gelderloos, J.A. and Kazlauskas, A. (1996) Phosphorylation of tyrosine 720 in the platelet-derived growth factor is required for the binding of Grb2 and SH-2 but not for activation of Ras or cell proliferation. *Mol. Cell. Biol.*, **16**: 6926–6936.
32. Farooq, A., Zeng, L., Yan, K.S., Ravichandran, K.S. and Zhou, M-M. (2003) Coupling of folding and binding in the PTB domain of the signaling protein Shc *Structure*, **11**: 905–913.
33. Laminet, A.A., Apell, G., Conroy, L. and Kavanaugh, W.M. (1996) Affinity, and kinetics of the interaction of the SHC phosphotyrosine binding domain with asparagine-*X*-*X*-phosphotyrosine motifs of growth factor receptors. *J. Biol. Chem.*, **271**: 264–269.

34. Benjamin, C.W. and Jones, D.A. (1994) Platelet-derived growth factor stimulates growth factor receptor binding protein-2 association with Shc in vascular smooth muscle. *J. Biol. Chem.*, **269**: 30911–30916.
35. Baxter, R.M., Secrist, J.P., Vaillancourt, R.R. and Kazlauskas, A. (1998) Full activation of the platelet-derived growth factor β-receptor kinase involves multiple events. *J. Biol. Chem.*, **273**: 17050–17055.
36. Irusta, P.M., Luo, Y., Bakht, O., Lai, C-C., Smith, S.O. and DiMaio, D. (2002) Definition of an inhibitory juxtamembrane WW-like domain in the platelet-derived growth factor β receptor. *J. Biol. Chem.*, **277**: 38627–38634.
37. Chan, P.M., Ilangumaran, S., La Rose, J., Chakrabartty, A. and Rottapel, R. (2003) Autoinhibition of the Kit receptor tyrosine kinase by the cytosolic juxtamembrane region. *Mol. Cell. Biol.*, **23**: 3067–3078.
38. Griffith, J., Black, J., Faerman, C., Swenson, L., Wynn, M., Lu, F., Lippke, J. and Kumkum, S. (2004) The structural basis for autoinhibition of FLT3 by the juxtamembrane domain. *Molecular Cell*, **13**: 169–178.
39. Tickenbrock, L., Schwable, J., Wiedehage, M., Steffen, B., Sargin, B., Choudhary, C., Brandts, C., Berdel, W.E., Muller-Tidow, C. and Serve, H. (2005) Flt3 tandem duplication mutations cooperate with Wnt signaling in leukemic signal transduction. *Blood*, **105**: 3699–3706.
40. Zheng, R. and Small, D. (2005) Mutant FLT3 signaling contributes to a block in myeloid differentiation. *Leuk. Lymphoma*, **46**: 1679–1687.
41. Kindler, T., Breitenbuecher, F., Kasper, S., Estey, E., Giles, F., Feldman, E., Ehninger, G. and Schiller, G. (2005) Identification of a novel activating mutation (Y842C) within the activation loop of FLT3 in patients with acute myeloid leukemia (AML). *Blood*, **105**: 335–340.
42. Yamamoto, Y., Kiyoi, H., *et al.* (2001) Activating mutation of D835 within the activation loop of FLT3 in human hematologic malignancies *Blood*, **97**: 2434–2439.
43. Gray, J.S., Wieczorowski, E., Wells, J.A. and Harris, S.C. (1942) The preparation and properties of urogastrone. *Endocrinology*, **30**: 129–134.
44. Gschwind, A., Fischer, O.M. and Ullrich, A. (2004) The discovery of receptor tyrosine kinases: targets for cancer therapy. *Nature Review*, **4**: 361–370.
45. Jorissen, R.N., Walker, F., Pouliot, N., Garrett, T.P.J., Ward, C.W. and Burgess, A.W. (2003) Epidermal growth factor receptor: mechanisms of activation and signalling. *Experimental Cell Research*, **284**: 31–53.
46. Olayioye, M.A., Neve, R.M., Lane, H.A. and Hynes, M.E. (2000) The ErbB signaling network: receptor heterodimerisation in development and cancer. *EMBO J.*, **19**: 3159–3167.
47. Hynes, N.E. and Stern, D.F. (1994) The biology of erbB2/neu/HER-2 and its role in cancer. *Biochim. Biophys. Acta.*, **1198**, 165–184.
48. Deb, T.B., Su, L., Wong, L., Bonvini, E., Wells, A., David, M. and Johnson, G.R. (2001) Epidermal growth factor (EGF) receptor kinase-independent signalling by EGF. *J. Biol. Chem.*, **276**, 15554–15560.
49. Jepson, S., Komatsu, M., Haq, B., Arango, M.E., Huang, D., Carraway, C.A. and Carraway KL. (2002) Muc4/sialomucin complex, the intramembrane ErbB2 ligand, induces specific phosphorylation of ErbB2 and enhances expression of p27(kip), but does not activate mitogen-activated kinase or protein kinaseB/Akt pathways. *Oncogene*, **21**: 7524–7532.
50. Carraway III, K.L. and Sweeney, C. (2001) Localization and modulation of ERBB receptor tyrosine kinase. *Current Opinion in Cell Biology*, **13**: 125–130.
51. Soltoff, S.P. and Cantley, L.C. (1996) p120^{cbl} is a cytosolic adaptor protein that associates with phosphoinositide 3-kinase in response to epidermal growth factor in PC12 and other cells. *J. Biol. Chem.*, **271**: 563–567.
52. Scaife, R.M., Courtneidge, S.A. and Langdon. A.Y. (2002) The multi-adaptor proto-oncoprotein Cbl is a key regulator of Rac and actin assembly. *J. Cell Science.*, **116**: 463–473.

53. Stamos, J., Sliwkowski, M.X. and Eigenbrot (2002) Structure of the epidermal growth factor receptor kinase domain alone and in complex with a 4-anilinoquinazoline inhibitor. *J. Biol. Chem.*, **277**: 64265–46272.
54. Landau, M., Fleishman, S.J., and Ben-Tal, N. (2004) A putative mechanism for downregulation of the catalytic activity of the EGF receptor via direct contact between its kinase and C-terminal domains. *Structure*, **12**: 2265–2275.
55. Aifa, S., Aydin, J., Nordvall, G., Lundstrom, I., Svensson, S.P.S. and Hermanson, O. (2005) A basic peptide within the juxtamembrane region is required for EGF receptor dimerization. *Exp. Cell Res.*, **302**: 108–111.
56. McLaughlin, S., Smith, S.O., Hayman, M.J. and Murray, D. (2005) An electrostatic engine model for autoinhibition and activation of the epidermal growth factor receptor (EGFR/ErbB) family. *J. Gen. Physiol.*, **126**: 41–53.
57. Garrett, T., McKern, N., Lou, M., Elleman, T., Adams, T., Lovrecz, G., Zhu, H., Walker, F., Frenkel, M., Hoyne, P., *et al.* (2002) Crystal structure of a truncated epidermal growth factor receptor extracellular domain bound to transforming growth factor α. *Cell*, **110**: 763–773.
58. Ferguson, K.M., Berger, M.B., Mendrola, J.M., Cho, H-S., Leahy, D.J. and Lemmon, M.A. (2003) EGF activates its receptor by removing interactions that autoinhibit ectodomain dimerization. *Molecular Cell*, **11**(2): 507–517.
59. Choowongkomon, K., Carlin, C.R. and Sonnichsen, F.D. (2005) A structural model for the membrane-bound form of the juxtamembrane domain of the epidermal growth factor receptor. *J. Biol. Chem.*, **280**: 24043–24052.
60. Poppleton, H.M., Wiepz, G.J., Bertics, P.J. and Patel, T.B. (1999) Modulation of the protein tyrosine kinase activity and autophosphorylation of the epidermal growth factor receptor by its juxtamembrane region. *Archives of Biochemistry and Biophysics*, **363**: 227–236.
61. Sun, H., Chen, Z., Poppleton, H., Scholich, K., Mullenix, J., Weipz, G.J., Fulgham, D.L., Bertics, P.J. and Patel, T.B. (1997) The juxtamembrane, cytosolic region of the epidermal growth factor receptor is involved in association with α-subunit of Gs. *J. Biol. Chem.*, **272**: 5413–5420.
62. Aifa, S., Frikha, F., Miled, N., Johansen, K., Lundstrom, I. and Svensson, S.P. (2006) Phosphorylation of Thr(654) but not Thr(669) within the juxtamembrane domain of the EGF receptor inhibits calmodulin binding. *Biochem. Biophys. Res. Commun.*, **347**: 381–387.
63. Zwick, E., Wallasch, C., Daub, H. and Ullrich, A. (1999) Distinct calcium-dependent pathways of epidermal growth factor receptor transactivation and PYK2 tyrosine phosphorylation in PC12 cells. *J. Biol. Chem.*, **274**: 20989–20996.
64. Li, W-P., Tsiokas, L., Sansom, S.C. and Ma, R. (2004) Epidermal growth factor activates store-operated Ca^{2+} channels through an insoitol 1,4,5-trisphosphate-independent pathway on human glomerular mesangial cells. *J. Biol. Chem.*, **279**: 4570–4577.
65. Yang, H., Megler, S., Sun, X., Wang, Z., Lu, L., Bonanno, J.A., Pleyer, U. and Reinach, P.S. (2005) TRPC4 knockdown suppresses EGF-induced store-operated channels activation and growth in human corneal epithelial cells. *J. Biol. Chem.*, **280**(37): 32230–32237.
66. San Jose, E., Benguria, A., Geller, P. and Villalobo, A. (1992) Calmodulin inhibits the epidermal growth factor receptor tyrosine kinase. *J. Biol. Chem.*, **267**: 15237–15245.
67. Li, H., Ruano, M.J. and Villalobo, A. (2004a) Endogenous calmodulin interacts with the epidermal growth factor receptor in living cells. *FEBS Letters*, **559**: 175–180.
68. Li, H., Sanchez-Torres, J., del Carpio, A., Sallas, V. and Villalobo, A. (2004b) The erbB2/Neu/Her2 receptor is a new calmodulin-binding protein. *Biochem. J.*, **381**: 257–266.

7

G proteins (I) – monomeric G proteins

Monomeric G proteins such as Ras are membrane-bound and alternate between activated GTP-bound and inactivated GDP-bound forms. Ras<GTP> activates its mitogenic effector Raf by causing its translocation to the membrane, where C-Raf is further activated by phosphorylation of its 'N-region', followed by *trans*-autophosphorylation (by C-Raf or B-Raf). B-Raf's 'N-region' does not require to be phosphorylated, and so B-Raf is more easily activated (*trans*-autophosphorylation only). Raf's large N-terminal regulatory region contains numerous inhibitory serine phosphorylation sites for PKA and Erk. Although PKA is usually inhibitory, in some circumstances PKA can activate B-Raf via the Ras relative, Rap. Activation of Raf leads to Erk-type MAPK activation and mitosis. The two forms of Ras (Ras<GDP> and Ras<GTP>) have different conformations in two mobile regions: switch I and II. The conformation change is triggered by loss of the γ-phosphate of GTP, breaking the bonds it makes with Thr35 (switch I) and Gly60 (switch II). In the active GTP-bound conformation, switch I acts as a binding surface for the effector enzyme Raf, which binds by β-strand addition. Ras<GTP> is inactivated by the intervention of the GTPase-activating protein, RasGAP, which inserts an activating arginine into the otherwise incomplete active site of Ras. Ras<GDP> is activated by docking with receptor-associated guanine nucleotide exchange factors such as Son-of-sevenless (Sos). In the GDP conformation, Ras switch II binds to Sos, which inserts an α-helix into the active site, destabilising the β/γ-phosphate binding scaffold and ejecting GDP, which is then replaced by GTP.

Structure and Function in Cell Signalling John Nelson

7.1 Classification

'G proteins' are so-called because they bind guanine nucleotides (GTP or GDP). Ras-like monomeric G proteins are related in both sequence and structure to a superclass of proteins known as 'P-loop GTPases' that are part of a wider grouping of 'P-loop NTPases'[1]. P-loop GTPases are distinguished from the other P-loop NTPases on the basis of their α/β fold, their distinct Walker A box/P-loop (explained below), and a conserved aspartate in their Walker B box.

The wider P-loop GTPase superclass is divided into two broad classes (Figure 7.1): those related to translation/elongation factors (TRAFAC class) and those related to signal sequence recognition particle GTPases (SIMIBI class). The TRAFAC class includes the elongation factor GTPases (such as EF-Tu), Septin GTPases, related motor ATPases (such as mysosin and kinesin) and the Ras-like superfamily of GTPases that are involved in signal transduction. The Ras-like superfamily encompasses both monomeric (Ras, Rac, Rho, Rap) and the alpha subunits of the heterotrimeric forms (Gαs, Gαi, Gαq). The monomeric small G proteins can be grouped into five subfamilies[2]:

- **Ras** group, (Ras, Rap and Ral) which are regulators of mitogenic response, gene expression and chemotaxis;

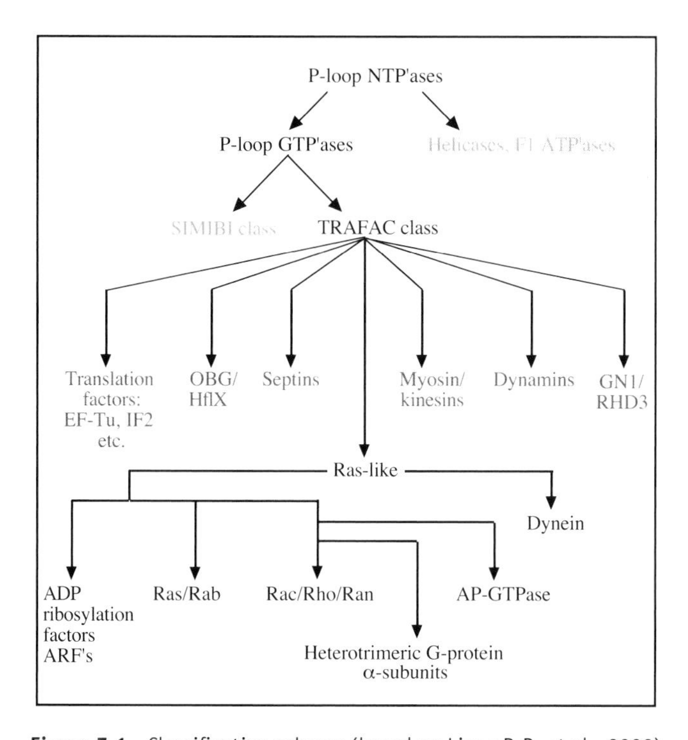

Figure 7.1 Classification scheme (based on Liepe D.D. et al., 2002)

- **Rab** group (the largest group with more than 60 members), which (like Arfs) regulate vesicular trafficking;
- **Rho** group (Rho, Rac and Cdc42), which are regulators of the actin cytoskeleton and gene transcription;
- **Ran** group, which regulate microtubules and transport of nucleocytoplasmic proteins;
- **Arf** group (Arf1-6; Arl1-7; Sar),which are regulators of vesicular trafficking and endocytosis.

7.2 ON and OFF states of Ras-like proteins

Ras and its relatives are active signal transducers, but only when GTP occupies their guanine nucleotide binding cleft/catalytic site. These monomeric G proteins may be classed as GTPases, but alone barely function as enzymes. The rate of basal GTP hydrolysis by Ras has been estimated to be $3{\cdot}4 \times 10^{-4}\ s^{-1}$ or 0·02 per minute[3]. To put it another way, it would take an average of 50 minutes to hydrolyse a bound GTP molecule. Furthermore, these 'enzymes' are decidedly unusual in that they do not readily release the product of hydrolysis (GDP). Basal GDP release rates ($4{\cdot}2 \times 10^{-4}\ s^{-1}$ or 0·025 per minute) are similar to the sluggish rate of hydrolysis.

In order to function, Ras needs input from accessory proteins, chiefly ***G***TPase ***a***ctivating ***p***roteins (GAPs) and ***g***uanine nucleotide ***e***xchange ***f***actors (GEFs). Interaction with a GAP is essential for the termination of Ras signalling because it dramatically increases GTP hydrolysis by Ras. The RasGAP protein neurofibromin increases the rate of GTP hydrolysis by HaRas from 0·01 to 1,380 per minute – such $\approx 10^5$-fold increases in catalysis are typical of the activating effects of other RasGAP proteins[4]. The result of such RasGAP intervention is the formation of Ras<GDP>, a stable resting form of the protein.

Classically, Ras<GDP> is activated downstream of single pass growth factor receptors such as EGFR and PDGFR. The ligand-occupied receptor signals its own activation status by recruitment of GEFs such as mSos (via the Grb2 adaptor). Ras<GDP> does not interact directly with the receptor but instead binds to mSos and this has the effect of causing a $\approx 10^5$-fold increase in the rate of GDP dissociation from Ras. A metastable complex of the GEF with the 'empty pocket' form of Ras is quickly followed by GTP binding and consequent de-binding of mSos. Activated Ras<GTP> is now free to interact with its effectors (Raf, RalGDS, PI-3-kinase) until it is turned off again by a subsequent RasGAP interaction (see Figure 7.2).

A third type of Ras-like regulator protein has been described, in the form of the ***g***uanine nucleotide ***d***issociation ***i***nhibitor (GDI) proteins for Rho and Rab family members. RabGDI and RhoGDI function primarily to sequester the small G proteins through binding and the masking of their prenylated *C*-termini. Shielding this membrane tether allows the G protein to be shuttled from the membrane to cytosol and on to other

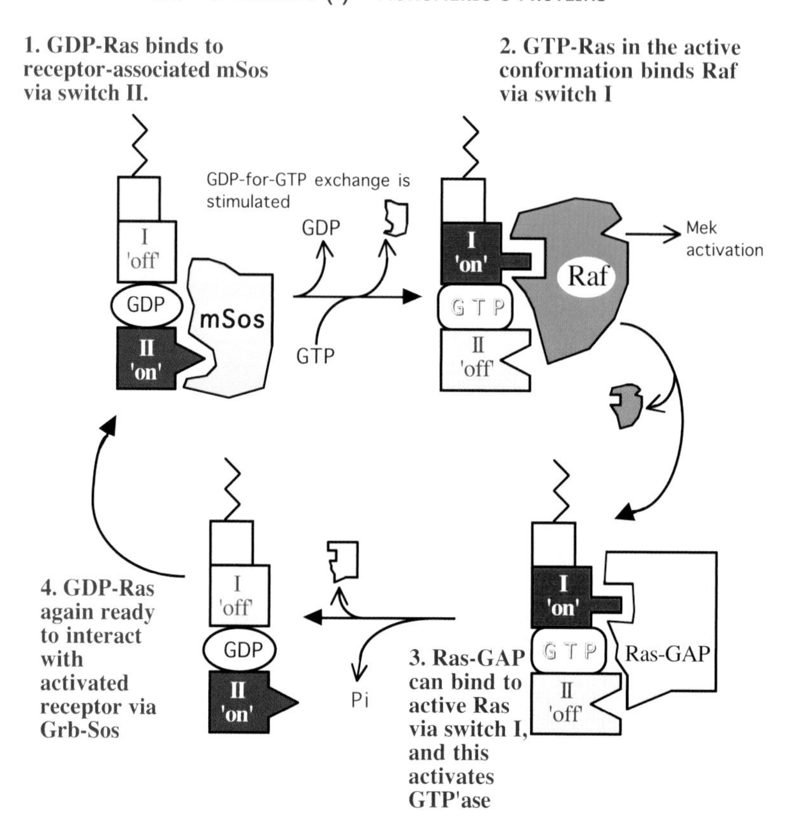

Figure 7.2 The Ras cycle

membrane compartments[5] – the inhibition of nucleotide exchange seems to be secondary to the sequestering effect. The effector Raf also has a 'GDI' effect while complexed with Ras<GTP> – the nucleotide is bound more tightly than in isolated Ras<GTP>.

7.3 Raf – a multi-domain serine/threonine kinase family of Ras effectors

Raf sits at the apex of the MAPK cascade and is the prime effector for Ras signals. Raf is a serine/threonine kinase that is present as three human isoforms with partially overlapping, but non-redundant, functions: ubiquitous Raf-1 (also referred to as C-Raf); A-Raf (also fairly ubiquitous); and B-Raf (high levels in neuronal tissue, lower levels elsewhere)[6]. All three share ***c***onserved homology ***r***egions, referred to as CR1, CR2 and CR3. The *C*-terminal half of each protein contains a serine/threonine kinase domain and this corresponds with CR3. In C-Raf, it stretches from amino acids 347 to 613.

The *N*-terminal half of the protein comprises a complex regulatory region that contains CR1 (aas: 51–194) and CR2 (aas: 254–269)[7]. CR1 contains both the ***R***as ***b***inding ***d***omain (RBD; aas: 55–131) and an adjacent ***c***ysteine-***r***ich ***d***omain (CRD;

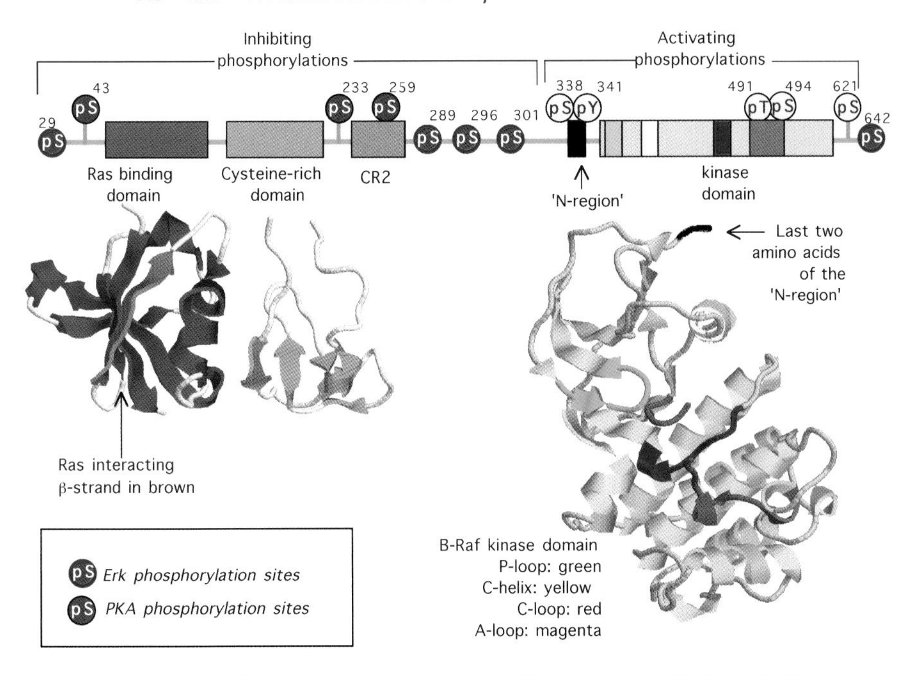

Figure 7.3 General domain organisation of Raf proteins (PDB file: 1UWH)

aas: 138–184), which functions as a 'zinc finger'. The CRD/zinc finger aids membrane association during Raf activation – it binds two molecules of zinc and a headgroup of a phosphatidylserine lipid of the plasma membrane inner leaflet[8]. The CRD is not sufficient for membrane binding alone – in the resting state, Raf is cytosolic and only translocates to the membrane in response to the appearance of Ras<GTP>. Although the primary interaction with Ras is via the RBD, the CRD also makes essential contacts with Ras.

Separate (truncated) structures are available for the isolated CRD of C-Raf (PDB file: 1FAR), the C-Raf RBD in complex with mSos (PDB file: 1C1Y), and the isolated kinase domain of B-Raf (PDB file: 1UWH). A schematic representation of the domain organisation and the isolated structures is shown in Figure 7.3. The numbering and alignment of subdomains is shown in Table 7.1.

The CR2 region is at the centre of a stretch of inhibitory serine phosphorylation sites. In contrast, the activating phosphorylations are either in the C-Raf A-loop (where threonine491 acts as the primary activating site, analogous to threonine197 of PKA) or in the 'negative-charge' regulatory region termed the 'N-region' (Figure 7.3, Table 7.1). A third activating site is found near the *C*-terminus.

7.3.1 Raf-Ras binding – translocation of Raf from cytosol to membrane

Raf regulation is extremely complex, unsurprisingly, given its importance in growth and development as evidenced by its common mutations in certain forms of cancer. Raf

Table 7.1 C-Raf and B-Raf sequences

```
C-Raf(Raf-1)
ACCESSION   P04049

  1 MEHIQGAWKT ISNGFGFKDA VFDGSSCISP TIVQQFGYQR RASDDGKLTD PSKTSNTIRV
 61 FLPNKQRTVV NVRNGMSLHD CLMKALKVRG LQPECCAVFR LLHEHKGKKA RLDWNTDAAS
121 LIGEELQVDF LDHVPLTTHN FARKTFLKLA FCDICQKFLL NGFRCQTCGY KFHEHCSTKV
181 PTMCVDWSNI RQLLLFPNST IGDSGVPALP SLTMRRMRES VSRMPVSSQH RYSTPHAFTF
241 NTSSPSSEGS LSQRQRSTST PNVHMVSTTL PVDSRMIEDA IRSHSESASP SALSSSPNNL
301 SPTGWSQPKT PVPAQRERAP VSGTQEKNKI RPRGQRDSSY YWEIEA...
                                           N-region
                                                       ... SEVM LSTRIGSGSF
361 GTVYKGKWHG DVAVKILKVV DPTPEQFQAF RNEVAVLRKT RHVNILLFMG YMTKDNLAIV
421 TQWCEGSSLY KHLHVQETKF QMFQLIDIAR QTAQGMDYLH AKNIIHRDMK SNNIFLHEGL
481 TVKIGDFGLA TVKSRWSGSQ QVEQPTGSVL WMAPEVIRMQ DNNPFSFQSD VYSYGIVLYE
541 LMTGELPYSH INNRDQIIFM VGRGYASPDL SKLYKNCPKA MKRLVADCVK KVKEERPLFP
601 QILSSIELLQ HSLPKINRSA SEPSLHRAAH TEDINACTLT TSPRLPVF

Regulatory region...
    Ras Binding Domain
    Cysteine-rich/zinc finger domain
    CR1

Kinase domain...
    P-loop
    C-loop
    A-loop

Phosphorylated residues...
    Activating phosphorylation sites
    Inhibiting phosphorylation sites
```

```
B-Raf
ACCESSION   P15056

  1 MAALSGGGGG GAEPGQALFN GDMEPEAGAG AGAAASSAAD PAIPEEVWNI KQMIKLTQEH
 61 IEALLDKFGG EHNPPSIYLE AYEEYTSKLD ALQQREQQLL ESLGNGTDFS VSSSASMDTV
121 TSSSSSSLSV LPSSLSVFQN PTDVARSNPK SPQKPIVRVF LPNKQRTVVP ARCGVTVRDS
181 LKKALMMRGL IPECCAVYRI QDGEKKPIGW DTDISWLTGE ELHVEVLENV PLTTHNFVRK
241 TFFTLAFCDF CRKLLFQGFR CQTCGYKFHQ RCSTEVPLMC VNYDQLDLLF VSKFFEHHPI
301 PQEEASLAET ALTSGSSPSA PASDSIGPQI LTSPSPSKSI PIPQPFRPAD EDHRNQFGQR
361 DRSSSAPNVH INTIEPVNID DLIRDQGFRG DGGSTTGLSA TPPASLPGSL TNVKALQKSP
421 GPQRERKSSS SSEDRNRMKT LGRRDSSDDW EIPD...
                                N-region
                                      ...GQITVG QRIGSGSFGT VYKGKWHGDV
481 AVKMLNVTAP TPQQLQAFKN EVGVLRKTRH VNILLFMGYS TKPQLAIVTQ WCEGSSLYHH
541 LHIIETKFEM IKLIDIARQT AQGMDYLHAK SIIHRDLKSN NIFLHEDLTV KIGDFGLATV
601 KSRWSGSHQF EQLSGSILWM APEVIRMQDK NPYSFQSDVY AFGIVLYELM TGQLPYSNIN
661 NRDQIIFMVG RGYLSPDLSK VRSNCPKAMK RLMAECLKKK RDERPLFPQI LASIELLARS
721 LPKIHRSASE PSLNRAGFQT EDFSLYACAS PKTPIQAGGY GAFPVH
```

interaction with Ras<GTP> is responsible for translocation of Raf from cytosol to membrane and is an essential first step in its kinase activation. At first sight, this appears to be a simple translocation of Raf from the cytosol to the membrane where its upstream activating kinases reside. However, translocation and Ras-binding is conditional upon Raf's prior phosphorylation status. Notably, in most cases, chronic PKA activation prevents Raf translocation.

7.3.2 cAMP inhibition of cell division *via* sequestration of Raf

In most cell types, sustained increases in cAMP lead to inhibition of mitosis. At least part of this suppressing effect is mediated by chronic activation of PKA. PKA phosphorylates C-Raf on three inhibiting sites *in vivo*: Ser43, Ser233 and Ser259. pSer43 is close to the RBD and inhibits Ras binding by steric hinderance. The other two pSer residues act as 14-3-3 binding sites. pSer259 of C-Raf acts as a major 14-3-3 protein binding site and when bound to 14-3-3, binding to Ras is inhibited[6]. Ser259 is also phosphorylated by PKB. pSer233 is a weaker 14-3-3 site but is theorised to collaborate with pSer259 to form an *N*-terminal high affinity bipartite binding site 'locking' C-Raf into an inactive state (Figure 7.4) that is incapable of Ras binding and membrane recruitment[6,9].

A structure of two pSer259 C-Raf peptides bound to a 14-3-3 dimer is shown in Figure 3.7 (Chapter 3).

An activating serine phosphorylation occurs on serine621 (*C*-terminal of the kinase domain) and this provides a second 14-3-3 binding site. Note that both pSer259 and pSer621 motifs conform to a high affinity 'mode 1' 14-3-3 binding site with serine at P^{-2} and proline at P^{+2}. The activating phosphorylation of Ser621 is controversial, with some studies finding that it is constitutively phosphorylated while others find it to be regulated[6]. For example, C-Raf serine 621 is reported to be rapidly (30 seconds) phosphorylated in response to nerve growth factor[7]. To add to the uncertainty, pSer621 has previously been implicated in inhibitory Raf-14-3-3 binding, working in combination (as an alternative to pSer233) with pSer259 to form a dimer binding platform[6,9,10].

RY[S]^{233}TP
ST[S]^{259}TP
SA[S]^{621}TP

7.3.3 Raf activation by translocation

14-3-3 regulation of Raf is theorised to shift between an inhibiting mode of binding to an activating mode of binding. The former is favoured by PKA phosphorylation of Ser259 and the latter is favoured by its dephosphorylation. Binding to the *N*-terminal PKA sites locks C-Raf into a 'closed' inhibited conformation that cannot dock with Ras[6,9]. If PKA is not activated, Raf is probably only phosphorylated on the activating C-terminal serine621, and 14-3-3 binding here allows membrane recruitment (if Ras is activated). At the membrane, negative charges are then imparted to the C-Raf 'N-region' by the action of the non-receptor tyrosine kinase Src, which phosphorylates tyrosine341. This action of Src is a prerequisite to the secondary phosphorylation of the proximate serine338 that follows. The kinase for this latter step may be PAK[11]. Activation of Src and PAK by growth factor receptors often parallels Ras activation (see Chapter 6).

It is unclear what part 14-3-3 binding to pSer621 plays in Raf's activating steps. There is some suspicion that 14-3-3 may act as a scaffold for activating kinases, thus facilitating efficient catalysis.

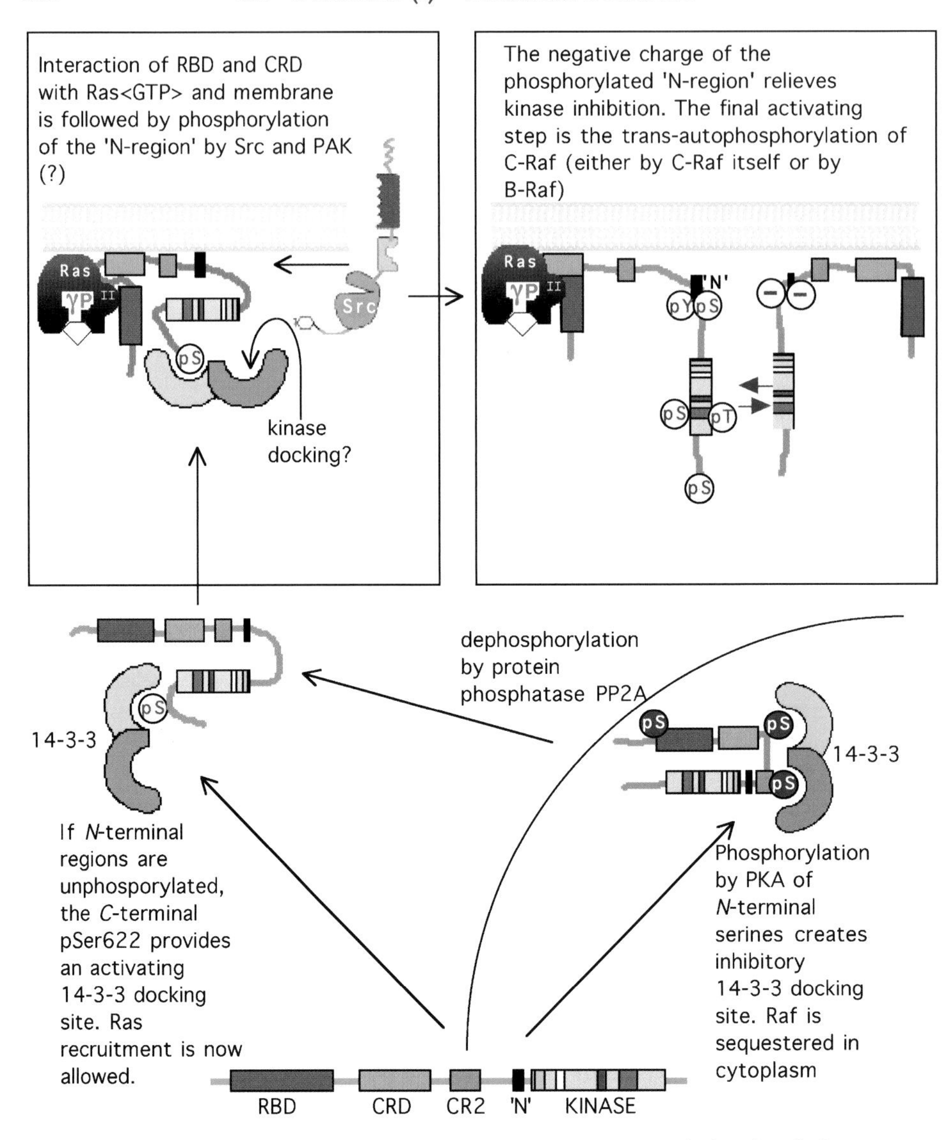

Figure 7.4 Activation of C-Raf by membrane recruitment and phosphorylation

7.3.4 B-Raf is less stringently inhibited than C-Raf

The regulation of A-Raf is less well researched but is thought to mirror that of C-Raf. B-Raf, on the other hand, has a distinct form of regulation. B-Raf differs from C-Raf in (i) having a higher constitutive (basal) activity and (ii) being independent of activation by Src. The reason for both differences lies in its 'N-region'. B-Raf lacks an equivalent to the activating pTyr341 of C-Raf. Instead it has a pair of negatively charged aspartate residues that, even in basal conditions, fulfil the need for activating negative charges.

Thus, Src does not, and cannot, activate B-Raf. Furthermore, the 'N-region' serine446 (equivalent to C-Raf pSer338) is constitutively phosphorylated. So it appears that B-Raf merely needs to be recruited to the membrane for the final step of activation[12]. (At this point it is worth noting that past numbering of B-Raf has been subject to a sequencing error – many papers have numbers one less than quoted here, as discussed in Reference 6.

7.3.5 Homologous or heterologous trans-autophosphorylation

The final step of Raf kinase activation is A-loop phosphorylation (Figure 7.4) – a step with which you will now be familiar. The main activating residue is threonine491 of C-Raf. A structure of B-Raf kinase domain indicates that its 'inactive' conformation is close to that of the ideal kinase active state, with only the A-loop DFG motif being in an inhibitory position[13]. Final activation is thought to occur through trans-autophosphorylation – upon cell activation, C-Raf forms homodimers. The Raf-Raf dimerisation, however, is independent of 14-3-3 proteins[10]. Notably, heterodimerisation of C-Raf and B-Raf also allows cross-phosphorylation and may explain why certain B-Raf mutants, with quite low kinase activity, can nevertheless produce strong stimulation of the MAPK pathway by subverting the normal C-Raf through inappropriate *trans*-phosphorylation[13].

7.3.6 Erk-1/2-type MAPK pathway activation

The Erk1/2 pathway is activated by Raf and eventually leads to the induction of immediate early response genes and initiation of cell division. The pathway was originally thought of as an enzyme 'cascade' in which a soluble kinase diffuses to downstream kinase targets activating several of them, these then diffuse to the next targets, each activating several more, and this was hypothesised to 'amplify' the signal. We now know that this is incorrect. In common with many other cytosolic 'cascades', the constituent kinases are scaffolded together so that each upstream kinase is right next to its downstream target. A 'cascade' or waterfall splashes everywhere, indiscriminately; but the MAPK pathway is actually more like a pipeline, directing its signals to discrete subcellular locations. Our current knowledge of the viscous and crowded cytoplasm makes clear that a diffusion-dependent protein 'cascade' is less efficient and therefore less 'amplifying' than an entrained cassette of kinases (Figure 7.5).

7.3.7 MAPK scaffolds

The best understood MAPK scaffolding protein is 'KSR'. It binds activated C-Raf, MAPK kinase (Mek) and MAPK (Erk)[14]. KSR was so named because at high levels it acts as a '***k***inase ***s***uppressor of ***R***as', supressing Ras/MAPK signals, but at low levels it

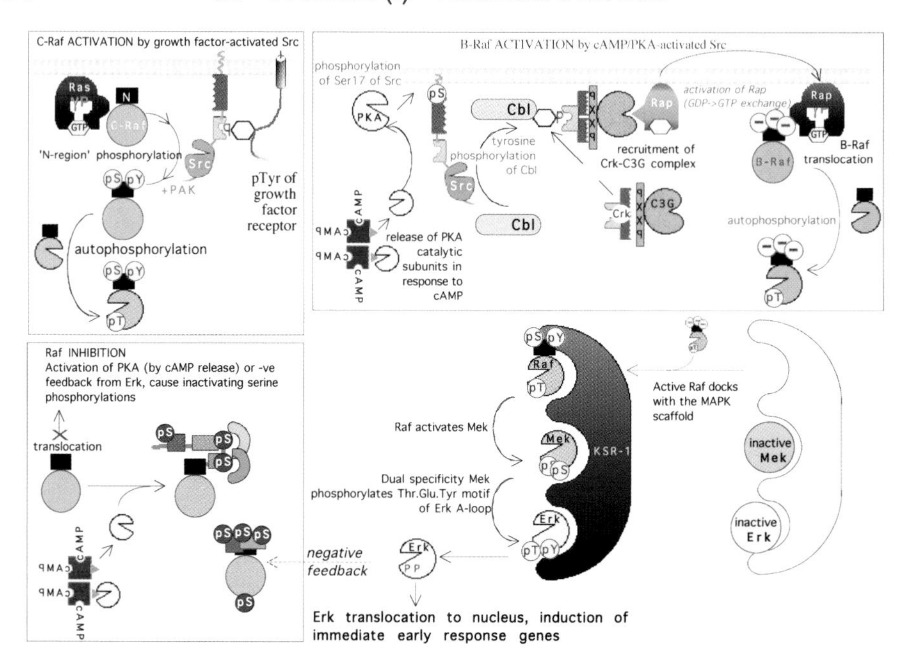

Figure 7.5 Differential regulation of C-Raf versus B-Raf

stimulates signalling. This is typical behaviour for a scaffold: at levels much higher than its client kinases, it disperses the signal because individual scaffolds bind just one client; at levels lower than the client, there is a much higher chance of achieving stoichiometric binding of all clients in each cassette. Interestingly, KSR is an inactive homologue of Raf.

7.3.8 Signal termination

The Erk1/2 pathway promotes cell division, but only if the pathway is transiently activated – persistent chronic stimulation has the opposite effect, leading to either differentiation or apoptosis[9]. Erk operates a negative feedback mechanism to shut off Raf activity if signalling persists too long, and it does this by phosphorylating C-Raf on multiple serines (Ser29; Ser289; Ser296; Ser301; Ser642) all of which are contained in 'proline-directed' motifs (Ser.Pro) that represent Erk substrate consensus sites (Figure 7.3, Table 7.1)[15]. C-Raf activity peaks 5 minutes after PDGF stimulation but declines to basal levels during the following 10 minutes, coincident with peak hyper-phosporylation. The hyperphosphorylated form of C-Raf can no longer bind to Ras and is released into the cytoplasm. The inhibited kinase is then re-set by dephosphorylation by protein phosphatase PP2A in collaboration with Pin1 (an isomerase that flips prolines from the *cis* to the *trans* conformation that conforms to PP2A substrate preference).

7.3.9 Other activating signals for Raf

A persistent, and occasionally controversial finding has been that activation of various protein kinase C isoforms also leads to the activation of the Erk-type MAPK pathway, either by direct activation of Raf or (the immediate) downstream MAPKK enzyme. Phosphorylation of C-Raf A-loop residues Ser497 and Ser499 by PKC has been reported to increase its kinase activity (as discussed in Reference 16). Cancer-promoting phorbol esters (e.g., TPA, also known as PMA) activate conventional PKC types and cause growth stimulation. TPA growth-promoting effects are mediated by activation of diacylglycerol-dependent PKCα, which in turn activates Raf – an activation that does not appear to occur at the membrane. The atypical PKCζ seems not to activate Raf, but can activate the Erk pathway by phosphorylating Mek[17]. In other studies, activated PKC-ζ has been found to be scaffolded to C-Raf by their dual binding to 14-3-3 protein and this complex is transitory because PKC-ζ phosphoryates 14-3-3, causing the complex to dissociate[18]. Whether this temporary liaison results in Raf activation or is actually a negative regulatory event is unclear.

7.4 Ras protein structure and function

There are three mammalian Ras genes (Ha-Ras, Ki-Ras and N-Ras). Their gene products were originally identified as hyperactive point-mutated oncoproteins commonly found in cancer. The proteins are ≈21 kDa in mass (p21^{Ras}), and are associated with the plasma membrane via long hydrocarbon chains, a result of fatty acylation reactions that occur during Ras processing.

Alignment of the primary amino acid sequences of Ras proteins from various species reveals two distinct regions:

- an *N*-terminal amino acid stretch (amino acids 1–165) that is highly conserved;
- a *C*-terminal amino acid stretch (amino acids 166–189) that is hypervariable.

All forms have a *C*-terminal CAAX motif that is farnesylated ('A' = aliphatic, 'X' = any amino acid); some are also prenylated. The farnesylation reaction is followed by proteolysis of the AAX and carboxymethylation of the (now) *C*-terminal Cys. The farnesyl group, however, needs an additional tether for Ras proteins to become membrane bound. K-Ras employs a polybasic patch of six lysines to achieve membrane tethering (reminiscent of myristoylated proteins see Chapter 3, Section 3.1.4), whereas H-Ras and N-Ras have additional fatty acid tethers in the form of cysteine-linked palmitoyl chains[19]. These post-translational modifications are not only essential to the membrane binding of Ras but also to the differential trafficking of the three isoforms to distinct microdomains of the plasma membrane and internal membranes. Drugs that inhibit Ras farnesylation lead to the synthesis of cytoplasmic inactive proteins (even in the case of Gln61 mutant Ras, which would otherwise be constitutively active). The

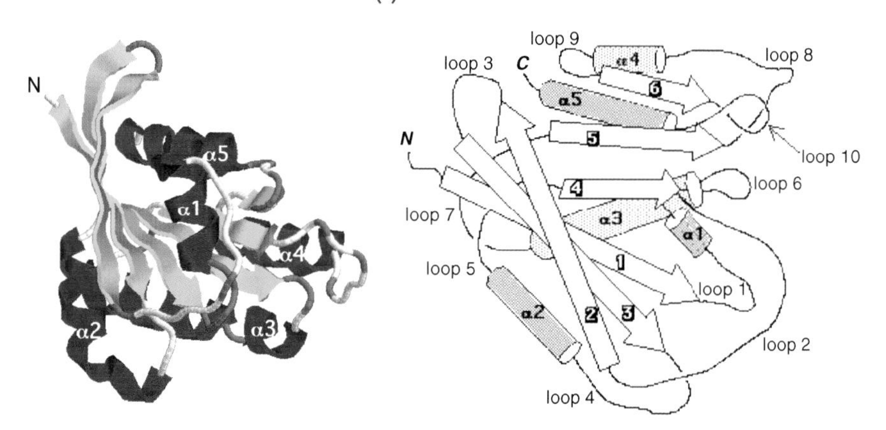

Figure 7.6 Schematic diagram and crystal structure of Ras showing alpha and beta structure

reason for this inactivity is that Ras must be membrane-associated in order to interact efficiently with its activators: the signal particles of integral membrane proteins of the single-pass receptor tyrosine kinase family.

Ras structures have been derived from *C*-terminally truncated proteins – even when full length Ras has been subjected to structural analysis, the *C*-terminal region is too disordered to be resolved[13]. The *N*-terminal 166 amino acids, which are well defined in X-ray crystal and NMR structures, comprise both the catalytic site and the binding surfaces for interaction with effectors and regulators. This 'G-domain' is highly conserved between species and is fully competent in GTP hydrolysis. The *N*-terminal domain is a tightly folded structure consisting of a central six-stranded β-sheet (five parallel strands and one anti-parallel) surrounded by five α-helices – the so-called 'α/β fold' (Figure 7.6). The β-strands and α-helices are connected by ten polypeptide loops, and it is within five of these loops that the catalytic site is centred, and where the switch-like mechanism of Ras is triggered.

7.4.1 The GTPase site of Ras: G-boxes and switch regions

In the Ras protein, the active site resides in the aforementioned five loops. These are the so-called 'G-regions' or 'G-boxes'. These G-regions sequences contain key residues that are highly conserved throughout the Ras family.

7.4.2 The P-loop (G-1)

The **G-1** region encompasses the glycine-rich pyrophosphate-binding loop (P-loop) and contains a conserved lysine (Lys16) and an adjacent conserved serine (Ser17) (Table 7.2). In some members of the superfamily, such as elongation factors (EFs) and ADP ribosylation factors (ARFs), this serine is replaced by a threonine[1].

Table 7.2 Harvey Ras catalytic domain sequence 1–166

	Met1	Thr2	Glu3	Tyr4	Lys5	Leu6	Val7	Val8	Val9	**Gly10**
G-1	**Ala11**	**Gly12**	**Gly13**	**Val14**	**Gly15**	**Lys16**	**Ser17**	Ala18	Leu19	Thr20
	Ile21	Gln22	Leu23	Ile24	Gln25	Asn26	His27	Phe28	Val29	Asp30
G-2	Glu31	**Tyr32**	**Asp33**	**Pro34**	**Thr35**	**Ile36**	**Glu37**	**Asp38**	Ser39	Tyr40
	Arg41	Lys42	Gln43	Val44	Val45	Ile46	Asp47	Gly48	Glu49	Thr50
	Cys51	Leu52	Leu53	Asp54	Ile55	**Leu56**	**Asp57**	**Thr58**	**Ala59**	**Gly60**
G-3	**Gln61**	**Glu62**	**Glu63**	**Tyr64**	**Ser65**	**Ala66**	**Met67**	**Arg68**	**Asp69**	**Gln70**
	Tyr71	**Met72**	**Arg73**	**Thr74**	**Gly75**	**Glu76**	Gly77	Phe78	Leu79	Cys80
	Val81	Phe82	Ala83	Ile84	Asn85	Asn86	Thr87	Lys88	Ser89	Phe90
	Glu91	Asp92	Ile93	His94	Gln95	Tyr96	Arg97	Glu98	Gln99	Ile100
	Lys101	Arg102	Val103	Lys104	Asp105	Ser106	Asp107	Asp108	Val109	Pro110
G-4	Met111	Val112	Leu113	Val114	Gly115	Asn116	Lys117	Cys118	Asp119	Leu120
	Ala121	Ala122	Arg123	Thr124	Val125	Glu126	Ser127	Arg128	Gln129	Ala130
	Gln131	Asp132	Leu133	Ala134	Arg135	Ser136	Tyr137	Gly138	Ile139	Pro140
G-5	Tyr141	Ile142	Glu143	Thr144	Ser145	Ala146	Lys147	Thr148	Arg149	Gln150
	Gly151	Val152	Glu153	Asp154	Ala155	Phe156	Tyr157	Thr158	Leu159	Val160
	Arg161	Glu162	Ile163	Arg164	Gln165	His166				

G-box regions are shaded yellow
P-loop is in green text underlined
Effector region is in red text (switch I is underlined)
Switch II is in blue text underlined

Inset shows Ras structure with G-boxes in yellow and GTP analogue in active site

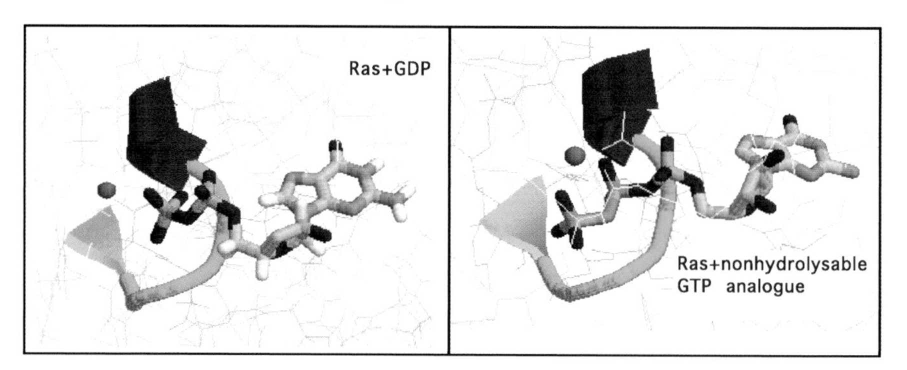

Figure 7.7 P-loop of Ras

In Ras-like monomeric G proteins, the **G-1** consensus sequence is GXGGXGKS and in the wider superfamily it is GXXXXGK(S/T). Such glycine containing motifs are common in dinucleotide- and mononucleotide-binding proteins[21,22]. The **G-1** region is sometimes referred to as a 'Walker A' box. In monomeric G proteins, P-loops extend from the first β-strand to the first α-helix (Figure 7.7). In the case of human Harvey Ras, the P-loop extends from residues 10–17 (Table 7.2).

The lack of sidechains in the glycine-rich turn allows the loop to wrap around the phosphates and the amide nitrogens provide a positively charged electrostatic field to accommodate the negatively charged phosphates. Glycines are essential here because no other amino acid could sterically accommodate the torsional angles experienced by Gly10 and Gly15 in the P-loop fold[23]. The sidechain of Lys16 hydrogen-bonds the β- and γ-phosphates and the sidechain hydroxyl of Ser17 coordinates the Mg^{2+} ion while its main chain amide hydrogen-bonds to the β-phosphate. During hydrolysis, the invariant lysine is thought to help stabilise the transition state by neutralising charges at the γ-phosphate and increasing the charge on the β-phosphate leaving group.

7.4.3 Switch I (G-2)

The **G-2** box is located in the loop ('loop-2') that connects the α-1 helix and β-2 strand in Ras (Figure 7.8). The **G-2** box contains an invariant threonine (Thr35 in Ras) – its sidechain hydroxyl group participates in coordination of the essential Mg^{2+} (Figure 7.9). The **G-2** box is contained within the **switch I** region of Ras (Table 7.2, Figure 7.8). This is one of two mobile elements in Ras (switches I and II) that are known to be involved in the binding of regulators and effectors. These two switches are the sites of the largest conformational changes seen when comparing Ras-GDP with Ras-GTP (Figure 7.10). As we shall see, the conformational changes are driven by the presence or absence of the γ-phosphate group of the bound guanine nucleotide. Both switch regions were identified in structural studies comparing GDP-occupied Ras with Ras bound to a non-hydrolysable GTP analogue (phosphoaminophosphonic acid-guanylate ester: GPPNHP or GNP); both switches were found to coincide with regions identified by independent

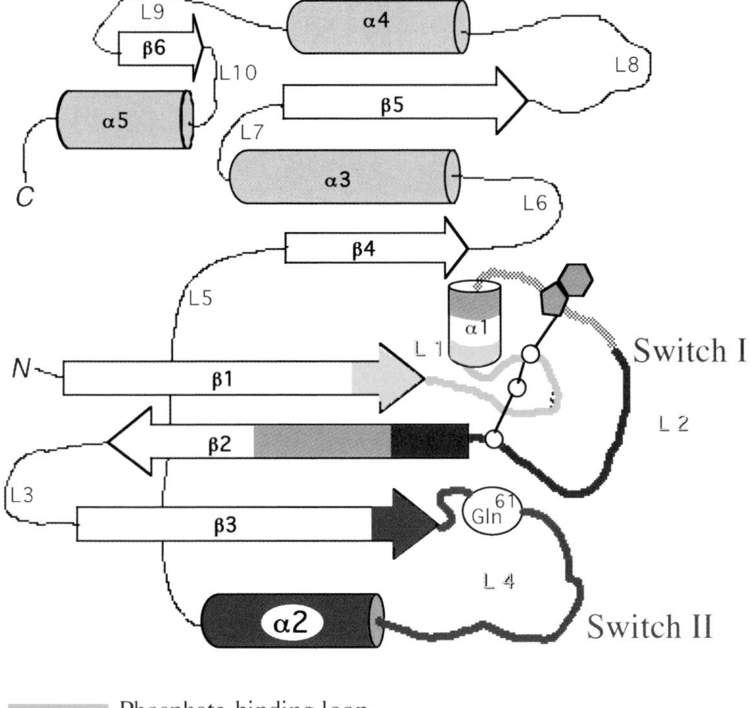

Figure 7.8 Schematic diagram of Ras

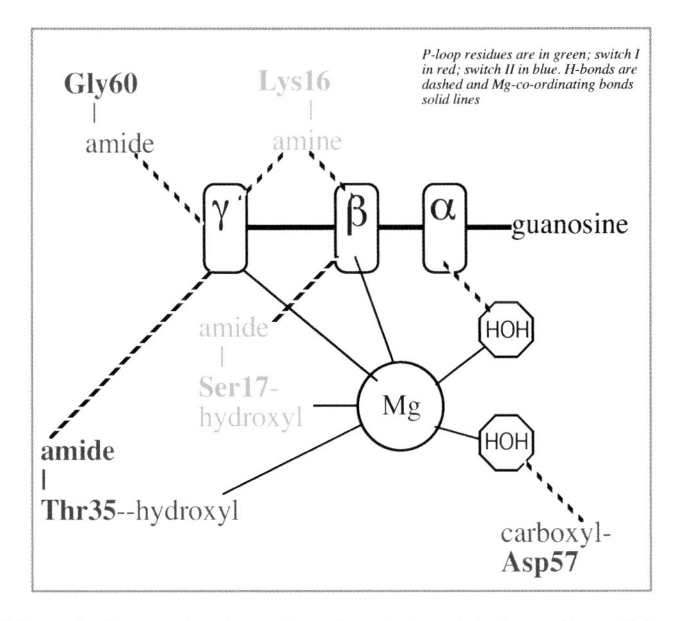

Figure 7.9 Schematic diagram showing points of contact made between Ras switch regions and the γ-phosphate of GTP

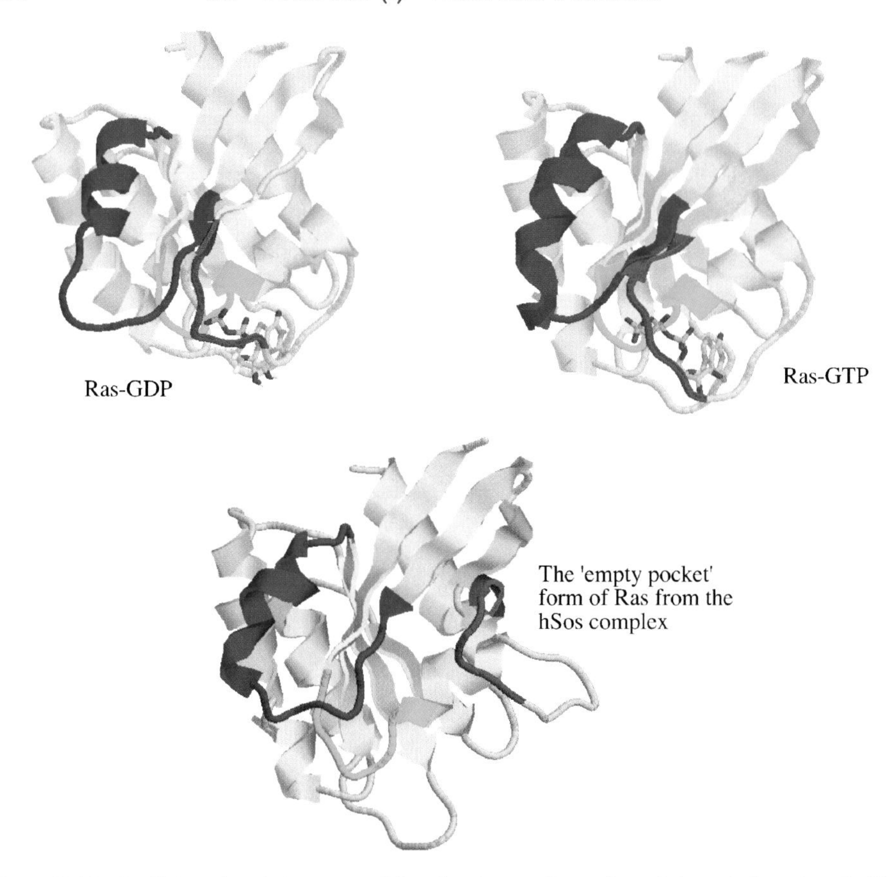

Figure 7.10 RasMol rendered structures of Ras showing conformation of the switch regions in GDP-bound (PDB file: 1CRQ) and GTP-bound (PDB file: 1CTQ) forms

biochemical investigations as being involved in binding to regulators such as mSos or RasGAP or effector enzymes such as Raf.

Switch I is part of the larger 'effector region', identified in solution studies of *oncogenic* Ras as the binding surface for the serine kinase Raf. Oncogenic Ras mutants cause cancer because they are less dependent upon activation by growth factors and tend to be constitutively active due to their inability to hydrolyse GTP – effectively their effector regions are stuck in the 'ON' position. Point mutations (of such oncogenic Ras proteins) that inactivate signalling *without affecting the GTPase* are centred on this region.

7.4.4 Switch II (G-3)

The **G-3** sequence (containing an invariant consensus DXXG) is found at the *N*-terminus of the α-2 helix and forms part of the **switch II** region of both monomeric and heterotrimeric subunits. The invariant aspartate coordinates the magnesium ion via a water molecule and the conserved glycine coordinates the γ-phosphate (in GTP-bound forms) (Figure 7.9). This glycine (Gly60 in Ras) is followed by a glutamine in both Ras and Gα-subunits. Gln61 of Ras (and the equivalent Gln in αs) is a common site of point mutation in human cancers. Almost all such mutations destroy GTPase activity, block GTPase activation by RasGAP and cause oncogenic conversion of Ras (reviewed in Reference 23; see also Chapter 10, Sections 10.3 and 10.17.1) and for this reason Gln61 is thought to be the most likely candidate for the prime catalytic residue.

Although conserved in Ras and Gα-subunits, this glutamine is not conserved in the wider TRAFAC class – it is replaced by histidine in EF-Tu and by a threonine in Rap-1A.

The **G-3** sequence (DXXG) is often referred to as a Walker B box. In Ras, Switch II runs from the *C*-terminal portion of β-strand 3 through loop 4 to α-helix 2.

Note: For the sake of simplicity, switch II numbering throughout the text is based upon homologies with switch regions identified in bovine transducin[24]. One should be aware that the span of switch regions is approximate and may vary from G protein to G protein and from author to author (see for example References 5 and 25).

The **G-4** sequence (containing an invariant consensus NKXD) is located in the loop between the β-5 strand and the α-4 helix and is partially responsible for binding the guanine base.

The **G-5** box runs between the β-6 strand and the α-5 helix and aids in guanine base recognition and binding.

7.5 The switch mechanism: hydrolysis-driven conformational change in Ras

The greatest conformational change noted in Ras-GDP versus Ras-GTP is a relaxation of the two switches when GTP is hydrolysed. This suggests a trap or spring-like mechanism that is 'cocked' in Ras-GTP and 'released' by removal of the γ-phosphate during hydrolysis[25]. The most easily discernible change is in switch II, where α-helix 2 has noticeably extended into loop-4 of Ras-GTP. In Ras-GDP, on the other hand, the helix is shorter, the loop longer, and the whole switch has swung away from the G-cleft (Figure 7.10). In some G protein structures, switch II is so disordered that it is not visible in the X-ray crystal (see GαiGDP, Chapter 8). Switch I is also more ordered in the GTP-bound conformation – the switch extends by an extra residue into β-strand 2 when GTP is bound – and swings away from the G-cleft in the GDP-bound form (Figure 7.10).

The question then is: how are these changes triggered? The answer appears to lie in the contacts made with the γ-phosphate of GTP, the essential Mg^{2+} ion, and key residues in both switch regions. To understand this, it is necessary to examine the active site in some detail.

Two informative structures are of human Harvey Ras: an X-ray crystallographic structure of Ras complexed with the GTP analogue GNP (PDB file: 1CTQ) shows the active site poised for hydrolysis[26]; whereas an NMR structure of Ras complexed with GDP (PDB file: 1CRQ) shows the active site and switch regions in a relaxed state[27]. Both studies used the truncated *N*-terminal domain (1–165 and 1–166, respectively).

GTP-bound Ras is a taut structure, because switch I and switch II are tethered to the γ-phosphate group. The schematic diagram (Figure 7.9) summarises the key contacts (<3 Å) that Ras makes with the γ-phosphate, the essential Mg ion and two water molecules[26]. The most obvious tethers are the Mg^{2+}-coordinating bond from the switch I Thr35 hydroxyl group, the H-bond between the γ-phosphate and the amide nitrogen of the switch II glycine60, the H-bond between the Mg^{2+}-coordinating water and the sidechain carboxyl group of switch II Asp57, and the Mg^{2+}-coordinating bond from the Ser17 sidechain hydroxyl group (Figure 7.11a). In an evaluation of a Ras structure complexed with a caged-GTP (and subsequently activated by photolysis followed by freezing) another H-bond was found from the backbone amide of Thr35 to the γ-phosphate[26].

7.6 GTP hydrolysis

As mentioned previously, isolated Ras has a very low rate of hydrolysis, taking ≈50 minutes to an hour to hydrolyse a single GTP[3,28]. Compared with Gαs subunits, which are themselves sluggish, hydrolysing a GTP in 15 seconds[29], Ras is a very poor enzyme indeed. In fact, it could be said that Ras has an incomplete active site. Ras lacks the active site arginine, conserved in switch I of heterotrimeric Gα-subunits, that is thought to help stabilise the transition state (Chapter 8, Section 8.2.3). Instead, Ras relies upon the GTPase-activating protein Ras-GAP to provide an 'arginine finger' that allows the Ras/RasGAP complex to hydrolyse GTP (see below).

No absolute consensus exists as to which residue acts as a general base during catalysis. It was originally proposed[30] that the carboxylate group of the conserved glutamate in the switch I/G-3 box (Ras: GQE**E**^{63}, Gαt: GQRS**E**^{203}) might act to activate a water molecule by abstracting a proton, thus serving as a general base much as a conserved glutamate does in P-loop NTPases such as the helicases (reviewed in Reference 1) or indeed the catalytic aspartate in protein kinases. However this glutamate residue has been ruled out because its substitution (by mutation) has no effect on hydrolysis[31].

On the other hand, natural and experimental point mutations of the conserved glutamine do have a profound effect upon hydrolysis – substitution of the Gln61 with almost any other amino acid (except glutamate) inactivates the active site.

The possible contribution of this Gln61 residue to catalysis is more easily understood by reference to the structures of Gα-subunits complexed with AlF_4^- and GDP (Chapter 8, Section 8.2.3), where the aluminium fluoride ion is thought to mimic the pentavalent transition state of the γ-phosphate group of GTP. Here, the Gα-subunit glutamine is within H-bonding distance of an AlF_4^--bound water (the presumed nucleophile). The sidechain guanidino group of the arginine of switch I (in Gα-subunits) coordinates the AlF_4^- (which mimics the γ-phosphate), and is thought to complete the active site by neutralising the developing negative charge on the γ-phosphate leaving group.

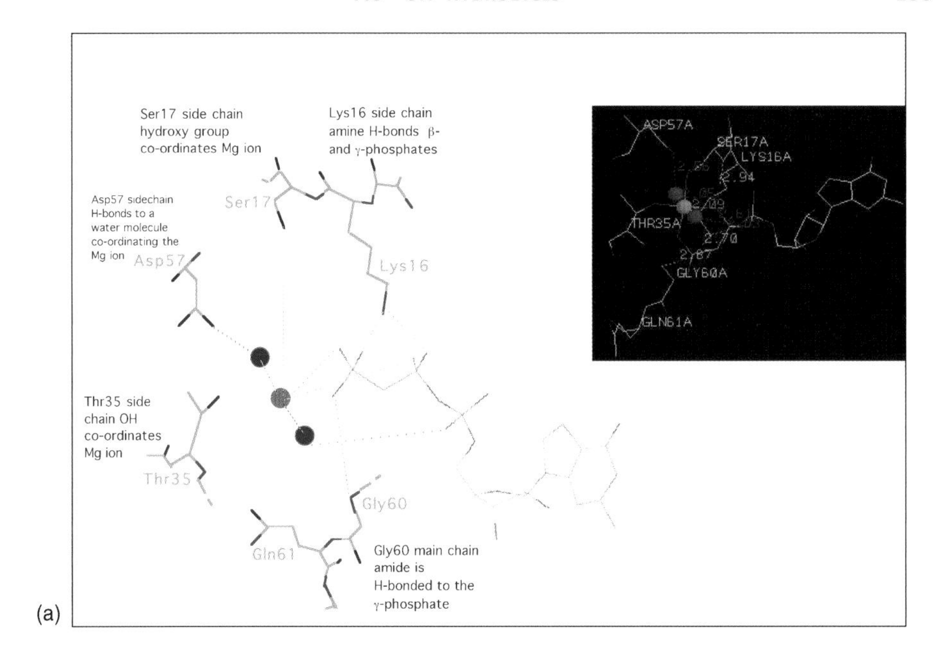

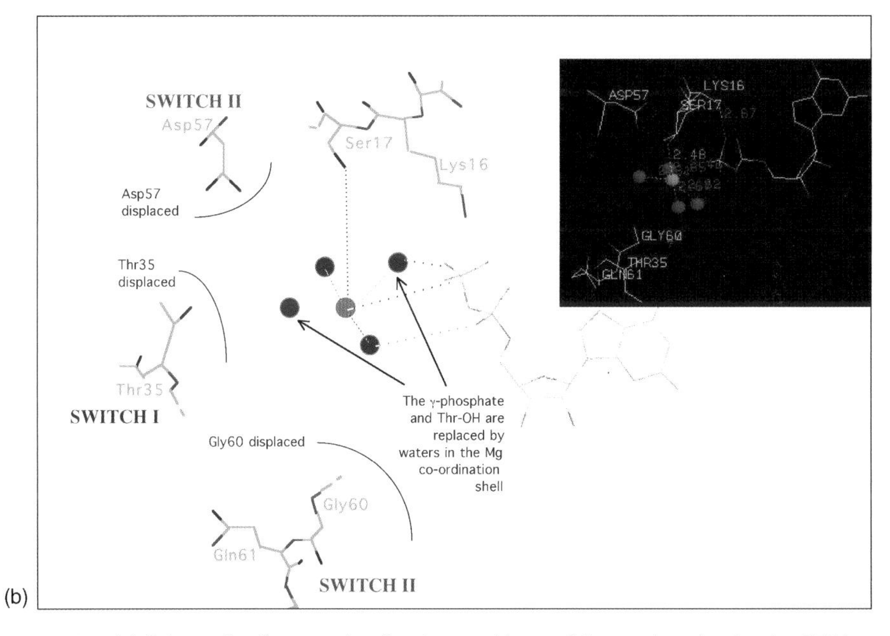

Figure 7.11 (a) Schematic diagram showing key residues of Ras active site in the GTP-bound conformation (insert shows the actual crystal structure rendered by RasMol, PDB file: 1CTQ); (b) (insert shows the actual NMR structure rendered by RasMol, PDB file: 1CRQ)

Interestingly, the complex of Ras with RasGAP is able to bind aluminium fluoride whereas Ras is not. It is likely that the contribution of an arginine (by RasGAP) completes the active site and thus allows Ras to more easily achieve the transition state seen in α-subunit structures.

An alternative, as a general base in catalysis, has been suggested to be the substrate itself. In this model, the γ-phosphate acts as the general base, abstracting a proton from a catalytic water molecule, thus creating a nucleophilic hydroxide ion that attacks (the now protonated) γ-phosphate.

7.6.1 Structural effects of loss of γ-phosphate

The loss of the γ-phosphate group breaks the bonds made by switch I Thr35 and switch II Gly60 and Asp57 (Figure 7.11a). The γ-phosphate group is replaced by water in the Mg co-ordination shell, as is the Thr35 hydroxyl (Figure 7.11b). The backbone H-bond from Gly60 is broken and Asp57 is displaced[23]. The net effect is the relaxation of both switch regions.

7.7 Effector and regulator binding surfaces of Ras

It is helpful to think of Ras switches being either 'ON' or 'OFF'. In this analogy switch I (which is primarily responsible for interaction with RasGAP and effector enzymes such as Raf, is 'ON' in the GTP-bound conformation and 'OFF' in Ras-GDP. Switch II (which predominates in interactions with the Ras-activating guanine nucleotide exchange factors such as mSos) is 'ON' when Ras is complexed with GDP and 'OFF' in Ras-GTP (see Figure 7.2). This prevents conflict between activating and inactivating factors – mSos only recognises switch II in the GDP conformation and Ras GAP only interacts with switches I and II in the GTP conformation.

7.7.1 RasGAP

$p120^{GAP}$ is a large multi-domain protein that contains two SH2 domains through which it is recruited to activated single-pass receptors such as PDGFR. The Ras-interacting region of RasGAP is composed of two alpha helical domains, the larger (*C*-terminal) of which contains the Ras binding cleft. A crystal structure of this portion – residues 718–1037 of RasGAP – complexed with residues 1–166 of Ras includes bound GDP, aluminium fluoride, magnesium and water molecules[32] (PDB file: 1WQ1).

Looking at the overall structure, one can see that RasGAP inserts a finger-like loop into the active site of Ras (Figure 7.12). This RasGAP loop contains the essential arginine (referred to above) that completes the active site of Ras. This catalytic loop is referred to as an 'arginine finger''. Although the main interaction is with the Ras effector region containing switch I (discussed in Chapter 6), the RasGAP *C*-domain also binds to, and stabilises, switch II as well as burying the P-loop in its binding cleft.

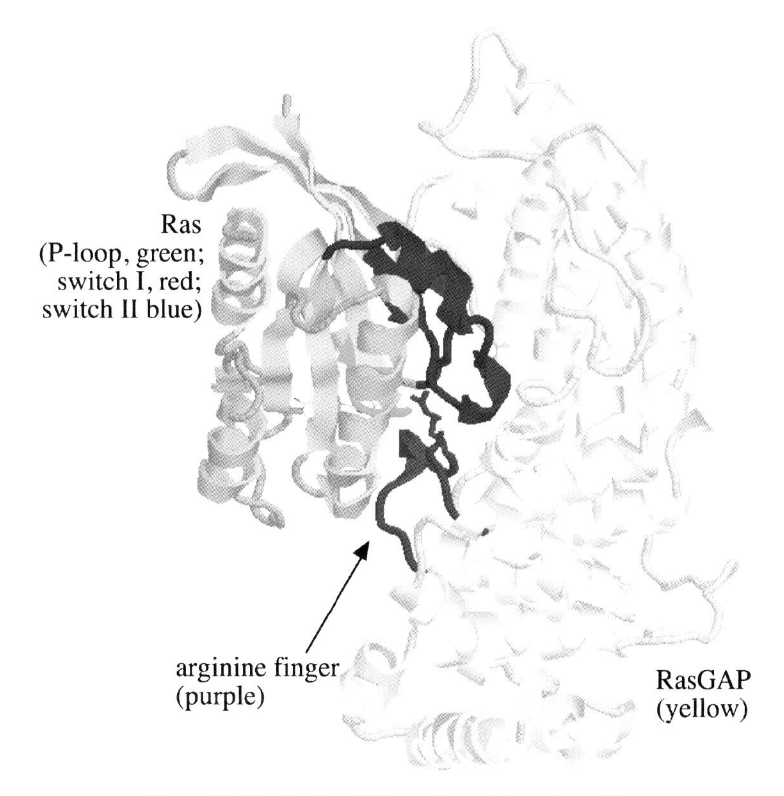

Figure 7.12 The RasGAP complex with activated Ras

Zooming in on the composite active site, one can see the transition state that is made possible by the arginine finger of GAP (Figure 7.13). In this model, the pentavalent transition state of the γ-phosphate group of GTP is mimicked by AlF_3 (rather than AlF_4^- in the α-subunit structures, Chapter 8). Arginine789 (supplied by RasGAP) is H-bonded via its backbone carbonyl to the amide of Gln61 sidechain (of Ras). This interaction moves Gln61 much closer to the active site than is the case in the ground state (compare Figure 7.13 with Figure 7.11). The carbonyl sidechain of Gln61 is now H-bonded to the catalytic water and the arginine guandino group (supplied by RasGAP) is H-bonded to the γ-phosphate. This produces a pentavalent transition state resembling the heterotrimeric AlF_4^- structures (reviewed in References 1 and 23).

The two residues can be thought of as collaborating in stabilising a Ras-RasGAP transition state that is much more difficult to achieve for Ras alone. In particular, the H-bond between the arginine finger and Gln61 moves its sidechain close enough to interact with the nucleophilic water (Figure 7.13), an interaction that is less favourable in the ground state[26,32]. Furthermore, the residues immediately adjacent to Gln61 (Glu62. Glu63.Tyr64) also interact with GAP residues and serve to further stabilise this important portion of switch II.

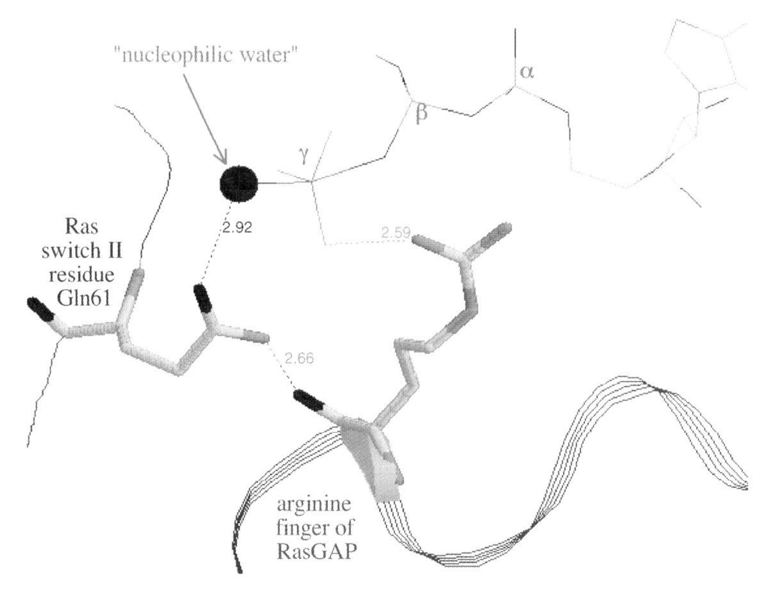

Figure 7.13 Active site of Ras complexed with GDP, aluminium fluoride and RasGAP

Specific GAP proteins exist for all members of the Ras-related monomeric G proteins, but sequence and structural homology only pertains within each of the GAP subfamilies. However, even though GAP protein structures vary, they do exhibit a common mechanism: arginine fingers have been identified in unrelated GAPs for most Ras superfamily members[33].

Note: In the RasMol/PDB file, the GAP fragment numbers the catalytic Arg789 as 'Arg76' and the finger loop runs from Asp64–Lys90 in RasMol/PDB file.

7.7.2 RasGEFs

Guanine nucleotide exchange factors (GEFs) serve two functions: (i) they enable Ras-GDP to dock with receptor-recruited adaptors such as Grb2/Shc (see Chapter 6) and (ii) they are fundamental to Ras activation, forcing exchange of GDP for GTP. GEF proteins appear quite unrelated to one another and are remarkably varied in the structures of their nucleotide exchange-catalysing domain. In contrast to the closely connected Ras family tree, there appears to be no evidence for a common GEF ancestor, nor is there any indication of convergent evolution. Consequently, it is not surprising that they do not share a common mechanism of action. The discussion will therefore be confined to the best known GEF for p21Ras: mammalian Sos.

We have seen above that the interactions of Ras with GDP or GTP are predominantly centred on the β- and γ-phosphates, with little α-phosphate interaction. For these reasons, Ras has a very low affinity for GMP – six orders of magnitude less than for GDP or GTP – and because a GEF functions only to eject GDP from the binding site of

Ras, it is unsurprising that GEF targets the P-loop of Ras (the primary binding site of the β-phosphate). The nucleotide-free Ras-Sos complex shows no selectivity for the guanine nucleotide that replaces GDP – either GTP or GDP will bind, but only GTP causes dissociation of the complex and thus preferentially replaces GDP (see Chapter 2, Section 2.4). The Ras-Sos complex is only stable in the absence of guanine nucleotide. Sos serves to stabilise the nucleotide-free ('empty pocket') form of Ras that is readied for GTP binding. The 'empty pocket' forms of G proteins are highly unstable except when associated with a GEF (in the case of Ras-like small G proteins) or with an activated 7-pass receptor (in the case of heterotrimeric G proteins).

A crystal structure shows $Ras_{1\text{-}166}$ complexed with the 'CDC-25' domain of hSos-1 in the absence of guanine nucleotide[34] (PDB file: 1BDK) – *Sos shares a stretch of homology with the yeast GEF, CDC-25, and this region represents the exchange-catalysing domain.* The catalytic domain of Sos is predominantly alpha helical (Figure 7.14).

Sos binds principally to the Ras-GDP-conformed switch II region via numerous sidechain interactions – a fact confirmed by mutagenesis experiments – and this docking site allows Sos to interact directly with the P-loop (reviewed in Reference 35). The key exchange-catalysing residues are centred on the Sos α-helix 'H', which extends from residue Gly931 to Glu942 (in the PDB file these are 326–337s). Two key residues appear to be the H-helix amino acids, $Glu942/^{PDB\ no:337}$ and $Leu938/^{PDB\ no:333}$, which can clearly be seen inserted into the active site of Ras.

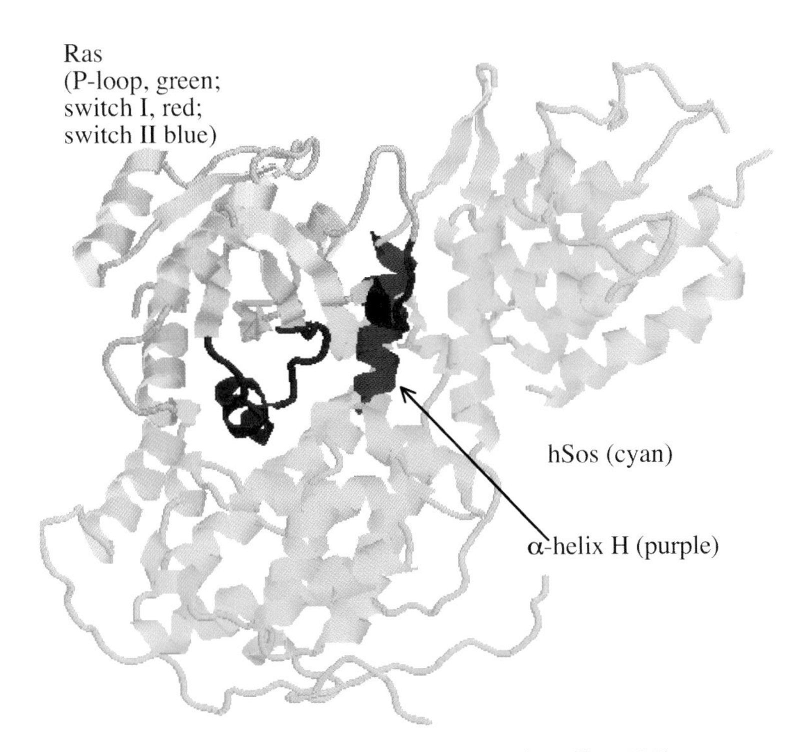

Figure 7.14 The Sos complex with Ras (PDB file: 1BDK)

The helical hairpin containing the α-helix H of Sos displaces switch I from the nucleotide binding site of Ras (Figure 7.14) and disrupts its structure, thus eliminating switch I's water-mediated, and direct, interactions with GDP.

The strong interaction of mSos with switch II is centred on hydrophobic binding of Ras residues Tyr64, Met67 and Tyr71 and, surrounding that, polar or ionic interactions with other residues of switch II[34]. In the nucleotide-free structure, switch II is in a different conformation and is less mobile than in Ras-GTP or Ras-GDP. In this Ras conformation, the amide of Gly60 is H-bonded to the Glu62 sidechain that also binds Lys16 of the P-loop. Furthermore, because the Sos α-H glutamate residue 942 interacts with the Ser17 residue of Ras P-loop (Figure 7.15) the net result is a distortion of the P-loop into a conformation that is incapable of interacting with the β-phosphate of GDP. Finally, Sos residue Leu938 collaborates with Glu942 by blocking the Mg^{2+} binding site.

This destruction of the β-phosphate scaffold provides a plausible explanation of how GDP is expelled from the Ras G-cleft, leaving it empty and exposed to the cytosol. GTP can enter the G-cleft (base end first), interacting with GMP-binding residues to form a low-affinity intermediate that eventually remodels switch II, turning it 'OFF' and dissociating Sos from the, now, GTP-bound Ras.

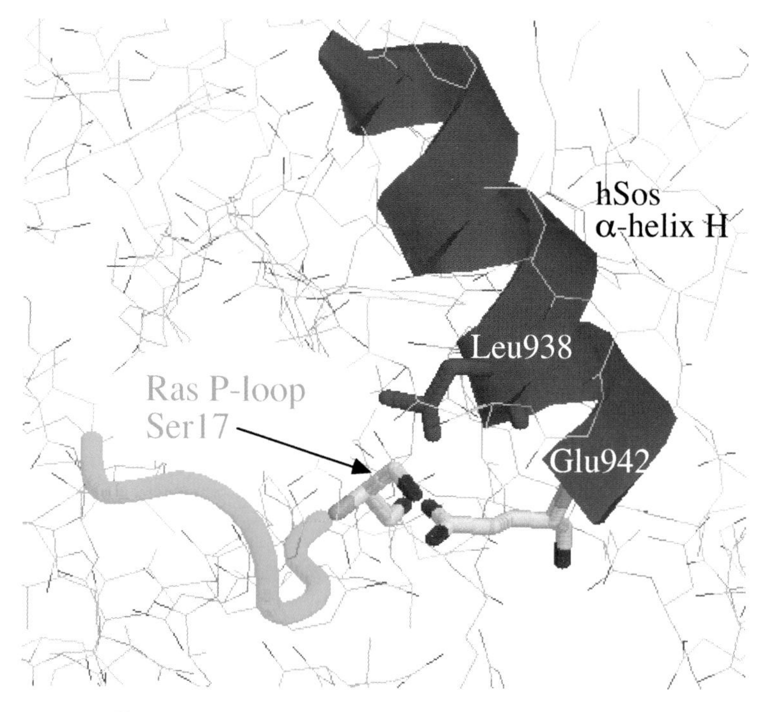

Figure 7.15 Interaction of Sos α-helix H with Ras P-loop

7.7.3 The Ras effector region and Raf binding

The interaction of Ras with downstream effectors was first detected indirectly as the ability to cause transformed growth of cultured fibroblasts. Oncogenic forms of Ras have damaged GTPase sites that do not respond to RasGAP intervention and therefore cause uncontrolled growth because they persist in the GTP-bound form. The 'effector region' was initially identified by deletion and substitution mutagenesis of oncogenic Ras – residues whose mutation blocked the transforming power of oncogenic Ras, without restoring GTPase activity, were considered to represent points of contact with an immediate downstream effector enzyme.

The oncogenic form of Ha-Ras was subjected to a series of deletion mutagenesis experiments, which showed that deletion or substitution of residues within a single stretch of amino acids (22–43) could inactivate the growth-promoting effects of oncogenic Ras; importantly these mutations did not rescue the inactive GTPase but instead prevented coupling to the downstream effector. Before Raf was identified as the Ras effector, it was at first thought that RasGAP may be the effector because its primary binding site was also the 'effector region' (switch I of Ras)[25]. In fact, Raf and GAP compete for binding to the effector region around switch I[36]. Support for the Ras effector region interacting with Raf came from a study showing that a synthetic peptide encompassing residues 17–44 of Ras is sufficient to block the binding of Raf to its target Ras and that the Ras point mutant $Asp^{38} \rightarrow Ala^{38}$ is inactive because it cannot bind Raf kinase[37].

Again, when crystal structures of a Ras-like protein complexed with the serine kinase Raf became available, the predictions from mutagenesis studies were largely confirmed. The Ras-binding domain (RBD) of Raf has been crystallised (PDB file: 1C1V) in complex with a non-hydrolysable GTP analogue and a homologue of Ras, Rap1A[38]. Rap1A, at least in some cell types, acts as an antagonist of Ras-Raf signalling[39]. Rap1 has complete identity with the effector region of Ras. Rap1A antagonises Ras-dependent activation of C-Raf (but not B-Raf) because Rap1 sequesters C-Raf in regions of the plasma membrane where a normally activating phosphorylation of C-Raf cannot occur. Ras-Raf complexes on the other hand are concentrated in 'lipid rafts' where the final activating phosphorylation of C-Raf on Ser338 occurs[40].

The first important point is that, unlike the Ras-Sos complex, Ras is not distorted by binding with Raf – the *overall* conformation of Rap1A is remarkably similar to the structure of Ras-GTP. However, it is worth noting one difference in the fine detail: the H-bonds linking the main chain amides of Thr35 and Gly60 to the γ-phosphate (see Figure 7.9) are much tighter (both are 0.25 Å shorter) than in isolated Ras-GTP and this may explain why Raf acts as a guanine nucleotide dissociation inhibitor (GDI) for Ras and Rap (discussed in Reference 38). The concept of GDI is discussed further in the next chapter (Section 8.5.2).

The crystal structure (Figure 7.16) reveals a Rap1A-Raf binding interface that is solely encompassed by the Ras/Rap effector region. The Ras/Rap β-2 strand forms an

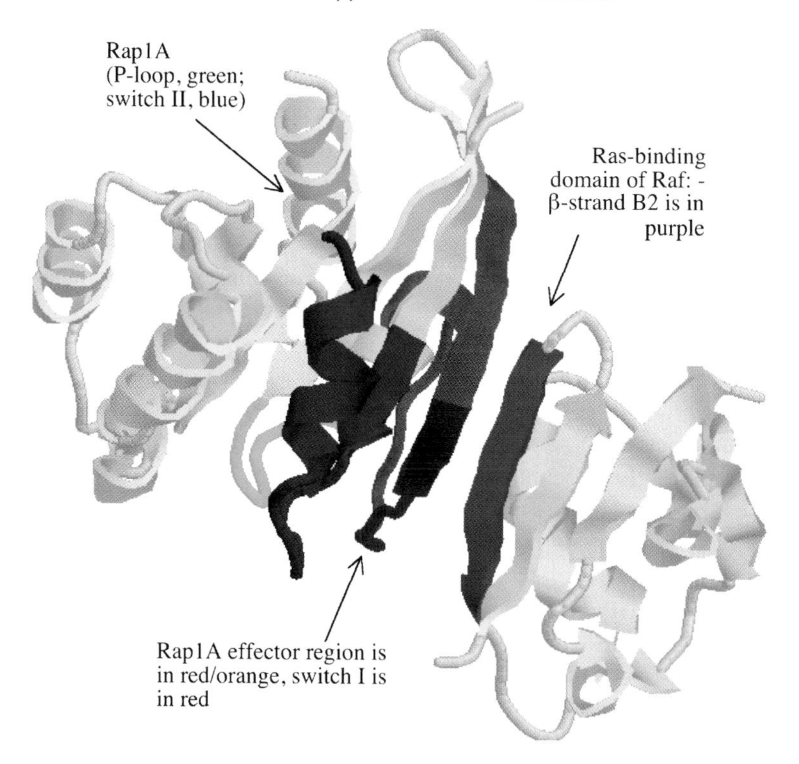

Figure 7.16 The complex of Rap1A with Raf RBP

anti-parallel interaction with a β-strand from Raf (B2 in Nassar's nomenclature). Effector region/switch I residues Glu37, Asp38, Ser39 and Arg41 in particular make strong polar interactions and H-bonds with complimentary residues of the Raf β-strand B2.

The above deals with the main interaction of Raf's RBD region with Ras switch II. However, Raf cystein-rich domain (CRD) also interacts with Ras, via residues in switch II that if mutated compromise Ras activity and Raf activation[41].

7.7.4 Rap1 and cAMP effects

To conclude this discussion of Ras orthologues, it is worth briefly reviewing the way that Rap can exert opposing effects on MAPK activation. As mentioned, Rap1 was originally known only as an antagonist of Ras-Raf signalling. Rap1 has two immediate effects on C-Raf: it sequesters C-Raf, and its very tight binding to the C-Raf CRD region blocks Ras binding[14]. Interestingly, if the C-Raf CRD is swapped with the (lower affinity) CRD of B-Raf, Rap1 now activates the C-Raf-(-B-Raf/CRD) chimera.

Rap1 is activated by a number of signals, particularly those that increase cAMP levels in the cell. In most cells, a sustained high level of cAMP is either toxic or growth-inhibitory. However, in some cell types (such as pituitary somatotrophs or ovarian carcinoma cells) high cAMP is mitogenic. There are undoubtedly gaps in our knowledge of how these events are regulated. However, the following summarises what is known thus far:

- Both growth-stimulatory and –inhibitory effects of cAMP are mediated by Rap1 and depend critically upon the ratio of C-Raf to B-Raf in the target cell (for the most part, at least). In cells with high levels of C-Raf and little B-Raf, cAMP activates PKA, which inhibits C-Raf by inhibitory phosphorylations that prevent its recruitment/activation by Ras. In cells that express B-Raf, cAMP causes activation of the Erk pathway and cell division. We have seen how tyrosine phosphorylation of (membrane-recruited) C-Raf's 'N-region' by Src is critical to its activation, whereas B-Raf does not need to be phosphorylated by Src. Strangely, the mode of action of Src is itself highly dependent upon *how* it was activated. Growth factor receptors activate Src by displacement of the intrachain pTyr527-SH2 bond of Src. In contrast, cAMP, via PKA, activates Src by phosphorylation of serine17 in its unique domain (Figure 7.5; see also Chapter 3, Section 3.1.4). Src is thereby switched from Ras/C-Raf activation (growth factor-activated Src) to Rap1/B-Raf activation (pSer17-activated Src).

- The Rap1-directed actions of pSer17-Src begin with construction of a membrane-associated scaffold through Src tyrosine-phosphorylating the ***C***asitas ***B***-***l***ineage (Cbl) protein[42]. pTyr-Cbl provides a platform for the docking of the adaptor protein Crk, which has a domain structure solely composed of SH2-SH3-SH3[43]. Crk is constitutively bound (via its central SH3 domain) to an unusually proline-rich region of a Rap-specific GEF called C3G. This cytoplasmic complex associates with membranes via SH2 docking to pTyr-Cbl. C3G then activates Rap1 by GDP-GTP exchange. Rap1 can then, dependent upon Raf isoform ratios, either sequester C-Raf (inhibiting Erk signals and cell division) or activate B-Raf (stimulating Erk signals, causing mitosis).

The net result of either C-Raf or B-Raf activation is transient stimulation of the Erk pathway and a downstream mitogenic response, which is conditional upon a variety of cellular conditions being favourable before it is permitted. This is discussed further in Chapter 10.

References

1. Leipe, D.D., Wolf, Y. I., Koonin, E.V. and Aravind, L. (2002) Classification and evolution of P-loop GTPases and related ATPases. *J. Mol. Biol.*, **317**: 41–72.
2. Bhattacharya, M., Babwah, A.V. and Ferguson, S.S.G. (2004) Small GTP-binding protein-coupled receptors. *Biochem. Soc. Trans.*, **32**: 1040–1044.
3. Neal, S.E., Eccleston, J.F., Hall, A. and Webb, M.R. (1988) Kinetic analysis of the hydrolysis of GTP by $p21^{N\text{-}ras}$. *J. Biol. Chem.*, **263**: 19718–19722.

4. Sermon, B.A., Lowe, P.N., Strom, M. and Eccleston, J.F. (1998) The importance of two conserved arginine residues for catalysis by the Ras GTPase-activating protein, neurofibromin. *J. Biol. Chem.*, **273**: 9480–9485.
5. Vetter, I.R. and Wittinghofer, A. (2001) The guanine nucleotide-binding switch in three dimensions. *Science*, **294**: 1299–1304.
6. Wellbrock, C. Karasarides, M. and Marais, R. (2004) The RAF proteins take centre stage. *Nat. Rev. Mol. Cell Biol.*, **5**: 875–885.
7. Hekman, M., Fischer, A., Wennogle, L.P., Wang, Y.K., Campbell, S.L. and Rapp, U.R. (2005) Novel C-Raf phosphorylation sites: serine 296 and 301 participate in Raf regulation. *FEBS Letters*, **579**: 464–468.
8. Avruch, J., Okhlatchev, A., Kyriakis, J.M., Luo, Z., Tzivion, G., Vavvas, D. and Zhang, X-F. (2001) Ras activation of the raf kinase: tyrosine kinase recruitment of the MAP kinase cascade. *Recent Prog. Horm. Res.*, **56**: 127–155.
9. Dumaz, N. and Marais, R. (2005) Integrating signals between cAMP and the RAS/RAF/MEK/ ERK signalling pathways. *FEBS Journal*, **272**: 3491–3504.
10. Tzivion, G., Luo, Z and Avruch, J. (1998) A dimeric 14-3-3 protein is an essential cofactor for Raf kinase activity. *Nature*, **394**: 88–92.
11. King, A.J., Sun, H., Diaz, B., Barnard, D., Miao, W., Bagrodia, S. and Marshall, M.S. (1998) The protein kinase Pak3 positively regulates Raf-1 activity through phosphorylation of serine-338. *Nature*, **396**: 180–184.
12. Carey, K.D., Watson, R.T., Pessin, J.E. and Stork, P.J.S. (2003) The requirement of specific membrane domains for Raf-1 phosphorylation and activation. *J. Biol. Chem.*, **278**: 3185–3196.
13. Wan, P.T.C., Garnet, M.J., Roe, S.M., Lee, S., Niculescu-Duvaz, D., Good, V.M., Cancer Genome Project, Jones, C.M., Marshall, C.J., Springer, C.J., Barford, D. and Marchais, R. (2004) Mechanism of activation of the RAF-ERK signaling pathway by oncogenic mutations of B-RAF. *Cell*, **116**: 855–867.
14. Kolch, W. (2000) Meaningful relationships: the regulation of the Ras/Raf/MEK/Erk pathway by protein interactions. *Biochem. J.*, **351**: 289–305.
15. Dougherty, M.K., Muller, J., Ritt, D.A., Zhou, M., Zhou, X.Z., Copeland, T.D., Conrads, T.P., Veenstra, T.D., Lu, K.P and Morrison, D.K. (2005) Regulation of Raf-1 by direct feedback phosphorylation. *Mol. Cell*, **17**: 215–224.
16. Mason, C.S., Springer, C.J., Cooper, R.G., Superti-Furga, G., Marshall, C.J. and Marais, R. (1999) Serine and tyrosine phosphorylations cooperate in Raf-1, but not B-Raf activation. *EMBO Journal*, **18**: 2137–2148.
17. Schonwasser, D.C., Marais, R.M., Marshall, C.J. and Parker, P.J. (1998) Activation of mitogen-activated protein kinase/extracellular signal-regulated kinase pathway by conventional, novel, and atypical protein kinase C types. *Mol. Cell. Bio.*, **18**: 790–797.
18. van der Hoeven, P.C.J., van der Wal, J.C.M., Ruurs, P., van Dijk, M.C.M. and van Blitterswijk, W.J. (2000) *Biochem. J.*, **345**: 297–306.
19. Roy, S., Plowman, S., Rotblat, B., Prior, I.A., Muncke, C., Grainger, S., Parton, R.G., Henis, Y.I., Kloog, Y and Hancock, J.F. (2005) Individual palmitoyl residues serve distinct roles in H-Ras trafficking, microlocalization, and signalling. *Mol. Cell. Biol.*, **25**: 6722–6733.
20. Milburn, M.V., Tong, L., deVos, A.M., Brunger, A, Yamaizumi, Z., *et al.* (1990) Molecular switch for signal transduction: structural differences between active and inactive forms of protooncogenic forms of Ras proteins. *Science*, **247**: 939–945.
21. Rossmann, M.G., Moras, D. and Olsen, K.W. (1974) Chemical and biological evolution of a nucleotide-binding protein. *Nature*, **250**: 194–199.
22. Walker, J.E., Saraste, M., Runswick, M.J. and Gay, N.J. (1982). Distantly related sequences in the alpha- and beta-subunits of ATP synthase, myosin, kinases and other ATP-requiring enzymes and a common nucleotide binding fold. *EMBO J.*, **1**: 945–951.

23. Sprang, S.R. (1997) G protein mechanisms: insights from structural analysis. *Ann. Rev. Biochem.*, **66**: 639–678.
24. Lambright, D.G., Noel, J.P., Hamm, H.E. and Sigler, P.B. (1994) Structural determinants for activation of the a-subunit of a heterotrimeric G protein. *Nature*, **369**: 621–628.
25. Marshall, M.S. (1993) The effector interactions of p21ras. *TIBS*, **18**: 250–254.
26. Scheidig, A.J., Burmester, C. and Goody, R.S. (1999) The pre-hydrolysis state of p21Ras in complex with GTP: new insights into the role of water molecules in the gtp hydrolysis reaction of Ras-like proteins. *Structure*, **7**: 1311.
27. Kraulis, P.J., Domaille, P.J., Campbell-Burk, S.L., Van Aken, T. and Laue, E.D. (1994) Solution structure and dynamics of Ras p21. GDP determined by heteronuclear three- and four-dimensional NMR spectroscopy. *Biochemistry*, **33**: 3515–3531.
28. Temeles, G.L., Gibbs, J.B., D'Alonzo, J.S., Sigal, I.S. and Scolnick, E.M. (1985) Yeast and mammalian ras proteins have conserved biochemical properties. *Nature*, **313**: 700–703.
29. Landis, C.A., Masters, S.B., Spada, A., Pace, A.M., Bourne, H.R. and Vallar, L. (1989) GTP'ase inhibiting mutations activate the α chain of Gs and stimulate adenylyl cyclase in human pituitary tumours. *Nature*, **340**: 692–696.
30. Noel, J.P., Hamm, H.E. and Sigler, P.B. (1993) The 2.2Å crystal structure of transducin-α complexed with GTPγS. *Nature*, **366**: 654–663.
31. Kleuss, C., Raw, A.S., Lee, E., Sprang, S.R. and Gilman, A.G. (1994) Mechanism of GTP hydrolysis by G-Protein α subunits. *Proc. Natl. Acad. Sci. USA*, **91**: 9828–9831.
32. Scheffzek, K., Ahmadian, M.R., Kabsch, W., Wiesmuller, L., Lautwein, A., Schmitz, F., and Wittinghofer, A. (1997) The Ras-RasGAP complex: structural basis for GTPase activation and its loss in oncogenic Ras mutants. *Science*, **277**: 333–338.
33. Bernards, A. (2003) GAPs galore! A survey of putative Ras superfamily GTPase activating proteins in man and Drosophila. *Biochimica et Biophysica Acta*, **1603**: 47–82.
34. Boriack-Sjodin, P.A. Margarit, M.S., Bar-Sagi, D. and Kuriyan, J. (1998) The structural basis of the activation of Ras by Sos. *Nature*, **394**: 337–343.
35. Cherfils, J. and Chardin, P. (1999) GEFs: structural basis for their activation of small GTP-binding proteins. *TIBS*, **24**: 306–311.
36. Schweims, T. and Wittinghoffer, A. (1994) GTP-binding proteins: structures, interactions and relationships. *Current Biology*, **4**: 547–550
37. Warne, P.H. *et al.* (1993) Direct interaction of Ras and the amino-terminal region of Raf-1 in vitro. *Nature*, **364**: 352–355.
38. Nassar, N., Horn, G., Herrmann, C., Scherer, A., McCormick, F. and Wittinghofer, A. (1995) The 2.2Å crystal structure of the Ras-binding domain of the serine/threonine kinase c-Raf1 in complex with Rap1A and a GTP analogue. *Nature*, **375**: 554–560.
39. Stork, P.J.S. and Dillon, T.J. (2005) Multiple roles of Rap1 in hematopoietic cells: complementary versus antagonistic functions. *Blood*, **108**: 2952–2961.
40. Carey, K.D., Watson, R.T., Pessin, J.E., and Stork, P.J.S. (2003) Requirement of specific membrane domains for Raf-1 phosphorylation and activation. *J. Biol. Chem.*, **278**: 3185–3196.
41. Drugan, J.K., Khosravi, F.R., White, M.A., Der, C.J., Sung, Y.J., Hwang, Y.W. and Campbell, S. L. (1996) Ras interaction with two distinct binding domains in Raf-1 may be required for Ras transformation. *J. Biol. Chem.*, **271**: 233–237.
42. Schmitt, J.M. and Stork, P.J.S. (2002) PKA phosphorylation of Src mediates cAMP's inhibition of cell growth via Rap1. *Mol. Cell*, **9**: 85–94.
43. Knudsen, B.S., Feller, S.M. and Hanafusa, H. (1994) Four proline-rich sequences of the guanine-nucleotide exchange factor C3G bind with unique specificity to the first Src homology 3 domain of Crk. *J. Biol. Chem.*, **269**: 32781–32787.

8 G proteins (II) – heterotrimeric G proteins

Heterotrimeric G protein α-subunits are structurally homologous with Ras, having a similarly-folded core containing the G-boxes and two conformationally fluid regions that are structurally equivalent (though non-homologous) to the switches I and II of Ras. In α-subunits, the Ras-like core has been further elaborated by extra inserted amino acid stretches plus a large *N*-terminal extension and a distinctive *C*-terminal tail. The largest insert (insert-1) is an independently-folded all-helix domain that (in combination with an arginine-containing linker peptide that connects it to the Ras core) acts like a tethered GAP and explains the higher catalytic rates of α-subunits compared with Ras. Although superficially similar in their movements, the switch regions in α-subunits function differently – switch II of Gα is the prime effector-binding region (rather than switch I, as in Ras). In α-subunits, switch II conformations are stabilised by a third switch region (switch III, insert-2) not found in Ras. The most significant difference from the Ras activation cycle is the obligate partnering of the GDP-bound Gα with a β/γ-dimer that (i) acts as an inhibitor of GDP release, (ii) sequesters the effector-binding switch II, and (iii) helps the heterotrimer to interpret the activation status of any cognate receptors it encounters.

Like Ras, Gα must enter a nucleotide-free ('empty-pocket') state after receptor activation, before it can go on to interact with effectors. The nucleotide-free Gα<EMPTY> monomer is very unstable except when complexed to both the agonist-occupied receptor plus the β/γ-subunit, just as Ras<EMPTY> is stabilised by the GEF, mSos. Nucleotide exchange begins when the occupied receptor (acting as a GEF) disturbs the switch-stabilising effect of the α-bound β/γ (acting as a GDI), with the result that GDP is released. The resulting ternary complex (Receptor-Gα<EMPTY>β/γ) has a similar affinity for GDP and GTP, but GTP-binding wins out because the

Structure and Function in Cell Signalling John Nelson

additional γ-phosphate has the effect of stabilising a new set of switch conformations that have no affinity for receptor or β/γ. This disintegrates the complex, causes closure of the G-cleft, allowing metastable monomeric Gα <GTP> (and/or freed β/γ) to move away to interact with effectors. Gα<GTP> interacts with effectors through a re-configured switch II that rotates in response to the presence of the γ-phosphate of GTP.

The activated Gα<GTP> is eventually 'de-fused' by its intrinsic GTPase, returning it to a Gα<GDP> state. The Gα<GDP> monomer has disordered switch regions and instead presents another binding surface made up of the *N*- and *C*-termini that signals to free β/γ that it is ready for another activation cycle.

8.1 Classification and structural relationship with Ras

Heterotrimeric G proteins were originally grouped according to the distinct activities of their respective α-subunits, properties such as receptor selectivity, effector enzyme specificity and varying susceptibility to bacterial toxins. The α-subunit is closely related to Ras and conserves the distinct features of the Ras-like G-domain. The variation in β/γ-subunits were historically ignored in classification of heterotrimeric G proteins, because there are relatively fewer β/γ-gene products and (at first) it was thought that the signal transducing power of the heterotrimer lay exclusively with the α-subunits. We now know that this is not the case. Many signals have been found to be transduced by free β/γ-subunits. Nevertheless, α- units are not generally selective about the β/γ dimer pair with which they associate[1], and so the α-subunit is still used as a means of classification of the different heterotrimers.

Table 8.1 shows a widely accepted classification scheme for α-subunits based upon sequence identity[2]. It is apparent that, with a few exceptions, closely homologous subunits display activities that are in common with the rest of the subfamily.

More recently, the structure of the α-subunit has been used to further refine the position of heterotrimeric G proteins within both the Ras-like superfamily and the wider superclass of 'P-loop GTPases'[3]. The advantage of sequence/structure-based classification is that the evolution of these proteins may be discerned. However, one should be aware that although ancestral forms of G protein are identifiable in present day lower organisms, their mode of action may be quite different. In mammals the monomeric G protein Ras is the prime activator of MAP kinase cascades, but in yeast it is the β/γ dimer of a heterotrimeric G protein that activates a MAP kinase pathway that is involved in pheromone response[4]. The corresponding yeast α-subunit appears not to act directly on such pathways, but instead acts as a sink to sequester β/γ dimers when extracellular pheromone signals are absent[5].

Gα-subunits are the product of 16 genes in the human genome. The presence of various splice variants adds to this complexity – around 20 different protein products are known. There are six β- and 12 γ-gene products[6].

G proteins act as heterotrimers in both their GDP-bound and 'empty pocket' forms but when activated by interaction with ligand-occupied receptor, the α-subunit exchanges GDP for GTP and the trimer separates into an α-GTP monomer and a free β/γ dimer. The

Table 8.1 Classification of G proteins based on α-subunit sequence identity

Class	Members	Toxin sensitivity	α-subunit-effector enzyme interaction	β/γ effects
Gs	αs, αolf	CTX-sensitive	stimulation of adenylyl cyclase	
Gi	αi αo αz αt (transducin),	all PTX-sensitive (except αz) (αt is PTX- & CTX-sensitive)*	all cause inhibition of adenylyl cyclase (except αt, which activates cGMP phosphodiesterase)	stimulation of phospholipase Cβ2§ inhibition of Ca^{2+} channels activation of K^+ channels‡
Gq	αq, α11, α14-16	PTX-resistant	stimulation of phospholipase Cβ1,3,4§	
G12	α12, α13	PTX-resistant	stimulation of phospholipase Cε†	

*see:- West, Jr., R.E., et al., 1988; Bornancin, F, and Chabre, M., 1991
† *see:* Kelley, G.G., et al., 2004
‡ *see:* Fernandez-Fernandez, J.M. et al., 2001
§ *see:* Jiang, H., et al., 1994; Smrcka, A.V. and Sternweiss, P.C., 1993

β/γ dimer itself does not disunite except under denaturing conditions and so always acts as a unit. The β and γ polypeptides dimerise via a coiled coil involving their respective *N*-terminal helices (see Figure 8.1) and are tethered to the membrane by a fatty acyl chain – either geranylgeranyl or farnesyl – attached to the *C*-terminus of the γ-subunit. The most complete structures available at the time of writing are crystal structures of heterotrimers containing either the rat Gαi-1 subunit[7] (Figure 8.1(A)), or a chimeric αt/i-subunit[8] (Figure 8.1(B)), complexed with a β/γ dimer.

In the first of these structures (Figure 8.1(A) and (C)), the rat Gαi1 subunit is complexed with bovine β1/γ2. The proteins are not truncated, but were modified to prevent any fatty acylation – the *N*-terminus of αi would normally be myristoylated, and the γ-subunit prenylated, to allow membrane tethering[7]. Despite the lack of resolution of the *C*-terminal 46 amino acids of the γ-subunit, the electron density mapping suggests it terminates proximate with the *N*-terminus of αi (only three *N*-terminal amino acids are unresolved, excluding the start methionine), pointing to a shared locus for membrane insertion of both the myristoyl moeity at the αi *N*-terminus and the prenyl of the γ *C*-terminus.

The most obvious feature of the conformations of both of the complexed α-subunits in the heterotrimer structures is that there is a long *N*-terminal α-helix (Figure 8.2(A)) – this is missing in all previous monomer structures of αi or αt proteins in isolation. In the case of monomeric αt (at least), this is because the subunits were proteolysed to remove the *N*-terminal 25 amino acids to aid crystalisation. In monomeric αi structures, however, the proteins are intact and show differences in the *N*- and *C*-termini. In particular, the *N*- and *C*-termini are completely disordered in the GTP-bound conformations (Figure 8.2(C)),

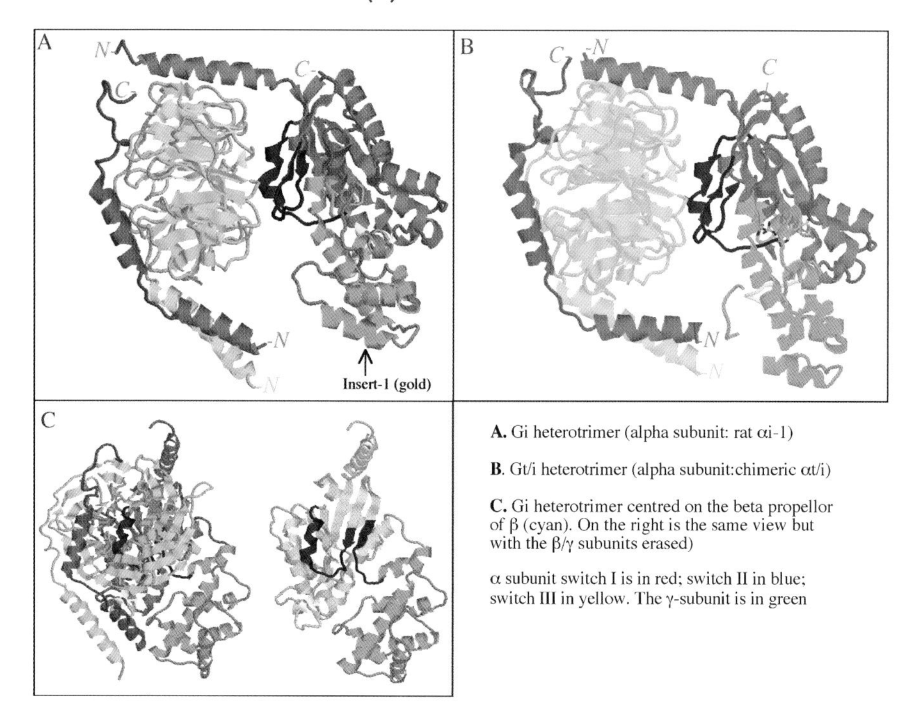

Figure 8.1 Heterotrimeric G protein structures

including the GTP-mimicking condition where GDP plus AlF_4^- is bound. In contrast, the GDP-bound form of αi (representing the de-fused transducer readied for re-combination with β/γ, having hydrolysed its bound GTP) has a clearly defined microdomain consisting of its *N*- and *C*-termini (Figure 8.2(B)), which presumably forms a recognition site for the free β/γ-subunit. The following list summarises the missing portions in each α-subunit (numbering does not include the start methionine).

- Heterotrimer-complexed αi <GDP> (PDB file: 1GP2) *N*-terminal 3 amino acids and *C*-terminal 6 amino acids are unresolved.

- αi<GTP> (PDB file: 1CIP) *N*-terminal 30 amino acids and *C*-terminal 7 amino acids are unresolved.

- αi <GDP plus AlF_4^-> (PDB file: 1GFI) *N*-terminal 31 amino acids and C-terminal 9 amino acids are unresolved.

- αi <GDP> (PDB file: 1BOF) only the 8 most *N*-terminal amino acids are unresolved: 8 aas missing from Switch III; 14 aas missing from Switch II.

In some models of membrane tethering of the heterotrimer, the extended *N*-helix is depicted as lying along the face of the membrane[6]. However, as has been recently

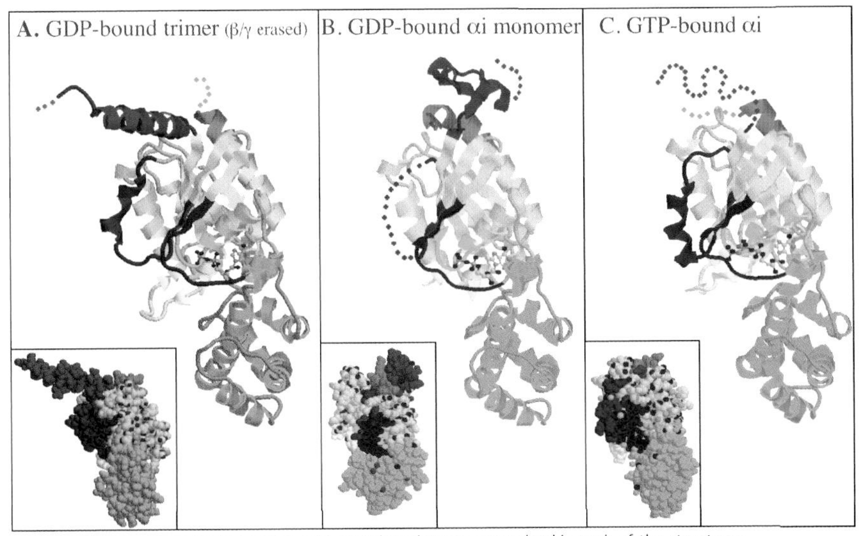

Dotted lines indicate the amino acid stretches that are unresolved in each of the structures. The N-terminal helix is purple; the C-terminal14 amino acid, darkgreen; P-loop, green; Switch I, red; Switch II, blue; Switch III, yellow. Inset is a spacefill model of each.

Figure 8.2 The conformations of the *N*- and *C*-terminal domains of αi

pointed out, this helix is actually hydrophilic, having a high proportion of charged residues (both basic and acidic). For this reason it is suggested that the *N*-helix hangs at an angle from its membrane-tethered fatty acid chain[9] (see Figure 8.3 and Table 8.2).

The second heterotrimer structure (Figure 8.1(B)) contains a chimeric αt/i-subunit engineered from transducin with the region between 216 and 294 replaced with the corresponding region from the α-subunit of Gαi (residues 220–298). That is to say, the chimeric α-subunit in this structure contains the *N*-terminal portions of transducin, including the P-loop, switches I and II, and the α-helical insert-1. The *C*-terminal third of the protein contains switch III from the αi-subunit followed by the *C*-terminal portion of transducin. Note that the final seven *C*-terminal amino acid stretch (a receptor-binding determinant) is missing as is the acylated extreme *N*-terminus; the γ-subunit is similarly truncated and is missing its *C*-terminal farnesylation site.

8.2 Gα-subunits: the Ras-like core, G-boxes and switch regions

As indicated above, the Gα subunits are related to monomeric G proteins such as Ras and share the highly conserved residues of Ras G-domains. Like the Ras protein, the active site of Gα subunits resides in five so called 'G-regions' or 'G-boxes'[10]. Certain key residues in these G-regions sequences are highly conserved throughout the Ras-like superfamily. In Gα subunits, the Ras-like core is interrupted by a number of inserted sequences that are not found in Ras itself (Table 8.2 and Figures 8.4 and 8.5). The largest of these inserts (insert-1) is almost as big as Ras itself.

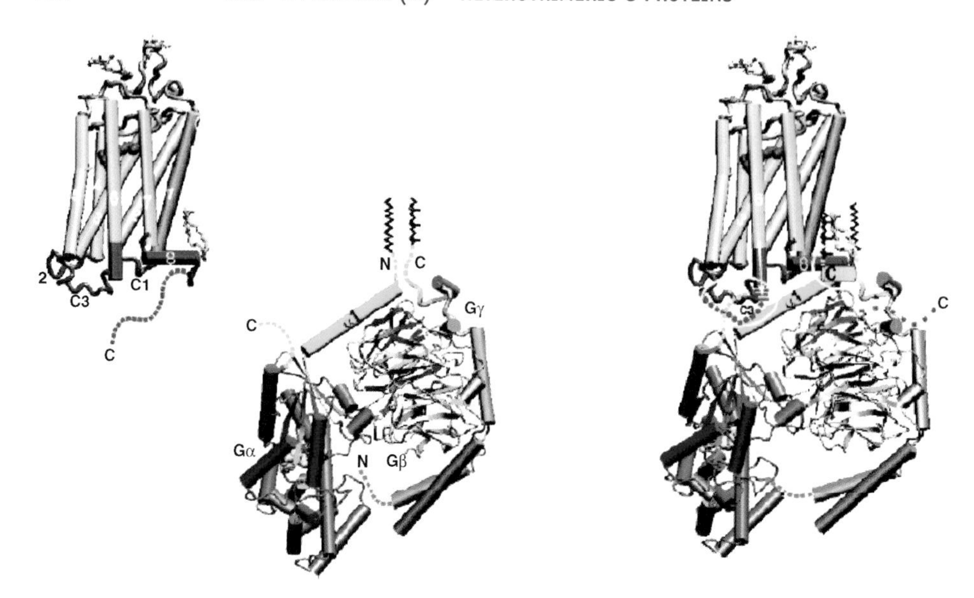

Figure 8.3 Model of 7-pass receptor coupling with a heterotrimeric G protein

8.2.1 The P-loop

The **G-1** region encompasses the glycine-rich pyrophosphate binding loop (P-loop) and contains a conserved lysine followed by a serine. The homology of P-loops in monomeric and heterotrimeric G proteins is high. In Ras and heterotrimeric α-subunits, the consensus sequence is GXGXXGKS and in the wider superclass it is GXXXXGK(S/T). Such glycine containing motifs were originally recognised in the Rossmann fold of dinucleotide and mononucleotide binding proteins[11]. Strictly speaking, the P-loop of G proteins is not a Rossmann fold, which classically defines a large domain made up of alternating α and β structure (βαβαβα) with a glycine-rich motif between the first β-strand and the following α-helix. Although G proteins do not follow the extended Rossmann motif, they do have P-loops extending from the first β-strand to the first α-helix. The **G-1** region is sometimes referred to as a 'Walker A' box, the glycine rich sequence having been found in certain ATP binding proteins along with a conserved 'Walker B' box[12].

8.2.2 Switch I/G-2

The **G-2** box is located in switch I of both Ras and heterotrimeric α-subunits. In Ras, switch I is the loop that connects the α-1 helix and β-2 strand in Ras ('loop-2'). In Gα, loop-2 is interrupted by the inserted large helical domain (insert-1) that precedes its switch I region. In α-subunits, switch I is found in the second 'linker peptide' connecting the Ras-like core to the independently folded alpha helical domain (Table 8.2 and Figure 8.5).

Table 8.2 Aligned sequences of Gα and Ras proteins in structural studies referred to in text

	1	11	21	31	41	51
Bov Gαs	MGCLGNSKTE	DQRNEEKAQR	EANKKIEKQL	QKDKQVYRAT	HRLLLLGAGE	SGKSTIVKQM
	1		10	20	30	40
Bov Gαt	MGA·GAS···	···AEE····	KHSRELEKKL	KEDAEKDART	VKLLLLGAGE	SGKSTIVKQM
	1		14	24	34	44
Rat Gαi	MGC·TLS···	···AEDKAAV	ERSKMIDRNL	REDGEKAARE	VKLLLLGAGE	SGKSTIVKQM
				1	4	14
Hum Ras				MTE	YKLVVVGAGG	VGKSALTIQL
	61	71	81	91	101	111
Bov Gαs	RILHVNGFNG	EGGEEDPQAA	RSNSDGEKAT	KVQDIKNNLK	EAIETIVAAM	SNLVPPVELA
	50	60		65	75	85
Bov Gαt	KIIHQDGYSL	E·········	······ECLE	FIAIIYGNTL	QSILAIVRAM	TTLNIQYGDS
	54	64		69	79	89
Rat Gαi	KIIHEAGYSE	E·········	······ECKQ	YKAVVYSNTI	QSIIAIIRAM	GRLKIDFGDA
	24					
Hum Ras	IQNH······	··········	··········	··········	··········	··········
	121	131	141	151	160	170
Bov Gαs	NPENQFRVDY	ILSVMNVPDF	DFPPEFYEHA	·KALWEDEGV	RACYERSNEY	QLIDCAQYFLD
	95	104	114	123	133	143
Bov Gαt	ARQDDAR·KL	MHMADTIEEG	TMPKEMSD·I	IQRLWKDSGI	QACFDRASEY	QLNDSAGYYLS
	99	108	118	127	137	147
Rat Gαi	ARADDAR·QL	FVLAGAAEEG	FMTAE·LAGV	IKRLWKDSGV	QACFNRSREY	QLNDSAAYYLN
Hum Ras	··········	··········	··········	··········	··········	···········
	181	191	201	211	221	231
Bov Gαs	KIDVIKQDDY	VPSDQDLLRC	RVLTSGIFET	KFQVDKVNFH	MFDVGGQRDE	**R**RKWIQCFND
	154	164	174	184	194	204
Bov Gαt	DLERLVTPGY	VPTEQDVLRS	RVKTTGIIET	QFSFKDLNFR	MFDVGGQRSE	**R**KKWIHCFEG
	158	168	178	188	198	208
Rat Gαi	DLDRIAQPNY	IPTQQDVLRT	RVKTTGIVET	HFTFKDLHFK	MFDVGGQRSE	**R**KKWIHCFEG
		28	32	42		65
Hum Ras	··········	······FVDE	YDPTIEDSYR	KQVVIDGETCLLDI	LDTAGQEEY	SAMRDQYMRT
	241	251	261	271	281	291
Bov Gαs	VTAIIFVVAS	SSY**NMVIRED**	**NQTNR**LQEAL	NLFKSIWNNR	WLRT ISVILF	LNKQDLLAEK
	214	224	234	244	254	264
Bov Gαt	VTCIIFIAAL	SAY**DMVLVED**	**DEVNR**MHESL	HLFNSICNHR	YFAT TSIVLF	LNKKDVFSEK
	218	228	238	248	258	268
Rat Gαi	VTAIIFCVAL	SDY**DLVLAED**	**EEMNR**MHESM	KLFDSICNNK	WFTD TSIILF	LNKKDLFEEK
	75	85		91	101	
Hum Ras	GEGFLCVFAI	NN········	······TKSF	EDIHQYREQI	KRVKDSDDVPMVLVGNKCDLAART	
	301	311	321	331	341	351
Bov Gαs	VLAGKSKIED	YFPEFARYTT	PEDATPEPGE	DPRVTRAKYF	IRDEFLRIST	ASGDGRHYCY
	274	282	292		299	309
Bov Gαt	I··KKAHLSI	CFPDYNGPNT	YEDA······	·······GNY	IKVQFLELNM	RRDVKE··IY
	278	286	296		303	313
Rat Gαi	I··KKSPLTI	CYPEYAGSNT	YEEA······	·······AAY	IQCQFEDLNK	RKDTKE··IY
	125				130	
Hum Ras	V·········	··········	···E······	·······SRQ	AQDLARSYG·	·······IP
	361	371	381	391		
Bov Gαs	PHFTCAVDTE	NIRRVFNDCR	DIIQRMHLRQ	YELL		
	317	327	337	347		
Bov Gαt	SHMTCATDTQ	NVKFVFDAVT	DIIIKENLKD	CGLF		
	321	331	341	351		
Rat Gαi	THFTCATDTK	NVQFVFDAVT	DVIIKNNLKD	CGLF		
	141	151	161	171	181	
Hum Ras	YIETSAKTRQ	GVEDAFYTLV	REIRQHKLRK	LNPPDESGPG	CMSCKCVLS	

KEY
XXXX – P-LOOP; XXXX – α-helical domain (≈ INSERT-1);
XXXX – SWITCH I; XXXX – SWITCH II; XXXX – SWITCH III (= INSERT-2)
(Important residues mentioned in the text are underlined)
switch residues that are paired in GTP-bound forms are in bold

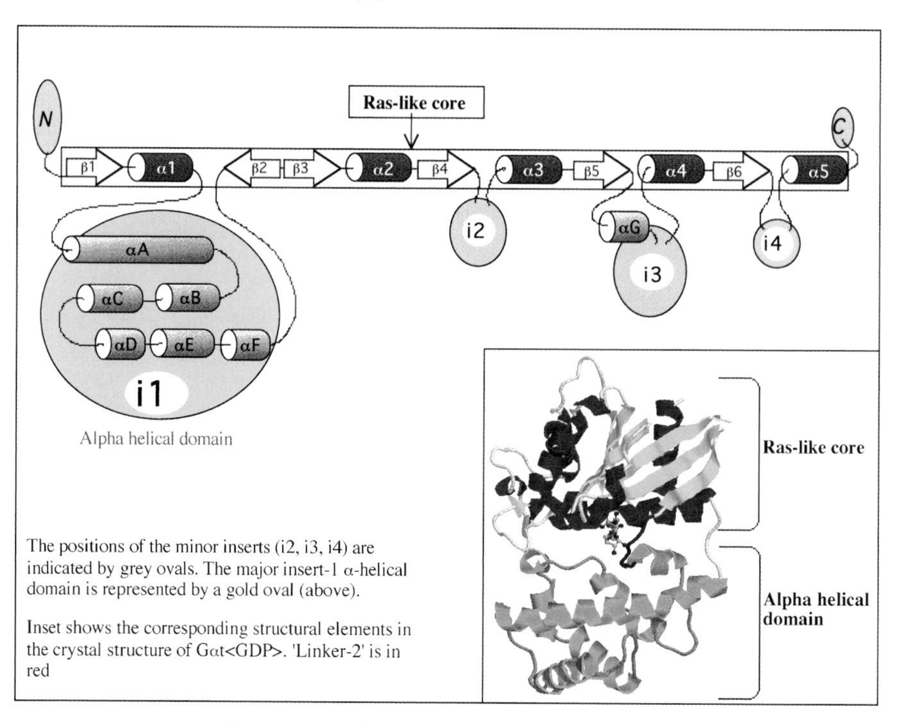

Figure 8.4 Schematic diagram of a Gα subunit

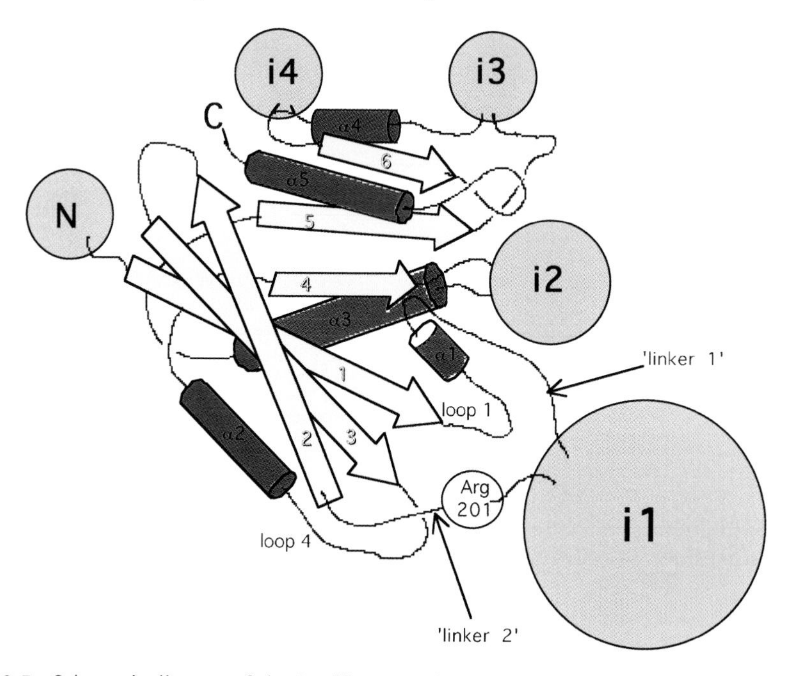

Figure 8.5 Schematic diagram of the Ras-like core of a Gαs subunit showing the position of extra peptide sequences (inserts and *N*-terminal extension) not found in Ras

In Ras, switch I contains an invariant threonine (Thr35 in Ras) involved in γ-phosphate-binding and Mg^{2+} coordination. Gα subunits also have a central threonine (see Figure 8.2) in their switch I loops, and the residue's functions in GTP hydrolysis-driven conformational rearrangements are similar to Ras. However, in other respects, switch I in Gα subunits is very different in structure and function. Most notably, switch I plays little or no role in contacting Gα-effectors; instead that role is taken over by switch II.

8.2.3 Switch I/insert-1: a tethered *G*TPase-*a*ctivating *p*rotein (GAP)

Apart from containing a common central threonine, switch I sequences (boxed in red, Table 8.2) in α-subunits show no homology with switch I in Ras. Switch I sequences of various α-subunit family members, on the other hand, are highly homologous with one another.

All Gα subunits contain an invariant arginine (Arg201 of αs, Arg174 of αi, Figure 8.5, underlined in Table 8.2) in switch I that is not found in Ras. This is the same arginine residue of αs that is ADP-ribosylated by cholera toxin. The covalent modification of Arg201 by cholera inactivates the catalytic site of αs and, because GTP is only very slowly hydrolysed by the damaged α-subunit, a persistent stimulation of adenylyl cyclase ensues in cholera-infected cells. We shall see later that the same arginine is also the site of (GTPase-disabling) spontaneous oncogenic point mutations in some cancers (Chapter 10, Section 10.17.1). Finally, it is the presence of this switch I arginine that allows GDP and AlF_4^-, together, to mimic GTP transition state in α-subunits (thereby causing receptor-independent and continuous activation of effector enzymes), an effect not seen with isolated Ras.

Gα switch I is considered to be part of insert-1: an independently-folded domain of five short helices surrounding a long sixth central helix. A significant observation was made when the insert-1 domain (71–214 of αs) and the Ras-like core domain (i.e., αs minus the insert) were expressed as separate proteins. Without the insert, the recombinant Ras-core had a very low rate of GTP hydrolysis (like Ras itself) but this could be restored to the normal (higher) GTPase rate of native Gαs by addition of the truncated insert-1-only protein; despite being entirely unstructured in the truncated insert-1-only protein, the switch I loop arginine (Arg201 in αs) is still able to exert its GAP-like effect[13]. The absence of this arginine may explain Ras's low intrinsic GTPase activity (and the activating power of Ras-GAP, see Chapter 7, Section 7.6). In heterotrimeric α-subunits, the arginine helps position the catalytic glutamine and its positive charge neutralises the growing negative charge of the γ-phosphate of GTP during hydrolysis thus stabilising the transition state, just as RasGAP's 'arginine finger' functions in Ras.

8.2.4 Switch II/G-3

The **G-3** sequence (containing invariant DXXG), often referred to as a Walker B box, is found at the *N*-terminus of the α-2 helix and is part of the **switch II** region of both monomeric G proteins and α-subunits. The main chain amide of the conserved glycine (226 in αs, 203 in αi) coordinates the γ-phosphate (in GTP-bound forms). The glycine is immediately followed by a glutamine in both Ras and Gα subunits. Point mutation of this Gln61 in Ras inactivates its GTPase and is a common site of point mutation in Ras in a variety of human cancers. Its equivalent residue in Gαs (Gln227, Gln204 in αi) is also a site of oncogenic point mutations in a smaller group of human cancers (see Chapter 10, Section 10.17.1). However, although it is an essential element in the catalytic site of Gα and Ras, this glutamine is not conserved in all members of the superfamily – it is replaced by histidine in EF-Tu.

G-4 The **G-4** sequence (containing consensus NKXD) spans the link between the β-5 strand and the 'α-G' helix and is partially responsible for binding the guanine base. In the αt/αi chimera structure, G-4 is Asn265–Asp268 and in αi it is Asn269–Asp272.

G-5 The **G-5** box runs between the β-6 strand and the α-5 helix and aids in guanine base recognition and binding. It contains a consensus sequence (TCA) in α-subunits and is found between Thr320–Ala322 in the αt/αi chimera structure and Thr324–Ala326 in αi.

8.2.5 Switch III

This region is not found in Ras. It is one of the insert regions (insert-2) of heterotrimeric α-subunits and is located between β-4 and α-3 (Asp227–Arg238 in bovine transducin and Asp231–Arg242 in rat αi1) (Table 8.2). The main function of switch III is to stabilise the extended helix-2 of switch II in the GTP-bound conformation, which is thus aligned for effector recognition. In the GTP-bound conformation, switch III residues form a series of interactions with cognate residues in switch II. A key interaction emanates from the conserved glutamate (Glu232 in αt; Glu236 in αi1) in the centre of switch III, whose carboxyl sidechain forms an H-bond with the backbone amide of the conserved arginine in switch II (Arg201 in αt; Arg205 in αi1)[14].

In the GDP-bound conformations, switch III moves away from switch II and no longer makes contact. The result is that the G-cleft is more open and the switch II helix is allowed to unwind to present a different surface that binds β/γ instead of effector enzymes. The rotation of the helix and cleft closure elicited by the binding of GTP is also stabilised by a second conserved glutamate just *C*-terminal of switch III, which also contacts switch II (discussed below).

8.3 GTP exchange, hydrolysis and switch movements

The key residues in the first two switch regions of α-subunits that are involved in nucleotide recognition are highly reminiscent of Ras. In α-subunits, the taut structure of the GTP-bound switches is primarily due to γ-phosphate contacts with the conserved switch I threonine and the invariant switch II glycine. When these primary contact points are broken by loss of γ-phosphate after hydrolysis, the structures relax. However, although the basic mechanism and structural consequences of GTP- versus GDP-coordinated conformational changes are broadly similar, there are major differences in the way the Gα switch regions organise effector, receptor or regulator binding interfaces.

8.3.1 GTP conformations

The GTP-bound forms of α-subunits can be thought of as metastable intermediates in catalysis. The very stability of such enzyme-substrate complexes allows sufficient time for the activated G protein to transduce and amplify the signal by virtue of the rapid turnover of its immediate effector enzyme.

As with Ras, the GTP-bound conformations of α-subunits are taut due to extensive interactions with the γ-phosphate. The conserved lysine and serine P-loop residues perform the same function in α-subunits as they do in Ras. In complex with nonhydrolysable GTP, analogue αi and αt structures[14–17] show the P-loop serine hydroxyl group (equivalent to Ser17 of Ras) coordinating the Mg^{2+} ion, while its amide nitrogen hydrogen bonds to the β-phosphate oxygen. The amine side chain of the P-loop lysine (equivalent to Lys16 of Ras) hydrogen-bonds to both β- and γ-phosphates.

In the switch I region of α-subunits, the invariant threonine (Thr35 in Ras) also coordinates the Mg^{2+} ion while its amide hydrogen-bonds with the γ-phosphate. Switch II contacts also are similar to Ras. The invariant switch II glycine (Gly60 in Ras) hydrogen-bonds via its backbone amide to the γ-phosphate. The invariant switch II aspartate (Asp223 in αs, Asp200 in αi; equivalent to Asp57 in Ras) hydrogen-bonds to one of the water molecules coordinating the Mg^{2+} ion. A simplified summary of γ-phosphate contacts is shown in the inset in Figure 8.6.

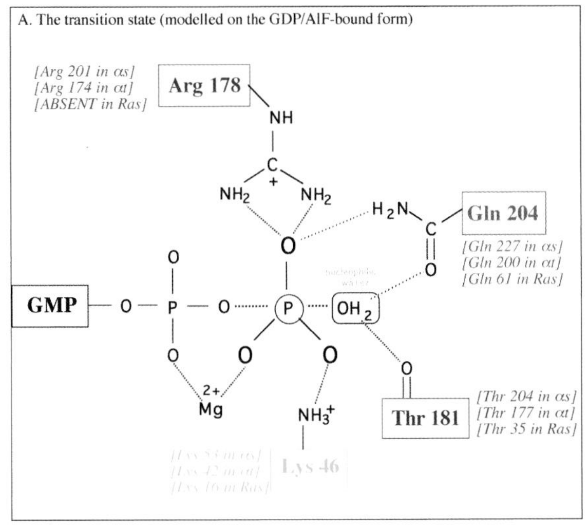

B. The GTP-bound state (the connections lost upon GTP hydrolysis are indicated by arced lines)

NH3 +
HN Gly203
HOH
Mg
Asp223
HOH
HO
Thr181

Residue numbers are from Rat αt, equivalent residues in other alpha subunits are in italics. P-loop residues in green; Switch I residues in red; Switch II residues in blue. Diagram in left panel adapted from :- Coleman, D. E., Berghuis, A. M., Lee, E., Linder, M. E., Gilman, A. G., and Sprang, S. R. (1994) Science 265, 1405 - 1412.

Figure 8.6 γ-phosphate contacts of the Gαi subunit

8.3.2 The transition state

Aluminium fluoride is a strong activator of GDP-bound Gα proteins. In crystal structures of αi and αt, in complex with GDP and AlF_4^-, the aluminium fluoride ion occupies the position expected for the γ-phosphate of GTP. The structure of this ion is thought to model the pentavalent transition state of the γ-phosphate – the extra valence representing the attack trajectory of the nucleophilic water molecule. When the catalytic glutamine is within H-bonding distance of this water, and the switch I arginine is close enough to help neutralise the growing charge on the γ-phosphate leaving group, the abstraction of a proton from the water creates a nucleophile that catalyses hydrolysis.

We have seen in Chapter 7 how Ras requires the arginine finger of RasGAP to complete its active site and help neutralise the charge on the γ-phosphate leaving group. Unlike α-subunits, Ras alone does not bind aluminium fluoride, but can if docked with RasGAP. Crucially, insertion of RasGAP's arginine also helps bring the catalytic glutamine into the correct conformation (that is captured by the presence of aluminium fluoride) through hydrogen-bonding to its side chain. In GTP-occupied Gα subunit structures, despite the presence of the arginine in switch I, the glutamine does not make contact with the catalytic water or the γ-phosphate except in α subunits occupied by GDP and AlF_4^-[18]. Indeed, both residues are far from the γ-phosphate in GTP-GαI, for example, compared with their positions in GDP-AlF_4^--Gαi (see Figure 8.7). Instead, it appears that the correct conformation (as modelled by GDP-AlF_4^-) simply assembles infrequently, explaining the sluggish rate of intrinsic hydrolysis. This is a reasonable assumption, given that even in the GTP-bound state, the switches are still highly mobile.

8.4 β/γ- and receptor-binding surfaces of α-subunits

8.4.1 The β/γ binding site of GDP-occupied α

The interaction between β/γ and α-subunits only occurs when the α-subunit is in the GDP-occupied state (but the β/γ-binding persists transiently with the nucleotide-free α-subunit during receptor-catalysed nucleotide exchange). Monomeric GDP-bound αi has disordered switch regions and instead presents a microdomain consisting of tightly folded *N*- and *C*-termini – this is probably a recognition tag for β/γ. β/γ-docking with α<GDP> remodels the *N*-terminus. The GDP-bound α-subunit in the trimeric crystal has a well-defined *N*-terminal helix structure that is not resolved in isolated α-subunit structures of GTP-bound α-subunits and this extended structure is the result of stabilising interactions with β. The *N*-terminal helix of the α-subunit binds at the edge of the β-propeller of the β-subunit[19]. The β-subunit also makes extensive contacts with the GDP-conformed switch I and II regions[20]. In particular, the parasol-like β-subunit effectively caps the potential effector binding site of switch II, precluding inappropriate activation of downstream signalling until nucleotide exchange remodels the conformation of the α-subunit (see Figure 8.1(c)). The β-subunit also traps the bound GDP, preventing inappropriate nucleotide exchange in the absence of a receptor signal.

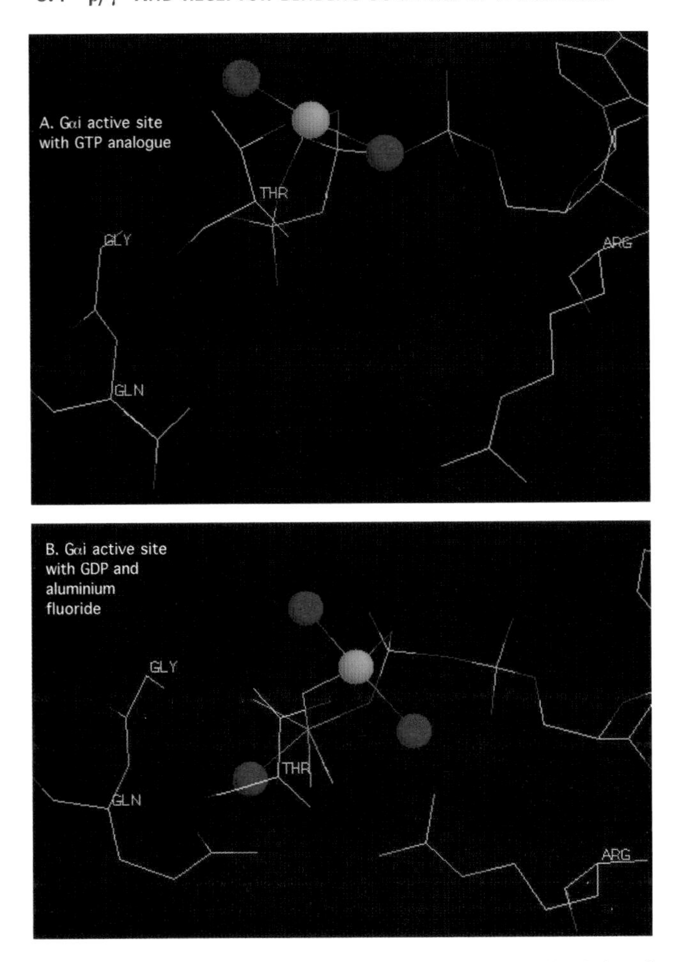

Figure 8.7 (a) GTP occupied alpha subunit (note Gln position); (b) Aluminium fluoride plus GDP mimics the GTP transition state

8.4.2 The receptor-binding interface of GDP-occupied G proteins

The dominant receptor-recognition domain of heterotrimer G proteins is found at the extreme *C*-terminus of the various α-subunits. The earliest indication of this came from studies on Gαi. When covalently modified by pertussis-toxin (PTX), Gαi is unable to couple to cognate receptors such as α2-AR. Thus, upstream receptors fail to couple with PTX-treated Gαi and can no longer inhibit the stimulation of adenylyl cyclase elicited by Gαs cognate receptors such as βAR. The reason is that pertussis ADP-ribosylates Gαi at a *C*-terminal cysteine residue in αi that is normally required for docking with the activated receptor – in human αi-1 this is cysteine-351 in the most *C*-terminal hexapeptide stretch (underlined) KD*C*GLF (Figure 8.3, Table 8.3)[21–23].

Table 8.3 Extreme C-terminal sequence motifs of human G protein α-subunits

α-sub	*C*-termini	ADP-ribosylated by Pertussis
Gαi-1	DVIIKNNLKDCGLF	YES (Cys351)
Gαi-2	DVIIKNNLKDCGLF	YES (Cys352
Gαi-3	DVIIKNNLKECGLY	YES (Cys351)
Gαo	DIIIANNLRGCGLY	YES (Cys351)
Gαz	DVIIQNNLKYIGLC^{355}	NO
Gαt	DIIIKENLKDCGLF	YES (Cys347)
Gαs	DIIQRMHLRQYELL^{394}	NO
Gαq	DTILQLNLKEYNLV^{359}	NO
Gα12	DTILQENLKDIMLQ^{381}	NO

A second indication of the importance of the *C*-terminus is found in the *unc* cell line, (Section 1.4.9) which contains a defective αs point mutant that fails to couple with AC because it contains a proline substitution in the final hexapeptide sequence (native αs = RQYELL^{394}; *unc* αs = PQYELL^{394}). In fact, the six most *C*-terminal amino acids represent distinct signatures for each of the functional classes of Gα subunit and are remarkably well conserved throughout metazoans. Gα1/2, for example, has an identical *C*-terminal hexapeptide receptor-binding sequence (KDCGLF) in mosquito, sea urchin, schistosome, rat, mouse, dog and human (the underlined cysteine is the target of pertussis toxin).

Although it is a ‘floppy’ structure, there is good evidence that the *C*-tail’s conformation (and/or the extent of its exposure to solvent) is altered in cycling between GDP-bound to a GTP-bound states. This change is apparently driven by switch II movements that are a consequence of the presence or absence of γ-phosphate (or AlF_4^-)[24]. In α<GTP>, the *C*-tail is tucked away, but in α<GDP> it is exposed to solvent and available for receptor interactions.

The *C*-terminal tail is insufficient, alone, to completely *specify* receptor interactions. Notably, Gαi-2 has an identical final 8 amino acid sequence to that of transducin (Table 8.3), but although the αi-subunit recognises α2-adrenergic receptors, transducin does not[25]. Further specificity and binding strength comes from receptor interactions with the *N*-terminal helix as well as the α4–β6 loop[6].

Isolated GDP-bound α-subunits have very low affinities for receptors. The receptor interaction is only possible when the α-subunit is reunited with a β/γ-dimer, because the latter contributes crucially to receptor coupling. The receptor makes essential contacts with the *C*-terminal region of the γ-subunit[9].

The receptor side of the interface is less well mapped. However, a definitive binding site for the *C*-terminus of α-subunit is located in the third intracellular loop, and a second at the end of the sixth transmembrane helix of 7-pass receptors. The γ-subunit is theorised to bind at the juxtamembrane region of the receptor’s *C*-terminal tail which, in rhodopsin at least, is folded as a small supernumerary cytosolic helix (Figure 8.3).

8.4.3 Receptor-induced GDP dissociation and nucleotide exchange

The *N*-terminal helix of Gα is a shared/overlapping binding site for both the β-subunit and the receptor. The receptor also contacts the γ-subunit directly[9] (Figure 8.3). It seems likely that the β/γ-dimer is used by the receptor to 'lever' open the guanine nucleotide pocket, facilitating nucleotide exchange, although it is by no means clear how exactly this is achieved. There has been much speculation in the past that the helical insert-1, which seems to effectively cap the G-cleft in all available α-subunit structures, may act as a trapdoor that must open to allow nucleotide exchange. Until the elusive 'empty pocket' form of the ternary complex is finally captured, this fascinating possibility remains just that[6].

No crystal structures of the empty pocket ternary complex, or the isolated Gα<EMPTY>, are available. In fact it is unlikely that isolated, monomeric, Gα<EMPTY> will be ever be crystalised as it aggregates, being either misfolded or denatured upon isolation without nucleotide[26]. In contrast, when it is bound to the ternary complex (i.e., agonist/receptor+α+β+γ), Gα<EMPTY> persists in an activatable condition. Very recent NMR data suggests that the ternary complex form of Gα<EMPTY> is in a conformationally dynamic state, quite different from all the stable Gα structures crystalised in the absence of activated receptor[27].

Activation of the α-subunit is completed when GTP enters the empty G-cleft and resets the switch regions. The β/γ binding interface of α<GTP> is thus lost, causing dissociation of the ternary complex.

8.4.4 Switch II helix rotation

In the monomeric structures of Gαt and Gαi bound to either a GTP analogue or the GTP-mimicking combination of GDP and aluminium fluoride, switch II adopts an extended structure that is stabilised by interaction with switch III. However, in these GTP-activated structures, a further point of contact forms with the conserved arginine of switch II (Arg204 in αt, *178 in Rasmol file*), which is H-bonded to the main-chain carbonyl of Gly199. In this conformation, the arginine also forms a salt-bridge with a conserved glutamate (Glu241 in αt, *215 in Rasmol file*) that is present at the *C*-terminal side of switch III. This latter ionic contact is theorised to form a latch or hasp that serves to trap GTP in the binding pocket. At the same time these contacts also contribute to the rotation of the switch II α-helix by 120° compared with the GDP-bound forms[20]. Compare Figures 8.8(A) and 8.9(A) with Figure 8.8A.

These contacts are not present in GDP-bound α-subunits because the three residues move apart after GTP hydrolysis as the helix unwinds. The Glu-to-Arg distance in αt<GTP> is $\approx$3Å but increases to more than 17 Å in αt<GDP>. The rotation of the helix can be appreciated from the position of the conserved tryptophan (207 in Gαt; 211 in Gαi).

Thus, the face of the switch II helix presented in the GTP-bound subunit (that interacts with effectors) is entirely different from the face of the GDP-bound subunit (that interacts with β/γ).

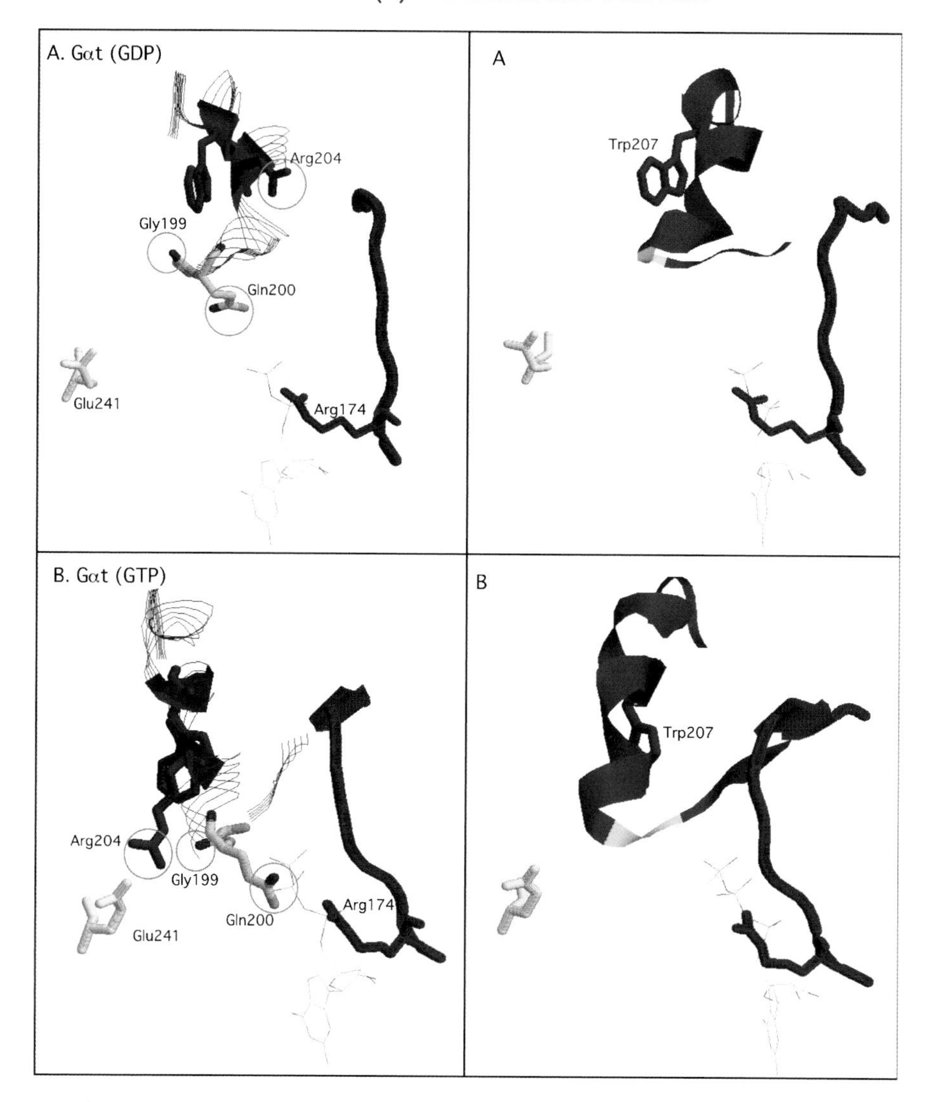

Figure 8.8 G-cleft 'hasp' closure and switch II rotation upon GTP-for GDP exchange

Because of the loss of contact between switch II and the region of switch III, the guanine-binding cleft of the monomeric GDP-bound Gα subunit is much more exposed to solvent and this allows GDP to dissociate slowly and spontaneously *in vitro*. By contrast, when Gα is capped by β/γ in the heterotrimer, GDP is almost irreversibly bound to the G-cleft. It is noteworthy that GDP-conformed switch II adopts a unique conformation in the heterotrimer (Figure 8.9(A)), which is different from that of monomeric αt<GDP> or αi<GDP> – the latter structure, of course, is completely disordered.

In monomeric GDP-bound Gα, the switch II helix is more flexible – in Gαi it is so floppy that it is unresolved in the X-ray structures. This is a result of the loss of

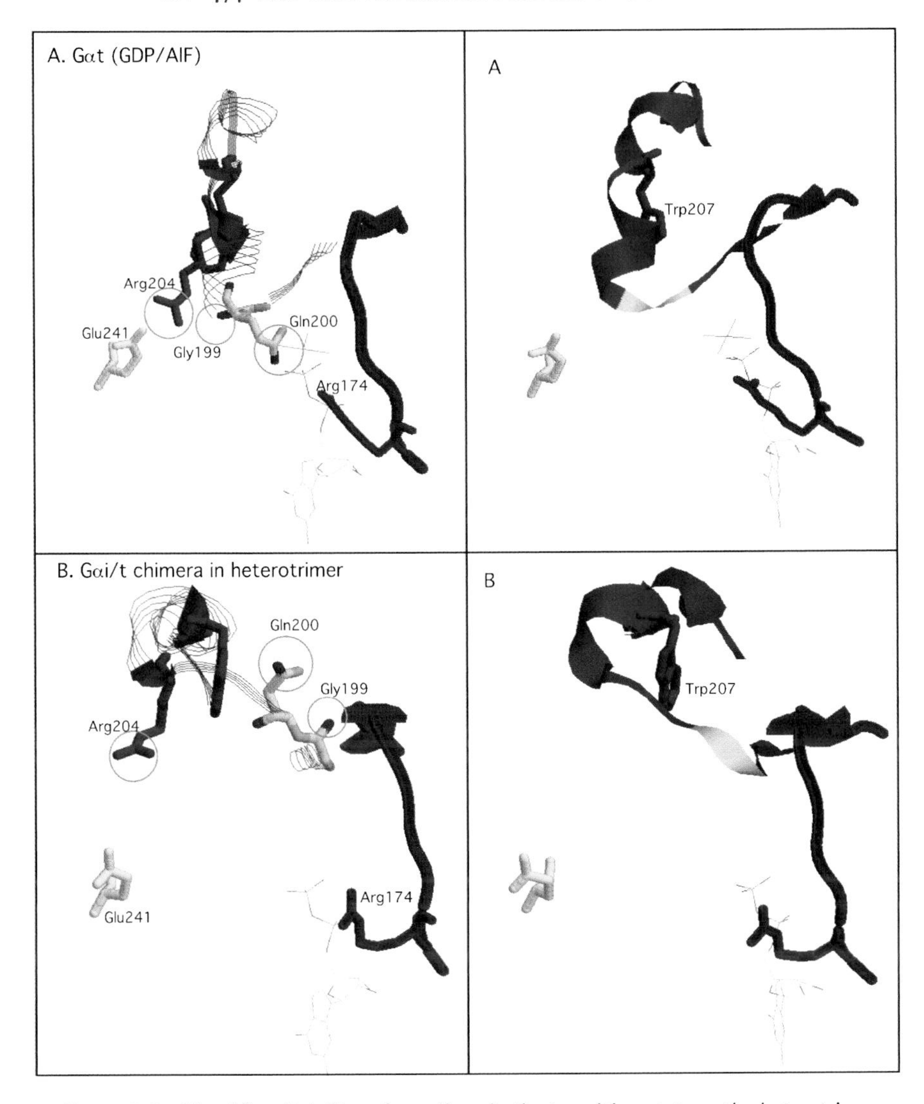

Figure 8.9 'Hasp' & switch II conformations in the transition state vs the heterotrimer

γ-phosphate contacts with switch I and II residues and this loss has the effect of opening the 'latch' of the G-cleft – Arg208 no longer forms the tripartite connection with Gly199 and Glu241. This further disturbs the positioning of the catalytic glutamine. Instead of being stabilised by the γ-phosphate of GTP, the inactive α<GDP>-subunit must now be stabilised by interaction with β/γ, which effectively acts as a GDP-dissociation inhibitor simply because it stabilises the switch regions. This arrangement persists until it is again disturbed by docking of an activated receptor, which acts like a guanine nucleotide exchange factor or GEF.

8.4.5 Switch II – the primary effector-binding surface of α-subunits

Effector enzymes for Gα-transduced signals include forms of adenylyl cyclase, PLCβ and cGMP-PDE. The best understood interaction is that of the stimulatory α-subunit of the G protein, Gs, with adenylyl cyclase, which has been long investigated, first biochemically and then structurally. The first crystal structure of the complex of Gαs with two halves of adenylyl cyclase's 'split' active centre confirmed many predictions from earlier studies, especially with regard to the effector-binding interface on αs, and for the first time elucidated the mechanism of forskolin[28] (see Figure 8.10). The first thing you will notice is that switch II of αs is sandwiched between a loop and an α-helix of the adenylyl cyclase C2 domain. Less obvious is a second interacting portion of αs: the loop connecting the α3-helix and β5-strand (coloured light blue), which interacts

A. Side view showing approximate orientation with respect to the plasma membrane

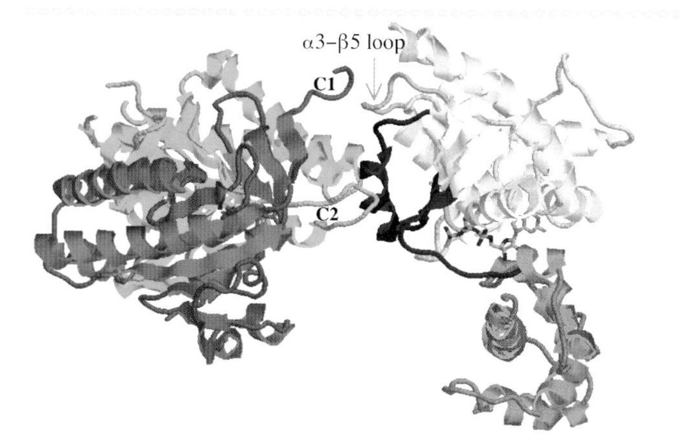

Gαs: switch I is in red; switch II in blue; switch III in yellow; P-loop in bright green; helical insert-1 in gold. α3–β5 loop light blue.

Adenylyl cyclase: C2 domain from AC-2 is in light blue-green; C1 domain from AC-5 dark green; forskolin rendered as pink sticks

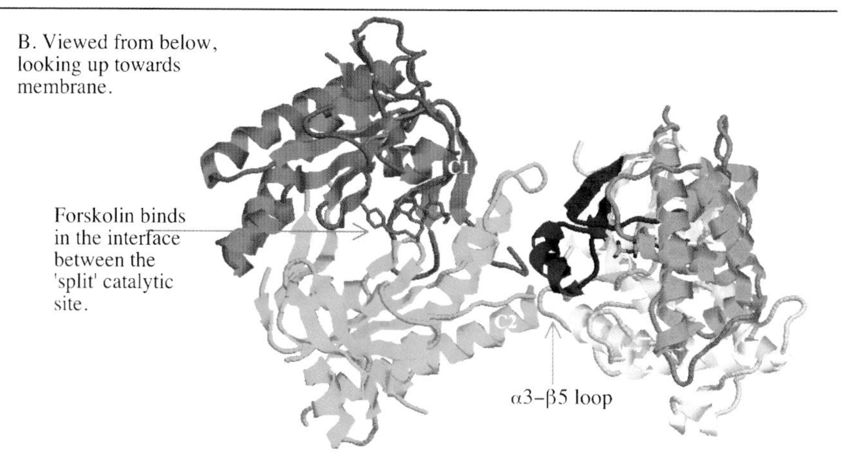

Figure 8.10 Crystal structure of Gαs in complex with adenylyl cyclase catalytic subunit pair

with portions of *both* C1 and C2. The dipterpine, forskolin, also makes contacts with both halves of adenylyl cyclase by a bridging action in the cleft between C1 and C2 (strictly speaking these are the truncated versions of C1 and C2 from AC-5 and AC-2, respectively (termed 'C1a' and 'C2a')).

The binding sites for forskolin and αs are distinct and act independently. Both are thought to activate the effector by uniting the split catalytic site – effectively gluing the two halves together. Incidentally, it is interesting to note that the two hydrophobic pockets that forskolin fits into are evolutionarily preserved, leading to speculation that an endogenous forskolin-like entity may exist, although none has so far been found.

Gαi, (and most likely all the other α-subunits) also binds to effectors predominantly through switch II [28]. In the case of adenylyl cyclase types 5 and 6, Gαi binding is non-competitive with Gαs[23]. Gαi switch II is proposed to bind at a site on the adenylyl cyclase C1 domain, directly opposite to that of Gαs. Both can bind at once, which explains Gαi's ability to block Gαs. However, Gαi can inhibit AC5/6 independently of Gαs – Gαi also blocks forskolin's ability, alone, to stimulate the enzyme[29].

8.5 Modulators of G protein activity – the 'RGS' protein family

The activity of some α-subunits' GTPase can be increased by ***r**egulators of **G** protein **s**ignalling* (RGS proteins). The GAP-type RGS proteins for heterotrimeric G proteins are unrelated to RasGAPs in both sequence and mechanism. RGS proteins share a ≈120 amino acid domain that represents their functional GAP module. They may be functionally classified into subfamilies dependent upon the presence or absence of other domains in their respective structures[30] and this broadly follows the classification based on RGS module structural relationships. The additional domains include PH-, PDZ-, DH/PH-, and PTB-domains among others, making these RGS proteins functionally diverse scaffolding proteins, whose functions are only now beginning to be revealed.

As we have seen, Gα has an inbuilt catalytic site arginine. The intrinsic catalytic rate of α-subunits is much higher than Ras due to the tethered GAP-like activity of the arginine in linker-2 connecting insert-1, and it is therefore unsurprising that the mechanisms of action of RGS proteins do not resemble that of RasGAP. The first indication that the intrinsic rate of Gα catalysis could be further enhanced came from studies on the interaction of Gαq with its effector, PLC-β1. Interaction with PLC-β1 dramatically increases the GTPase rate of Gαq, an effect not seen with Gαs and adenylyl cyclase (although modest increases are reported)[31].

8.5.1 RGS proteins and GTPase activation

The 'standard model' of G proteins as fixed timers of 7-pass receptor transduction, derived from *in vitro* biochemical studies, still holds true for the αs-containing G proteins that stimulate adenylyl cyclase. The rates of the GTP hydrolysis and nucleotide exchange at αs are unaltered by the addition of effectors or other regulators[32,33].

However, the fixed-rate-GTPase paradigm was never quite able to account for the variable (usually faster) kinetics observed in *in vivo* situations where Gi/o or Gq proteins were the transducers. The paradox was resolved when it was realised that, just like Ras-type monomeric G proteins, the GTPase activity of αi-subunits (in particular) could also be accelerated by RGS proteins (reviewed in Reference 34). Such RGS proteins downregulate 7-pass receptor signalling by shortening the life-time of α<GTP>. At least 37 RGS proteins are encoded in the human genome, most are Gi/o interactors, with a smaller number of Gq interactors[34]. The RGS fold encompasses 9 α-helices with the α-subunit-interacting core being an antiparallel 4-helix bundle that presents loops that interact with switches I, II and III. This is thought to stabilise switches I and II in a conformation favourable to achieving the transition state.

Why have Gi/o proteins been so targetted? Perhaps it is because Gi proteins are highly active in neural transmission through their coupling with ion channels and it is likely that their additional level of control by RGS proteins is necessary to allow for flexibility and rapid responses to neuronal signals – a situation that does not apply to the more leisurely metabolic signalling transduced by Gs.

8.5.2 RGS proteins: inhibition of nucleotide exchange; crosstalk with other pathways

Latterly, it has become apparent that certain RGS proteins manifest other regulatory properties because of their additional modules. The R12 subfamily of RGS proteins, for example, acts as ***g***uanine nucleotide ***d***issociation ***i***nhibitors (GDIs), by virtue of the presence of a 'GoLoco' domain that causes downregulation of G protein transduction, this time by inhibiting GTP entry into the α-subunit. The GoLoco domain binds only to Gαi<GDP> and its interaction site is centred on switch II. GoLoco can be thought of as functionally usurping the β/γ-subunit's role as a Gα-GDI while at the same time blocking β/γ-binding at switch II and thereby blocking the GEF-action of receptors. R12 proteins also possess Ras binding domains and PTB- and PDZ-domains, suggesting potentially complex roles in crosstalk with single-pass receptor pathways as well as scaffolding functions.

A different subfamily of RGS proteins contains tandem DH/PH-domains that function as a RhoGEFs. In this case, the RhoGEF-containing RGS protein acts as an effector allowing Gα12/13 and Gαq to activate the monomeric G protein Rho[34]. Finally, R7-type RGS proteins have PH-domains and a Gγ-like domain that specifically binds the neuronal-specific Gβ subunit, β5. The possibility exists that R7 not only acts as a Gα-GAP but also (in the form of R7/β5) might functionally replace conventional β/γ in facilitating the uncoupling of Gα to 7-pass receptors.

Confusingly, the dual specific AKAP scaffolding protein (D-AKAP-2) that binds PKA regulatory subunits is also an RGS-containing protein, whereas other AKAPs are not. An evolutionary dendrogram illustrates the structural relationships and classification of these proteins, based upon the RGS modules they harbour[34] (Figure 8.11).

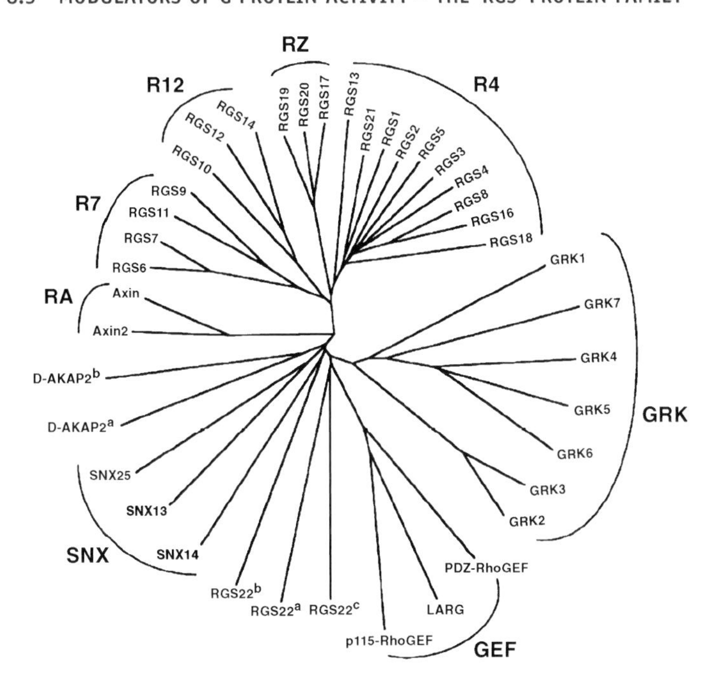

Figure 8.11 Dendrogram showing relationships and classification of RGS proteins (Siderovski, D.P. & Willard, F.S. (2005) *Int. J. Biol. Sci.*, **1**: 51–66)

8.5.3 Gαi/o/q GEF proteins – unrelated to RGS

Reminiscent of Ras regulation, there appear to be a small number of ***g**uanine nucleotide* ***e**xchange* ***f**actors* (GEFs) specific for Gαi/o/q subunits. The known proteins ('Ric-8A and B'; named for '***r**esistant to* ***i**nhibitors of* ***c**holinesterase*) are unrelated to RGS proteins and have no discernable modular structure resembling any other signal molecules. They act *in vitro* by binding to monomeric Gα<GDP> only (like GoLoco, they do not bind to heterotrimers) and thereby cause ejection of GDP, yielding a stable Ric-8 complex with the now *nucleotide-free* α-subunit[33]. The Ric-8/ Gα<EMPTY> complex only dissociates when GTP is subsequently supplied. This would be expected to upregulate signalling by speeding up GTP-for-GDP exchange, analogous to the action of mSos, but it differs in that it appears to act independently of receptors. However, although a physiological role for Ric-8-homologues in aspects of cell division in *Drosophila* and *C. elegans* has been established by knockout, it is fair to say that the biochemistry and structure/functional mechanisms of Ric-8 are yet to be fully explored.

8.5.4 GRKs – RGS domain-containing S/T-kinases

Oddly, members of the RGS family had unknowingly been discovered in the late 1980s in the form of serine/threonine kinase enzymes, collectively known as

G protein-coupled ***r***eceptor ***k***inases (GRKs), although their relationship to the wider RGS-containing family was unrecognised at the time (see reviews in References 34 and 35).

The first GPCR-kinase to be identified was rhodopsin kinase, now known as 'GRK-1'. It is activated by docking with the third intracellular loop of rhodopsin when the receptor is in the light-activated conformation, and it desensitises rhodopsin by phosphorylating several sites in the receptor's *C*-terminal tail. These phosphorylation events have the effect of creating a binding site for an accessory protein known as 'arrestin', which binds to the phosphorylated *C*-tail and thus prevents Gαt coupling by steric hinderance. Arrestin binding also initiates internalisation of receptors, which are either dephosphorylated and recycled or degraded.

Soon after the discovery of rhodopsin kinase, another serine/threonine kinases was found to play a key role in the desensitisation of 7-pass receptor signalling in conditions of chronic stimulation with agonists. The second GRK was originally named 'β-***a***dreneric receptor ***k***inase' (or βARK) because it was found to phosphorylate serine residues in the βAR receptor's *C*-terminal tail after long-term stimulation with βAR agonists. βARK-1 and the closely related βARK-2 are now known as GRK-2 and GRK-3, respectively. Like GRK-1, GRK-2/3 also acts in concert with accessory proteins (in this case, β-arrestins) that bind the receptor's *C*-tail after βARK has phosphorylated it.

Unlike GRK-1, which is constitutively membrane-associated through a *C*-terminal farnesyl post-translational modification, GRK-2/3 are cytosolic enzymes that can only be recruited to the membrane through interaction with free β/γ-subunits. An excess of free β/γ-subunits is an indication that a G protein coupled receptor is being chronically stimulated. βARK has a PH-domain. PH-domains predominantly bind phosphoinoside headgroups but some (like those in GRKs) are capable of binding to the centre of the β-propeller of the *free* β/γ subunit – the site otherwise occupied by the switch II region of Gα<GDP>-subunits. When tethered to the membrane by β/γ, βARK is in a position to collision-couple with receptors and will phosphorylate any 7-pass receptors that are ligand-occupied – like GRK-1, GRK-2/3 only binds to the agonist-occupied receptor conformation.

A recent structure of GRK-2 in complex with a Gαq-containing heterotrimer, reveals another way that GRKs can disrupt G protein signalling. That is, by sequestering the chronically activated Gα subunits. GRK-2 does this via its *N*-terminal RGS domain, which binds to the switch II effector-binding site of the Gα subunit, masking it, and thus disrupting signal transduction (see Figure 8.12). In the crystal, GRK-2 can be seen scaffolding a Gαq subunit through its RGS module and, simultaneously, a β/γ-dimer through its PH domain. Whether such simultaneous scaffolding pertains *in vivo* is uncertain at present.

8.6 Signal transduction by β/γ subunits

β/γ-subunits were at first thought to play no activating role in signal transduction, other than the inhibitory role of sequestering the effector-binding region of the α-subunit while the cell is unstimulated. However, after some initial resistance to the idea, it

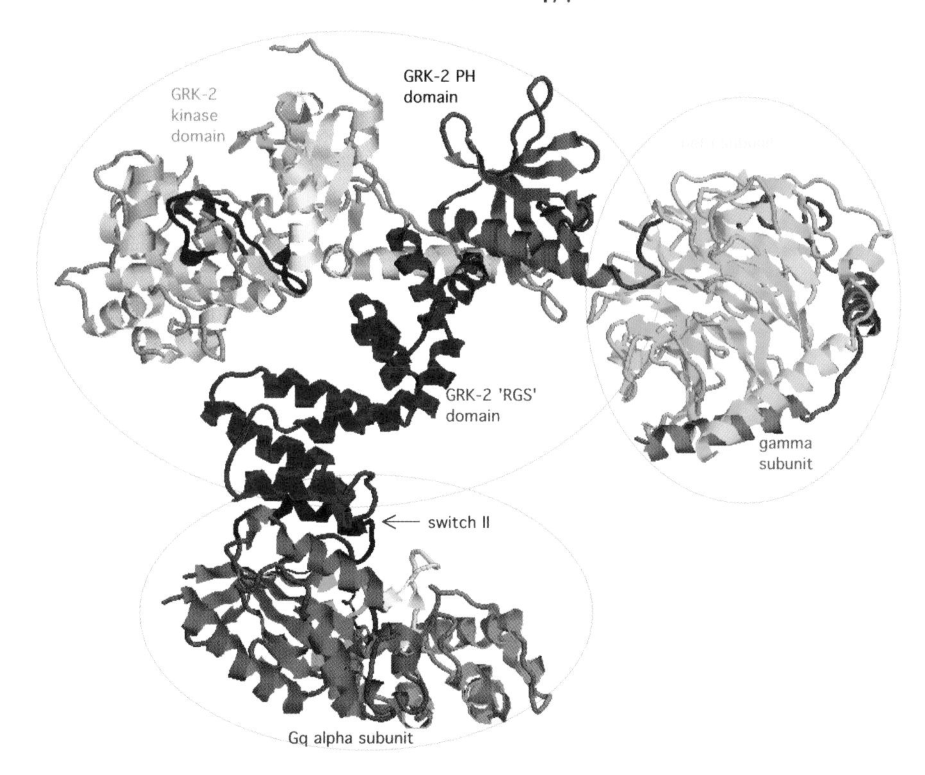

Figure 8.12 The complex of GRK-2 with Gαq and β/γ

gradually became accepted that free β/γ dimers could also activate certain effectors. Actually, in primitive organisms β/γ seems to be the primordial transducer, with the α-subunits acting as mere 'sinks' to sequester β/γ in resting cells. In metazoans, the 'sequestration' is mutual – β/γ masks the α-subunit's switch II while the α-subunit, in a reciprocal fashion, masks the effector-binding face of the β-subunit's β-propeller. Table 5.2 (Chapter 5) summarises the modifying effects that β/γ-subunits potentially exert on certain isoforms of adenylyl cyclase, both inhibitory and stimulatory. For example, β/γ-subunits released from Gq heterotrimers (downstream of α1-AR) can co-activate adenylyl cyclase 2/4 when Gαs subunits are also activated (downstream of β-AR)[36].

The finding that β/γ subunits are transducers in their own right has explained a long-existing paradox. As we saw in Chapter 5, adrenergic stimulation of male rat liver activates glycogenolysis through an IP3-dependent release of calcium that is mediated by Gq activated PLC-β. Because Gq is involved, PLC-β activation in liver is generally pertussis toxin insensitive. However, in other tissues IP3 production and calcium release provoked by agonist stimulation is pertussis-toxin sensitive, pointing to a Gαi-containing G protein. For example, PTX was found to block the production of IP3 in leukaemic HL60 cells, normally stimulated by chemotactic peptide fMet-Leu-Phe, a 7-pass receptor agonist[37].

Such observations were later explained by the finding that subtypes of PLC-β were more, or less, sensitive to activation by free β/γ subunits (Chapter 5, Table 5.3). β/γ subunits are excellent activators of the PLC-β2 isoform, where they are equipotent with Gαq, but are 50–100-fold less able to activate PLC-β1, -β3 or -β4. Generally, free β/γ subunits must be in excess to have this effect and this appears to be why this alternative IP3 signal is pertussis toxin-sensitive – Gαi/o-containing G proteins are much more abundant than other G proteins and are the predominant source of excess amounts of free β/γ-subunits[21].

GRK-2 (βARK) is recruited to the membrane by free β/γ subunits. Like the GRK family of receptor kinases, PLC-β subtypes also contain PH domains. Given that the PH domain of GRK clearly binds to β/γ in an 'effector-like' manner, there may be reason to expect that PLC-β subtypes also bind β/γ via PH interactions, possibly varying due to differences in PH domains between subtypes. Equally, however, there is evidence that the 'X'–'Y' linker of PLC-β2 may be an interaction site for β/γ[31].

One final caveat: although we have been used to thinking of the different β- and γ-isoforms as interchangeable, there are notable exceptions to this general rule. For example, heterotrimers containing β_1/γ_1 subunits couple more effectively with rhodopsin than any other combinations, and β_2/γ_2 subunits are selective inhibitors of T-type calcium channels[30].

References

1. Graf, R., Mattera, R., Codina, J., Evans, T., Ho, Y.K., Estes, M.K. and Birnbaumer, L. (1992) Studies on the interaction of alpha subunits of GTP-binding proteins with beta gamma dimers. *Eur J Biochem.*, **210**: 609–619.
2. Simon, M.I., Strathmann, M.P. and Gautam, N. (1991) Diversity of G-proteins in signal transduction. *Science*, **252**: 802–808.
3. Leipe, D.D., Wolf, Y. I., Koonin, E.V. and Aravind, L. (2002) Classification and evolution of P-loop GTPases and related ATPases. *J. Mol. Biol.*, **317**: 41–72.
4. Whiteway, M.S., Wu, C., Leeuw, T., Clark, K., Fourest-Lieuvin, A., Thomas, D.Y. and Leberer, E. (1995) Association of the yeast pheromone response G-protein beta gamma subunits with the MAP kinase scaffold Ste5p. *Science*, **269**: 1572–1575.
5. Wang, Y. and Dohlman, H.G. (2004) Pheromone signaling mechanisms in yeast: a prototypical sex machine. *Science*, **306**: 1508–1509.
6. Hamm, H.E. (1998) The many faces of G-protein signaling. *J. Biol. Chem.*, **273**: 669–672.
7. Wall, M.A., Coleman, D.E., Lee, E., Iniguez-Lluhi, J.A., Posner, B.A., Gilman, A.G. and Sprang, S.R. (1995) The structure of the G-protein heterotrimer Giα1β1γ2. *Cell*, **83**: 1047–1058.
8. Lambright, D.G., Sondek, J., Bohm, A., Skiba, N.P., Hamm, H. and Sigler, P.B. (1996) The 2.0Å crystal structure of a heterotrimeric G-protein. *Nature*, **379**: 311–319.
9. Chabre, M. and le Maire, M. (2005) Monomeric G-protein-coupled receptor as a functional unit. *Biochemistry*, **44**: 9395–9403.
10. Sprang, S.R. (1997) G-protein mechanisms: insights from structural analysis. *Ann. Rev. Biochem.*, **66**: 639–678.
11. Rossmann, M.G., Moras, D. and Olsen, K.W. (1974) Chemical and biological evolution of nucleotide-binding protein. *Nature*, **250**: 194–199.

12. Walker, J.E., Saraste, M., Runswick, M.J. and Gay, N.J. (1982). Distantly related sequences in the alpha- and beta-subunits of ATP synthase, myosin, kinases and other ATP-requiring enzymes and a common nucleotide binding fold. *EMBO J.*, **1**: 945–951.
13. Benjamin, D.R., Markby, D.W., Bourne, H.R. and Kuntz, I.D. (1995) Solution structure of the GTPase activating domain of as. *J. Mol. Biol.*, **254**: 681–691.
14. Lambright, D.G., Noel, J.P., Hamm, H.E. and Sigler, P.B. (1994) Structural determinants for the activation of the α-subunit of a heterotrimeric G-protein. *Nature*, **369**: 621–627.
15. Noel, J.P., Hamm, H.E. and Sigler, P.B. (1993) The 2.2Å crystal structure of transducin-α complexed with GTPγS. *Nature*, **366**: 654–663.
16. Coleman, D.E., Berghuis, A.M., Lee, E., Linder, M.E., Gilman, A.G. and Sprang, S.R. (1994) Structures of active conformations of Gi alpha 1 and the mechanism of GTP hydrolysis. *Science*, **265**: 1405–1412.
17. Mixon, M.B., Lee, E., Coleman, D.E., Berghuis, A.M. and Gilman, A.G. (1995) Tertiary and quaternary structural changes in Gi alpha 1 induced by GTP hydrolysis. *Science*, **270**: 954–960.
18. Sunahara, R.K., Tesmer, J.J.G., Gilman, A.G. and Sprang, S.R. (1997) Crystal structure of the adenylyl cyclase activator Gsa. *Science*, **278**: 1943–1947.
19. Milligan, G. and Kostenis, E. (2006) Heterotrimeric G-proteins: a short history. *Br. J. Pharmacol.*, **147**: S46–S55.
20. Iiri, T., Farfel, Z and Bourne, H.R. (1998) G-protein diseases furnish a model for the turn-on switch. *Nature*, **394**: 35–38.
21. Gilman, A.G. (1994) G-proteins and the regulation of adenylyl cyclase. *Nobel Prize Lecture (Physiology or Medicine)*, pp. 182–212.
22. Bokoch, G.M., Katada, T., Northrup, J.K., Ui, M. and Gilman, A.G. (1984) Purification and properties of the inhibitory guanine nucleotide-binding regulatory component of adenylate cyclase. *J. Biol. Chem.*, **259**: 3560–3567.
23. Taussig, R., Tan, W-J., Hepler, J.R. and Gilman, A.G. (1994) Distinct patterns of bidirectional regulation of mammalian adenylyl cyclase. *J. Biol. Chem.*, **269**: 6093–6100.
24. Yang, C-S., Skiba, N.P., Mazzoni, M.R. and Hamm, H.E. (1999) Conformational changes at the carboxyl terminus of Gα occur during G-protein activation. *J. Biol. Chem.*, **274**: 2379–2385.
25. Kostenis, E., Waelbroeck, M. and Milligan, G. (2005) Promiscuous G-alpha proteins in basic research and drug discovery. *Trends Pharmacol. Sci.*, **26**: 595–602.
26. Zellent, B., Veklich, Y., Murray, J., Parkes, J.H., Gibson, S. and Liebman, P.A. (2001) Rapid irreversible G-protein alpha subunit misfolding due to intramolecular kinetic bottleneck that precedes Mg^{2+} 'lock' after GTP/GDP exchange. *Biochemistry*, **40**: 9647–9656.
27. Abdulaev, N.G., Ngo, T., Ramon, E., Brabazon, D.M., Marino, J.P. and Ridge, K.D. (2006) The receptor-bound 'empty pocket' state of the heterotrimeric G-protein a-subunit is conformationally dynamic. *Biochemistry*, **45**: 12986–12997.
28. Tesmer, J.J.G., Sunahara, R.K., Gilman, A.G. and Sprang, S.R. (1997) Crystal structure of the catalytic domains of adenylyl cyclase in a complex with Gsα. GTPγS. *Science*, **278**: 1907–1916.
29. Sunahara, R.K. and Taussig, R. (2002) Isoforms of mammalian adenylyl cyclase: multiplicities of signaling. *Molecular Interventions*, **2**: 168–184.
30. McCudden, C.R., Hains, M.D., Kimple, R.J., Siderovski, D.P. and Willard, F.S. (2005) G-protein signaling: back to the future. *Cellular and Molecular Life Sciences*, **62**: 551–577.
31. Morris, A.J. and Malbon, C.C. (1999) Physiological regulation of G-protein-linked signaling. *Physiological Reviews*, **79**: 1373–1430.
32. Berman, D.M and Gilman, A.G. (1998) Mammalian RGS proteins: barbarians at the gate. *J. Biol. Chem.*, **273**: 1269–1272.
33. Tall, G.G., Krumins, A.M. and Gilman, A.G. (2003) Mammalian Ric-8A (synembryn) is a heterotrimeric Gα protein guanine nucleotide exchange factor. *J. Biol. Chem.*, **278**: 8356–8362.

34. Siderovski, D.P and Willard, F.S. (2005) The GAPs, GEFs, and GDIs of heterotrimeric G-proteins. *International Journal of Biological Sciences*, **1**: 51–66.
35. Penn, R.B., Pronin, A.N. and Benovic, J.L. (2000) Regulation of G-protein-coupled receptor kinases. *Trends in Cardiovascular Medicine*, **10**: 81–89.
36. Defer, N., Best-Belpomme, M. and Hanoune, J. (2000) Tissue specificity and physiological relevance of various isoforms of adenylyl cyclase. *Am. J. Physiol. Renal Physiol.*, **269**: F400–F416.
37. Brandt, S.J., Dougherty, R.W., Lepatina, E.G. and Niedel, J.E. (1985) Pertussis toxin inhibits chemotactic peptidestimulated generation of inositol phosphates and lysosomal enzyme secretion in human leukemic (HL60) cells. *PNAS*, **82**: 3277–3280.

9 The insulin receptor and the anabolic response

There are three members of the human insulin receptor family: the insulin receptor, the insulin-like growth factor-I (IGF-I) receptor and the insulin receptor related (IRR) protein, which is an 'orphan' receptor with no known ligand. The insulin receptor and its close homologue, the IGF-I receptor, differ from other RTKs in (i) being persistently dimerised and (ii) not providing stable docking sites for phosphotyrosine-binding adaptors. Instead, the insulin receptor uses an intermediary substrate protein as a surrogate for its signal transduction particle assembly. This substrate, IRS, is phosphorylated by the receptor kinase on multiple tyrosines while transiently associated, before leaving the receptor to act as an isolated anchorage point for SH2 domain proteins. The temporary nature of the liaison means that the insulin receptor may potentially modify several IRS molecules before its kinase inactivates again, thus providing an unusual amplification mechanism.

9.1 The insulin receptor – a pre-dimerised RTK with a unique substrate

The insulin receptor (InsR) follows the general protein kinase convention of requiring A-loop phosphorylation for activation, but differs in already being dimerised. Even in the absence of its ligand insulin, interchain disulphide bonds between the partner monomers ensures persistent dimerisation. The 'monomeric' InsR is derived from a single gene but its protein product is cleaved during

Structure and Function in Cell Signalling John Nelson

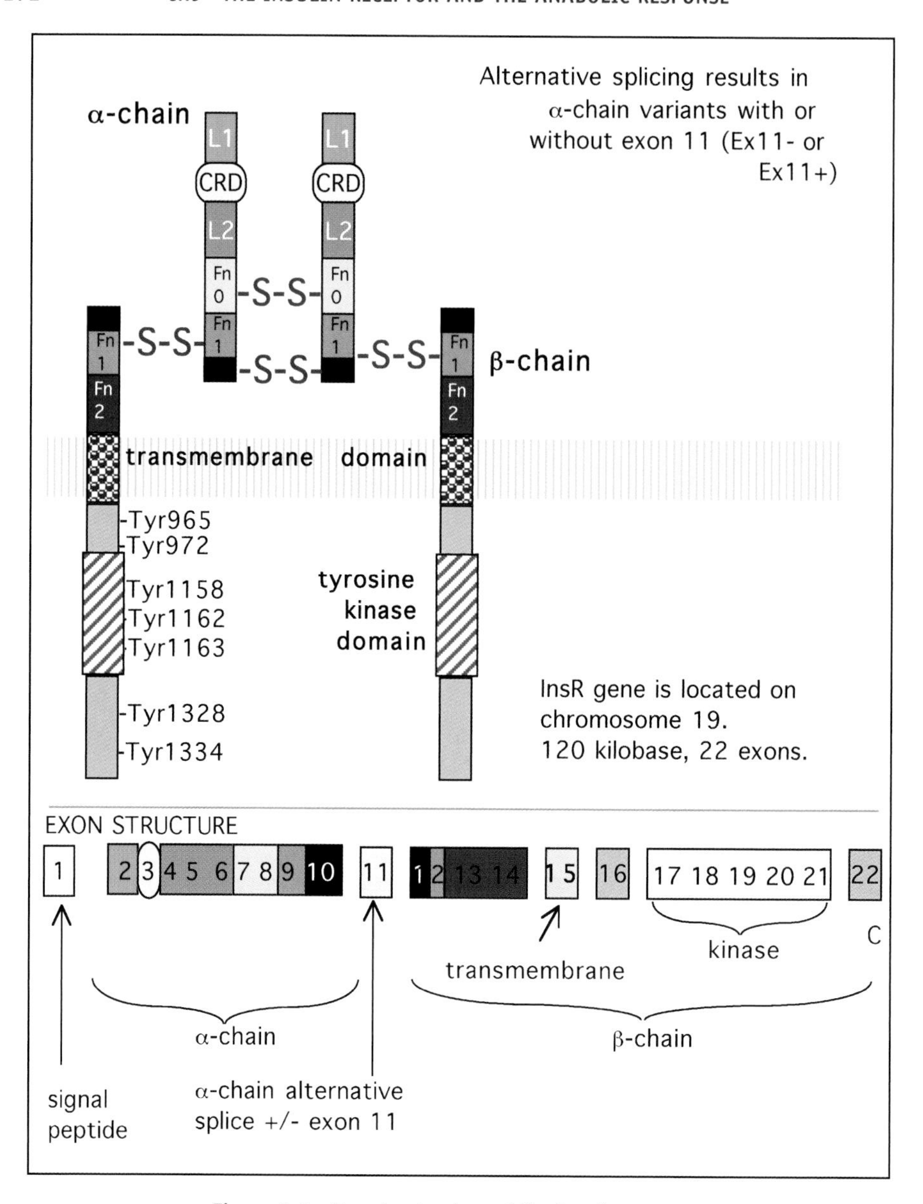

Figure 9.1 Domain structure of the insulin receptor

processing, leaving two polypeptides (the 'α-chain' and the 'β-chain') that remain tethered to one another by an intrachain disulphide bond (see Figure 9.1). The α-subunit contains the ligand-binding domain and is entirely extracellular. The two dimerising disulphide bonds form between the two α-chains. The β-chain spans the membrane with a small, extracellular *N*-terminal portion where the α–β disulphide linker is anchored. The intracellular portion of the β-chain contains the tyrosine kinase domain, and regulatory regions that are juxtamembrane- or *C*-terminal of the kinase domain.

There are two human InsR splice variants that differ in the presence or absence of 12 amino acids near the *C*-terminus of the mature α-subunit. Apart from in the foetus (see Section 9.10.5 later in this chapter), the following summarises their general expression. The InsR-A (*minus* exon 11) has a relatively low affinity for insulin and is the only form expressed in leukocytes. InsR-B (*plus* exon 11) has a high affinity for insulin (twice that of InsR-A) and is expressed at elevated levels in insulin-responsive tissue like skeletal muscle, liver and adipose (up to 200,000 receptors per cell). Other cell types have a mixture of both isoforms. Practically all vertebrate tissues contain insulin receptors, which can be as few as 40 receptors per cell in erythrocytes.

9.1.1 Insulin receptor residues numbering

The numbering of insulin receptor residues varies in the literature and has the potential to be confusing. The numbering system used here (see Table 9.1) omits the signal sequence but includes the alternatively spliced mini exon (exon 11), which codes for an additional 12 amino acids[1]. This means that amino acid numbers used here ($Ex11^+$) will be 12 higher than papers using the numbering system of Ullrich[2], which is based upon the short splice variant ($Ex11^-$). The X-ray crystallography papers tend to use the long splice variant numbering system, whereas many of the biochemical papers use short splice variant numbering.

9.1.2 Three clusters of autophosphorylated tyrosines in the InsR intracellular region

The InsR β-chain tyrosine residues that become autophosphorylated after ligand binding are found in three clusters (*here and throughout this chapter, the* $Ex11^-$ *numbers are in italics, in parentheses*):

- Juxtamembrane: $Ex11^+$ tyrosines: 965, **972** and 984 *($Ex11^-$:953,* ***960*** *and 972)*
- Activation loop: $Ex11^+$ tyrosines: 1158, 1162 and 1163 *($Ex11^-$: 1146, 1150 and 1151)*
- *C*-terminal: $Ex11^+$ tyrosines: 1328 and 1334 *($Ex11^-$: 1316 and 1322)*

Although three juxtamembrane domain tyrosines may be phosphorylated *in vitro*, it is the central pTyr972 (*$Ex11^-$: 960*) that is predominantly autophosphorylated *in vivo* and acts as the prime activator of the majority of the receptor's downstream metabolic effects[3]. Tyr965 (*$Ex11^-$: 953*) is possibly also phosphorylated and may act as a receptor internalisation signal.

Table 9.1 The complete human insulin receptor sequence. Note that numbering varies between studies depending on whether the signal sequence and/or the splice variant regions are counted

```
                   signal peptide  27|
        1 MGTGGRRGAA AAPLLVAVAA LLLGAAG

                   mature alpha-chain 1    4
                                     HLY PGEVCPGMDI RNNLTRLHEL ENCSVIEGHL
  34    61 QILLMFKTRP EDFRDLSFPK LIMITDYLLL FRVYGLESLK DLFPNLTVIR GSRLFFNYAL
  94   121 VIFEMVHLKE LGLYNLMNIT RGSVRIEKNN ELCYLATIDW SRILDSVEDN HIVLNKDDNE
 154   181 ECGDICPGTA KGKTNCPATV INGQFVERCW THSHCQKVCP TICKSHGCTA EGLCCHSECL
 214   241 GNCSQPDDPT KCVACRNFYL DGRCVETCPP PYYHFQDWRC VNFSFCQDLH HKCKNSRRQG
 274   301 CHQYVIHNNK CIPECPSGYT MNSSNLLCTP CLGPCPKVCH LLEGEKTIDS VTSAQELRGC
 334   361 TVINGSLIIN IRGGNNLAAE LEANLGLIEE ISGYLKIRRS YALVSLSFFR KLRLIRGETL
 394   421 EIGNYSFYAL DNQNLRQLWD WSKHNLTTTQ GKLFFHYNPK LCLSEIHKME EVSGTKGRQE
 454   481 RNDIALKTNG DKASCENELL KFSYIRTSFD KILLRWEPYW PPDFRDLLGF MLFYKEAPYQ
 514   541 NVTEFDGQDA CGSNSWTVVD IDPPLRSNDP KSQNHPGWLM RGLKPWTQYA IFVKTLVTFS
 574   601 DERRTYGAKS DIIYVQTDAT NPSVPLDPIS VSNSSSQIIL KWKPPSDPNG NITHYLVFWE
 634   661 RQAEDSELFE LDYCLKGLKL PSRTWSPPFE SEDSQKHNQS EYEDSAGECC SCPKTDSQIL
 694   721 KELEESSFRK TFEDYLHNVV FVPRKTSSGT GAEDPRPSRK RR


                                                   736        744
                   mature beta-chain                SLGDVGNV TVAVPTVAAF
 754   781 PNTSSTSVPT SPEEHRPFEK VVNKESLVIS GLRHFTGYRI ELQACNQDTP EERCSVAAYV
 814   841 SARTMPEAKA DDIVGPVTHE IFENNVVHLM WQEPKEPNGL IVLYEVSYRR YGDEELHLCV
 874   901 SRKHFALERG CRLRGLSPGN YSVRIRATSL AGNGSWTEPT YFYVTDYLDV PSNIAKIIIG
 934   961 PLIFVFLFSV VIGSIYLFLR KRQPDGPLGP LYASSNPEYL SASDVFPCSV YVPDEWEVSR
 994  1021 EKITLLRELG QGSFGMVYEG NARDIIKGEA ETRVAVKTVN ESASLRERIE FLNEASVMKG
1054  1081 FTCHHVVRLL GVVSKGQPTL VVMELMAHGD LKSYLRSLRP EAENNPGRPP PTLQEMIQMA
1114  1141 AEIADGMAYL NAKKFVHRDL AARNCMVAHD FTVKIGDFGM TRDIYETDYY RKGGKGLLPV
1174  1201 RWMAPESLKD GVFTTSSDMW SFGVVLWEIT SLAEQPYQGL SNEQVLKFVM DGGYLDQPDN
1234  1261 CPERVTDLMR MCWQFNPKMR PTFLEIVNLL KDDLHPSFPE VSFFHSEENK APESEELEME
1294  1321 FEDMENVPLD RSSHCQREEA GGRDGGSSLG FKRSYEEHIP YTHMNGGKKN GRILTLPRSN
1354  1381 PS

                    grey = numbering of pro-form (before processing)

-27-0:-      Signal peptide (cleaved off during ER processing)       1-27
1-735:-      Mature α-chain (ligand-binding, extracellular)          28-758
718-729:-    Splice variant region (missing in short isoform)        745-756
736-1355:-   Mature β-chain (catalytic and regulatory, cytosolic)    763-1382

Transmembrane domain (boxed)

Catalytic domain: 996-1267
1002-1010:- P-loop
1150-1179:- Activation loops
1130-1137:- Catalytic loop
1038-1051:-"C-alpha helix"
= residues that are tyrosines in the IGF-IR sequence
```

The tyrosine clusters are phosphorylated in an ordered sequence. First the tyrosines in the activation loop of the kinase domain are phosphorylated, with the juxtamembrane region being phosphorylated very soon after. The *C*-terminal domain tyrosines are phosphorylated last, and after some delay[4]. As we discussed in Chapter 4, the three tyrosines of the A-loop need to be phosphorylated for full activity and there is evidence of a graded activation as each one is modified – the monophospho- form is almost inactive, the bisphospho- form is partially active and the trisphospho-kinase is fully active[5]. Tyr1163 is phosphorylated last, resulting in the single biggest elevation in activity especially towards exogenous substrates[6]. This accords with crystallographic evidence that pTyr1163 plays a key role in the correct assembly of the A-loop into a platform for peptide substrate recognition[7]. There is some debate over which residue is phosphorylated first, but there is agreement that tyrosine 1162 phosphorylation is the first effective step in de-repression of active site inhibition through the electrostatic repulsion from the catalytic aspartate that results from addition of the negatively charged phosphoryl moiety. Point mutation of Tyr1158, however, suggests it plays little part in kinase activation.

9.2 InsR and IGF-IR: differentiation leads differential tissue effects

The effects of insulin are widespread and varied, with many specialised tissues set up to respond in very different and distinct ways – for example, liver is stimulated to synthesise glycogen and adipose stimulated to synthesise fat. Some effects of insulin are independent of other signals: mobilising glucose transporters to increase glucose uptake, for example. However, many of insulin's signalling pathways lead to de-activation of downstream targets previously activated by glucagon- or adrenaline-pathways. In effect, insulin reverses glucagon or adrenaline effects, re-setting these metabolic switches.

The diverse effects of insulin and the IGFs on mature vertebrate tissues are 'pre-ordained' by differentiation programmes that have terminated in target cells with differing, and characteristic, patterns of gene expression/repression. This spectrum of response-control varies from broad to quite discrete. Overarching control can be exerted through receptor expression – tissues that need to be acutely insulin responsive (muscle, liver, adipose) have very high levels of InsR; others have very much lower levels. In the same vein, liver has no IGF-I receptors[8]; probably because it is the main source of secreted IGF-I[9] – lack of receptors would circumvent autocrine growth stimulation. More discrete control is exerted through expression patterns of downstream proteins that change as differentiation progresses, but many of the precise molecular mechanisms remain obscure.

A frequently used experimental model was developed from a murine fibroblast clone (the 3T3-L1 cell line) that spontaneously differentiates into insulin-responsive adipocytes after several weeks in culture. The spontaneous differentiation programme can be accelerated by high doses of serum or insulin, or by treatment with isobutylmethylxanthine plus dexamethasone. Differentiation results in morphological changes (formation of intracellular lipid droplets, for example) that are accompanied by an upregulation of anabolic enzymes (such as fatty acid synthetase) as well as dramatic changes in signalling programmes. The adipocytes become more responsive to ACTH- and β-adrenergic-stimuli, and become increasingly insulin sensitive. The latter effect is due, in part, to an enormous 35-fold increase in InsR expression (although insulin sensitivity peaks before InsR reaches maximum levels)[10]. Significantly, differentiation leads to induction of a hormone-sensitive phosphodiesterase, PDE-3B (discussed in Section 9.8.2 and 9.8.3), that is absent in the preadipocyte[11].

Similarly, as brown preadipose cells differentiate into adipocytes, the expression of InsR rises six-fold while IGF-IR only increases by 40%, resulting in a ratio of 120,000 InsR:100,000 IGF-IR molecules per adipocyte, reversing their original relative abundance[12]. Furthermore, although IRS-1 is virtually undetectable in preadipocytes, its expression and phosphorylation levels rise dramatically upon differentiation[13]. This is accompanied by an increase in the activation of protein phosphatase-1 (PP-1), downstream of PI-3-kinase, at the expensive of a relative decrease in PKB activation (also downstream of PI-3-kinase)[14].

9.3 Features of metabolic control in key tissues

In the adult, a constant supply of glucose is absolutely essential for brain function as it is the only adult tissue that cannot survive on alternative fuel sources alone (free fatty acids, amino acids). Glucose-6-phosphatase has a very limited tissue distribution. It is found chiefly in the liver, the kidney cortex, intestinal mucosa and β-cells of the endocrine pancreas. However, liver is the only glycogen storage organ expressing the enzymes and thus is the main source of glucose from glycogendysis during starvation/stress.

Glucose can diffuse slowly in and out of tissues, but its influx is greatly increased by a family of glucose transporter proteins, the GLUT family. When glucose is internalised, it is phosphorylated by hexokinase or glucokinase, and the negatively-charged product (glucose-6-phosphate) is no longer membrane-permeant and is effectively trapped inside the cell.

After a meal, it is essential to remove glucose quickly from the circulation and store it. The main receiving centres are the liver and skeletal muscle, with a lesser uptake into adipose tissue. The bulk of insulin-stimulated glucose disposal is due to skeletal muscle uptake and subsequent synthesis of glycogen[15], with the remaining 25% of glucose going the liver[16]. Unlike peripheral tissues, glucose disposal into the liver is not only controlled by insulin but is also responsive to the concentration of blood glucose. Like insulin-secreting pancreatic β-cells, liver acts as a glucose sensor. This property results from expression of liver-specific isoforms of metabolic enzymes and regulators. Striated muscle can also control its glucose uptake and glycogen in an insulin-independent manner – vigorous exercise, for example, stimulates glucose uptake and glycogenesis in skeletal muscle[17]. Both muscle and liver are also sensitive to the size of their glycogen depots – glycogen depletion facilitates upregulation of glycogen synthase activity. A major difference between muscle and liver glycogen metabolism lies in their distinct modes of regulation of protein phosphatase-1 activity.

In adipose tissue, energy is stored in the form of triglycerides and the size of the store is controlled by the actions of adrenaline/glucagon *versus* insulin upon the hormone-sensitive lipase (HSL). HSL is a homodimeric protein with a monomeric structure of *N*-terminal fat-binding domain and *C*-terminal catalytic domain. The latter domain contains a large regulatory loop that is subject to serine-phosphorylation by PKA. This activates the enzyme and results in lipolysis, depletion of the triacylglycerol store, and export of free fatty acids to the circulation[18].

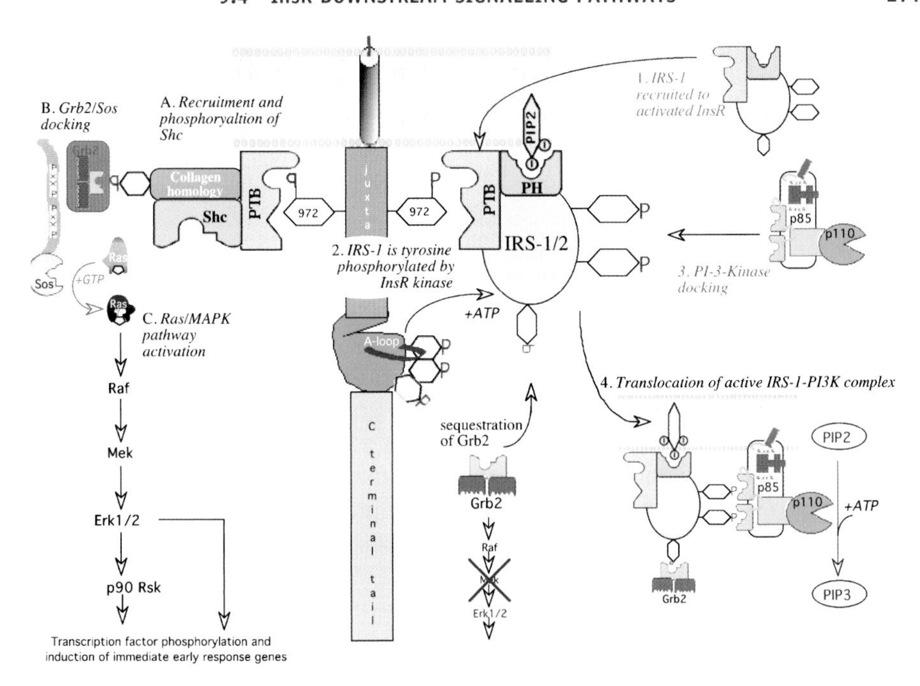

Figure 9.2 Two pathways from the insulin receptor juxtamembrane tyrosine 972 (*Ex11 960*)

9.4 InsR downstream signalling pathways

Two prominent pathways were first identified in insulin signal transduction: the Ras-Mek-Erk1/2-p90Rsk pathway and the IRS1/2-PI3K-PKB pathway (Figure 9.2). The former is familiar as the classic mitogenic pathway of the growth factors – in fact, MAP kinase was first discovered as an insulin-activated protein in 3T3-L1 cells[19]. It was originally thought that MAPK mediated some of insulin's anabolic effects through activation of the p90 Rsk isoform – Rsk is a serine/threonine ***r***ibosomal protein ***S***6 ***k***inase.

9.4.1 MAPK/p90Rsk pathway only mediates growth effects

Shc, like IRS-1, binds to InsR via its PTB domain and the association leads to Shc phosphorylation. Shc binds to the same pTyr972 (*Ex11*$^{-}$*:960*) that IRS-1 binds but, despite their common preference for the minimal NPEpY motif, they display differences in preferences to amino acids further away from the motif[20].

Downstream of Erk1/2 activation, p90Rsk is also activated by phosphorylation by Erk1/2. p90Rsk was shown to phosphorylate 'site 1' of the PP1 glycogen targeting subunit (G^{SUB}) and this phosphorylation activates protein phosphatase-1 (PP1), *in vitro*. However, a number of experimental challenges to this hypothesis have all but ruled out a role for the MAPK pathway in anabolic functions. For example, inhibiting MAPK-kinase (Mek) has no effect on insulin's ability to stimulate glucose uptake, glycogen synthesis, or lipogenesis[21]. Furthermore, the defective glucose uptake seen in type 2 (insulin-resistant) diabetic muscle was found to be associated with an inactive PI-3-kinase pathway – in all cases, the MAPK pathway functioned normally[22].

9.4.2 PI-3-kinase is the prime anabolic effector – is there a second (non-MAPK) anabolic pathway: (CAP-Cbl-Crk)?

Although it is generally accepted that the Erk1/2-type MAPK pathway is solely concerned with growth effects, and the PI-3-kinase pathway is primarily responsible for anabolic effects, there remains a suspicion that a second, alternative pathway may exist for insulin stimulation of glucose uptake, independent of PI-3-kinase. A Cbl pathway (involving the associated CAP and Crk proteins) has been proposed as an alternative to PI-3-kinase in activating translocation of glucose transporters to the cell surface. But this too is challenged by the finding that RNAi-silencing of Cbl, CAP or Crk has no effect on insulin's ability to activate glucose transport[23]. Cbl is recruited to the InsR, as it is to many other growth factor receptors, and is also phosphorylated in adipocytes. But Cbl is usually associated with negative regulation of such receptors (via ubiquitation-directed degradation).

Cbl-b is a major susceptibility gene in experimental type I diabetes but its disruption does not lead to perturbed GLUT4 translocation, rather its dysfunction aids lymphocyte infiltration and immune destruction of pancreatic β-cells[24]. Furthermore, a Cbl-knockout mouse showed improved insulin sensitivity in peripheral tissues[25] – exactly the opposite effect one would expect if Cbl were truly essential for insulin-stimulated glucose uptake. The evidence for and against was recently reviewed[26,27].

In summary, the effects of insulin on anabolism and glucose-uptake are heavily (if not solely) dependent upon activation of the Class 1A PI-3-kinase via $p85\alpha^{PI3K}$, $p50\alpha^{PI3K}$ or $p55\alpha^{PI3K}$ adaptors[28]. The predominant second messenger for insulin's anabolic effects is, therefore, the membrane-bound lipid, PIP3.

9.5 The insulin receptor substrate – a surrogate signal transduction particle

A major difference from more conventional RTKs is that InsR does not (stably) recruit many signalling molecules; instead it uses a separate protein to assemble its signal transduction particle: the insulin receptor substrate (IRS) (Figure 9.2). There are four IRS isoforms (IRS-1, -2, -3, -4), each with a similar make up of an *N*-terminal PH module, next to a PTB module, which together comprises the 'targeting' domain. A more variable *C*-terminal region is referred to as the 'activator' domain[29]. This presents multiple substrate motifs for the InsR kinase, which favours a consensus sequence of

Tyr.Met.Xxx.Met. Four out of the (at least) eight IRS-1 tyrosines phosphorylated by InsR in turn provide perfect docking sites for $p85^{PI3K}$, whose SH2 domains favour **pTyr**. Met/Φ.Xxx.Met. IRS proteins are also multiply serine/ threonine phosphorylated in this *C*-terminal region, and this modification leads to inhibition of their activity[13].

The main two IRS proteins have an overlapping tissue distribution but differ in their relative importance in certain tissues. In skeletal muscle, IRS-1 is the primary transducer of insulin signals; in liver, IRS-2 predominates. In pancreatic β-cells, IRS-1 is a part of the glucose-sensing apparatus whereas IRS-2 is critically concerned with β-cell growth[30].

The transient nature of its receptor interaction means that several IRS proteins can associate with a single activated InsR, one after another, each becoming tyrosine phosphorylated during the association, and each then going on to form an independent signalling complex. Thus, the receptor can use IRS proteins to amplify its signal[31]. More importantly, activated IRS proteins are free to relay the signal by physically moving to other parts of the cell, taking their cargo of activated PI-3-kinase with them.

The other proteins that associate directly with InsR are mostly PTB-domain-containing, and the majority of these also contain a PH domain and are IRK substrates. These include Shc, Gab1 and Cbl, among a small number of others[26].

9.5.1 IRS protein targetting

The subcellular locations, and the translocations of IRS proteins are thought to hold the key to unravelling insulin's enigmatic signal transduction. But the problem has proven difficult to resolve due to the fleeting nature of the interactions and there is still much to learn.

Most researchers agree that in the resting state IRS proteins are located in the cytoplasm or associated with cytoskeletal elements; then, upon stimulation with insulin, they translocate to the plasma membrane[32], before partially redistributing to specific intracellular compartments where they act as activated signalling nodes. Insulin, for example causes activated IRS-1/PI-3-kinase complexes to translocate to internal vesicles – derived from the tubulovesicular endosomal membrane system – that are enriched in glucose transporters[26].

After stimulation with IGF-I on the other hand, IRS-1 (activated by the IGF-IR) translocates to the nucleus – this movement is critically dependent upon IRS-1's PTB domain[33].

The precise contributions that IRS-1/2 PH- and/or PTB-modules make to signalling outcomes and translocations are still debated. The PH domain is believed to be a prime mediator of IRS-1 membrane localisation, but it also plays a distinctive role in regulating activation – foreign PH domains (with similar phosphoinositide specificity) do not restore IRS functionality when swapped with the native domain[34]. IRS-1 PH domain (like that of β-ARK) can also bind to protein, specifically the PH domain-interacting protein (PHIP) that mediates certain InsR signals – a dominant-negative version of PHIP blocks InsR mitogenic effects[35].

What is certain is that, when the receptor is present at normal levels, phosphorylation of Tyr972 (*Ex11⁻ 960*) of InsR is an essential prerequisite for IRS proteins themselves to be phosphorylated. Furthermore, both PH- and PTB-domains are needed for the efficient IRS-InsR coupling that allows this to happen[31]. However, when InsR or IRS is separately over-expressed, the tyrosine phosphorylation of IRS-1 seems less dependent upon the integrity of its PTB or PH domains[31,32].

9.5.2 IRS-interacting proteins – Class 1A PI-3-kinases

In contrast to the receptor, with its preference for PTB proteins, IRS-1/2 pTyr sites recruit SH2 domains only. And, importantly, it is the p85 regulatory/p110 catalytic heterodimer of the Class 1A PI-3-kinase that is predominantly recruited. Peptide binding studies imply that the $p85^{PI3K}$ tandem SH2 pair bind cooperatively, forming a high affinity complex with a pair of pYMXM motifs on an activated IRS[13].

9.6 IRS-1/2 phosphorylation and PI-3-kinase activation

IRS-1, like the other family members, interacts via its PTB domain with the autophosphorylated InsR, assisted by its membrane-binding PH domain. A crystal structure of the PH-PTB 'targeting' domain of IRS-1 (PDB file: 1QQG) shows that both the modules share the expected PH-superfold and, in the same study, binding studies on the truncated IRS-1 reveal their binding preferences[29].

IRS-1's *N*-terminal tandem PH/PTB module pair[29] are tightly packed against each other and in an orientation that appears designed to permit the PH domain to engage the membrane while the PTB domain (binding site 45° rotated) can dock with the InsR juxtamembrane pTyr972 (Figure 9.3). The PH domain of IRS-1 binds either PIP3 or PIP2 with almost equal affinity, but unlike many PH domains its lipid binding is unaffected by soluble phosphoinositide headgroups – IP3, for example, does not compete for its binding of PIP2.

PTB binding to pTyr motifs is lower affinity than SH2 binding and the weaker affinity underlies the transitory nature of IRS-1 interaction with the receptor. The upshot is that it is difficult to co-immunoprecipitate activated IRS-1 with an InsR or IGF-IR receptor antibody, in stark contrast to SH2 proteins associating with autophosphorylated EGFR, which are easily co-immunoprecipitated with anti-EGFR antibodies. To circumvent this, InsR protein—protein interactions have for the most part been studied in yeast 2-hybrid systems[36].

PTB domains have a limited range of ligand motifs. Compared with the wide diversity of pTyr motifs that SH2 domains have evolved to recognise, PTB domains bind almost exclusively to Asn.Pro.Xxx.pTyr motifs, which adopt the reverse turn that PTB and PDZ domains also favour. However, beyond the NPXY motif, more distant residues can affect specificity.

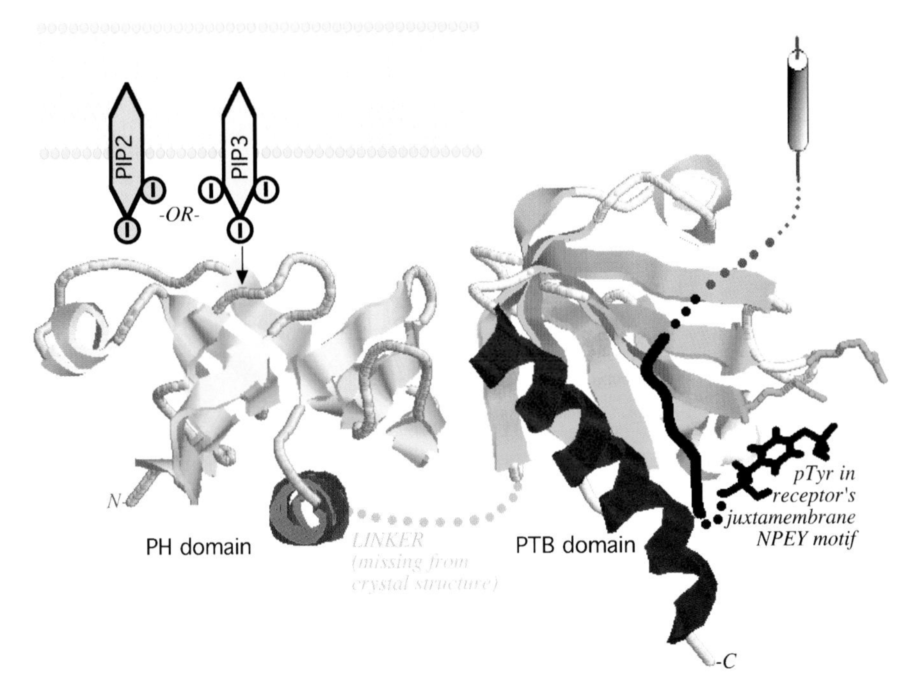

Figure 9.3 The tandem PH-PTB domain of IRS-1 indicating approximate membrane orientation

9.7 Protein phosphatase-1 (PP-1)

PP-1 is a ubiquitous protein phosphatase with a specificity for phosphoserine or phosphothreonine. Other than the phospho-amino acid preference, PP-1 shows little discrimination in the surrounding amino acid sequence context. PP-1 would therefore be extremely promiscuous, were it not for its essential interaction with targeting subunits – partner proteins that recognise substrates on PP-1's behalf. The glycogen-targeting subunits, the G^{SUB} group, are of interest here (the complex area of PP-1 interacting proteins has been comprehensively reviewed in Reference 37).

9.7.1 Glycogen granule targetting of PP-1

Glycogen granules are found both in the cytosol and bound to the endoplasmic reticulum, and the enzymes responsible for both synthesis and degradation are themselves bound to the granules. Regulatory enzymes are also scaffolded to the granule by G^{SUBs}, which are essential partners of the ***p***rotein ***p***hosphatase catalytic subunits, in particular PP-1 (Table 9.2). Liver and skeletal muscle express tissue-specific isoforms: $G_L{}^{SUB}$ (liver only) and $G_M{}^{SUB}$ (skeletal and heart muscle only), while two others are widely distributed. They are protein targeting to glycogen (PTG) and the less-well understood R6[38]. The former subunits' activities and expression are subject to hormonal control, whereas PTG activity appears to be more constitutive[39,40,41].

Table 9.2 Tissue-specific differences in gene expression

glycogen-targetting	Liver	Skeletal muscle	Adipose	β-cells
G_M^{SUB}	0	high	0	0
G_L^{SUB}	higher levels after feeding/insulin	0	0	high
PTG	equal to fasting levels of G_L^{SUB}	present	high	present
cAMP-opposing				
PDE-3B	present	0	present	present

Sources[39, 38, 40, 41]

Adrenaline-stimulated glycogenolysis in muscle The muscle G_M^{SUB} isoform is (uniquely) hormone-regulated by phosphorylation and also differs from the other regulatory subunits in being membrane-bound to the sarcoplasmic reticulum by a *C*-terminal transmembrane extension not present in the rest of the family members. cAMP-elevating signals, such as adrenergic stimulation of adenylyl cyclase, probably activate discrete pools of scaffolded PKA, which release their catalytic subunits locally, and this results in the direct phosphorylation of G_M^{SUB}. PKA phosphorylates Ser48 ('site 1') and Ser67 ('site 2') of G_M^{SUB} at similar rates, but 'site 1' is de-phosphorylated much more slowly (by PP2A or PP2B) when the cAMP signal subsides[42]. When doubly-phosphorylated, G_M^{SUB} can no longer bind PP-1, which is excluded from its substrates by being released from the glycogen granule into the cytoplasm where it is further inhibited by 'inhibitor-1', a protein activated by PKA phosphorylation[38]. 'Site 1'-singly phosphorylated G_M^{SUB}–PP-1 complex is actually even more active than non-phosphorylated G_M^{SUB}, and persistence of phosphoSer48 is thought to ensure speedy return of glycogen synthesis after adrenaline stimulation is over.

As mentioned earlier, an insulin-induced phosphorylation of 'site 1' by p90Rsk (downstream of MAPK) that is seen *in vitro*, is not believed to occur *in vivo*.

9.7.2 p70Rsk – inducer of GS dephosphorylation?

A second form of Rsk exists: p70Rsk. This is activated by PDK-1 phosphorylation downstream of InsR, but can also be activated by amino acids, or mTOR (mammalian target of rapamycin). Recently, physiological doses of insulin given to human subjects were found to cause activation of p70Rsk and, at the same time, activate GS without discernable effect upon GSK-3 activity[43]. This suggests that p70Rsk directly activates G_M^{SUB} and this causes activation (i.e., de-phosphorylation) of GS. However, it remains controversial whether G_M^{SUB} phosphorylation is a cause of GS activation: insulin is reported to either have no effect on G_M^{SUB} phosphorylation status[41] or to cause phosphorylation of both 'site 1' and 'site 2' of G_M^{SUB} (via p90Rsk and p70Rsk), but mutation of the target amino acids (Ser48 and Ser67) has no effect upon insulin's GS activating action[44].

9.8 Insulin reverses effects of adrenaline and/or glucagon

9.8.1 Insulin's reversal of adrenaline-induced glycogenolysis in muscle

Glycogen synthase in both muscle and liver is subject to continuous, inhibiting phosphorylation by a constitutively active serine/threonine kinase that has come to be known as ***g**lycogen **s**ynthase **k**inase*-3 (GSK-3). At present, the only form of control that is agreed upon is that insulin simply switches GSK-3 off. Insulin does this through activation of PKB, which phosphorylates GSK-3, inactivating it. Constitutive protein phosphatase activity is unopposed and returns GS to its active, unphosphosphorylated state.

Note that an insulin-activatable phosphodiesterase, PDE-3B, that is present in liver and adipose, is absent from striated muscle. The competing pathways in muscle are summarised in Figure 9.4.

9.8.2 Insulin's reversal of adrenaline- and glucoagon-induced glycogenolysis in liver

Liver-specific G_L^{SUB} is not phosphorylated but instead is controlled by a unique allosteric inhibition by activated glycogen phosphorylase (GPa). We discussed earlier how PKA and PhK activate GP and inactivate GS, both by phosphorylation. An additional effect of PKA activation of GP is that it prevents GS re-activation because

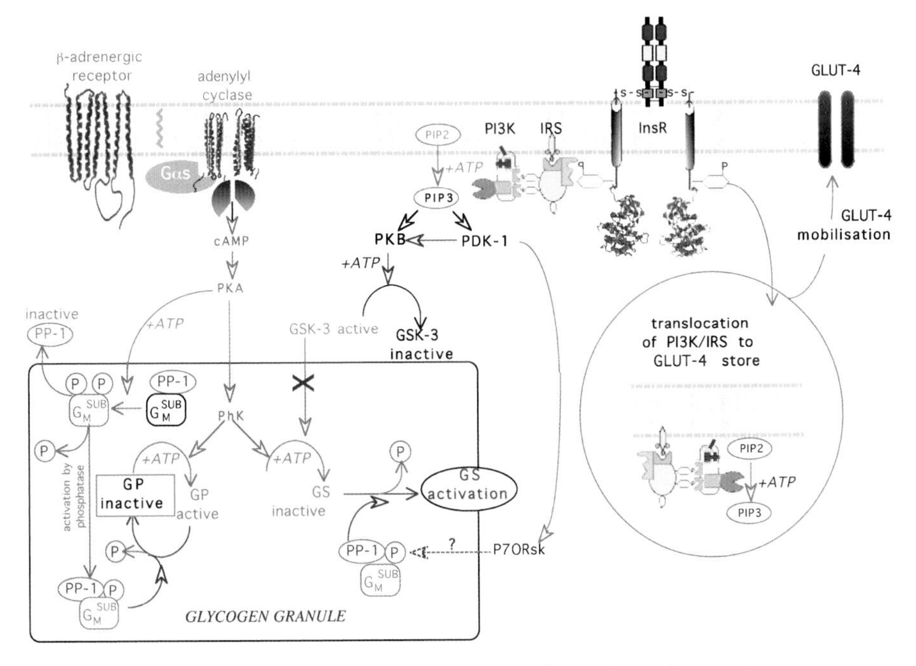

Figure 9.4 Competing and non-competing pathways in muscle

G_L^{SUB}/PP-1 (GS-directed) phosphatase activity is blocked by even nanomolar amounts of GPa[16].

The liver isoform of GP differs from that of muscle (or brain) in being exquisitely sensitive to glucose. Due to constitutive expression of membrane localised GLUT2, the concentration of glucose in liver is the same as that in the blood. In high glucose, the liver isoform of GPa binds glucose at an allosteric site, forcing a conformational change that exposes Ser14 and thereby greatly increases its rate of de-phosphorylation. It has been suggested that insulin synergises with high glucose in activating GS through its activation of PKB and inhibition of GSK-3.

PKB has been shown to phosphorylate and activate a novel form of cAMP ***p****hospho*-***di****esterase*: PDE-3B. Activated PDE-3B eliminates the cAMP pool, inactivates PKA and reverses the phosphorylation status of GP and GS[45].

Liver G_L^{SUB} differs from other isoforms in being allosterically modified by glucose-6-phosphate, and specifically that produced by glucokinase (a form of hexokinase unique to the liver)[16]. Binding of G-6-P to G_L^{SUB} activates its GS-directed phosphatase activity, but only after the levels of GPa have diminished. Glucokinase differs from hexokinase in that it translocates from nucleus to cytoplasm in response to high glucose, suggesting the compartmentalised production of G-6-P is essential for G_L^{SUB} activation. The hepacocyte pathways are shown in Figure 9.5.

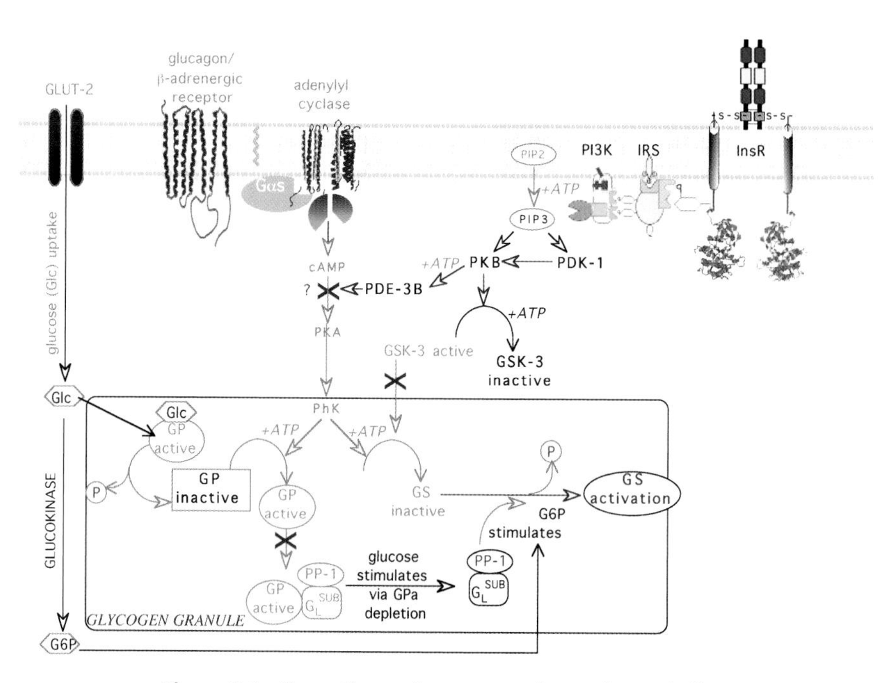

Figure 9.5 Competing and non-competing pathways in liver

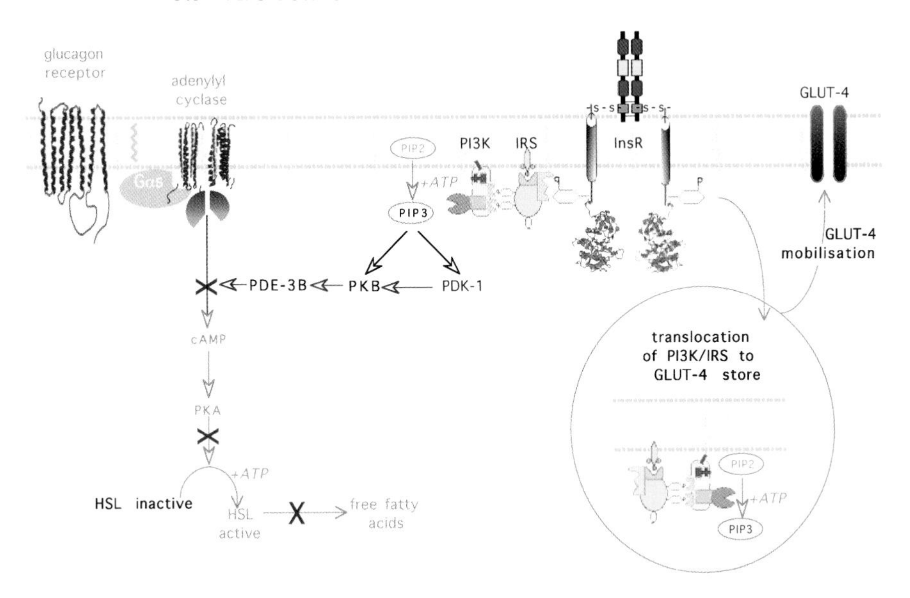

Figure 9.6 Competing and non-competing pathways in adipocyctes

9.8.3 Insulin's reversal of adrenalin/glucagon-induced lipolysis in adipose tissue

Adipocyte ***h***ormone-***s***ensitive ***l***ipase (HSL) is acutely activated by serine phosphorylation by PKA and this results in breakdown of triglyceride stores and the release of free fatty acids into the blood stream. Again, the major anti-lipolytic effects of insulin are thought to be mediated primarily through insulin's activation of PKB and its phosphorylation/activation of PDE-3B (Figure 9.6). This suppression of cAMP production effectively blocks the lipolytic effect of adrenaline or glucagon[46].

9.9 PIP3 downstream effects – glycogen synthesis

The appearance of PIP3 at the plasma membrane causes translocation and serial activation of a number of serine/threonine kinases, and culminates in the activation of glycogen synthase and inactivation of glycogen phosphorylase. There are three main serine/threonine protein kinases activated by insulin-induced PIP3 signals: ***P***IP3-***d***ependent protein ***k***inase (PDK), ***p***rotein ***k***inase ***B*** (PKB, also known as 'Akt') and the 'atypical' protein kinase Cζ (PKCζ). All three enzymes are cytosolic in resting cells but are membrane-recruited and activated by the appearance of PIP3. Both PDK and PKB have PH domains, whereas PKCζ (which is without a PH domain) binds PIP3 via its 'atypical' C1 domain[47].

PDK is constitutively active before binding to PIP3 lipid but, being in the cytosol, is unable to signal – it appears to autophosphorylate its A-loop on a serine (Table 9.3)[48].

Table 9.3 Identical PDK-1 substrate sequences in the A-loop PKB and PKC-ζ

```
PKB
241 RERVFSEDRA RFYGAEIVSA LDYLHSEKNV VYRDLKLENL MLDKDGHIKI TDFGLCKEGI
301 KDGATMKTFC GTPEYLAPEV LEDNDYGRAV DWWGLGVVMY EMMCGRLPFY NQDHEKLFEL
361 ILMEEIRFPR TLGPEAKSLL SGLLKKDPKQ RLGGGSEDAK EIMQHRFFAG IVWQHVYEKK
421 LSPPFKPQVT SETDTRYFDE EFTAQMITIT PPDQDDSMEC VDSERRPHFP QFSYSASGTA

PKC-ζ
361 ERGIIYRDLK LDNVLLDADG HIKLTDYGMC KEGLGPGDTT STFCGTPNYI APEILRGEEY
```

The A-loop activating threonines (bold red text) and identical *C*-terminal amino acids (boxed in yellow) are specific for PDK-1. The 'priming' site serine of PKB is in bold white text, boxed in green. The autophosporylation site of PDK-1 is shown below for comparison.

```
PDK-1
181 CTRFYTAEIV SALEYLHGKG IIHRDLKPEN ILLNEDMHIQ ITDFGTAKVL SPESKQARAN
241 SFVGTAQYVS PELLTEKSAC KSSDLWALGC IIYQLVAGLP PFRAGNEYLI FQKIIKLEYD
```

Translocation of PDK and its two mains substrates, PKB and PKC-ζ, occurs as soon as PIP3 is generated at the plasma membrane. PDK can then dock with and phosphorylate membrane-bound PKB and PKC-ζ, fully activating both. PDK-1 phosphorylates A-loop threonine residues (equivalent to PKA Thr197) in both enzymes and this leads to reversal of autoinhibition. Both substrate kinases (although otherwise unrelated) have identical sequences adjacent to the activating threonine (Table 9.3).

PDK-1 is unusual in that it requires its substrates to be 'primed' by serine phosphorylation before it can phosphorylate the A-loop. In the case of PKB, this 'priming' phosphorylation is at serine473 in its extreme *C*-terminal tail[49]. This 'priming' phosphorylation is carried out by PDK-2, another PIP3-dependent ST-kinase[28]. The recent crystal structure of PDK-1 reveals a hydrophobic-binding pocket (the 'PIF pocket') adjacent to a phosphate-binding pocket, and these make up a docking site that allows PDK-1 to dock with serine-phosphorylated *C*-terminal hydrophobic stretches found in substrates such as PKB and PKC-ζ[48]. PDK-1 (possibly aided by PDK-2) also phosphorylates the p70 isoform of Rsk.

9.9.1 PKB and GSK-3 inactivation

***G**lycogen **s**ynthase **k**inase-3 (GSK-3) is a constitutively active ST-kinase that continuously phosphorylates **g**lycogen **s**ynthase (GS). Serine phosphorylation of GS by GSK-3, or by adrenaline/glucagon-activated phosphorylase kinase (PhK), serves to inhibit the enzyme and blocks the synthesis of glycogen. Insulin turns GS on again by lowering its phosphorylation status. One way it achieves this is by blocking the inhibitory actions of GSK-3. GSK-3 is, itself, inactivated by phosphorylation, and PKB is the kinase that does this. PKB appears to interact with GSK-3 at the membrane where a complex can be detected after cell transfection with an oncogenic form of PKB – an inactive form of PKB can block insulin's ability to inactivate GSK-3[49]. This InsR-induced

PKB-inactivation of GSK-3 is a pathway to activation of GS in both liver and skeletal muscle (Figures 9.4 and 9.5).

It has been noted that GSK-3 inactivation is the prime route for insulin's stimulation of GS in fibroblast-like 3T3-L1 preadipocytes, but when they are differentiated into adipocytes, protein phosphatase-1 activation becomes the dominant way for insulin to upregulate GS[14]. Furthermore, in liver, GSK-3 inhibition only accounts for an estimated 20% of insulin's activation of GS and it is proposed that the remainder is through acute upregulation of a hormone-sensitive GS-directed phosphatase, rather than the GSK-3 inhibition route, which presumably relies on constitutive phosphatase to return GS to an unphosphorylated state[50]. The subject has been controversial for many years, and remains so.

9.9.2 PKC-ζ – negative feedback control

As alluded to earlier, the atypical PKCζ is translocated to the plasma membrane in response to PIP3 generation, and is activated by PDK-1, which phosphorylates PKCζ on its activating A-loop threonine410 (equivalent to PKA Thr197). PKCζ has been suggested to collaborate with Ras and Raf to activate the MAPK pathway[51]. Conventional PKCs can partially activate Raf by serine phosphorylation (as discussed in Chapter 5). PKCζ has been suggested to play a positive regulatory role in insulin-induced glucose transport, although this is controversial[27]. Perhaps the best evidence suggests that PCKζ activation exerts a negative feedback inhibition of InsR signalling – PKCζ serine/threonine-phosphorylates IRS-1, causing IRS-1 to dissociate from InsR, preventing its further tyrosine phosphorylation, and leading to its inactivation through protein tyrosine phos-phatase action[51].

9.9.3 PIP3 downstream effects – GLUT4 mobilisation

One of the first noticeable effects in the gradual acquisition of type 2 diabetes is development of peripheral resistance to insulin that manifests as impaired glucose uptake. The impairment is due to an inability of insulin to induce the mobilisation of glucose transporters to the plasma membrane. It is well established that GLUT4 translocation to the plasma membrane (Figure 9.7) is the predominant mechanism by which insulin loads glucose into cells after a meal and it is this response that is impaired in acquired insulin resistance. GLUT4 is rapidly translocated to the plasma membrane within minutes of insulin-stimulation, as discussed earlier.

Insulin, for example, causes activated IRS-1/PI-3-kinase complexes to translocate to internal vesicles – derived from the tubulovesicular endosomal membrane system – that are enriched in glucose transporters[26]. IRS-1/PI3K becomes rapidly co-localised with internal GLUT4 vesicles after insulin stimulation[52] – the production of PIP3 at internal membranes coincides with IRS-2 co-localisation[53]. It seems that physical delivery of the IRS/PI3K complex to the cell's interior is the only way that PIP3 can be efficiently

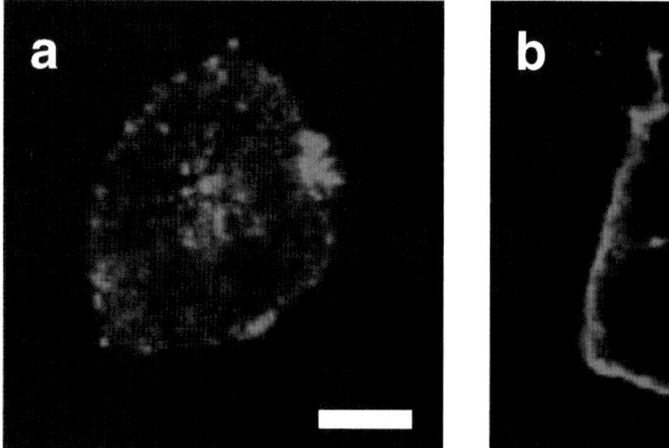

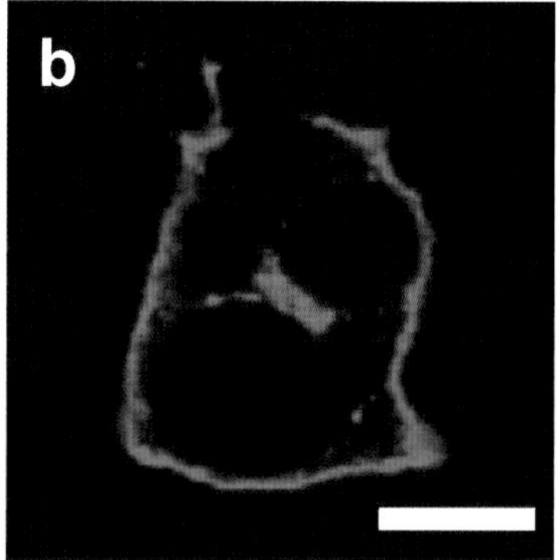

Adipocytes expressing GLUT4 labelled with green fluorescent protein in (a) basal conditions, or (b) after treatment with insulin

Figure 9.7 Translocation of GLUT 4

generated at such internal membranes. The consequence of the local increase in PIP3 concentration is that GLUT4 vesicles bud-off from the tubulovesicular endosomal membrane system and translocate to the plasma membrane. A large proportion of the cell's internal GLUT4 store can be seen to move to the plasma membrane within minutes of insulin stimulation[52].

The details of vesicular transport are beyond the scope of this book and insulin's precise pathway(s) in this context is still under investigation. Therefore, the following is merely an indication of the possible mechanisms.

Basal (unstimulated) vesicle trafficking moves GLUT4 back and forth from internal stores to the plasma membrane, but the protein spends most time in the internal depot. These basal GLUT4 vesicle movements are from an endosomal recycling compartment (ERC) that differs from a large subset of insulin-induced vesicles, which are dependent on a protein known as VAMP-2 for fusion with the plasma membrane – ERC movements are VAMP-2 independent[27]. Basal recycling is from transferrin-containing endosomes and is independent of microtubules. A proportion of activated GLUT4 vesicles are dependent upon microtubule integrity and are kinesin motor-driven along these 'highways' in response to insulin[54]. This insulin-induced pool of GLUT4 is composed of specialised GLUT4 storage vesicles (GSV).

Translocation of GLUT4 from the GSV is activated by the intracellular arrival of the IRS/PI-3-kinase complex and consequent local PIP3 generation. This results in fission of GLUT4-containing vesicles, which are then driven to a parking position just below the plasma membrane before a fusion step completes their delivery to the surface (Figure 9.8). Fission from the GSV is not well understood, but may involve Arf-like monomeric G proteins. Fusion of the vesicle with the plasma membrane is somewhat better understood and has much in common with vesicle exocytosis at synapses, including the requirement for NSF (***N***-ethylmaleimide-***s***ensitive ***f***actor) SNAP (***s***oluble ***N***SF ***a***ttachment ***p***rotein), v- and t-SNARE (***v***esicle- and ***t***arget-membrane ***SNA***P ***re***ceptor) and VAMP-2 (***v***esicle-***a***ssociated ***m***embrane ***p***rotein-***2***)[55]. It is thought that

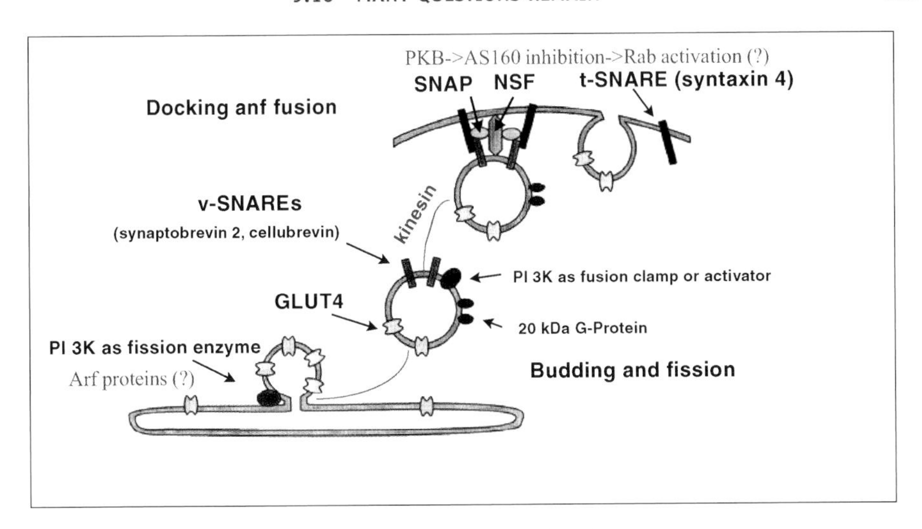

Figure 9.8 GLUT 4 translocation

the GLUT4 vesicle translocation and assembly of the multimeric fusion complex is regulated by a member of the Rab family of monomeric G proteins, probably Rab4[55].

The control of this whole process by insulin is exerted by intracellular PIP3 generation, but is also dependent upon PKB activation. Recently, a new PKB (or 'Akt') target was found and named ***A***kt ***s***ubstrate of ***160***kDa (AS160). AS160 is serine/threonine phosphorylated by PKB on five sites – mutation of four of these sites to alanine has the effect of blocking a substantial proportion of GLUT4 transport. AS160 contains a Rab-GAP domain and it is tempting to speculate that its phosphorylation by PKB may destroy its GTPase activating potential and that this, in turn, may release a 'brake' on Rab's activity[27]. Interestingly, it has been estimated that the PKB-activated step in the pathway is at the late stage of translocation that mediates fusion[56].

To summarise, insulin-induced GLUT4 comes from two pools: (i) the GSV pool that requires PI-3-kinase, PKB and VAMP-2; and (ii) the ERC pool that only requires PI-3-kinase activation[54].

9.10 Many questions remain

Having reviewed the accumulated data on insulin signals, one is left with several puzzles that are only beginning to be resolved, and we examine some of these in the following sections.

9.10.1 Insulin activates the Erk1/2 MAPK pathway – why, then, is the insulin receptor not as mitogenic as the PDGF receptor

In the recent past, it was usually considered that although insulin *might* be mitogenic, its receptor *was not*. This apparent paradox was easily explained, as follows.

High, supraphysiological, concentrations of insulin (≈ 5-10 μg/ml) can keep many types of tissue-cultured cells dividing even when serum is withdrawn from the medium. Supplementation with high dose insulin forms the basis of many serum-replacement formulae, 'HITES' for example (which contains ***h***ydrocortisone, high-dose ***i***nsulin, ***t***ransferrin, o***e***stradiol and ***s***elenite)[57]. However, the insulin concentration used (around 1 μM), is an order of magnitude higher than needed to activate fully the InsR (whose Kd is in the low nanomolar range). Here, the mitogenic effect of insulin is mediated by the IGF-I receptor, to which insulin binds with low affinity. But it was always difficult to rule out mitogenicity of InsR because of co-expression of IGF-IR in the same cells. And it has been shown in a few rigorous studies that InsR is (at least in certain important circumstances) weakly mitogenic, alone[58], a finding definitively confirmed in murine knockouts.

One explanation for the insulin receptor's weak mitogenic properties may be that the main conduit for its activation of MAPK, Shc, is in competition with IRS-1 for a limited pool of Grb2[59]. It appears that formation of the phosphoShc-Grb2 complex is the predominant way for InsR or IFG-IR to activate MAPK. Opposing that, the IRS-1–Grb2 interaction sequesters Grb2 and prevents MAPK activation. In addition, both IRS-1 and Shc must compete for the same NPEpY sequence of InsR in order to be phosphorylated[36] – of course the same is true of IGF-IR. Finally, compared with mitogenic growth factor receptors, InsR only weakly phosphorylates Shc[60] – Shc tyrosine phosphorylation is, of course, a necessary prerequisite for interaction with Grb2.

9.10.2 PDGR-β activates PI-3-kinase but does not exert anabolic effects like the insulin receptor Why?

Two explanations have been offered[28] in answer to this question. PDGFR activates PLCγ in parallel with PI-3-kinase and PLCγ is thought to damp-down the production of PIP3 because both enzymes compete for the same substrate, PIP2. In support of this, it has been determined experimentally that eliminating the PLCγ docking sites in PDGFR greatly enhances the production of PIP3 from the stimulated receptor mutant.

Insulin, in contrast, does not activate PLCγ and so its upregulation of PIP3 production is unopposed. A further difference lies in deployment of IRS-1/2. First, the sequential activation of multiple IRS proteins by an insulin receptor greatly amplifies the PIP3 signal because multiple PI-3-kinases are thereby activated. Second, the mobility of activated IRS-PI3K complexes allows them to reach cell subcompartments that the PDGFR-PI3K complex is barred from entering.

9.10.3 The insulin receptor and the IGF-I receptor are homologues – why is one anabolic and the other mitogenic?

Not only is IGF-IR more mitogenic than InsR, but it also stimulates transformation (anchorage independent growth). Paradoxically, IGF-IR and InsR induce similar levels of Shc tyrosine-phosphorylation and Shc-Grb2 complex formation[61]; but IGF-IR activates the MAPK pathway more efficiently than InsR. This may be because the inhibitory serine/threonine phosphorylation of Sos by IGF-IR (which attenuates its activity) is much delayed compared with the quicker inactivation by InsR – meaning IGF-IR-induced Shc-Grb2-Sos complex is stable for longer than that from InsR[62].

9.10.4 Do differing *C*-terminal tails cause differing regulation of growth responses in InsR *versus* IGF-IR?

Although the shared kinase domains of both receptors are 84% homologous, the *C*-terminal domains of the β-chains show only limited homology in phosphorylation motifs – perhaps this explains the differences in mitogenicity? For example, Tyr1334 (*Ex11*$^{-}$ *1322*) of InsR is conserved in IGF-IR (at position 1316), but an equivalent to InsR's Tyr1328 (*Ex11*$^{-}$ *1316*) is not found in IGF-IR; instead, IGF-IR has a phenylalanine at this position (1310 in IGF-IR) (Tables 9.1 and 9.4). Contrariwise, IGF-IR has a pair of tyrosines (at 1250 and 1251) that are missing from InsR (it has Phe.His instead). IGF-IR mutants with both the tyrosine pair and the phenylalanine substituted with the corresponding InsR residues, displayed defective growth stimulation and reduced transforming ability, with the bulk of this loss in activity being attributable to the substituted tyrosine pair[63].

Another divergence in their *C*-tails is the presence of a quartet of serine residues in IGF-IR (aas: 1280–1283) that are absent in InsR. Mutation of these serines to alanines has no effect on the mutant receptor's ability to stimulate mitogenesis in response to IGF-I, but the mutant receptor loses its transforming ability[61]. It has been shown that phosphoserine-binding 14-3-3 proteins interact with the *C*-tail of IGF-IR at Ser1283 but do not bind to InsR; 14-3-3 proteins are tyrosine-phosphorylated by IGF-IR and have been suggested to play a role in regulation of both mitogenesis and apoptosis[64].

Table 9.4 The human IGF-I receptor β-chain sequence

```
871   901  YTARIQATSL SGNGSWTDPV FFYVQAKTGY ENFIHLIIAL PVAVLLIVGG LVIMLYVFHR
931   961  KRNNSRLGNG VLYASVNPEY FSAADVYVPD EWEVAREKIT MSRELGQGSF GMVYEGVAKG
991   1021 VVKDEPETRV AIKTVNEAAS MRERIEFLNE ASVMKEFNCH HVVRLLGVVS QGQPTLVIME
1051  1081 LMTRGDLKSY LRSLRPEMEN NPVLAPPSLS KMIQMAGEIA DGMAYLNANK FVHRDLAARN
1111  1141 CMVAEDFTVK IGDFGMTRDI YETDYYRKGG KGLLPVRWMS PESLKDGVFT TYSDVWSFGV
1171  1201 VLWEIATLAE QPYQGLSNEQ VLRFVMEGGL LDKPDNCPDM LFELMRMCWQ YNPKMRPSFL
1231  1261 EIISSIKEEM EPGFREVSFY YSEENKLPEP EELDLEPENM ESVPLDPSAS SSSLPLPDRH
1291  1321 SGHKAENGPG PGVLVLRASF DERQPYAHMN GGRKNERALP LPQSSTC
```

grey = numbering of pro-form (before processing)
F = tyrosine autophosphorylation site in insulin recptor missing from IGF-IR
SSSS = serine quartet not found in insulin receptor
Transmembrane domain (boxed)
P-loop
Activation loops
Catalytic loop
Y = tyrosine phosphorylation sites

9.10.5 IFG-II, insulin receptor-A and 'half receptors'

Recently it has been found that the insulin homologue, IGF-II, binds with a high affinity to the short isoform of the insulin receptor (InsR-A)[65]. This is somewhat surprising, given that it has a very low affinity for the predominant long isoform (InsR-B). Perhaps more surprising is that insulin and IGF-II produce distinctly different signalling outcomes from the same InsR-A; IGF-II being more mitogenic, with insulin biased toward metabolic effects. So far, no major differences in signalling routes have been found. Significantly, InsR-A is at much higher expression levels than InsR-B in fast-growing foetal and malignant tissues. Knockouts in mice have proven that expression of IGF-II (which is at high levels in the foetus, but declines dramatically post-natally) and co-expression of *either* IGF-IR *or* InsR are essential for foetal development. Apparent reliance upon an IGF-II–InsR-A axis in early development hints at how dysregulation of InsR splicing (i.e., changing the ratio of InsR-A:InsR-B) may lead to foetal growth defects and cancer progression[8,65].

The mitogenic and metabolic selectivity is further extended by the presence of 'half receptors', or 'hybrid receptors', which are InsR monomers heterodimerised with IGF-IR partner monomers. Such hybrid receptors are widespread in a variety of tissues and their differential expression patterns play a modulatory role in foetal growth. The various possible homodimers and heterodimers have distinct affinities for the three possible ligands (insulin, IGF-I and -II) as judged by radiolabelled-IGF-I ligand displacement assays (see Table 9.5, reproduced from Reference 66). The InsR-A/IGF-IR hybrid binds all three ligands with high affinity; InsR-B/IGF-IR only binds IGF-I with high affinity. The InsR-A/IGF-IR hybrid, uniquely, allows insulin to induce *trans*-autophosphorylate the IGF-IR β-chain, something it cannot do to IGF-IR homodimers except at very high concentrations. Such a high affinity for insulin and IGF-II suggests an important route for stimulation of IGF-IR-like signals during foetal growth.

Table 9.5 Displacement of ^{125}I-labelled IGF-1 by related cold ligands – affinities of IGF-IR, short ('IR-A') and long ('IR-B') isoforms of the insulin receptor, and IGF-IR hybrids with InsR-A ('R^{A}') or InsR-B ('R^{B}')

R^{-} cells	EC$_{50}$ of unlabeled ligand		
	Insulin	IGF-I	IGF-II
		nM	
Hybrid-R^{A}	3.7 ± 0.9	0.3 ± 0.2	0.6 ± 0.1
Hybrid-R^{B}	> 100	2.5 ± 0.5	15.0 ± 0.9
IGF-IR	> 30.0	0.2 ± 0.3	0.6 ± 1.0
IR-A	0.2 ± 0.2	> 30.0	0.9 ± 0.4
IR-B	0.3 ± 0.4	> 30.0	11.0 ± 5.0

Reproduced with kind permission from: -Pandini, G., Frasca, F., Mineo, R., Sciacca, L., Vigneri, R., and Belfiore, A.(2002) Insulin/insulin-like growth factor i hybrid receptors have different biological characteristics depending on the insulin receptor iso form involved. J. Biol. Chem., 277: 39684–39695.

In brown adipose cells, the ability of IGF-I to stimulate PKB is strictly dependent upon co-expression of InsR with IGF-IR (presumably, operating as a hybrid); insulin-stimulated tyrosine phosphorylation of IRS-1 is lowered by co-expression of IFG-IR with InsR, whereas IRS-2 activation is unaffected[12]. The authors found, however, that insulin's actions (in these cells, at least) were primarily mediated by InsR-A (and not by InsR-A/IGF-IR half receptors). Interestingly, the number of hybrid receptors in brown adipocytes clones stayed constant, even when one or other of the receptors was over-expressed. The supporting evidence for the role of hybrids in rodent and human growth and development is beyond the scope of this book, but is the subject of an excellent review by Nakae[8].

References

1. Ebina, Y., Ellis, L., Jarnagin, K., Edery, M., Graf, L., Clauser, E., Ou, J-H., Masiarz, F., Kan, Y.W., Goldfine, I.D., Roth, R.A. and Rutter, W.J. (1985) The human insulin receptor cDNA: the structural basis for hormone-activated transmembrane signalling. *Cell*, **40**: 747–758.
2. Ullrich, A., Bell, J.R., Chen, E.Y., *et al.* (1985) Human insulin receptor and its relationship to the tyrosine kinase family of oncognes. *Nature*, **313**: 756–761.
3. Feener, E.P., Backer, J.M., King, G.L., Wilden, P.A., Sun, X.J., Kahn, C.R. and White, M.F. (1993) Insulin stimulates serine and tyrosine phosphorylation in the juxtamembrane region of the insulin receptor. *J. Biol. Chem.*, **268**: 11256–11264.
4. Tornqvist, H.E. and Avruch, J. (1988) Relationship of site-specific β subunit tyrosine autophosphrylation to insulin receptor (tyrosine) protein kinase activity. *J. Biol. Chem.*, **263**: 4593–4601.
5. Wei, L., Hubbard, S.R., Hendrickson, W.A. and Ellis, L. (1995) Expression, characterization, and crystallization of the catalytic core of the human insulin receptor protein tyrosine kinase domain. *J. Biol. Chem.*, **270**: 8122–8130.
6. Dickens, M. and Tavare, J.M. (1992) Analysis of the order of autophosphorylation of human insulin receptor tyrosines 1158, 1162 and 1163. *Biochem. Biophys. Res. Commun.*, **186**: 244–250.
7. Hubbard, S.R. (1997) Crystal structure of the activated insulin receptor tyrosine kinase in complex with peptide substrate and ATP analog. *EMBO J.*, **16**: 5572–5581.
8. Nakae, J., Kido, Y. and Accili, D. (2001) Distinct and overlapping functions of insulin and IGF-I receptors. *Endocrine Reviews*, **22**: 818–835.
9. Sjogren, K., Liu, J-L., Blad, K., Skrtic, S., Vidal, O., Wallenius, V., leRoth, D., Tornell, J., Isaksson, O.G.P, Jansson, J-O. and Ohlsson, C. (1999) Liver-derived insulin-like growth factor I (IGF-I) is the principal source of IGF-I in blood but is not required for postnatal body growth in mice. *Proc. Natl. Acad. Sci. USA*, **96**: 7088–7092.
10. Rubin, C.S., Hirsch, A., Fung, C. and Rosen, O.M. (1978) Development of hormone receptors and hormonal responsiveness *in vitro*. *J. Biol. Chem.*, **253**: 7570–7578.
11. Taira, M., Hockman, S.C., Calvo, J.C., Belfrage, P. and Manganiello, V.C. (1993) Molecular cloning of the rat adipocyte hormone-sensitive cyclic GMP-inhibited cyclic nucleotide phosphodiesterase. *J. Biol. Chem.*, **268**: 18573–18579.
12. Entingh-Pearsall, A. and Kahn, C.R. (2004) Differential roles of the insulin and insulin-like growth factor-I (IGF-I) receptors in response to insulin and IGF-I. *J. Biol. Chem.*, **279**: 38016–38024.
13. White, M.E. (1997) The insulin signalling system and the IRS proteins. *Diabetologia*, **40**: S2–S17.

14. Brady, M.J., Bourbonais, F.J. and Saltiel, A.R. (1998) The activation of glycogen synthase by insulin switches from kinase inhibition to phosphatase activation during adipogenesis in 3T3-L1 cells. *J. Biol. Chem.*, **273**: 14063–14066.
15. Ruan, H. and Lodish, H.F. (2003) Insulin resistance in adipose tissue: direct and indirect effects of tumor necrosis factor-a. *Cytokine and Growth Factor Reviews*, **14**: 447–455.
16. Bollen, M., Keppens, S. and Stalmans, W. (1998) Specific features of glycogen metabolism in the liver. *Biochem. J.*, **336**: 19–31.
17. Yeaman, S.J. Armstrong, J.L., Bonavaud, S.M., Poinasamy, D., Pickersgill, L. and Halse, R. (2001) Regulation of glycogen synthesis in human muscle cells. *Biochem. Soc. Trans.*, **29**: 537–541.
18. Yeaman, S.J. (2004) Hormone-sensitive lipase – new roles for an old enzyme. *Biochem J.*, **379**: 11–22.
19. Ray, L.B. and Sturgill, T.W. (1987) Rapid stimulation by insulin of a serine/threonine kinase in 3T3-L1 adipocytes that phosphorylates microtubule-associated protein 2 *in vitro*. *Proc. Natl. Acad. Sci. USA*, **84**: 1502–1506.
20. He, W., O'Neill, T.J. and Gustafson, T.A. (1995) Distinct modes of interaction of SHC and insulin receptor substrate-1 with the insulin receptor NPEY region via non-SH2 domains. *J. Biol. Chem.*, **270**: 23258–23262.
21. Lazar, D.F., Wiese, R.J., Brady, M.J., Corley Mastick, C. Waters, S.B. Yamauchi, K., Pessin, J. E., Cuatrecasas, P. and Saltiel, A.R. (1995) Mitogen-activated protein kinase inhibition does not block the stimulation of glucose utilization by insulin. *J. Biol. Chem.*, **270**: 20801–20801.
22. Cusi, K., Maezono, K., Osman, A., Pendergrass, M., Patti, M.E., Pratipanawatr, T., DeFronzo, R.A., Kahn, C.R. and Mandarino, L.J. (2000) Insulin resistance differentially affects the PI 3-kinase- and MAP kinase-mediated signaling in human muscle. *J. Clin. Invest.*, **105**: 311–320.
23. Mitra, P., Zheng, X. and Czech, M.P. (2004) RNAi-base analysis of CAP, Cbl, and CrkII function in the regulation of GLUT4 by insulin. *J. Biol. Chem.*, **279**: 37431–37435.
24. Thien, C.B.F. and Langdon, W.Y. (2005) cCbl and Cbl-b uquitin ligases: substrate diversity and the negative regulation of signalling responses. *Biochem. J.*, **391**: 153–166.
25. Molero, J.C. Jensen, T., *et al.* (2004) Cbl-deficient mice have reduced adiposity, higher energy expenditure, and improved peripheral insulin action. *J. Clin. Invest.*, **114**:1326–1333.
26. Saltiel, A.R. and Pessin, J.E. (2002) Insulin signaling pathways in time and space. *Trends in Cell Biology*, **12**: 65–71.
27. Thong, F.S.L., Dugani, C.B. and Klip, A. (2005) Turning signals on and off: GLUT4 traffic in the insulin-signaling pathway. *Physiology*, **20**: 271–284.
28. Shepherd, P.R., Withers, D.J. and Siddle, K. (1998) Phosphoinositide 3-kinase: the key switch mechanism in insulin signalling. *Biochem. J.*, **333**: 471–490.
29. Dhe-Paganon, S., Ottinger, E.A., Nolte, R.T., Eck, M.J. and Shoelson, S.E. (1999) Crystal structure of the pleckstrin homology-phosphotyrosine binding (PH-PTB) targeting region of insulin receptor substrate 1. *Proc. Natl. Acad. Sci. USA*, **96**: 8378–8383.
30. Kido, Y., Nakae, J. and Accili, D. (2001) The insulin receptor and its cellular targets. *J. Clin. Endocrinol. Metab.*, **86**: 972–979.
31. Yenush, L., Makati, K.J., Smith-Hall, J., Ishibashi, O., Myers Jr, M.G. and White, M.F. (1996) The plextrin homology domain is the principle link between the insulin receptor and IRS-1. *J. Biol. Chem.*, **271**: 24300–24306.
32. Jacobs, A.R., LeRoith, D. and Taylor, S.L. (2001) Insulin receptor substrate-1 pleckstrin homology and phosphotyrosine-binding domains are both involved in plasma membrane targetting. *J. Biol. Chem.*, **276**: 40795–40802.
33. Prisco, M., Santini, F., Baffa. R., Liu, M., Drakas, R., Wu, A. and Baserga, R. (2002) Nuclear translocation of insulin receptor susbtrate-1 by the simian virus 40 T antigen and the activated type 1 insulin like growth factor receptor. *J. Biol. Chem.*, **277**: 32078–32085.

34. Burks, D.J., Pons, S., Towery, H., Smith-Hall, J., Myerws, M.G.J., Yenush, L. and White, M.F. (1997) Heterologous pleckstrin domains do not couple IRS-1 to the insulin receptor. *J. Biol. Chem.*, **272**: 27716–27721.
35. Falang-Fallah, J., Randhawa, V.K., Nimnual, A., Klip, A., Bar-Sagi, D. and Rozakis-Adcock, M. (2002) The pleckstrin homology (PH) domain-interacting protein couples the insulin receptor substrate 1 PH domain to insulin signaling pathways leading to mitogenesis and GLUT4 translocation. *Mol. Cell. Biol.*, **22**: 7325–7336.
36. Isakoff, S.J., Yu, Y-P., Su, Y-C., Blaikie, P., Yajnik, V., Rose, E., Weidner, K.M., Sachs, M., Margolis, B. and Skolnik, E.D. (1995) Interaction between the phosphotyrosine binding domain of Shc and the insulin receptor is required for Shc phosphorylation by insulin in vivo. *J. Biol. Chem.*, **271**: 3959–3962.
37. Ceulemans, H. and Bollen, M. (2003) Functional diversity of protein phosphatase-1, a cellular economizer and reset button. *Physiol. Rev.*, **84**: 1–39.
38. Cohen, P.T.W. (2002) Protein phosphatase 1 – targeted in many directions. *J. Cell Science*, **115**: 241–256.
39. Newgard, C.B., Brady, M.J., O'Doherty, R.M. and Saltiel, A.R. (2000) Organising glucose disposal. Emerging roles of the glycogen targeting subunits of protein phosphatase-1. *Diabetes*, **49**: 1967–1977.
40. Harndahl, L., Jing, X-J., Ivarsson, R., Degerman, E., Ahren, B., Manganiello, V.C., Renstrom, E. and Stenson Holst, L. (2002) Important role of phospodiesterase 3B for the stimulatory action of cAMP on pancreatic β-cell exocytosis and release of insulin. *J. Biol. Chem.*, **277**: 37446–37455.
41. Movesesian, M.A., Komas, N., Krall, J. and Manganiello, V.C. (1996) Expression and activity of low Km, cGMP inhibited cAMP phosphodiesterase in cardiac and skeletal muscle. *Biochem. Biophys. Res. Commun.*, **225**: 1058–1062.
42. Walker, K.S., Watt, P.W. and Cohen, P. (2000) Phosphorylation of the skeletal muscle glycogen-targetting subunit of protein phosphatase 1 in response to adrenaline in vivo. *FEBS Letters*, **466**: 121–124.
43. Liu, Z., Wu, Y., Nicklas, E.W., Jahn, L.A., Price, W.J. and Barrett, E.J. (2003) Unlike insulin, amino acids stimulate p70S6K but not GSK-3 or glycogen synthase in human skeletal muscle. *Am. J. Physiol. Endocrinol. Metab.*, **286**: E523–E528.
44. Liu, J. and Brautigan, D.L. (2000) Insulin-stimulated phosphorylation of the phosphatase-1 striated muscle glycogen-targeting subunit and activation of glycogen synthase. *J. Biol. Chem.*, **275**: 15940–15947.
45. Zhao, A.Z., Shinohara, M.M., Huang, D., Shimizu, M., Eldar-Finkleman, H., Krebs, E.G., Beavo, J.A. and Bornfeldt, K.E. (2000) Leptin induces insulin-like signaling that antagonises cAMP elevation by glucagon in hepatocytes. *J. Biol. Chem.*, **275**: 11348–11354.
46. Rahn, T., Ridderstrale, M., Tornqvist, H., Manganiello, V., Friedrikson, G., Belfrage, P. and Degerman, E. (1994) Essential role for phosphatidylinositol 3-kinase in insulin-induced activation and phosphorylation of the cGMP-inhibited cAMP phosphodiesterase in rat adipocytes. *FEBS Letters*, **350**: 314–318.
47. Chou, M.M., Hou, W., Johnson, J., Graham, L.K., Lee, M.H., Chen, C.S., Newton, A.C., Schaffhausen, B.S. and Toker, A. (1998) Regulation of protein kinase C zeta by PI 3-kinase and PDK-1. *Curr. Biol.*, **8**: 1069–1077.
48. Biondi, R.M., Komander, D., Thomas, C.C., Lizcano, J.M., Deak, M., Alessi, A.R. and van Aalten, D.M.F. (2002) High resolution crystal structure of the human PDK1 catalytic domain defines the regulatory phosphpeptide docking site. *EMBO J.*, **21**: 4219–4228.
49. van Weeren, P.C., de Bruyn, K.M.T., de Vries-Smits, A.M.M., van Lint, J. Burgering, B.M.Th. (1998) Essential role for protein kinase B (PKB) in insulin-induced glycogen synthase kinase 3 inactivation. *J. Biol. Chem.*, **273**: 13150–13156.

50. Aiston, S., Coghlan, M.P. and Agius, L. (2003) Inactivation of phosphorylase is a major component of the mechanism by which insulin stimulates hepatic glycogen synthesis. *Eur. J. Biochem.*, **270**: 2773–2781.
51. Liu, Y-F., Paz, K., Herschkovitz, A., Alt, A., Tennenbaum, T., Sampson, S.R. Ohba, M., Kuroki, T., LeRoith, D. and Zick, Y. (2001) Insulin stimulates PKCζ-mediated phosphorylation of insulin receptor substrate-1 (IRS-1). *J. Biol. Chem.*, **276**:14459–14465.
52. Heller-Harrison, R.A., Morin, M., Guilherme, A. and Czech, M.P. (1996) Insulin-mediated targeting of phosphatidylinositol 3-kinase to GLUT4-containing vesicles. *J. Biol. Chem.*, **271**: 10200–10204.
53. Niswender, K.D., Gallis, B., Blevins, J.E., Corson, M.A., Schwartz, M.W. and Baskin, D.G. (2003) Immunocytochemical detection of phosphatidylinositol 3-kinase activation by insulin and leptin. *J. Histochem. Cytochem.*, **51**: 275–283.
54. Fletcher, L.M., Welsh, G.I., Oatey, P.B. and Tavare, J.M. (2000) Role for microtubule cytoskeleton in GLUT4 vesicle trafficking and in the regulation of insulin-stimulated glucose uptake. *Biochem. J.*, **352**: 267–276.
55. Pessin, J.E., Thurmond, D.C., Elmendorf, J.S. Coker, K.J. and Okada, S. (1999) Molecular basis of insulin-stimulated GLUT4 vesicle trafficking. *J. Biol. Chem.*, **274**: 2593–2596.
56. van Dam, E.M., Govers, R. and James, D.E. (2004) Akt is required at a late stage of the insulin-induced GLUT4 translocation to the plasma membrane. *Mol. Endocrinol.*, **19**: 1067–1077.
57. Carney, D.N., Bunn Jr, P.A., Gazdar, A.F., Pagan, J.A. and Minna, J.D. (1981) Selective growth in serum-free hormone-supplemented medium of tumor cells obtained by biopsy from patients with small cell carcinoma of the lung. *Proc. Natl. Acad. USA*, **78**: 3185–3189.
58. Van Obberghen, E. (1994) Signaling through the insulin receptor and the insulin-like growth factor-I receptor. *Diabetologia*, **37**: S125–S134.
59. Yamauch, K. and Pessin, J.E. (1994) Insulin receptor substrate-1 (IRS1) and Shc compete for a limited pool of Grb2 in mediating downstream signaling. *J. Biol. Chem.*, **269**: 31107–31114.
60. Virkamaki, A., Ueki, K. and Kahn, R. (1999) Protein-protein interaction in insulin signaling and the molecular mechanisms of insulin resistance. *J. Clin. Invest.*, **103**: 931–943.
61. Li, S., Resnicoff, M. and Baserga, R. (1996) Effect of mutations at serines 1280–1283 on the mitogenic and transforming activities of the insulin-like growth I factor. *J. Biol. Chem.*, **271**: 12254–12260.
62. Sasaoka, T., Ishiki, M., Sawa, T., Ishihara, H., Takata, Y., Imamura, T., Usui, I., Olefsky, J.M., and Kobayashi, M. (1996) Comparison of the insulin and insulin-like growth factor 1 mitogenic intracellular signaling pathways. *Endocrinology*, **137**: 4427–4434.
63. Esposito, D.L., Blakesley, V.A., Koval, A.P., Scrimgeour, A.G. and LeRoth, D. (1997) Tyrosine residues in the C-terminal domain of the insulin-like growth factor-I receptor mediate mitogenic and tumorigenic signals. *Endocrinology*, **138**: 2979–2988.
64. Furlanetto, R.W., Dey, B.R., Lopaczynski, W. and Nissley, S.P. (1997) 14-3-3 proteins interact with the insulin-like growth factor receptor but not with the insulin receptor. *Biochem.*, **327**: 765–771.
65. Frasca, F., Pandini, G., Scalia, P., Sciacca, L., Mineo, R., Costantino, A., Goldfine, I.D., Belfiore, A. and Vigneri, R. (1999) Insulin receptor isoform A, a newly recognized, high-affinity insulin-like growth factor II receptor in fetal and cancer cells. *Mol. Cell. Biol.*, **19**: 3278–3288.
66. Pandini, G., Frasca, F., Mineo, R., Sciacca, L., Vigner. R. and Belfiore, A. (2002) Insulin/insulin-like growth factor I hybrid receptors have different biological characteristics depending on the insulin receptor isoform involved. *J. Biol. Chem.*, **277**: 39684–39695.

10

Mitogens and cell cycle progression

Most adult human cell types do not divide, but may be recruited back into the cell cycle by wounding or cancer. Oncogenes are cancer-causing genes, many of which were first identified in oncogenic animal retroviruses. These genes were picked up from the vertebrate host and incorporated into the viral genome in a disabled form. Retroviral oncoproteins are truncated or point-mutated and are frequently homologues of oncoproteins associated with human cancer. Their normal counterparts are frequently protein components of single-pass receptor tyrosine kinase (RTK) pathways. A number of these oncogenic/RTK-pathways are de-regulated versions of mitogenic pathways deduced from fibroblast growth factor requirements, *in vitro*.

Studies in yeast first revealed the cell cycle 'clock' that times each phase of mitosis by the synthesis and destruction of cyclins, the essential partners of 'cyclin-dependent kinase' enzymes. This gave us the cyclin/cdk model of cell division control.

The above findings, and the careful elucidation of development in model organisms, have gradually coalesced into a consensus model of mitogenesis that applies at least to human malignancy. But, it is not without its controversy and challenges.

Much past research has focused on what triggers replication of the cells genome. This set of triggering events occurs in G1, which is the only part of the cycle that growth factors influence. Growth factors induce an 'early gene' transcription programme. The earliest transcription factors (TFs) induced downstream of RTK activation are the 'immediate early response genes'. Many of these RTK-activated TFs are also induced/activated by certain 7-pass receptors or environmental stress (such as UV or ionising radiation). TFs are also activated by loss of pathway control in cancer, where oncogenic mutations render certain mitogenic pathways ligand-independent and constitutively active. The TFs themselves can

Structure and Function in Cell Signalling John Nelson

be oncoproteins. Oncogenic and mitogenic 'immediate early response' TFs then activate transcription of certain 'delayed early response gene products', including the cyclins that drive the G1/S-phase transition.

Opposing this, other TFs actively suppress mitogenic gene transcription. Loss of function of these 'tumour suppressors' allows uncontrolled cell division by default. Other tumour-suppressing gene products, whose loss-of-function also leads to cancer, are not TFs. These include growth-inhibitory extracellular ligands and their receptors, GTPase activating proteins, 'pocket proteins' that sequester growth-promoting TFs, and proteins involved with degradation pathways, among others.

10.1 The mitogenic response and the cell division cycle

Cellular replication is the fundamental property of all life and, microscopically, is the most dramatic endpoint of concerted cell signalling in metazoans. As a cell physiological and biochemical event, the mitogenic response is probably the most extensively researched area in cell signalling. The central paradigms in our understanding of the mitogenic response owe a great deal to cancer research. Indeed one could argue that we know more about aberrant cell division in human malignancy than we do about normal cell division control in the adult. And our understanding of cell growth control in embryonic development leans heavily upon genetic manipulation of non-human organisms (yeast, worms, flies, *Xenopus*, rodents). Thus, one must bear in mind that much of what we know was deduced from 'abnormal' cell division in human malignancy, or from developmental biology studies in non-human organisms.

10.1.1 Large scale biophysical events in the cell division cycle

The only visually obvious landmarks in the cell cycle occur during the brief period of mitosis (Figure 10.1) when condensed sister chromatids become visible. Mitosis can be divided into prophase, metaphase, anaphase and telophase on the basis of steps in chromatid separation. During the rest of the cell cycle (interphase) chromosomes are invisible. However, large scale biochemical events that leave no visual fingerprints do occur during interphase leading to the idea that interphase itself could be divided into phases. The most obvious biochemical event is in S-phase when DNA is replicated. This is followed by a gap (G2) where the cell rests before committing to mitosis. Before S-phase is another gap (G1), where a characteristic set of proteins must be made including: first, the products of the 'immediate early response' genes, followed by 'delayed early response' gene products. G1 is the only period of the cell cycle in which growth factors can exert mitogenic effects.

10.1.2 The cyclin model

Cyclin was first discovered in 1983 by Tim Hunt as a protein that appears and disappears in synchrony with phases of the division cycle of early sea urchin embryos[1]. Prior to that,

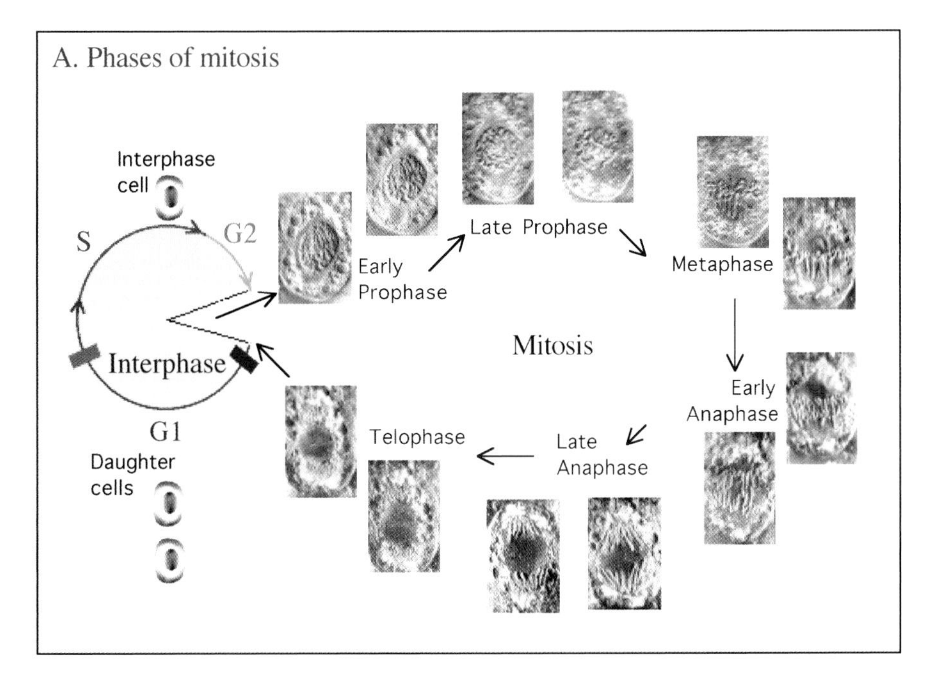

B. Idealised mammalian cell cycle

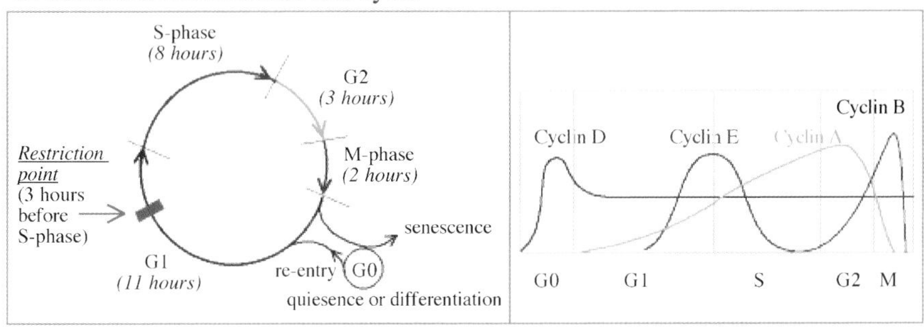

Figure 10.1 Phases of mitosis

Paul Nurse had identified a set of genes required for the fission yeast cell division cycle (the cdc genes), among which Cdc2 was predominant in controlling mitosis (Table 10.1)[2]. A maturation-promoting factor (MPF) was identified as a factor that promoted mitosis and meiosis and MPF turned out to be a complex of Cdc2 (a cyclin-dependent kinase, the metazoan form of which is now referred to as Cdk-1) and cyclin B[3,4]. As discussed in Chapter 4, Cdk enzymes are activated by A-loop phosphorylation as in many other protein kinases, but unlike other families a Cdk's activation is only achieved when it is united with the appropriate cyclin partner – an interaction that also has modifying effects on the Cdk's substrate specificity. As more cyclins and Cdks were discovered and their upstream and downstream effectors elucidated, cyclin-Cdk complex formation and periodic cyclin destruction became

Table 10.1 *C*ell *d*ivision *c*ycle proteins and other division controlling factors

	Mammals	Budding yeast (*Saccharomyces cerevisae*)	Fission yeast (*Schizosaccharomyces pombe*)
Cyclin-dependent kinases (Cdk's)	Cdk-1	Cdc28	Cdc2*
	Cdk-2, Cdk-4/6	*absent*	*absent*
G1/S cyclins	Cyclin D/E/A	Cln proteins	*'Cdc' nomenclature: e.g., Cdc-13*
M phase cyclins	Cyclin B	Clb proteins	Cdc13†
DNA-dependent protein kinase	ATM/ATR	Mec1	Rad3
Checkpoint kinases	Chk-1	Chk1	Chk1
	Chk-2	Rad53	Cds1¥
Dbf4-dependent kinase activator	ASK ('activator of S-phase kinase') ‡	Dbf4§ (Dna52) *f*	Dfp1 (Him1)
Dbf4-dependent kinase	Ddk, HuCdc7	Cdc7p§	Hsk1p#
Mini Chromosome Maintenance helicase components	Mcm2	Mcm2	Nda1§ (Cdc19) *f*
	Mcm4	Cdc54	Cdc21 *f*
	Mcm5	Cdc46	Nda4§
	Mcm6	Mcm6	Mis5
	Mcm7	Cdc47 *f*	Mcm7
Origin binding proteins	Orc1-6 ('origin recognition complex')	Orc1-6	Orp1-6 *f*
	Mcm10	Mcm10	Mcm10
Licensing- & S-phase-promoting factors	Cdt1	Tah11 (Sid2) *f*	Cdt1 ('Cdc10 dependent transcript 1') ¶
	Cdc6	Cdc6	Cdc18
	Cdc45	Cdc45	Cdc45

*Sherr, C.J. and Roberts, J.M. (2004) Living with or without cyclins and cyclin-dependent kinases. *Genes & Development,* **18**: 2699-2711.

†Nurse, P. (1994) Ordering S phase and M phase in the cell cycle. *Cell*, **79**: 547-550.

§Toyn, J.H., Toone, W.M., Morgan, B.A. (1995) The activation of Dna replication in yeast. *TIBS*, **20**: 70-73.

‡Kumagai, H. Sato, N., Yamada, M., Mahony, D., Sghezzi, W., Lees, E., Arai, K-I and Masai, H. (1999) A novel growth- and cell cycle-regulated protein, ASK, activates human Cdc-7-related kinase and is essential for G1/S transition in mammalian cells. *Mol. Cell. Biol.*, **19**: 5083-5095.

*f*Forsburg, S.L. (2004) Eukaryotic MCM proteins: beyind replication intitiation. *Microbiology and Molecular Biology Reviews,* **68**: 109-131

¶Nishitani, H. and Lygerou, Z. (2002) Control of DNA replication licensing in a cell cycle. *Genes to Cells,* **7**: 523-534.

Sclafani, R.A. (2000) Cdc7p-Dbf4p becomes famous in the cell cycle. *J. Cell Science*, **113**: 2111-2117.

¥ Rhind, N. and Russell, P. (2000) Chk1 and Cds1: linchpins of the DNA damage and replication checkpoint pathways. *J. Cell Biol.,* **113**: 3889-3896.

the dominant paradigm for cell division control during the last quarter of a century – a view that was supported by discoveries of aberrant cyclin expression and signalling in cancer cells. (Hunt, Nurse and Leland Hartwell were awarded the 2001 Nobel Prize for their discoveries of key regulators of the cell cycle).

10.1.3 Summary of the budding yeast cell cycle

Both budding and fission yeast possess only one cyclin-dependent kinase, equivalent to mammalian Cdk-1, and progression through the cell cycle is directed by its interaction with a number of cyclin partners. In *S. cerevisae*, a complex of Cdk-1 with the G1 cyclin Cln3 stimulates induction of Cln1 and Cln3 synthesis. Cln1/3–Cdk-1 complexes build up, leading to progression through 'START' – 'START' is a point in late G1, roughly equivalent to mammalian R-point, which cells must traverse before they can enter S-phase. In S-phase, the Cln proteins are degraded and expression of Clb5 and Clb6 increases. These Clb gene products are homologous with the sea urchin cyclin B discovered by Hunt. Clb5/6–Cdk-1 complexes accumulate and drive cells through the DNA replication of S-phase. Clb5/6 is destroyed at the end of S-phase and Cdk-1 is then free to associate with M-phase-promoting Clb family members to form the budding yeast equivalent of maturation-promoting factor (MPF, often referred to as 'mitosis-promoting factor'). After division, the M-phase Clbs are destroyed and the cycle 'clock' is re-set to G1 (Figure 10.2).

10.1.4 Mammalian cyclin cycle model

Post-embryonic metazoan cell cycles are subject to much more complex control. In particular, there is more than one Cdk involved in regulation and the cyclin partners include extra subfamilies that fulfil specialised roles in G1 and S. As in the yeast cyclin model, mammalian Cdks are present throughout the cell cycle but their cyclin partners are subject to periodic synthesis-destruction cycles that result in the partner kinase activity being turned on and off.

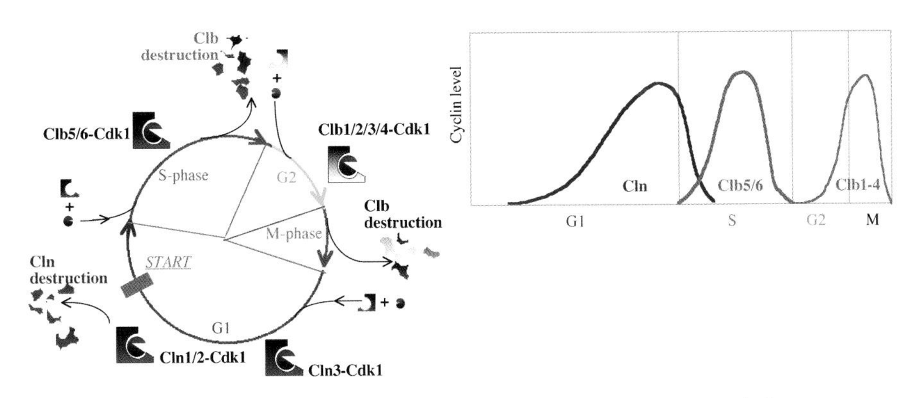

Figure 10.2 Budding yeast cell cycle and periodic cyclin accumulation

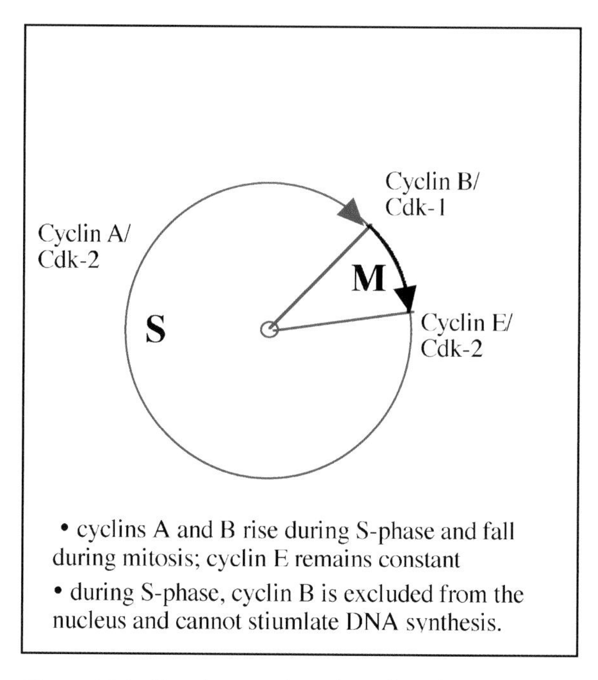

Figure 10.3 Vertebrate early embryonic cell division cycle

10.1.5 Embryonic cell cycle has no 'gaps'

The very earliest cell division cycles of the vertebrate embryo are quite different from normal adult cell divisions (Figure 10.3). Early embryonic cell divisions are rapid and synchronous. In early *Xenopus* oocytes, cells divide every 30 minutes. Cells proceed rapidly through the cell cycle without a gap (G1) between the end of mitosis (M-phase) and the beginning of a new round of DNA replication (S-phase), and without a gap (G2) before the next division[5]. However, once cellular differentiation begins, divisions are interrupted by long gaps and cells become asynchronous. In *Xenopus* oocytes, this occurs around the 13th cell division just after the midblastula transition (MBT), which is when zygotic genes begin to be expressed. In vertebrates, early embryonic cell division is controlled by only three cyclins (Cyclin A, B and E) and two cyclin dependent kinases (Cdk-1 and 2)[4], but post-embryonic cells acquire a new set of controls superimposed upon the embryonic ones. Of particular interest over the years has been the G1 cyclins (Cyclin D1,2,3) and their cyclin-dependent kinase partners (Cdk-4 and -6), which are absent in early embryonic cells.

Pathways that are involved in embryonic development have often been held to be recapitulated in cancer (the 'onco-foetal hypothesis') and, although certain embryonic pathways that should be dormant in adults, are re-activated in cancer (such as the Wnt/β-catenin axis), in other respects the embryonic cell cycle is very different from the de-regulated and varied cancer cell cycles.

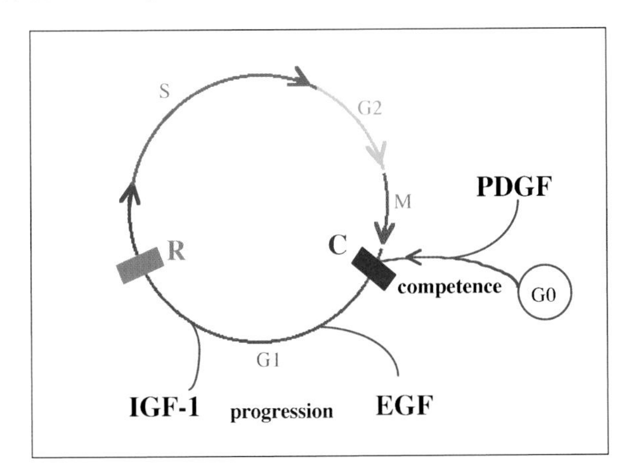

Figure 10.4 Growth factors in a fibroblast cell cycle

In general, adult cells (except stem cells) do not divide. They are terminally differentiated, 'quiescent', and removed from the cell cycle, 'resting' in a state called 'G0'. Terminated cells can be benignly recruited back into the cell cycle, for example during wound-healing – a processes set off by platelets releasing PDGF at the wound site. This can be mimicked in the lab by withdrawing serum from fibroblasts (thus pushing them into G0), then re-treating with PDGF, which makes them 'competent' to divide (G0 → G1), followed by EGF and IGF-1 treatment, which induces cell cycle 'progression' (G1 → S) and results in a round of synchronised cell divisions (Figure 10.4). Carcinogenesis also begins with terminated G0 cells being brought back into the cell cycle, but through some pathological event(s).

10.2 G0, competency, and the point of no return in G1 – the 'R-point'

The R-point marks the boundary between growth factor dependency and independency The Restriction-point ('R-point') has been defined as the last point in G1 at which growth factors are required, and it coincides with the termination of the early protein synthesis programme (early gene induction) some two to three hours before S-phase (Figure 10.5). This early and middle phase of G1 is therefore sensitive to amino acid deprivation and protein synthesis inhibitors[6,7].

10.2.1 What is G0?

Fibroblast cells that are suddenly deprived of serum arrest in G1. However, in long term serum-deprivation they enter a state of quiescence called G0. Although there has been controversy over the existence of G0 as a separate state from G1[8], it has become clear that G0 cells can be identified because they uniquely lack key components of the pre-replicative complexes found in cycle-competent G1 cells[9].

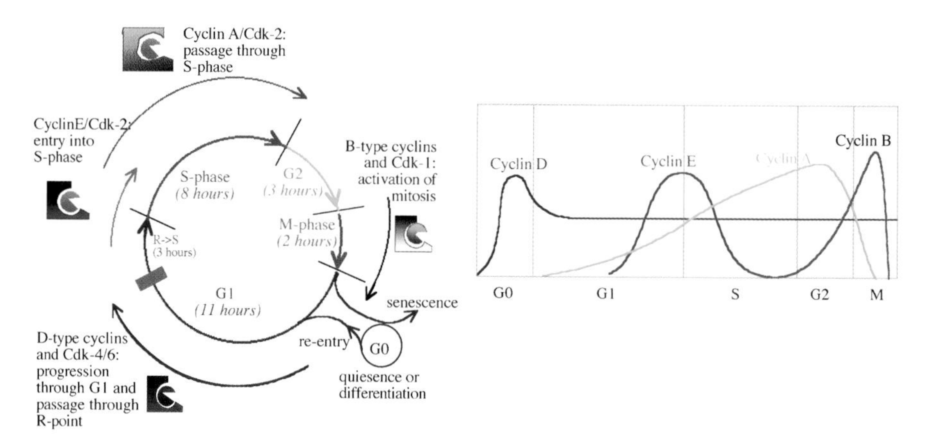

Figure 10.5 Cyclins and the mammalian cell cycle

10.2.2 The commitment point and competency factors

Analysis of serum provided clues to the factors driving cells through G1/0 checkpoints.

Serum contains the polypeptide growth factors that isolated cells require for stimulation of cell division. Plasma does not substitute for serum and does not support growth of cells in culture (plasma is blood from which all blood cells have been removed before platelet lysis and clotting can take place). The type of growth factor required varies according to the cell type – fibroblasts are taken as a general example, only.

10.2.3 Growth factors and the fibroblast cell cycle

Serum is blood that has naturally clotted and then been clarified by centrifugation. During clotting of serum, platelets burst and release platelet-derived growth factor (PDGF). It was found that a combination of PDGF and platelet-poor plasma could substitute for serum and supported growth. Plasma contains the growth factors EGF (epidermal growth factor) and IGF-I (insulin-like growth factor-I, also known as somatomedinC), but does not contain PDGF. It was found that the sequence of addition was important. A brief exposure to PDGF (2 hours) was sufficient to render G0 cells *competent* for up to 12 hours after the removal of PDGF. In other words, if plasma was added at any time during the following 12 hours, the cells would divide; plasma added after that time was ineffective. However, exposure to plasma first and PDGF second had no effect.

It was theorised that serum-starved G0 cells are re-activated by PDGF and become able to pass the 'commitment point' – a barrier between G0 and the start of G1 (Figure 10.1). Such cells re-enter the cell cycle at G1 and are said to be 'competent'. The result is the appearance of 'pre-replicative complexes' at origins of replication on

DNA in early G1. However, before passing into the DNA synthesis phase (S-phase) the cells need to be exposed to the plasma '*progression* factors' EGF (in early G1) followed by IGF-I (in late G1). This stimulates passage through the R-point, after which point cells are fully *licensed* for replication have no further need of growth factor stimulation to complete DNA synthesis and mitosis.

10.3 Oncogene products derived from growth factor pathway components

The earliest oncogene products whose mechanism was understood were the polypeptide growth factors. The Sis retrovirus causes sarcoma in woolly monkeys (hence its name: '***Si**mian **S**arcoma virus*') by the action of a single gene. The product of the *v*Sis retroviral oncogene was found to be a polypeptide growth factor. In 1983, Doolittle showed that the gene product was identical to mammalian B-type PDGF[10]. Sarcoma is a cancer of fibroblasts, which are cells that *do not* normally produce PDGF but *do* express the receptor. This inappropriate expression of PDGF in virus-infected fibroblasts results in them becoming 'self-stimulating'. A cell producing a ligand that stimulates its cognate receptor on itself is said to be operating an 'autocrine loop'. Such cells are then subject to chronic autonomous stimulation of the PDGFR pathway. It is significant that, although both PDGF-A and-B chains are mitogens, only PDGF-B produces transformed growth[11].

Other components of these growth factor pathways are potential oncogenes (termed 'proto-oncogenes') and we have discussed some earlier (see Table 10.2). EGFR family members may be truncated and, having lost the auto-inhibitory ligand-binding domain, become constitutively active (Chapter 6, Section 6.7.1). In human cancers, over-expression of erbB1 and erbB2 is common[12]. vSrc, for example, is truncated and constitutively active because it lacks the regulatory tyrosine527 (Chapter 3, Section 3.1.5). The oncoprotein also causes sarcoma in chickens.

Oncogenic versions of Ras include two point mutants that have disabled GTPases and are locked in a constitutively active state. Transformation of the foci of cells by the carcinogen 7,12-dimethylbenzanthracene (DMBA) is caused by a single adenine to thymine transversion at codon 61 of the Ras gene. This results in an oncogene product with arginine substituted for the catalytic Gln at position 61 in switch II[13]. A P-loop point mutation resulting in Gly12 being replaced with a lysine or arginine also results in an oncoprotein with an inactivated GTPase rate some 6-fold slower than normal[14].

There is a vast body of literature showing indisputably that PDGF or serum stimulation in G0/1 upregulates both cyclin D levels and Cdk-4/6 activity, and that the ultimate downstream target is the *retinoblastoma protein* (R_B) and the transcription factors that are part of the immediate early response gene expression pattern.

Interestingly, the full 'mitogenic response' of human fibroblasts to serum includes not only a mitogenic gene transcriptional programme but also a set of genes involved in the wound healing programme – genes for various coagulation-, inflammation-, and angiogenic-factors are induced in parallel in the same 15 minutes to 24 hour timeframe[15].

Table 10.2 Classes of oncogene

	Protein Type	**Proto-oncogene**	**Oncogene**
Class 1	**Growth Factors**	PDGF B-chain	Sis (abnormally expressed)
Class 2	**Tyrosine Kinases**		
	a) Receptor TK's	EGF receptor (c-Erb B1) cErb B2	v-Erb B (truncated) ErbB1/2 (overexpressed)
	b) Non-receptor TK's	c-Src Abelson gene	v-Src (truncation) Abl (Translocation-Philidelphia chromosome)
Class 3	**7-pass Receptors**	Angiotensin III Receptor	MAS
Class 4	**G proteins**		
	Heterotrimeric G proteins	Gs - α_S subunit Gi - α_i subunit	Pituitary Gsp (point mutation) Ovarian/adrenal Gip-2 (point mutation)
	Monomeric G proteins	Ras	H-Ras, K-Ras, N-Ras (point mutations)
Class 5	**Serine/threonine kinases**	c-Raf	v-Raf (truncation or point mutations)
Class 6	**Ser/Thr and Tyr kinases**		STY
Class 7	**Adaptor proteins** (non-catalytic, containing SH2 and SH3)	c-CrkII	v-Crk (translocation)
Class 8	**Cell Cycle Regulators**	Cyclin D1	Bcl-1 (or PRAD-1) (translocation, overexpression)
Class 9	**Transcription Factors**	Thyroid Receptor	vErb A (truncated)
			Myc (overexpressed) Jun (truncation or point mutation) Fos (overexpressed)
Class 10	DNA repair enzymes		MSH-2
Class 11	Mitochondrial membrane factor		Bcl-2
Class 12	Unknown		LCO

Adpated from Hesketh, R (1995) The Oncogenes FactsBook

10.4 Transcription and cyclins

For decades, the central dogma or consensus view of division control has been the cyclin/Cdk centric view. In this model, the periodic synthesis and destruction of cyclins turns on, and off, respectively, the activity of a series of cyclin dependent kinases (Cdks) and this cyclical activity is seen as the engine that drives cell division[16]. Whereas recent knockout studies have questioned the absolute requirement for individual G1 cyclins and Cdks in embryonic development (for example), there is no doubt that G1 cyclin complexes are predominantly involved in human carcinogenesis. For that reason, the present discussion is confined to signalling pathways known to contain a high preponderance of proto-oncogenes.

'Early response' genes are transcribed during G1 up to the restriction point. This is the period where amino acid deprivation can 'restrict' growth. During this phase of the cell cycle, a uniquely large amount of (specific) mRNA is made, including the messages for cyclin D and cyclin E.

10.5 Cyclin dependent kinases

Although the primary/primordial control of cyclin-dependent kinase activity is (as the name suggests) the simple appearance and disappearance of its activating cyclin partner, phosphorylation of the cyclin also plays an important switch-like role in control of cyclin-dependent kinase activity – B-type cyclins in particular.

10.5.1 Activating and inactivating phosphorylations

Cdks, like many other protein kinase enzymes, require A-loop phosphorylation for full activity. The cyclin H-Cdk-7 complex (***C****dk-****a****ctivating* ***k****inase*, or 'CAK') is a universal activator as it phosphorylates the A-loop (= threonine160 of Cdk-1) of cyclin-complexed Cdk-1,2,4,6, and this aligns the A-loop with the cyclin interface to form a peptide substrate-binding surface that is determined in part by the cyclin (Chapter 4)[17] (Figure 10.6). The cyclin not only facilitates activation but also dictates substrate preference of the bound Cdk – a Cdk may phosphorylate different substrates if it switches cyclin partners.

10.5.2 Inactivating phosphorylations of Cdks

Cdk-1 and Cdk-2 are subject to inhibitory P-loop phosphorylations on tyrosine14 and threonine15 (see Table 4.5, Chapter 4); the D-type Cdk-4/6 lack the threonine but retain the tyrosine. Prior to mitosis, the primed cyclinB/Cdk-1 complex is kept in check by two phosphorylation events. Wee-1 is a tyrosine kinase and it phosphorylates Cdk-1 on tyrosine15 in its P-loop[18]. This prevents ATP binding – even in otherwise activated cyclin/Cdk. Wee-1 knockouts divide during S-phase producing progressively smaller ('wee') cells. Cdk-1 is at the same time phosphorylated as the adjacent threonine14 is by the ST-kinase, Myt-1[19,20]. Thus, Wee-1/Myt-1 restrains cyclinB/Cdk-1 complexes from prematurely activating mitosis, while their levels build up during S-phase and G2.

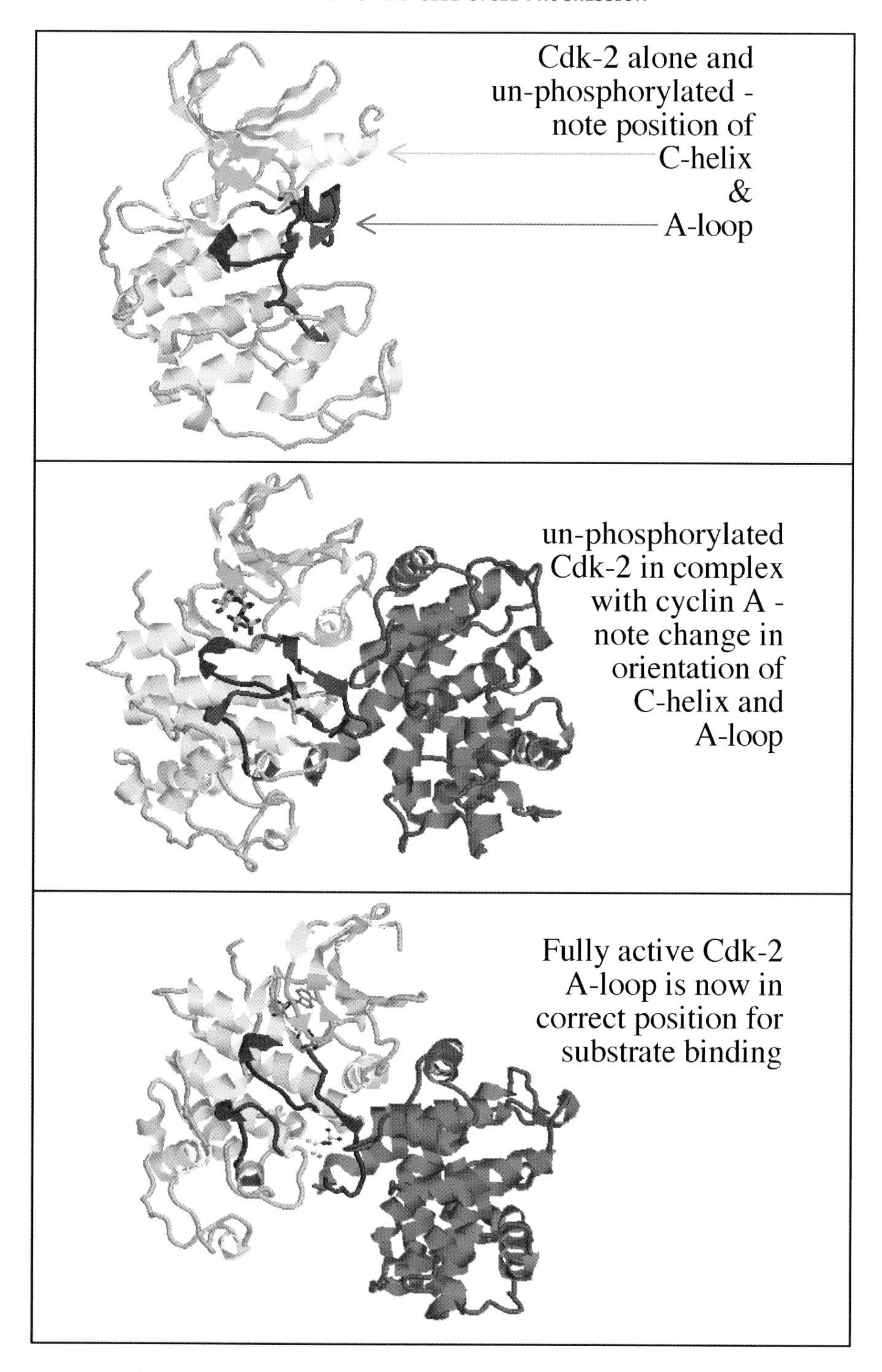

Figure 10.6 Structures of Cdk-2 in active and inactive conformations

Cyclin E/Cdk-2 activity is similarly restrained by P-loop phosphorylation in the run-up to S-phase[21].

10.5.3 Activating dephosphorylations of Cdks

De-phosphorylations of the cyclin-bound Cdk occurs *en masse* at respective transition points and this is accomplished by the dual-function protein phosphatase Cdc25, which is capable of removing phosphates from both tyrosines and threonines[22].

- Cyclin E/Cdk-2 is de-phosphorylated by the action of Cdc25A at the G1/S-phase transition.
- CyclinB/Cdk-1 is de-phosphorylated by the action of Cdc25B/C at the G2/M-phase transition.

10.5.4 DNA damage prevents dephosphorylation of Cdks

Cdc25s are important targets during DNA damage control by the ATM pathway. Radiation damage leads to activation of the checkpoint kinases Chk-1 and Chk-2. Their activation results in the phosphorylation of Cdc25A and Cdc25C, which are thus inhibited; Cdc25C by nuclear exclusion through 14-3-3 binding and translocation to cytoplasm, and Cdc25A through phosphorylation-induced proteolysis[23]. Cdc25 phosphorylation leads to cell cycle arrest at either the G1/S- or G2/M-transition.

10.6 Deactivation by cyclin destruction

The termination of Cdk activity is also (primarily) brought about by proteolytic degradation of the cyclin partner.

We have seen above that as a cyclin's level builds, so too do cyclin/Cdk complex levels, but with enzymic activity held in check. Then, at a critical level of cyclin/Cdk, the kinase activity is abruptly switched on at a transition point. However, for this to be allowed the preceding cyclin/Cdk complex (with the exception of D-types) must have first been efficiently switched off to allow approach to transition. This is simply done by removing the activating cyclin by destruction. With the exception of cyclin D, the cyclins are all degraded at cell cycle-specific time points – cyclin D levels stay constant as long as growth factors are present, but fall to zero in their absence (Figure 10.5). The exquisite control of degradation pathways is subject to many complex signalling mechanisms that will only be touched on here.

10.6.1 APC/cyclosome (APC/C) and SCF – E3 ubiquitin ligase complexes

All cyclins are eventually destroyed by the proteosome (or 'proteasome'), a multi-protein complex containing diverse proteolytic enzymes that degrade any protein that is 'tagged' for destruction by poly-ubiquitination. Ubiquitin is a highly conserved 76 amino acid peptide that is added (via isopeptide bonds) to lysine sidechains of the target protein by the combined activities of three types of proteins: a ubiquitin-activating enzyme ('E1'), a ubiquitin-conjugating enzyme ('E2') and a ubiquitin ligase ('E3')[24] – their activities are opposed by a set of de-ubiquitinating enzymes. The cyclical destruction of cyclins is under the control of two oligomeric protein complexes containing E3-ubiquitin ligase activity: APC/C (when active, binds cyclin B) and SCF (constitutively active, binds phosphorylated substrates).

10.6.2 APC/C

The complex that ends mitosis, by targetting cyclin B for destruction, is the aptly-named '***a***naphase ***p***romoting ***c***omplex' (APC) or '***c***yclosome', referred to here as 'APC/C'. (This should not be confused with the '***a***denomatous ***p***olyposis ***c***oli' protein, which unfortunately has acquired the same 'APC' abbreviation[25]). The APC/C is activated at the start of anaphase and causes the destruction of cyclin B, bringing an end to mitosis, and preventing another division (disallowed until after another round in of DNA replication). For example, mitotic exit and reassembly of the nuclear envelope cannot begin until cyclin B has gone and lamin B phosphorylation ceases. The non-cyclin kinase Polo is also degraded by the APC/C toward the end of anaphase[26]. The APC/C complex is activated by serine/threonine phosphorylation around the beginning of anaphase[4]. The complex is extensively phosphoryated *in vitro* by either cyclin B/ Cdk-1 or Polo-like kinase, which are both at peak activity at that time. However, careful analysis suggests that cyclin B/Cdk-1 is responsible for APC/C phosphorylation/ activation *in vivo*, in effect signing its own death warrant[27].

10.6.3 SCF

The Skp1-Cullin1-F-box complex (SCF) is responsible for directing G1 and S-phase cyclin destruction – the 'Skp' component was first discovered as a cyclin A/Cdk-2 binder named '***S***-phase ***k***inase associated ***p***rotein'[24]. In the case of cyclin D, its phosphorylation on threonine286 by GSK-3 triggers its recognition by SCF [28] – GSK-3, of course, re-activates once growth factors are withdrawn, see Section 10.8.7. Cyclin E is degraded via SCF after phosphorylation of its equivalent threonine. The licensing factor Cdt1 is also targetted by SCF in a cell cycle-dependent manner after it has been phosphorylated by cyclin A/cdk-2[29].

The destruction of cyclin A is mediated by the APC/C at the start of M-phase[30], but cyclin A also associates with components of the SCF in S-phase[31]. In S-phase cyclin A/ Cdk-2 enzymic activity is specifically inhibited by Skp1 and Skp2 (also an F-box protein); Skp1/2 binding also prevents A-loop phosphorylation/activation of inactive cyclin A/Cdk-2 by CAK.

10.7 Cyclin dependent kinases – activation through cyclin synthesis

10.7.1 Two sets of early genes – immediate and delayed

Growth factor stimulation of G1 cells drives the cell through the R-point towards the transition to S-phase. Within 30 minutes of growth factor stimulation, *immediate* early response genes are transcribed, including the well known proto-oncogenes cMyc, cFos and c-Jun; and other transcription factors such as STAT proteins, β-catenin, NF-κB, Egr-1, Ets-proteins and CREB may be transcribed, dependent upon cell type and stimulus.

Cell synchronisation experiments showed that the immediate early gene transcription programme preceded the synthesis of delayed early response genes. Delayed early response gene products include the D-type cyclins and cyclins E and A.

10.8 Mitogenic pathway downstream of single pass tyrosine kinase receptors

10.8.1 Transcription factor families involved in triggering the mitogenic response

Fos Fos appears to be a primary target of single pass receptor tyrosine kinase growth factor pathways and associated oncogenes. Fos is absent in non-dividing cells. Induction of Fos transcription in resting or quiescent cells is a trigger that eventually leads to cyclin D1 induction. However, the Fos gene promoter is also the target of certain 7-pass receptor pathways that also converge on Fos activation. The Fos promoter contains 'response elements' (short DNA stretches recognised by specific transcription factors) that correspond to the serum response element (SRE: single-pass and protein kinase C targets), the cAMP response element (CRE: cAMP/PKA target)[32]; Fos also contains an AP-1 site through which it can stimulate its own synthesis via positive feedback.

The Fos transcription programme leading to cyclin D1 production can be bypassed by factors binding directly to the cyclin D1 promoter, which not only contains an AP-1 site (Fos/Jun binding) but also has an SRE, a CRE, STAT-binding-, NF-κB-binding- and lymphoid Tcf/β-catenin-binding sites[33]. Cyclin D1 is commonly over-expressed in cancers, particularly breast cancer[34]. The oncogenic forms of cyclin D1 are also known as 'Bcl-1' or 'PRAD-1'.

10.8.2 Myc

Cyclin D2, on the other hand, is directly induced by another immediate early gene, Myc. The cyclin D2 gene has two Myc-binding sites on its promoter[35]. Myc itself is induced by STAT transcription factors that are activated by the tyrosine kinase Src. STAT proteins have a single SH2 domain and they associate with activated EGFR – a liaison that allows activated Src to tyrosine-phosphorylate STAT[36]. Tyrosine phosphorylation causes dimerisation of STAT through its SH2 domain binding to the pTyr of its partner and vice versa. Myc is a member of the bHLH-zip family, which has an *N*-terminal ***b***asic domain (DNA-binding), a ***H***elix-***L***oop-***H***elix domain and a ***zip***per domain. Activating Myc sites are also found on the Cdc25 phosphatase promoters[37].

10.8.3 Induction of Fos by 'serum response element' binding

A direct effect of MAPK pathway activation is the immediate activation of Elk1-like 'ternary complex transcription factors'. This is caused, in the nucleus, by their phosphorylation by activated Erk1/2. Activation of 'ternary complex transcription factors' leads to an amplified production of potent AP-1 transcription factors. Although Jun is present in resting cells, and could potentially form AP-1 homodimers, it is constitutively held in an inactive state by inactivating phosphorylations by glycogen synthase-3 (GSK-3)[38].

10.8.4 The Ets family of 'ternary complex transcription factors' – Elk-1, Sap-1/2

The '***T***ernary ***C***omplex transcription ***F***actors' are collectively known as TCF proteins[39]. This abbreviation is avoided here because, unfortunately, the same acronym has more recently become a standard term for lymphoid ***T***-***c***ell transcription ***f***actors (Tcfs) that are involved in Wnt/β-catenin signalling[40].

Ternary complex factors are a subgroup of the Ets family that have a shared structure of an *N*-terminal Ets domain (93 aas) that binds to DNA and a central 'B box' (21 aas) that mediates binding to SRF homodimers and, thereby, facilitates 'ternary complex formation'. All of these ternary complex factors are controlled by a 'switchable' *C*-terminal transactivation domain with conserved 'proline-directed' MAPK phosphorylation sites. Phosphorylation of the *C*-terminal sites is required to activate these transcription factors[41].

The ternary complex factors include Elk-1, Sap-1 and Sap-2.

10.8.5 The 'serum response factor' – MADS box-containing transcrition factors

The obligate partners of ternary complex factors such as Elk-1 are the ***s***erum ***r***esponse ***f***actors (SRFs), a ubiquitous group of proteins that contain a conserved sequence motif: the MADS box[10]. SRF is required to enable Elk-1 (activated downstream of MAPKs) to induce the transcription of Fos. The Fos gene contains a binding site for SRF in its promoter, known as a ***s***erum ***r***esponse ***e***lement (SRE). SRF binds as a homodimer to the DNA of the SRE sequence in the Fos gene promoter and this allows activated Elk-1 to dock – the ternary complex factor only binds to the SRE if it is occupied by SRF. Elk-1 binds both to the SRF proteins and to a short 5′ sequence of the SRE's DNA[42].

10.8.6 Signalling sequence of single-pass tyrosine kinase receptors leading to cyclin D induction – serum response element

The cannonical RTK pathway outlined in Chapter 7 starts with Ras activation and leads to Raf activation and consequent Mek activation. This dual-specificity MAPK kinase threonine- and tyrosine-phosphorylates the MAPK, Erk1/2, activating it and releasing it from the KSR scaffold. Activated Erk1 causes serine-phosphorylatation of the p90 form of ribosomal S6 protein kinase (p90Rsk), activating it. Activation of p90Rsk is complex, as Rsk has two kinase domains. The A-loop of its *C*-terminal kinase domain is threonine-phosphorylated by Erk and this leads to serine-phosphorylation of the A-loop of its N-terminal kinase (possibly by autophosphorylation)[43]. This event is facilitated by a physical association between Erk and Rsk. Both active Erk1 and p90Rsk can translocate to the nucleus where they phosphorylate transcription factors that constitute the 'ternary complex'. Erk1/2 phosphorylates Elk-1, activating it[44]; p90Rsk phosphorylates SRF (Figure 10.7)[39].

Erk1 phosphorylation at the *C*-terminal domain of Elk-1 is thought to cause a conformational change resulting in unmasking of the B-box and Ets domains. Phospho-Elk-1-like factors can bind to SRF dimers on serum response element of the Fos promoter. The complex of Sap-1 with an SRF dimer docked with SRE DNA (based upon the Fos promoter) has recently been solved and illustrates the structure and function of these 'ternary complexes[45]. The truncated Sap-1 is missing the phosphorylatable *C*-domain. The B-box-linker-Ets domain of Sap-1 is complexed with the core MADS domains of SRF. The B-box is responsible for binding to the MADS domain of one of the SRF pair only, while the flexible linker (invisible in structure) twists around the DNA strand to allow the Sap-1 Ets domain to bind with the other. The SRF MADS box interacts with the 'CArG' sequence of the minor groove and the Ets domain of Sap-1 binds to the 'Ets-binding site' sequence of the major groove on the opposite face of the DNA (Figure 10.8).

Fos synthesis commences and this activating signal is reinforced by stabilising and activating phosphorylations of the *C*-terminal domain of the Fos protein. First, the protein is stabilised by two 'priming' serine phosphorylations by p90Rsk and Erk. This is followed by two activating phosphorylations of threonine residues by Erk[46].

10.8.7 Activation of Jun (and Myc) by inactivation of glycogen synthase kinase-3 (GSK-3)

Unlike Fos, its partner Jun is constitutively expressed in resting cells but is inactive. This is due to a continuous, constitutive phosphorylation by GSK-3. Jun's suppression is reversed by p90Rsk, which phosphorylates GSK-3, inactivating it. In the absence of the inactivating GSK-3 signal, active Jun levels rise. Fos can now dimerise with Jun to form the highly potent Fos/Jun AP-1 heterodimer (Figure 10.9). AP-1 sites are present on both Fos and Jun genes[47]. Therefore, the low levels of Fos/Jun effectively self-stimulate their own synthesis as long as GSK-3 is suppressed.

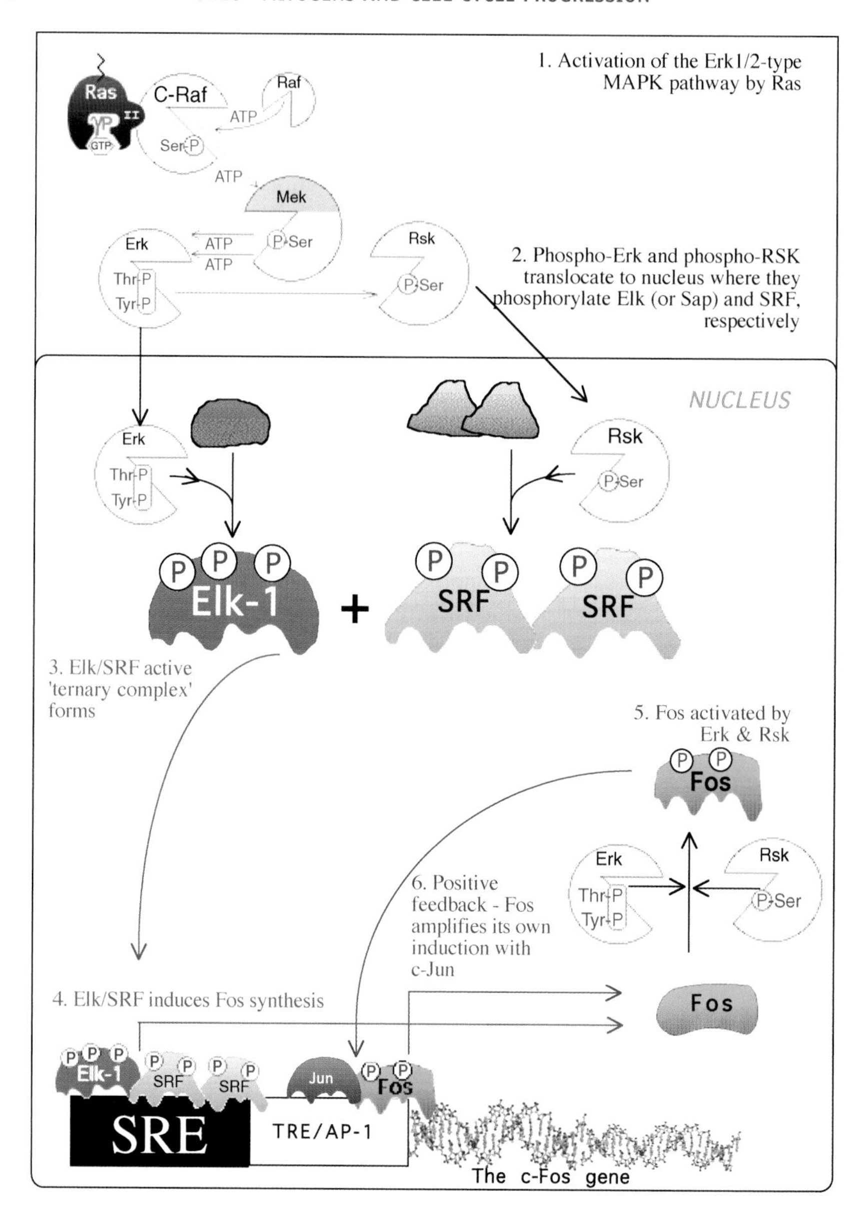

Figure 10.7 The activation of Fos synthesis by the Ras/MAPK pathway

GSK-3 is also inactivated through phosphorylation by PKB (also known as Akt), which is also activated downstream of growth factors, either by RTK-recruitment of $p85^{PI3K}/p110^{PI3K}$ or via direct activation of $p110^{PI3K}$ by Ras[48]. Myc would also be subject to an inhibiting phosphorylation by GSK-3, were it not inhibited by PKB or Rsk. Myc is also subject to a stimulating phosphorylation by Erk, which leads to full activation (Figure 10.10).

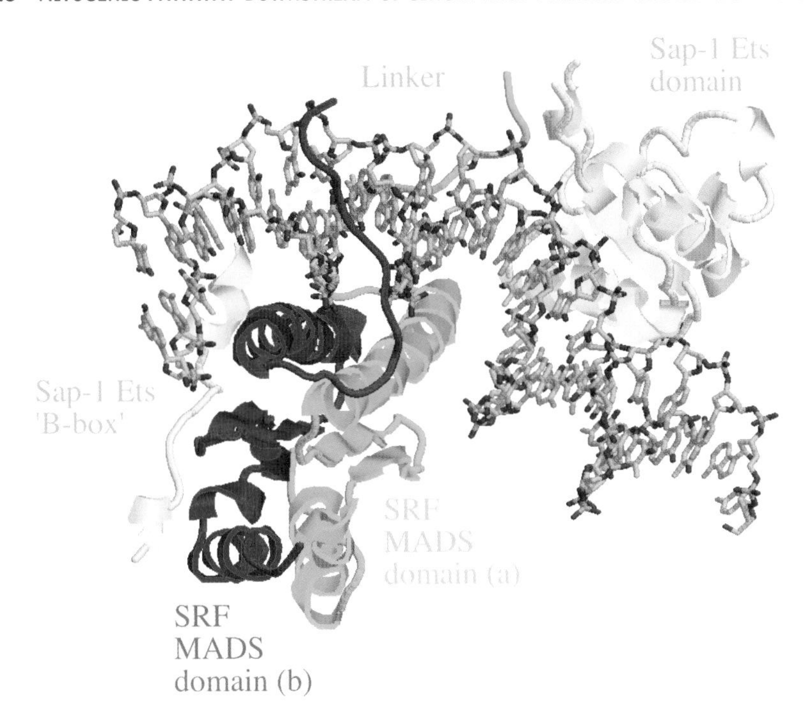

Figure 10.8 Sap-1 complexed with a SRF dimer docked with the Fos gene promoter DNA (PDB File: 1HBX)

Phylogenetically, GSK-3 most closely resembles cyclin dependent kinase enzymes such as Cdk-1 and Cdk-2[38], which is why structural studies use the Cdk terminology of 'T-loop' ('threonine-containing activation loop') instead of A-loop (activation loop)[49] (see Chapter 4, Section 4.4). Cdk enzymes (like most ST kinases) require an A-loop threonine residue to be phosphorylated to allow substrate access and to form the peptide-binding surface. GSK-3 is very different – it has a pair of phosphotyrosines in its active A-loop, but no phosphothreonine. The phosphothreonine is instead supplied by its 'primed' substrates such as glycogen synthase, which needs to be phosphorylated by casein kinase II before it can be efficiently phosphorylated by GSK-3[49]. The consensus GSK-3 substrate sequence is S/T.X.X.X.pS/pT. The inactivating serine phosphorylations of GSK-3 occur at a glycine-rich *N*-terminal extension (beyond the kinase core) and this creates a novel pseudosubstrate that binds intramolecularly to the active site and inhibits the enzyme.

10.8.8 AP-1 complexes – bZip transcription factors

AP-1 (***a****ctivator* ***p****rotein*-***1***) is a term used to describe homo- or heterodimers of the bZIP transcription factors (***b****asic region leucine* ***zip****per*), so-called for the specialised α-helix that aids their dimerisation via the formation of a coiled coil. Coiled coils result when

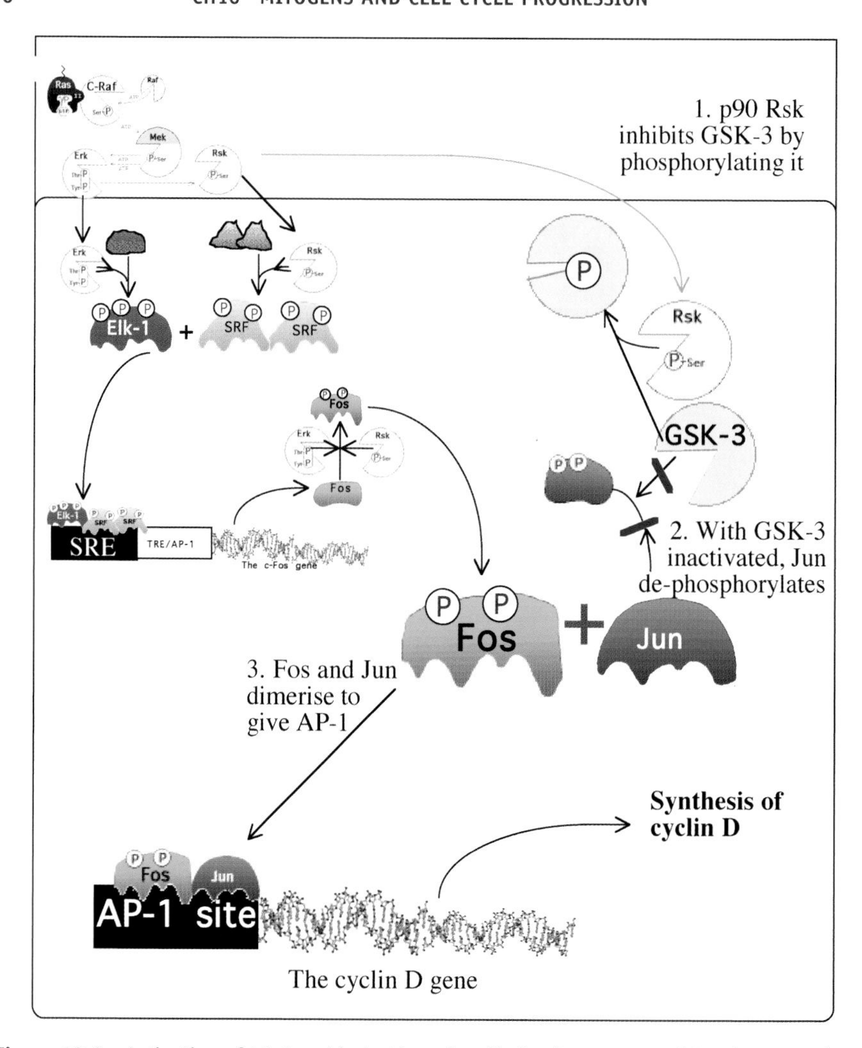

Figure 10.9 Activation of AP-1 and induction of cyclin D1 downstream of Ras /MAPK pathway

two or more right-handed alpha-helices wind around each other in a left-handed supercoil. An α-helix that can participate must have a seven amino acid periodicity such that it positions a leucine at positions 'a' and 'd' (see Figure 10.11). The non-polar leucines are paired along one face of the helices allowing them to 'zip' together – the leucines being seen as the teeth of a zip.

AP-1-forming bZIP proteins are:

- Jun family: c-Jun, JunB and JunD
- Fos family: c-Fos, FosB, Fra-1 and Fra-2

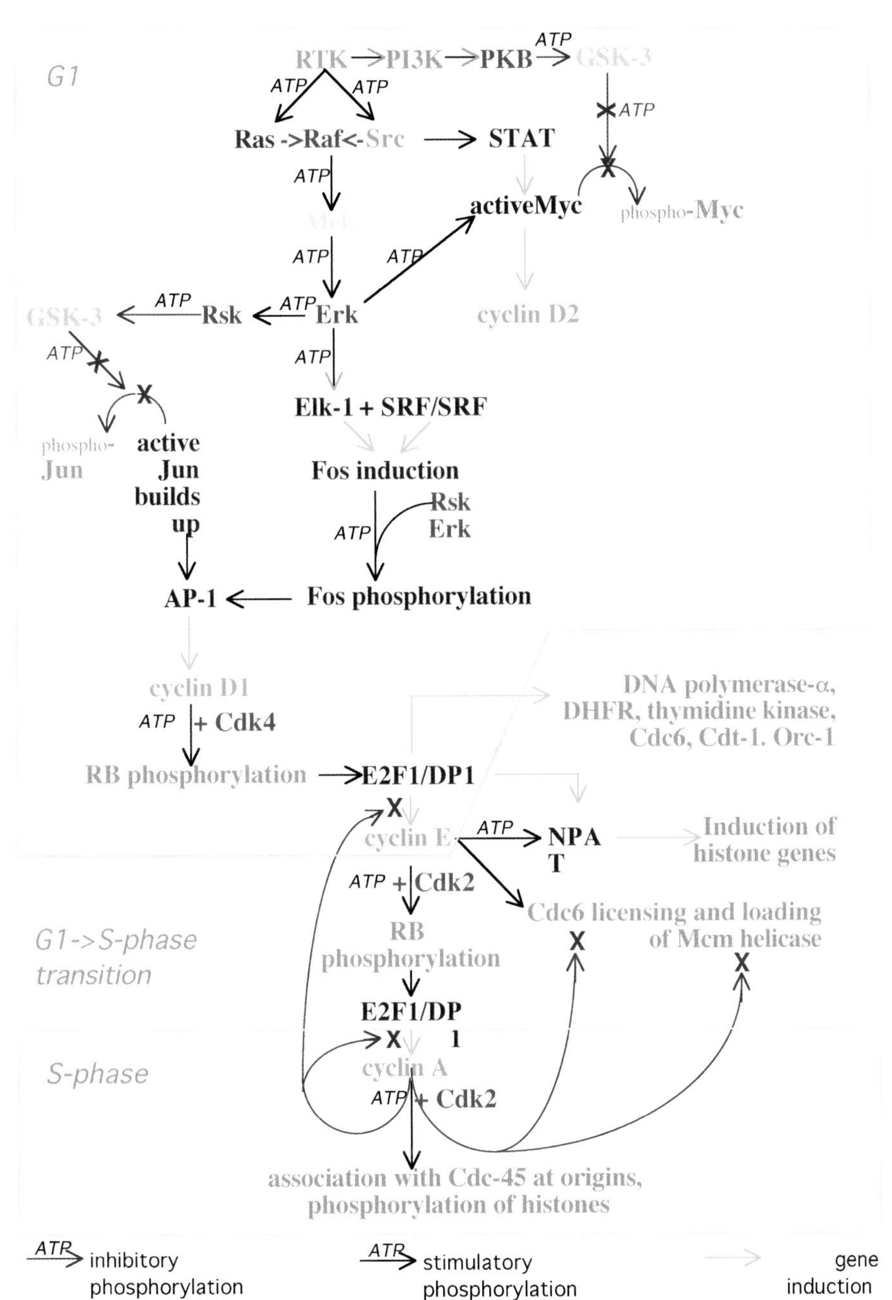

Figure 10.10 Summary of transcriptional activation of progression to S-phase

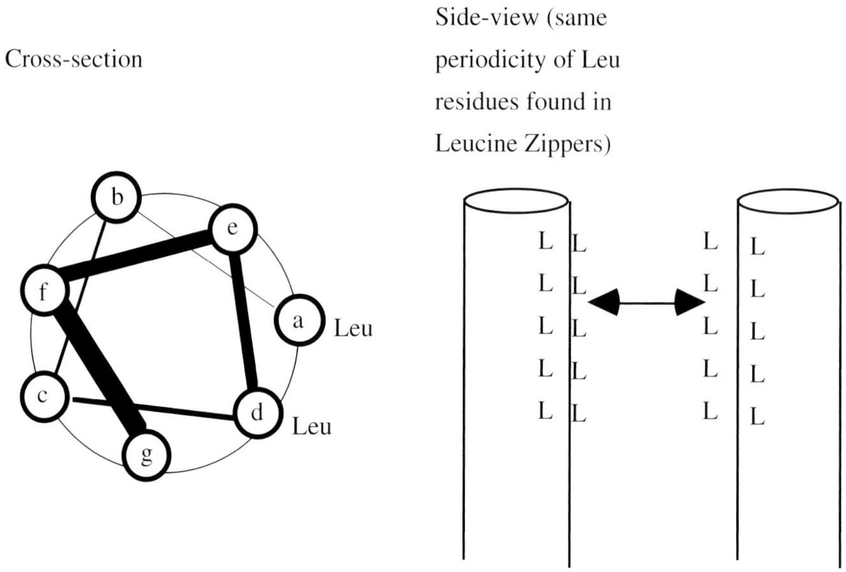

Cross-section

Side-view (same periodicity of Leu residues found in Leucine Zippers)

b e f a Leu c d Leu g

L L L L L L L L L L

L L L L L L L L L L

Figure 10.11 Leucine zipper

- ***J**un **d**imerisation **p**artners*: JDP1 and JDP2
- ***A**ctivating **t**ranscription **f**actor family*: ATF2, ATF3/LRF-1 and B-ATF

10.8.9 AP-1 response elements on DNA

Jun and ATF proteins can form stable homodimers, but Fos proteins cannot. However, Jun/Fos heterodimers are more stable than Jun/Jun dimers. c-Jun containing dimers bind to AP-1 recognition sites on a select group of gene promoters. Heterodimerisation of c-Jun with Fos increases the transcriptional capacity of c-Jun at these sites, whereas dimerisation with Jun-B inhibits c-Jun – Jun-B antagonises the transcriptional activity of c-Jun. The regions on promoters that bind such AP-1 factors were originally termed '***T**PA **r**esponse **e**lements*' (TREs) because they were found to be responsible for transducing the tumour-promoting effects of phorbol esters.

ATF2-like proteins heterodimerise with c-Jun and instead direct it to ***c**AMP **r**esponse **e**lements* (**CRE**s), so-named because they are involved in cell-type-specific gene transcription programmes downstream of cAMP generation and PKA activation. In particular, the CRE is involved in unusual pituitary tumours caused by constitutive activating point mutations of the Gs α-subunit (Section 10.17.2)[47].

10.8.10 AP-1 and cyclin D1 induction

The cyclin D1 promoter contains both a TRE site (TPA-response element) and a CRE site (cAMP-response element), and both of these promoter elements are targetted by the AP-1 complex (in the form of C-Jun/Fos dimers or C-Jun/ATF2 dimers, respectively).

Other AP-1 dimers are also found, including a transcriptional repressor made up of Fos and JunB.

In certain cells, the cyclin D1 CRE site is activated by CREB (***c***AMP ***r***esponse ***e***lement ***b***inding protein) downstream of protein kinase A (see the Section 10.17.1). D-type cyclins are also induced by a number of other immediate early gene products and transcription factors, including STAT proteins, β-catenin, NF-κB, Egr-1 and Ets-proteins.

10.9 CyclinD/Cdk-4/6 – only important substrate is R_B

The key G1 progression factor is a D-type cyclin complex with cyclin dependent kinase-4/6, which forms as cyclin D levels rise (D-type cyclin levels differ from the other cyclins in remaining high throughout the cell cycle if growth factors are present). A paradoxical factor in the formation of the active Cyclin D/Cdk-4 complex is that an 'assembly factor' is required – the assembly factor is an inhibitor of Cdk-2 (discussed in Section 10.15.2).

Cyclin D/Cdk-4's main purpose is to phosphorylate the retinoblastoma (R_B) 'pocket protein', which sequesters the transcription factors (E2F and DP1) needed for the next stages of the cell cycle – i.e., transition from G1 to S-phase. R_B is unphosphorylated in G0 and early to mid-G1 but becomes hyperphosphorylated from late G1. Hyperphosphorylation destroys its E2F binding capacity (Figure 10.12).

10.10 Retinoblastoma-related 'pocket proteins' – negative modulators of E2F

The R_B family of proteins bind E2F/DP1 dimers and inactivate them. R_B has two regions that bind the transcription factors (Table 10.3): the central 'pocket domain' and C-terminal domain (R_BCD). Recent crystal structures of complexes of truncated R_B and E2F show the interaction in detail and help explain how phosphorylation of R_B leads to E2F release. In one structure, the transactivation domain peptide from E2F is bound to the pocket domain of R_B (Figure 10.13)[50]. The R_B pocket domain is made up of two cyclin box folds and the transactivation peptide of E2F binds in the interface between them. This effectively sequesters the transactivation domain, thus blocking its ability to cause gene activation – five hydrophobic residues of E2F that are required for transactivating activity are buried in the 'pocket'. This explains how R_B can suppress gene transcription by direct masking of E2F's transactivating segment without the requirement for histone modifying enzymes.

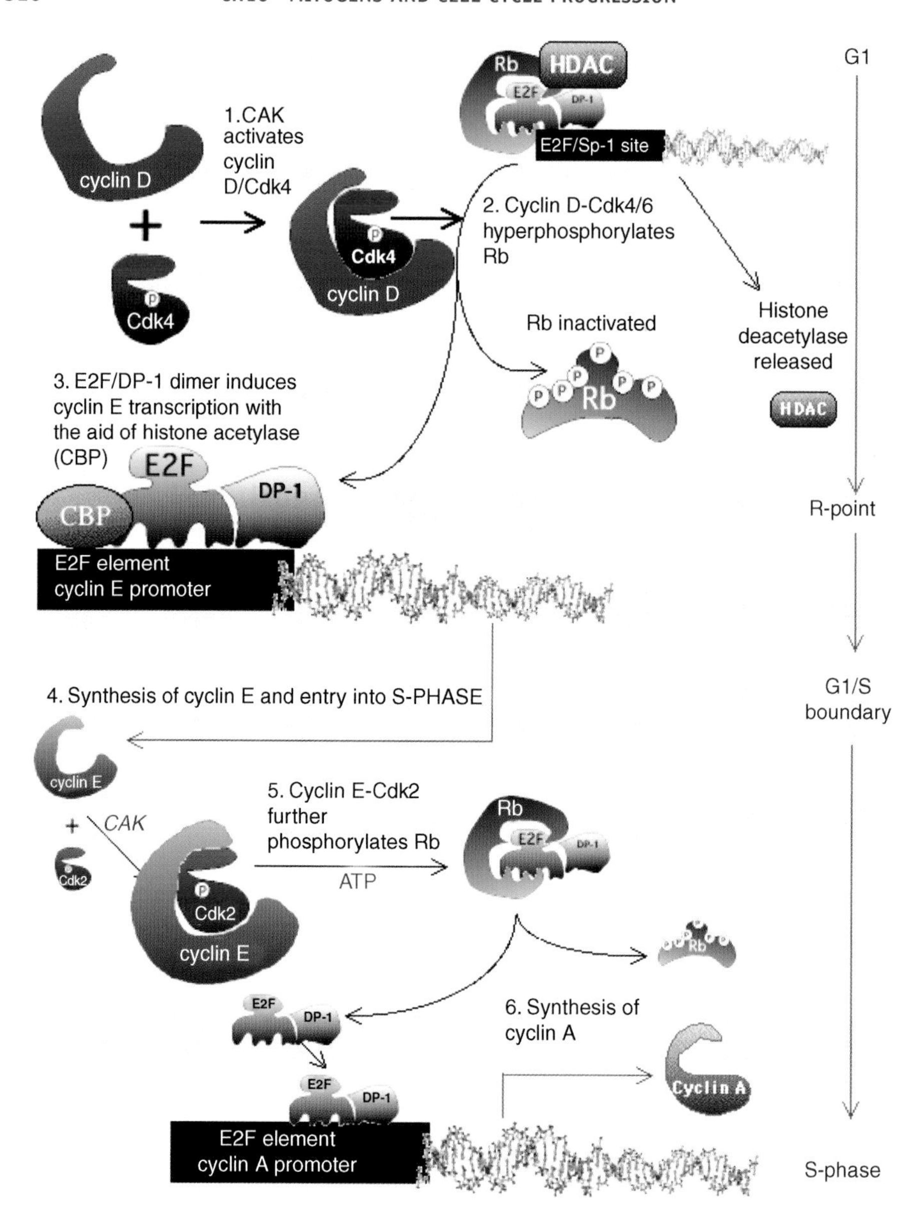

Figure 10.12 Cyclin E and cyclin A induction

The R_B pocket region has a second protein-binding site in the form of a shallow groove at the edge of the cyclin box 'B' (Figure 10.13). This site binds proteins with an 'LxCxE' motif (such as D-type cyclins and PP-1, endogenous proteins that normally control activity), but it also binds viral proteins that deregulate its activity[51] (viral proteins such as the papilloma E7 protein function to dislodge E2F). The shallow groove

Table 10.3 Schematic of E2F-like proteins and the RB pocket protein

E2F-1	1→120	DNA-Binding →199	Coiled-Coil →245	Marked Box →300	367	Transactivation →[illegible]
DP-1	1→104	DNA-Binding →197	Coiled-Coil →266	Marked Box →349	410	

R_B	1 →379	Pocket Domain →785	R_B C-Domain →928
P^n sites	T S S T T T	S S S S	SS SS TT826

Table information sources: -

Rubin, E., Tamrakar, S. and Ludlow, J.W. (1998) Protein phosphatase type 1, the product of the retinoblastoma susceptibility gene, and cell cycle control. *Frontiers in Bioscience*, **3**: d1209-1219.

Rubin, S.M., Gall, A-L., Zheng, N. and Pavletich, N.P. (2005) Structure of the Rb C-terminal domain bound to E2F1-DP1: a mechanism for phosphorylation-induced E2F release. *Cell*, **123**: 1093-1106.

also binds the histone deacetylase enzymes HDAC-1 and -2 that are employed by R_B to form repressor complexes with E2F[51].

10.10.1 Phosphorylation/inactivation mechanism

A second structure shows truncated R_BCD-only bound to an E2F-1/DP-1 dimer (Figure 10.14), and this study offers an explanation as to how phosphorylation of R_B

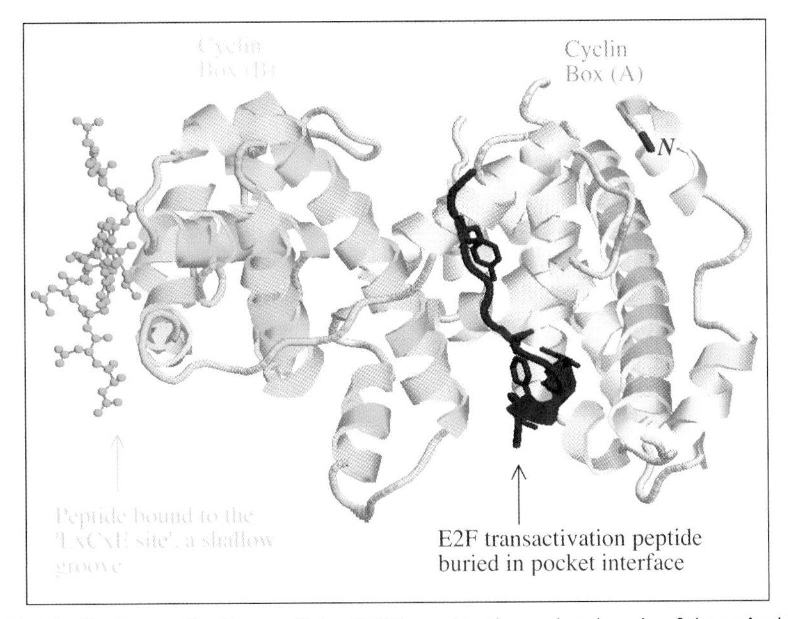

Figure 10.13 The transactivation peptide of E2F bound to the pocket domain of the retinoblastoma protein

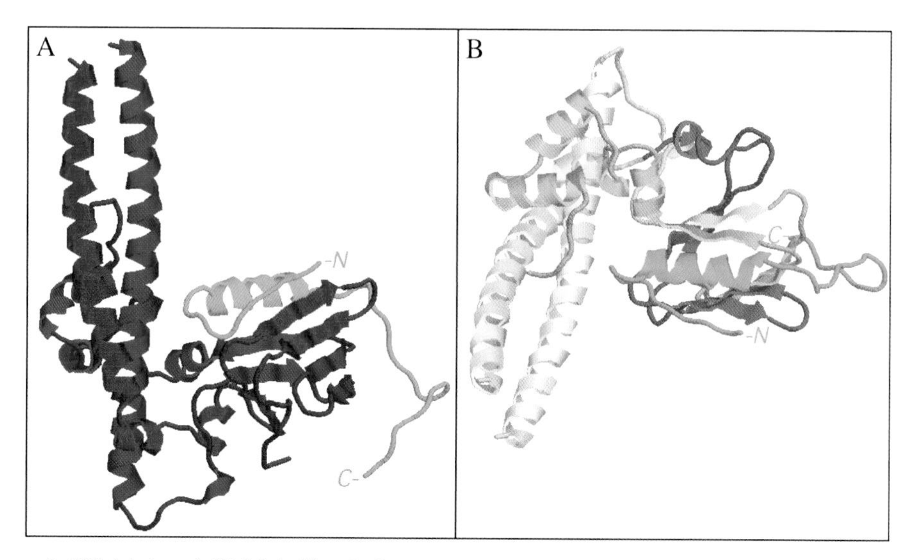

A. E2F-1 is in red, DP-1 is in blue, R_BC Domain is in green. B. E2F coiled-coil is in yellow, that of DP-1 is in light yellow. The marked box of E2F is magenta, that of DP-1 is light pink. Note that both Marked Box domains contact R_B.

Figure 10.14 Coiled-coil & Marked Box regions of E2F-1/DP-1 dimer complxed with the *C*-terminal domain of the retinoblastom protien

by Cdk-4/6 and Cdk-2 causes dissociation of the complex. The E2F dimer consists of truncated proteins encompassing the coiled-coil and Marked Box regions only[52]. The E2F/DP dimer is held together by the coiled-coil interaction of their long helices and by a β-sandwich constructed from shared strands of the two Marked Box domains; the truncated R_BCD forms a high affinity complex with both Marked Box domains. Missing from the structure is the *N*-terminal half of the R_BCD, which also binds E2F/DP and contains six Cdk consensus phosphorylation sites (Table 10.2). No single cyclin/Cdk can phosphorylate all 16 potential sites along the full length of the protein; both Cdk-4/6- and Cdk-2-containing complexes are needed. Evidence suggests that the *N*-terminal half of R_BCD is phosphorylated first and this causes destablisation of the protein, allowing further phosphorylations in the pocket (leading to unmasking of E2F transactivation domain). Further phosphorylation of the paired threonines in the *N*-terminal part of R_BCD (Table 10.2) turns the domain into a ligand for its own LxCxE site, and there is a shift of binding of the R_BCD from E2F/DP to an intramolecular association with the pocket that results in release of E2F/DP (Figure 10.15). It is thought that phosphorylations at the R_B *N*-terminal region play a similarly facilitating role[52].

In some cases, R_B binding converts the E2F/DP dimer into an active repressor of gene transcription. As well as specifying the gene target, the E2F component also influences which of the retinoblastoma (R_B) family of 'pocket proteins' sequesters the heterodimer in resting conditions. The R_B family has three members: R_B itself, p107 and p130. These pocket proteins can only bind transcription factors when they are de-phosphorylated.

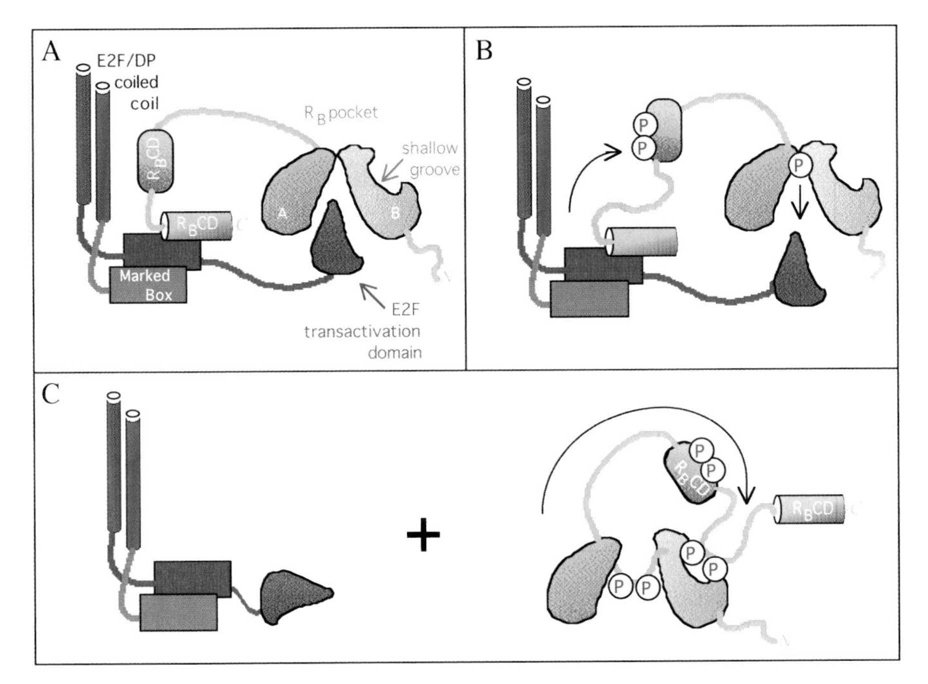

A. The C-terminal domain of R_B binds E2F/DP via the Marked Boxes, the pocket binds and masks the transactivation domain. B. Initial phosphorylations of the *N*-terminal part of R_BCD causes destabilisation allowing pocket phosphorylation. C. Further phosphorylations of the threonine pair in the *N*-terminal part of R_BCD leads to an intramolecular bond with the shallow groove and disolution of the complex.

Figure 10.15 Phosphorylation-driven release of E2F/DP from R_B

10.10.2 The E2F family of transcription factors – the targets of the 'pocket proteins'

E2F ('***E2***a transcription ***f***actor') was originally discovered as a host protein that binds to the viral E2a promoter in adenovirus infected cells. Almost simultaneously, it was identified in a different context as a differentiation-regulated transcription factor, hence its alternative name: DRTF1[53].

The E2F family form two groups that partner each other in forming active heterodimers. They differ in presence or absence of conserved domains. All members have a central DNA binding domain and an adjacent dimerisation domain. Only the E2F group contain a 'Marked box' domain that is *C*-terminal of the dimerisation domain and an *N*-terminal pocket protein-binding domain. Both of these domains are missing in the DP-1/2 group. The *N*-terminal pocket protein-binding domain is located in the middle of the *C*-terminal transactivation domain of the E2F types – pocket protein binding and gene transactivation activities are mutually exclusive.

All E2F types contain a single DNA binding domain except E2F7, which has two.

E2F group:	E2F1, E2F2 and E2F3 contain an *N*-terminal cyclin A binding domain E2F4 contains a unique serine-rich motif towards *N*-terminus. E2F5, like E3F4, lacks cyclin A binding domain. E2F6 lacks pocket protein domain, but contains Marked box'. E2F7 lacks pocket protein domain, lacks DP-dimerisation domain.
DP group	DP1 and DP2 contain DNA-binding and dimerisation-domains only.

10.10.3 R_B – a DNA-binding, E2F protein-binding tumour suppressor

Familial retinoblastoma is caused by mutation of the 'retinoblastoma protein' (R_B), a tumour suppressor gene product implicated in spontaneous cancers as well. Pocket proteins (R_B, p107 and p130) are named for their protein-binding pocket that is selective for E2F transcription factors. R_B-like proteins inhibit E2F's transactivating activity – either by passively sequestering the E2F/DP complex or converting it to a gene repressor.

Any combination of E2Fs 1–5 with either DP1 or 2 is possible. The DP part of the heterodimer is nonspecific (only serving to stabilise the active complex) and is constitutively expressed – DP1 is ubiquitous, DP2 is heart-specific[54]. The E2F component is the specifying and controlling partner. E2F1-3a are strong transcriptional activators that associate with R_B. They are at low levels in quiescent cells but rise to high levels in late G1. E2F4 and 5 are weak activators that lack the nuclear localisation motif of E2F1-3, and are expressed throughout the cell cycle, but in G0 and early G1 are recruited to the nucleus via the other pocket proteins, p107 and p130, which results in formation of active repressor complexes[55]. E2F3b isoform is also thought to form repressor complexes with R_B in G0 cells.

10.10.4 E2F targets – genes for DNA replication and licensing, delayed early response genes (cyclin E and A), and NPAT

Given their expression in the part of the cell cycle that leads up to the S-phase, it is not surprising that E2F binding sites are found in the promoters of many genes directly involved in DNA synthesis, including dihydrofolate reductase, thymidine kinase and DNA polymerase-α. E2F response elements are also present on the promoters of the 'delayed early response genes': Cdk-1, cyclin E and cyclin A. The E2F member of the dimer determines specificity in some cases – E2F-1-containing dimers are potent and specific inducers of Cyclin A; E2F-3-containing dimers induce Cdk-1 preferentially. E2F-1 also induces the synthesis of NPAT at the G1/S-phase boundary (see Section 10.11.1)[56]. Finally, E2F transcription factors also control the synthesis of licensing factors (Cdc6 and Cdt1) and components of the pre-replicative complex (Mcm subunits and Orc1, Figure 10.10)[57].

10.11 De-repression of the cyclin E gene by cyclin D/Cdk-4/6

The effects of D-type cyclin dependent kinases upon pocket protein effectors only extend as far as cyclin E induction. E-type cyclin dependent kinases cause further R_B phosphorylations that trigger a further set of effects.

Cyclin E is only expressed in late G1, and its gene contains two E2F sites. One is responsible for repression of the gene in G0 and early G1 (via E2F-4 plus p107/130 complexes) and this repression is lifted by cyclin D/Cdk-4/6 phosphoryation R_B[58]. The other E2F site (a 'bi-partite E2F-Sp1 site') is occupied by a repressor complex made up of E2F-1 (or -2 or -3), R_B, and other proteins, one of which is histone deacetylase (HDAC), an activity that is also associated with CREM repressor.

Histones are acetylated in regions of chromatin being actively transcribed. HDAC, recruited by the dephospho-pocket proteins, may be key to this repressional activity because de-acetylation of histones causes local condensation, preventing access for the transcriptional machinery. Cyclin D/Cdk-4/6 phosphorylation of R_B causes it and HDAC to be released from the promoter, leaving behind an active E2F-1/DP1 complex that is further activated by recruitment of histone acetylases such as p300/CBP, which serve to de-condense the chromatin by hyperacetylation of histones[59]. p300/CBP is the '***C***REB ***b***inding ***p***rotein' and is a common co-activator of transcription factors (Figure 10.12) see also Section 10.17.1.

10.11.1 Cyclin E/Cdk-2 substrates – R_B, NPAT, nucleophosmin

E2F1/DP1 dimers, activated by phospho-R_B de-binding, result in the synthesis of cyclin E, peaking at the G1/S-phase transition. Cyclin E preferentially associates with Cdk-2 and thus forms an active holoenzyme. Cyclin E/Cdk-2 causes further phosphorylation of R_B proteins, thereby releasing more E2F/DP dimers that are thought to lead to the subsequent induction of cyclin A as the cell passes into S-phase (Figure 10.12) – phosphorylation of R_B by D-type Cdk is insufficient for cyclin A induction. It is thought that this may be due to differences in the mode of repression of the cyclin A gene (that is inhibited by RB recruited nucleosome remodelling complexes) compared with the cyclin E gene (inhibited by RB recruited HDAC). Cyclin D/Cdk-4/6 activity leaves the former repressors intact, whereas subsequent cyclin E/cdk-2 induction is necessary to disassemble the latter repressor, which controls cyclin A[60].

But what else does cyclin E/Cdk-2 do to cause the G1/S-phase transition? Recently, a cell cycle-specific substrate of cyclin E/Cdk-2 has been discovered in the form of a serine-rich protein named NPAT (for '***n***uclear ***p***rotein mapped to the ***AT*** locus' – its gene is located next to the **A**taxia **T**elangiectasia locus on chromosome 11). Over-expression of NPAT alone causes S-phase entry, an effect that is accelerated by co-expression of cyclin E (but not cyclin A)[61]. NPAT physically associates with cyclin E/Cdk-2 and is phosphorylated by it at the time of the G1/S-phase boundary. Phosphorylated NPAT then promotes the induction of histone synthesis in preparation for assembly of new chromosomes in S-phase (Figure 10.10)[62] – histones make up 50% of chromatins mass and are essential for DNA packaging[60]. In cells, NPAT is associated with multi-functional nuclear organelles called 'Cajal bodies' (or 'coiled bodies') that contain high concentrations of small ribonucleoproteins involved in histone precursor mRNA processing, along with a FADD-like enzyme[63]. Interestingly, NPAT is dispersed during mitosis but can be visualised as two nuclear spots in G1 and G2 cells, that become four in S-phase – the G1/2 spots are associated with the histone cluster on chromosome 6, and the additional two spots in S-phase are associated with the second histone cluster on chromosome 1[64].

Another key cyclin E/Cdk-2 substrate is nucleophosmin, which causes centrosome duplication[65].

10.11.2 Cyclin E – licensing and loading of helicase

A few of the substrates/binding partners of cyclin/Cdk complexes that seem to be involved in G1 → S → G2 transitions are known – and these nuclear substrates (that are responsible for DNA replication) are discussed in the following sections.

Synchronised fibroblasts, released from G0, enter S-phase at 19 hours. Three hours prior to this, the cells pass the 'R-point' and this last phase of G1 when the progression to S-phase becomes independent of further growth factor stimulation. This commitment to replicate occurs when pre-replicative complexes become 'licensed' (Figure 10.16). Licensing allows loading of the DNA helicase, and this is the responsibility of two

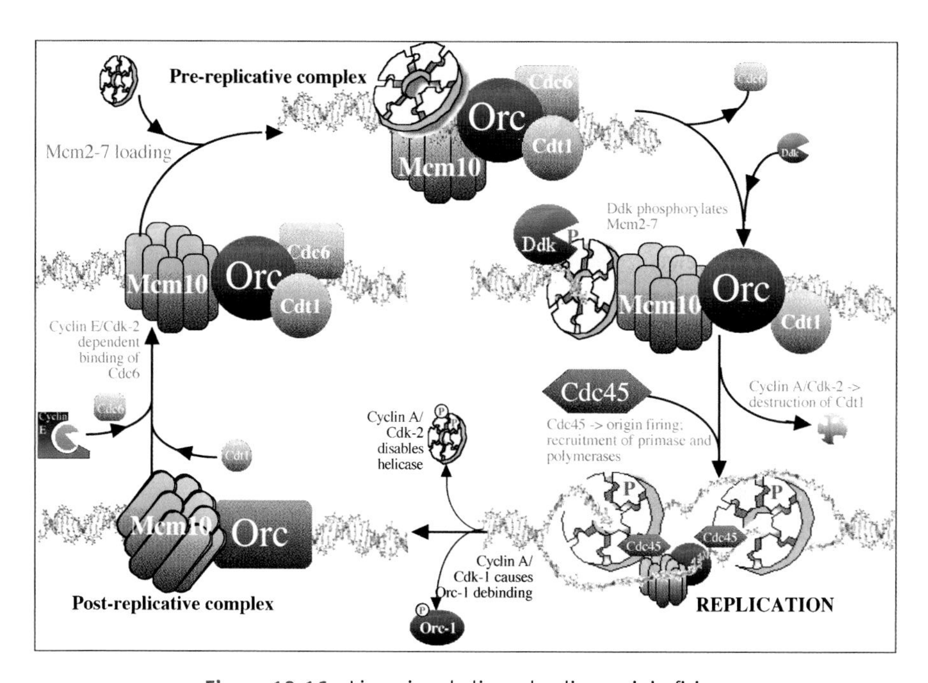

Figure 10.16 Licensing, helicase loading, origin firing

proteins, Cdc6 and Cdt1. The binding of Cdc6 to post-replicative complexes has recently been shown to be activated by cyclin E/Cdk-2 – the cyclin E/Cdk-2/Cdc6 complex is associated with chromatin at this time, where it stimulates the loading of Mcm-2 (a member of the helicase)[66].

10.12 Cyclin A/Cdk-2 – S-phase progression and termination

Cyclin E's main responsibility is to push cells through the G1/S-phase boundary, whereas cyclin A's main function is to maintain S-phase progression and, at the same time, ensure that DNA is replicated – *but only once*. Cyclin A/Cdk-2 appears to both aid DNA replication at origins and yet also prevent the same origin being re-used – a seeming contrary pair of activities not yet fully understood.

Cyclin A/Cdk-2 is associated with chromatin throughout S-phase. At least part of cyclin A's ability to maintain DNA replication appears to be due to its phosphorylation of histone H1 at active transcription forks[67]. This occurs after origin firing caused by the loading of Cdc45, which then acts to recruit Cdk-2/cyclin A. The localised cyclin A/Cdk-2 phosphorylation of histone causes decondensation of chromatin that then allows access for the transcription machinery.

10.12.1 Cyclin A/Cdk-2 – prevention of origin re-firing

It is significant the cyclin A/Cdk-2 has the opposite effect to cyclin E upon the licensing factor, Cdc6. Cyclin A/Cdk-2 phosphorylates Cdc6, and this disables its helicase-loading activity. Thus cyclin A/Cdk-2 acts to prevent DNA re-replication by 'de-licensing' origins.

Cyclin A/Cdk-2 also phosphorylates and deactivates the initiating activity of the DNA polymerase α-primase (the only enzyme that can initiate DNA replication *de novo*) – interestingly cyclin E/Cdk-2 was reported to do the opposite, activating the enzyme's initiating activity[68]. Such an effect of cyclin A/Cdk-2 would again contribute to an inhibition of inappropriate re-initiation.

Cyclin A/Cdk-2 is also thought to inhibit the helicase activity of Mcm2-7. Cyclin A/Cdk-2 phosphorylates Mcm-4 during S-phase, and this disables the helicase (Figure 10.16)[69]. Again, this is thought to prevent an origin being used more than once.

10.12.2 Terminating S-phase – cyclin A effects

Cyclin A has roles to play in the termination of S-phase. As noted earlier, the potent E2F transcription factors are a key force in driving S-phase and several of the cell cycle dependent forms have specific cyclin A binding sites at their *N*-terminal ends. Cyclin A/Cdk-2 associates with E2F-1/2/3 via the cyclin A-binding domains, with the result that the DP-1 protomer in the E2F/DP-1 dimer is phosphorylated. Cyclin A/Cdk-2 phosphorylation of DP-1 inhibits the transcription factor and shuts down the E2F transcription programme[54].

As well as Cdk-2, cyclin A can also pair up with Cdk-1. It has been found that cyclin A/Cdk-1 complexes prevent re-replication by hyperphosphorylation one of the 'origin recognition complex' members, Orc-1[70]. The hyperphosphorylated Orc-1 can no longer bind to chromatin, is released, ubiquitinated and degraded (Figure 10.16).

10.13 The controlled process of mammalian DNA replication

10.13.1 How does a cell know when to dvivide?

During G1, the interphase cell processes cues relating to nutrient supply (is it sufficient to complete division?), cell size (am I big enough to divide?), as well as information from other cells (hormones, growth factors, cell-to-cell communication) that either delay or permit the cell to proceed with the energy-costly business of replicating its chromosomes. Cells that pass the various tests in G1, cross the R-point, and are said to be 'licensed' to replicate.

10.13.2 DNA replication

In S-phase, the most important constraint is that each chromosome must be faithfully replicated, completely, and once only. Each replicated chromosome must also have a centromere and terminating telomeres. Unlike prokaryotic chromosomes that are replicated in one single operation, eukrayotic (and surprisingly archaea) chromosomes are replicated in non-overlapping segments called 'replicons'[71]. How do eukaryotes ensure their replicons are not re-replicated?

10.13.3 Pre-replicative complex formation begins in G1

Origin binding proteins initiate the pre-replicative complex formation Along each eukaryotic chromosome are a number of regions of DNA known as 'origins of replication' and these origins are where each replication segment begins (a circular bacterial chromosome only has a single origin). DNA replication begins at one origin and continues up to the next origin, making a single replicon.

Eukaryote origins were first identified in budding yeast, which contain short stretches of DNA called 'autonomously replicating sequence' elements (ARS) that cause plasmids containing them to replicate once during S-phase[72]. An ARS is A/T rich and contains an 11bp consensus sequence: (A/T)TTTAT(A/G)TTT(A/T), known as the 'A' element, or core element. The A element core sequence and adjacent 3 sequences (B elements) are required to bind a protein complex called the origin recognition complex (ORC). In bacteria, a single origin-binding protein DnaA is present, which although unrelated, shares structural similarities to eukaryotic Orc proteins. Archaea have a single origin-binding protein Orc1 that exhibits amino acid similarities with both the eukaryotic Orc1 and the helicase loader Cdc6 – whether the archaea Orc1/Cdc6 fulfils the functions of origin recognition, helicase loader or both is unclear. Eukaryotes have a more complicated way of marking origins; instead of a single origin binding protein, they have a six-member Orc consisting of Orc1–6[71]. Orc1, Orc4, Orc5, Cdc6, archaeal Orc1/Cdc6 protein and the bacterial DnaA protein are all members of the AAA^+ superfamily of ATPases.

The Orc complex is made up of six subunits: Orc1–6, and this holocomplex is responsible for locating and marking each origin during G1. Orc2–6 are expressed throughout the cell cycle but Orc1 expression is cycle-regulated, and although Orc2–6 remains chromatin-bound throughout the cell cycle in proliferating cells, Orc1 undergoes hyperphosphorylation by cyclin A/Cdk-2 resulting in its dissociation from chromatin throughout mitosis (Figure 10.17)[9]. M-phase Orc can no longer recruit helicase and this disabled post-replicative state of the complex is one way that re-replication of the genome is prevented until the next S-phase[73]. Resynthesis of Orc-1 by E2F induction on the next G1/S-phase transition reforms Orc1-6, which then acts a landing pad for the other proteins that make up the pre-replicative complex, aided by Mcm10 proteins.

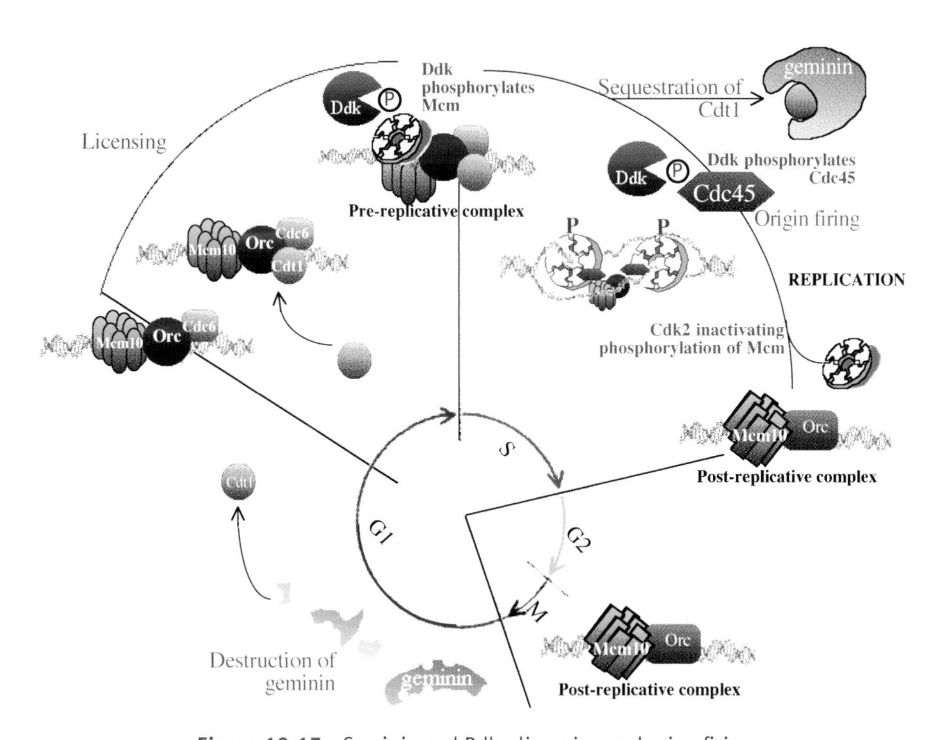

Figure 10.17 Geminin and Ddk - licensing and orign firing

10.13.4 Helicase loading

The eukaryotic helicase is a ring-shaped hexameric complex made up of six Mcm proteins (Mcm2–7). Mini chromosome maintenance (Mcm) proteins were first discovered in yeast due to their mutant forms being unable to support correct initiation of DNA synthesis in ARS-containing plasmids[72]. Mcm proteins are absent in bacteria but a single Mcm homologue is present in archaea[71]. Like members of the Orc, Mcm proteins are also members of the AAA$^+$-ATPase family whose members all form heterohexamers with 1+1+1+1+1+1 stochiometry[74]. Mcm proteins display two typical ATPase signature sequences: the Walker A box containing a P-loop and a Walker B box with a motif (IDEFDKM) that is conserved in Mcms and defines this subfamily. About 70 amino acids *C*-terminal of the Walker B box is an 'arginine finger' motif that is involved in complex assembly. The Mcm proteins are only competent as ATPases when assembled, and this occurs by interaction of the arginine finger of one Mcm inserting into the P-loop' of the next Mcm (see Figure 10.18).

Although Mcm2 and 3 have nuclear localisation signals (NLS) and therefore might chaperone the others (which lack an NLS) in nuclear translocation, it has been found that in cycling cells Mcm proteins are actually present in the nucleus at all parts of the cycle. However, as we shall see, Mcm proteins are absent from the cell in G0 and must be re-synthesised at commitment – presumably this is where Mcm2/3 NLS comes into play.

The Mcm complex is loaded onto the Orc-marked origins on chromatin at the G1 → S-phase transition and gradually dissociates again as S-phase progresses. In G2 and M-phase the Mcm complex is no longer bound to DNA origins[75]. Binding of the Mcm2–7 helicase is dependent upon another 'mini chromosome maintenance' protein, Mcm10, which acts as a bridge between Orc and the helicase. Note that, although Mcm10 was identified by the same genetic screen as Mcm2–7, Mcm10 is biochemically unrelated to the Mcm2–7 proteins that make up the helicase[73].

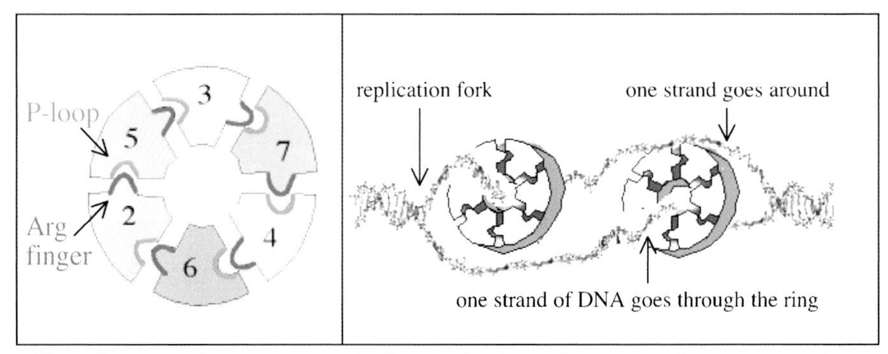

The helicase functions to open up the DNA replication forks by unwinding them enough to let the replicative gyrases, polymerases and primases gain access

Figure 10.18 The structure of the Mcm2-7 helicase complex

Mcm10 proteins, like Orc, remain bound to chromatin throughout the cycle and bind to the helicase and Orc when the helicase loaders, Cdc6 and Cdt1, have joined the complex to license it for replication.

10.13.5 Geminin control of helicase loading and licensing

Before the Mcm helicase can bind to Orc, the initiator proteins Cdc6 and Cdt1 must bind first, and this occurs during G1. Cdc6 appears to act as the helicase loader in synergy with the Cdt1 protein[76]. Cdt1 plays an important part in the timing of licensing the origin for replication and also in preventing re-replication of the replicon after it has been copied once.

In metazoans, Cdt1 licensing activity is controlled by a small (25 kDa) nuclear protein called geminin. It was originally identified in *Xenopus* as a mitotically degraded protein that is capable of inhibiting DNA replication[77]. Geminin contains a 9-amino acid 'destruction box' motif and its presence is cell cycle-regulated, being ubiquitinated and degraded by the APC/C during the metaphase-anaphase transition. Geminin's inhibitory mode of action upon DNA replication was found to be dependent upon its ability to bind and sequester Cdt-1 after each origin is licensed at the end of G1[78]. Geminin sequesters Cdt1 in S-phase to prevent origins being re-licensed and continues to sequester Cdt1 through G2 and mitosis (during which period, Cdt1 synthesis drops) and thereby prevents any inappropriate re-association with Mcm2–7 (Figure 10.17)[74]. Human Cdt1, like yeast Cdt1 homologues, reaches peak levels in the G1 phase before disappearing after the transition to S-phase[76]. Cdc6, in contrast, is present throughout the cell cycle, only disappearing in G0 cells[9].

Geminin, itself, is expressed from the S-phase to the end of mitosis – a period when licensing is disallowed – and its destruction after mitosis allows newly synthesised Cdt1 to act unopposed in licensing origins once more in preparation for a new round of replication. Geminin also seems to play a role in metazoan differentiation – it activates neuronal differentiation in embryos and its depletion in *Xenopus* embryos results in an unusual lethal phenotype in which early division cycles are normal but cells arrest and die after the 13th round of division, the point at which the midblastula transition (MBT) occurs, marking the beginning of differentiation of cells[4].

10.13.6 Origin firing – Ddk and Cdc45

Cdc45 is a conserved protein that acts to trigger the pre-replicative complex (Orc1–6; Mcm2–7; Cdc6; Cdt1) at the G1/S transition. Cdc45 acts at the same point as the unusual cell cycle kinase, Ddk.

Ddk is the latest name given to a protein kinase (see Table 10.1) – the '***D***bf 4-***d***ependent ***k***inase' – that is activated at the G1/S-phase transition, peaking at S-phase, and remains high until the end of mitosis. Ddk is related to casein kinase II but behaves like a Cdk, in that it needs a partner protein (acting like a cyclin) to become active. The human form of this latter protein is ASK ('***a***ctivator of ***S***-phase ***k***inase')[79]. Ddk activity is also controlled like a cyclin/cdk complex – the levels of kinase protein (Ddk) are fairly constant throughout the cell cycle but its activating partner (ASK) is degraded during G1 (probably by the APC/C) and must be resynthesised for the next S-phase promotion event[80].

Origin firing appears to be triggered by appearance of ASK/Ddk, which phosphorylates the Mcm2 portion of the origin-bound Mcm2–7 helicase complex. This phosphorylation creates a binding site for Cdc45, which displaces Cdc6 and Cdt1[81]. The binding of Cdc45 activates the helicase activity and transcription can begin. It is noteworthy that, until recently, helicase activity of eukaryotic Mcm complexes was only demonstrated for Mcm-4,6,7; Mcm2,3,4,5,6,7 complexes were inactive. Very recently, however, a holocomplex of Cdc45–Mcm2,3,4,5,6,7 from embryonic tissue proved to have helicase activity, confirming the need for Cdc45 in origin firing[82].

Cdc45 binding to the origin causes debinding of Cdc6 and Cdt1, thus remodelling the pre-replicative complex such that its helicase activity is activated[82]. Cdc45-dependent unwinding occurs followed by binding of RPA (replication protein A) proteins that bind the single stranded DNA and help keep the nascent replication fork open and maintain recruitment of DNA polymerase-α and primase[9]. The Mcm2-7 helicase complexes gradually dissociate from chromatin during S-phase (cyclin A/Cdk-2 phosphorylation) as each origin fires and this ensures each one fires once only. Removal of helicase produces an unlicensed post-replicative complex consisting of Mcm10 and the Orc complex minus Orc-1 (removed due to cyclin A/Cdk-2 phosphorylation).

10.14 Cyclin B translocations and M-phase

Cyclin B is synthesised from the end of S-phase, through G2, peaking in M-phase, but cyclin B/Cdk-1 activity is restrained by P-loop phosphorylations, as mentioned earlier. As a further safeguard, the complex is kept out of the nucleus (where its substrates reside) by active transport until required. Cyclin B gene promoters include sites for '***U***pstream ***s***timulatory ***f***actor' (USF)[83] and forkhead transcription factors[84]. It is suggested that the PI3K/PKB pathway must be shut off in S-phase to allow cyclin B to accumulate – PKB, activated in G1, phosphorylates forkheads and causes their nuclear exclusion and inactivation to allow S-phase entry; the PKB pathway is then shut off in S-phase, allowing activated forkheads to build up in the nucleus and activate the synthesis of cyclin B. More induction is caused by the USF element in G2.

10.14.1 What triggers mitosis?

As discussed earlier, cyclin B/Cdk-1 enzymic activity is held in check by P-loop phosphorylation until the G2/M transition. This is backed-up by another restraining mechanism: nuclear exclusion. The ability to exclude cyclin B/Cdk-1 from the nucleus is thought to be important not only in the normal cycle but also as a response to DNA

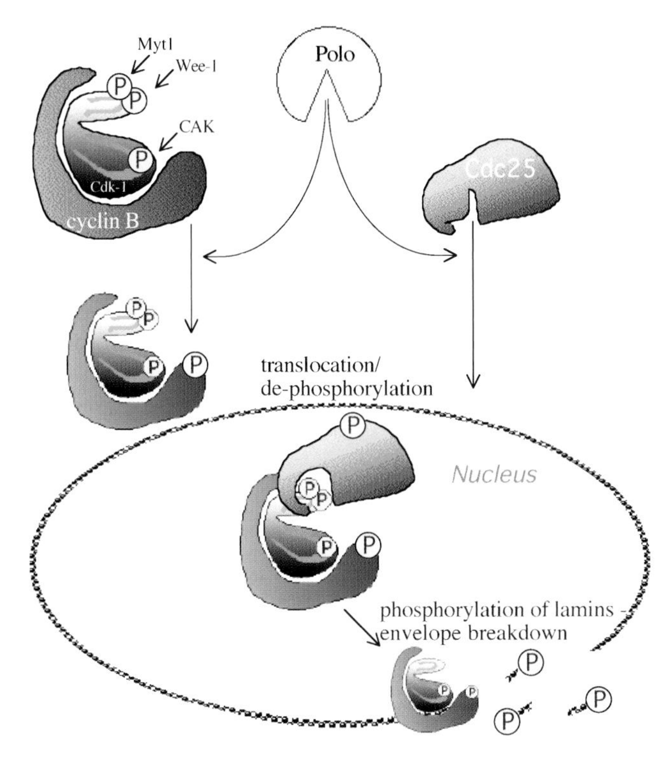

Figure 10.19 Polo - the ultimate mitotic trigger?

damage – keeping cyclin B/Cdk in the cytoplasm is one way the cell uses to block mitosis until it is ready (Figure 10.19)[85].

- Cyclin B contains a cytoplasmic retention signal (CRS) that is lacking in the other cyclins, which are always nuclear[86].
- The CRS is a 42 amino acid motif at its *N*-terminal that, when unphosphorylated, is recognised and bound by Crm-1, a nuclear exporter.
- At G2/M, when the CRS is multiple serine-phosphoryated, the cyclin B CRS is converted into a nuclear localisation signal (NLS) and cyclin B/Cdk-1 can now enter the nucleus.
- Prior to nuclear import, the kinase partner Cdk-1 subunit is tyrosine15 de-phosphorylated by Cdc25C.

10.14.2 POLO – the ultimate mitotic trigger?

Cdc25 is cytoplasmic and is activated (and also permitted to move to the nucleus) by phosphorylation by a very recently discovered ser/thr kinase called 'Polo' – mammalian polo-like kinases are called Plk-1, 2 and 3[87]. Interestingly, and significantly, it appears that Polo is also responsible for phosphorylating the CRS motif of cyclin B and causing its translocation[88]. Polo, itself, appears to be activated by the small amounts of active cyclin B/Cdk-1 that are present at G2/M – Polo activates more cyclin B/Cdk-1 and a positive feedback amplifies the transition signal. Accumulating nuclear cyclinB/Cdk-1 is activated by the nucleus-localised phospho-Cdc25C.

Fully active cyclin B/Cdk-1 immediately begins to phosphorylate its major substrates, including (for example) the nuclear lamins of the nuclear envelope, causing depolymerisation and breakdown of the structure – nuclear envelope breakdown is essential to allow chromosome separation. Finally, at anaphase, cyclin B is ubiquitinated by the APC/C and destroyed by the proteasome. The cell physiology and biomechanics of mitosis are beyond the scope of this text but are covered in numerous reviews.

10.15 Cdk inhibitors

There are two families of proteins that act as Cdk inhibitors.

- The INK family bind to Cdk-4/6 and form inhibited binary complexes that are unable to bind (and be activated by) cyclin D. They have no effect on the other Cdks.

- The WAF/Cip family, by contrast, bind to cyclin/Cdk dimers to form a ternary (or higher) complexes.

The binding of INK or WAF/Cip proteins to Cdks is mutually exclusive, and this is explained by X-ray crystallographic structural studies of various complex combinations[89]. The corollary is that the induction of high levels of INK may shift WAF/Cip from Cdk-4/6 (which it stimulates) to Cdk-2 (which it inhibits). A crystal structure shows the disabling effect that $p19^{INK4d}$ binding exerts upon the catalytic site of Cdk-6 and illustrates the ankyrin-repeat structure of the inhibitor (Figure 10.20)[89].

10.15.1 The INK proteins

The INK family members are between 15 and 19 kDa in size and in humans include the proteins $p16^{INK4a}$, $p15^{INK4b}$, $p18^{INK4c}$ and $p19^{INK4d}$. A source of possible confusion lies in the choice of 'ARF' as an acronym for an '***a***lternative ***r***eading ***f***rame' protein of the INK4a gene[90]. Here, this protein is referred to as $p19^{Arf}$, to avoid confusion with the ADP-ribosylation factors (similarly shortened to 'Arf'); although '$p19^{Arf}$' is how it is commonly described, it should be noted that this is actually the murine form – the human form is $p17^{Arf}$[90].

The INK family proteins are particularly potent inhibitors of the D-type Cdk-4 and -6 kinase activities. They are often absent in transformed cells, are present in normal cells at low levels, and are at increased levels in senescent cells. Certain INK proteins respond

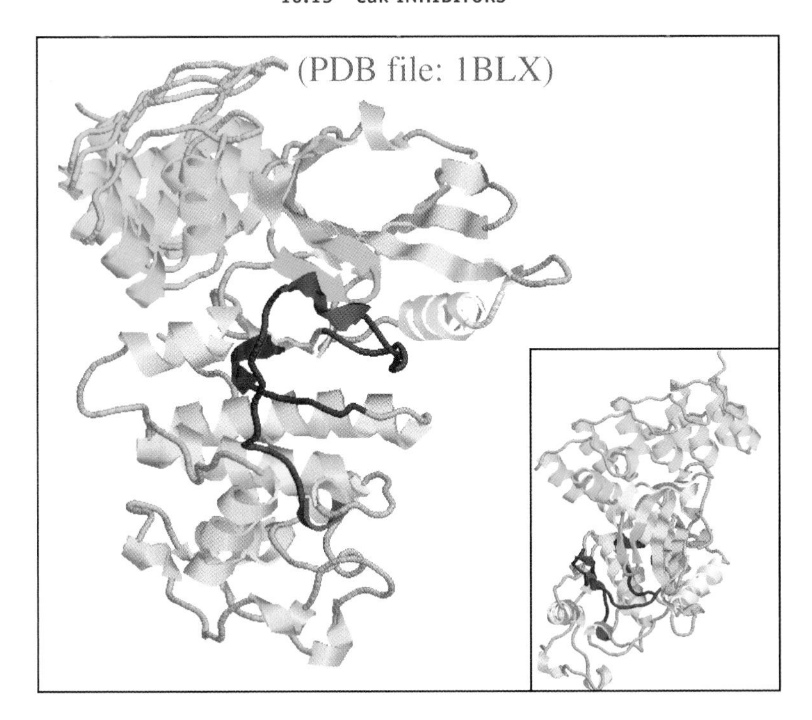

INK is in cyan. Cdk-6 (grey): P-loop, green; C-loop, red; A-loop, purple; C-helix, yellow. Inset: different viewpoint shows the ankyrin repeat structure of INK more clearly.

Figure 10.20 The complex of Cdk-6 and $p19^{INK4d}$

to external cytostatic signals. For example, ***t**ransforming **g**rowth **f**actor*-β (TGF-β, unrelated to TGF-α) is a growth-inhibitory peptide that binds to an unusual single-pass serine/threonine kinase receptor. TGF-β stimulation of adult cells causes induction of the $p15^{\mathbf{INK4b}}$ gene product, which in turn results in G1 arrest *via* inhibition of the D-type kinases activities[91].

The $p19^{\mathbf{Arf}}$ protein, on the other hand, plays a role in the control of apoptosis by p53 (discussed below).

10.15.2 The Cip/WAF family

The second Cdk inhibitor family (the Cip/Kip/WAF family) has a rather confusing nomenclature; the human family includes $p21^{\mathbf{Cip1}}$ (also known as $p21^{\mathbf{Waf1}}$), $p27^{\mathbf{Kip1}}$ (also known as $p27^{\mathbf{Cip}}$), and $p57^{\mathbf{Kip2}}$ (also known as $p57^{\mathbf{Cip}}$)[90]. These proteins are upregulated after DNA damage. $p21^{\mathbf{WAF}}$ was named from ***w**ild-type p53 **a**ctivated **f**ragment* and was found to be identical to $p21^{\mathbf{Cip}}$ (***C**dk-**i**nteracting **p**rotein*)[92].

Cip/WAF was first thought to act solely as a Cdk-2 inhibitor that functions to block G1/S transition. More recently, its enabling effect upon D-type cyclin/Cdk activation has been recognised[93] – one Cip/WAF is found to be associated with single active cyclin D/Cdk-4/6 complexes and its purpose appears to be to aid cyclin D/Cdk-4/6 formation, acting as an 'assembly factor'. It is suggested that in late G1, cyclin D/Cdk-4/6 acts as a 'sink' to sequester Cip/WAF away from cyclin E/Cdk-2, which it would otherwise efficiently inhibit[94]. However, this assembly-enabling function may extend to cyclin E/ Cdk-2 as well, because it too is present in an active form complexed with one molecule of Cip/WAF[95]. Cip/WAF inhibits cyclin E/Cdk-2 and cyclin A/Cdk-2 at higher than equal levels of stochiometry, but even higher levels of Cip/WAF are probably needed to inhibit cyclin D/Cdk-4/6.

10.16 p53 cell cycle arrest and apoptosis

P53, the 'guardian of the genome', is an unusual transcription factor that acts as a tumour suppressor protein by preventing DNA damage being duplicated to daughter cells. Its actions are primarily controlled by synthesis and destruction of the protein. However, the protein activity is also modulated by post-transcriptional modifications including phosphorylation, acetylation and ubiquitination. Induction of p53 synthesis is a normal response to DNA damage caused by UV- or ionising radiation, but dysregulation of p53 expression and function is also very common in sporadic human cancers (around 50% show such aberrations). Loss of p53 function is also the cause of heightened cancer susceptibility in the familial 'Li Fraumeni' syndrome, whose sufferers have an inherited germline mutation in one of the p53 alleles.

The wild type p53 protein product is 393 residues in length and consists of an *N*-terminal transactivation domain, followed by a proline-rich PXXP repeat domain, a central DNA binding domain, a tetramerisation domain and a *C*-terminal regulatory domain. The control of p53 function is largely exerted through control of stability and degradation of the protein. In normal cells, p53 is expressed at low levels that are precisely controlled during the cell cycle. Cellular stress or DNA damage leads to much higher p53 levels. Generally, phosphorylation of p53 stabilises the protein by preventing its ubiquitination/degradation, and also increases its DNA binding affinity at its target sites. Its stability is also controlled by acetylation of its *C*-terminal regulatory domain – again, this is induced by cell stress/DNA damage. Here, acetylation of multiple lysine residues (by histone acetylases like CBP) results in stabilisation and increased transcriptional activity of p53[96]. These acetylation sites are lysines that are also subject to ubiquitination by Mdm2; acetylation protects them from ubiquitination and thus prevents p53 degradation.

10.16.1 p53 and Cip/WAF

Ionising radiation (γ-rays) leads to double-stranded breaks in DNA and this is sensed by a complex mechanism involving the ATM protein kinase (for ***a***taxia-***t***elangiectasia ***m***utated), whose activation leads to an increase in cellular p53 levels. The central

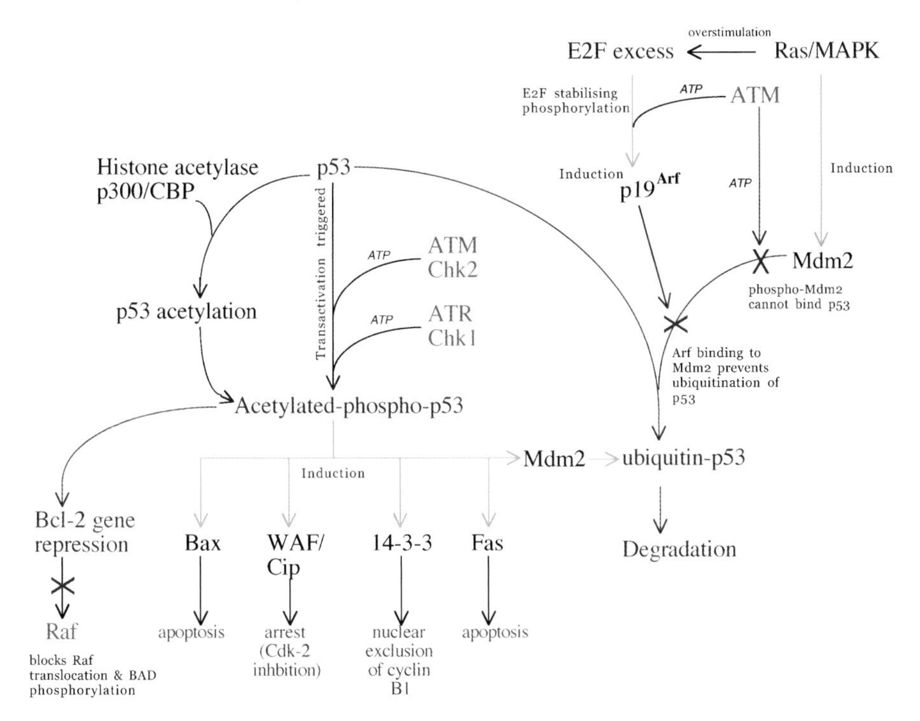

Figure 10.21 Summary of pathways controlling p53

actions follow from activation of ATM, which phosphorylates/activates the checkpoint kinase Chk2 (also known as Cds1); ATM and Chk2 both phosphorylate p53. The ***AT***M ***r***elated kinase (ATR) is activated by UV-induced DNA and replication blockade and it phosphorylates/activates the checkpoint kinase Chk1; ATR and Chk1 then phosphorylate/activate p53[97]. ATM also phosphorylates E2F-1, which results in its stabilisation and increases E2F-dependent induction of p19^{Arf}[90].

Both DNA damage systems result in phosphorylation of the *N*-terminal transactivation domain of p53 at serines-15 and -20. This has two direct effects: (i) it causes dissociation of the negative regulator, Mdm2, from the transactivation domain; (ii) it leads to recruitment of histone acetyltransferases (like CBP). The unmasking of the transactivation domain and *C*-terminal acetylation stabilise p53 and convert it into an activated transcription factor/repressor and this triggers a p53 gene transcription/repression programme. The best documented effect is induction of high levels of WAF/Cip to build up, which causes G1 arrest through inhibition of cyclin E/Cdk-2 (Figure 10.21)[96].

10.16.2 Mdm2 and p19^{Arf} – control of p53

The control of p53 expression is modulated by a protein called Mdm2 (for ***m***urine ***d***ouble ***m***inute (the human form is correctly, though rarely, referred to as 'HDM2'). Mdm2 is a p53-inducible gene and probably serves to balance the p53 response by negative feedback in normal cells. However, synthesis of the Mdm2 protein is also induced by growth factors, independently of p53 – Mdm2 is an immediate early response gene product downstream of PDGF – and Mdm2 in this setting ensures suppression of p53 to allow mitogenesis to proceed unopposed[98]. Mdm2 can do this because it has ubiquitin ligase activity and, when active, results in ubiquitination and degradation of wild type p53. Oncogenic activation of the Ras/MAPK pathway can also induce continual Mdm2-dependent destruction of p53.

As was alluded to earlier, the stimulation of proliferation normally requires *transient* stimulation of the Ras/MAPK pathway. In normal cells, *sustained* stimulation of this pathway leads to growth arrest or cell death.

In normal cells with wild-type p53, oncogenic effects are balanced and opposed by a tumour surveillance system, which responds to *hyper*-stimulation of mitogenic pathways (Ras or Myc oncogenes) by induction of the INK family homologue, $p19^{\mathbf{Arf}}$. $p19^{\mathbf{Arf}}$ has an E2F1 site on its promoter and its induction is thought to be due to a hyperabundance of free E2F/DP dimers caused by the continual phosphorylation of $R_{\mathbf{B}}$ by cyclin dependent kinases[98]. $p19^{\mathbf{Arf}}$ inhibits Mdm2, prevents it from ubiquitinating p53 for destruction, and this allows p53 to arrest the division of such mutation-carrying cells[90]. Chronic Ras stimulation also induces $p15^{\mathbf{INK4b}}$ expression[99], which may function to displace WAF/Cip from Cdk-4/6, forcing its binding and inhibition of cyclin E/Cdk-2, further contributing to arrest.

10.16.3 Apoptosis

Apoptosis is a controlled mode of programmed cell death, first described by Andrew Wyllie in 1972[100] – sadly, this important observation was neglected for several years. Apoptosis is very different from the 'accidental', toxin-induced, inflammation-provoking, lytic, cell death termed 'necrosis'. Apoptosis is an ATP-consuming programme that avoids the inflammatory response by being non-lytic. Apoptosis begins with disappearance of the nuclear envelope, nuclear condensation, and membrane blebbing, which eventually leads to budding-off of vesicles, whereby the entire cell contents is gradually dissipated by the 'falling away' of intact membrane-bounded cell fragments – apoptosis is Greek for 'falling away'. *In vivo*, these fragments are gobbled up by phagocytes; *in vitro*, apoptotic bodies from tissue-cultured cells undergo secondary necrosis and lyse.

10.16.4 Apoptosis or cell cycle arrest – majority verdict by a jury of Cdk inhibitors, survival factors, and pro-apoptotic factors

The control and execution pathways of apoptosis are exceedingly complex and unfortunately beyond the scope of this book. The discussion is therefore confined to some of the effects of p53.

The result of wild-type p53 upregulation/activation is either cell arrest or cell death. The outcome of this stark choice is executed by common mechanisms but (depending upon cell type, type of cellular insult and type of damage) the actual decision may have been arrived at in many different ways. It seems likely that the duration or extent of p53 upregulation, combined with the summation of anti-apoptotic/survival signals are deciding factors in favouring cell arrest (to allow damage repair) over cell death. It is not always the same set of factors that form the majority that decides in favour of cell death each time.

10.16.5 BH domain proteins and mitochondrial outer membrane permeabilisation

The p53 cell death programme is induced primarily through the 'intrinsic pathway' that leads to mitochondrial outer membrane permeabilisation (MOMP)[101]. Genes induced by p53 include those for a variety of proteins with BH domains – pro-apoptotic members of the Bcl-2 family (see Chapter 3)[96]. The Bcl-2 family includes: (i) anti-apoptotic members with BH-1,2,3,4 domains; (ii) multi-domain pro-apoptotic members with BH-1,2,3 domains; and (iii) pro-apoptotic BH3-only proteins (see Table 3.2, Chapter 3).

BH domain proteins are involved in regulating the release of calcium and apoptotic factors from the mitochondrion and are also involved in calcium homeostasis linked with the endoplasmic reticulum. In the absence of death signals, pro-apoptotic BAX is monomeric and cytoplasmic, whereas anti-apoptotic Bcl-2 is bound to the outer membrane of the mitochondrion. Bcl-2 (and Bcl-X_L) serve to sequester BH3-only proteins. Pro-apoptotic BH3 domains of BH3-only proteins are 'warheads' that need to be shielded. This prevents their binding to the BH1-2-3 pro-apoptotic proteins (such as BAX). This is how Bcl-2 suppresses apoptosis – by acting as a sink for BH3-only factors[102].

10.16.6 p53 and apoptosis

p53 activation of apoptosis is mediated by transcriptional and non-transcriptional means. In general, p53 causes downregulation of anti-apoptotic BH proteins and upregulation of pro-apoptotic ones. For example, p53 is well established as an inducer of BAX gene expression. But it also induces faster responses by non-transcriptional interaction with existing Bcl-2. p53 binds directly to Bcl-2 on mitochondria and causes release of BH3-only proteins such as BAD. These then shift to BAX-binding and cause translocation and a chain-reaction of oligomerisation at the mitochondrion membrane. The BH3 domain in resting monomeric BAX is thought to be shielded, but is exposed after binding the BH3-only activators. BAX BH3 can then bind to other BAX monomers and an oligomeric pore is thus built up. The pore-forming action of BAX (or Bak) at the mitochondrion is thought to cause MOMP, and leads to calcium leakage, depolarisation and eventual release of pro-apoptotic factors such as cytochrome C.

Cytochrome C and other released mitochondrial factors lead to activation of the caspase pathway that results in DNA fragmentation and nuclear envelope destruction. However, in many cases, p53-induced apoptosis is caspase-independent. Caspases are proteases with catalytic ***c***ysteines and ***asp***artate substrate specificity. They exist as inactive zymogens that are activated by internal cleavage mechanisms induced downstream of Fas ligand signals in the 'extrinsic pathway' or cytochrome C in the 'intrinsic pathway'.

10.16.7 Survival factors opposing induction of apoptosis

One of the more surprising findings in cancer research is that transfection of normal cells with a single oncogene often causes increased apoptosis rather than proliferation – this occurs upon Myc over-expression. Over-expression of E2F-1 also causes p53-induced apoptosis[54]. However, if Myc over-expression is combined with over-expression of anti-apoptotic Bcl-2 or Bcl-X_L, or is coupled with the loss of p19^{Arf} or p53 expression, then hyperproliferation results[101]. The potential oncogenic effects of viral Myc (vMyc) are also opposed by 'survival factors' from the Ras pathway. Transfection of human cells with oncogenic vMyc *alone*, leads to apoptosis. However, co-transfection of vMyc with the constitutively active oncogene vRaf, results in hyperproliferation because Raf acts as a survival signal to prevent apoptosis. Raf does this by interaction with the BH4 domain of mitochondrion-bound Bcl-2 (BH4 is a hallmark of anti-apoptotic BH proteins). Bcl-2 mediated recruitment of Raf to the mitochondrion allows it to phosphorylate the pro-apoptotic BH3-only BAD[103]. Phospho-BAD becomes bound by 14-3-3 proteins and is removed from the nucleus to the cytoplasm.

A second survival pathway is downstream of PI3K/PIP3, which results in PKB/Akt activation. PKB similarly phosphorylates BAD and represses apoptosis[101]. Finally, Cdk inhibitors have a part to play in the decision making process. In colorectal cancer cells, p53 activation normally causes WAF/Cip induction, but if WAF/Cip induction is blocked (by Myc) apoptosis ensues instead[96].

10.17 7-pass receptors and mitosis

10.17.1 The Gsp oncogene

A prototypical example of cAMP-induced cell proliferation was discovered in the Gsp oncogene that is responsible for rare human pituitary tumours[104].

Most cells are growth-inhibited by high levels of intracellular cAMP, but a few cell types naturally proliferate in response to increased cAMP levels, including ovarian carcinoma cells, thyroid and anterior pituitary somatotrophs[105]. Indeed, cholera toxin is often used to isolate initially ovarian cancer cell lines, because it selectively stimulates their growth over that of contaminating stromal cells. In normal pituitary somatotrophs, growth hormone-releasing hormone (GHRH) binds to a Gα-coupled 7-pass receptor and stimulates cAMP production, which in turn causes growth hormone release.

Table 10.4 Properties of mutated Gαs

Effects on GTPase activity

	wild type	**Arg**->Cys201	**Gln**->Arg227	Cholera
kcat (min^{-1})	4.1	0.12	0.12	0.17

The decreased rate of GTP hydrolysis allows the mutant to stimulate the adenylate cyclase effector for 10-20 times longer than normal.

Effects on Adenylate Cyclase activity (pmol cAMP.mg^{-1}.min^{-1})

	Wild type	**Arg**->Cys201	**Gln**->Arg227
Basal	13	170	180
Stimulated	170	130	120

However, this rise in cAMP also stimulates cell division during normal development, and can be reasserted in tumours associated with acromegaly (characterised by excessive production of growth hormone via GHRH overstimulation). Landis discovered a dominant oncogenic transformation of the heterotrimeric G protein, Gαs, in pituitary tumours of acromegaly patients[104] – the oncogene was named gsp for 'Gs protein'. Two point mutations were found.

One mutation point is in the same position as that commonly mutated in Ras: the catalytic glutamine of switch II Gln227 of as (corresponding to Gln61 of Ras) is substituted with an arginine. This disables its GTPase and locks it into a hyperactive GTP-bound state. A second mutation point is the switch I arginine (not present in Ras) that acts as part of the 'tethered GAP' of insert-1. Mutation of this residue to cysteine or histidine was found to have a similar inhibitory effect upon the GTPase and consequent constitutive activation (see Table 10.4).

In pituitary cells with a Gsp oncogene, cAMP is maintained at constantly high levels. This causes chronic stimulation of PKA, leading to increased levels of free PKA catalytic subunit translocating to the nucleus. The result is activation of a transcription factor named CREB (cAMP response element binding protein). The CREB family includes CREM and ATF-1. CREB can bind to a cAMP response element (CRE) on the Fos promoter and CREB is activated by serine phosphorylation by PKA, which creates a docking site for the histone acetylase, CBP[106]. The consequence is the activation of cFos synthesis and stimulation of synthesis of the cell division cycle genes (Figure 10.22).

10.17.2 Wnt/β-catenin

Wnt-1 is the founder member of a family of secreted glycoprotein hormones (19 in humans) that signal via binding to 7-pass receptors of the 'frizzled' family (10 in humans). Wnt was found independently as a mutated gene causing ***w***ingless fruit flies

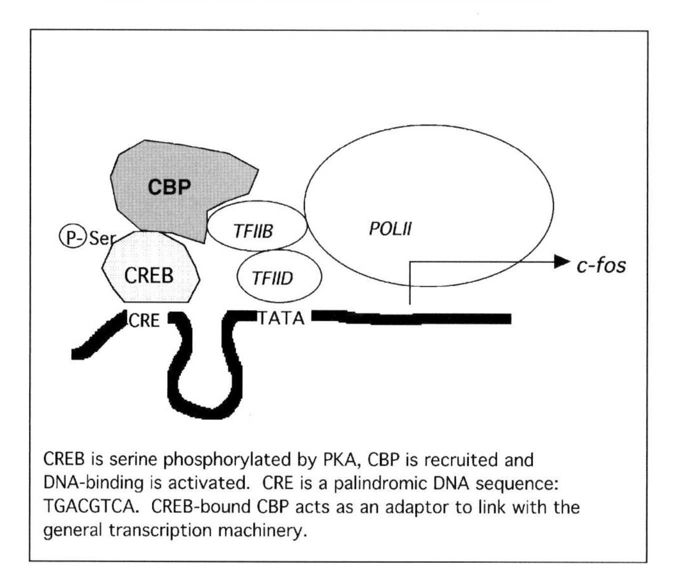

Figure 10.22 A cAMP response element is present in the c-fos promoter and activation of PKA leads to induction of c-fos

and as the murine '*Int*-1' gene activated by nearby insertion of the murine mammary tumour virus. Wnt signalling is part of an important developmental programme that can be reactivated in cancer, particularly colorectal cancers where it is implicated in 85% of sporadic cases[107]. There are three forms of signalling from Wnt/frizzled 7-pass receptors: (i) a calcium- mobilising pathway (probably mediated by heterotrimeric G proteins); (ii) a 'planar cell polarity'-organising pathway (involving RhoA); and (iii) the 'canonical' pathway[108]. As well as that, the downstream target of the latter pathway, β-catenin, has an additional (non-signalling?) role as part of the adherens junctions between cells. This, then, is a highly complex area that will only be touched upon where relevant to the present discussion.

Wnt/β-catenin and cyclin D One agreed result of oncogenic Wnt pathway deregulation is a sustained increase in levels of cyclin D1. This is the combined result of two actions: (i) Wnt signalling causes induction of cyclin D1 and Myc gene transcription; and (ii) Wnt signals result in decreased degradation of cyclin D1[109,110]. The 'canonical' Wnt pathway primarily controls degradation events through the formation of variable multi-protein complexes, whose functions reflect the presence/absence or activation/inactivation of proteins scaffolded at any one time. The activated pathway aims to increase levels of stabilised β-catenin, which acts as a co-activator of gene transcription in combination with proteins of the lymphoid T-cell transcription factor family (Tcfs).

β-catenin has an *N*-terminal domain (≈130 aas) containing four regulatory serine/threonine phosphorylation sites, a central region (≈535 aas) containing 12 tri-helical

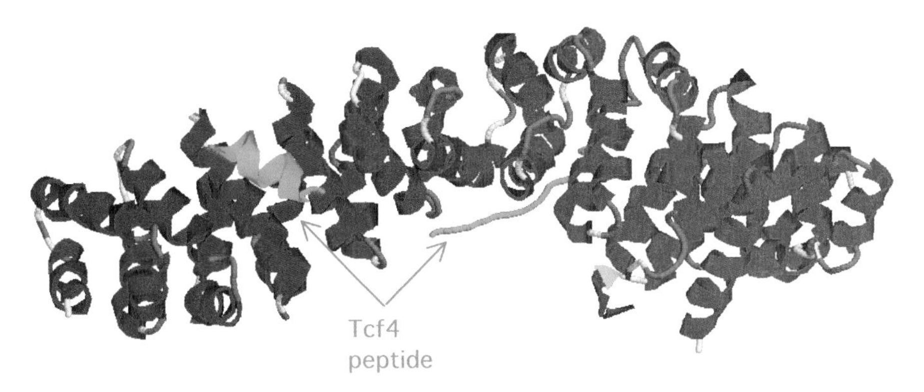

Figure 10.23 Complex of β-catenin and Tcf4 peptide

repeats (termed armadillo or 'arm' repeats) that twist into a rigid superhelix, and a *C*-terminal (≈100 aas) transactivation domain that recruits transcriptional co-activators. A crystal structure shows the central armadillo repeat region of β-catenin bound to an interacting peptide from the Tcf-4 transcription factor (Figure 10.23)[111].

Scaffolding in the canonical Wnt pathway Ligation of the Wnt ligand to the frizzled receptor recruits co-receptors of the low-density lipoprotein receptor family and this leads to activation of a cytoplasmic protein called Dishevelled (Dvl). The downstream target of Dvl is GSK-3, which becomes inhibited. GSK-3 would otherwise constitutively phosphorylate β-catenin, leading to its continual removal.

The pathway centres upon the scaffolding function of Dvl and axin (Figure 10.24). Dvl contains a central PDZ domain that binds casein kinase Iϵ (CK-Iϵ) and/or an oncogene product termed Frat; Dvl can also bind casein kinase II[112]. Axin acts as a β-catenin destruction-promoting scaffold in resting cells. Axin has an *N*-terminal RGS (regulator of G protein signalling) domain, which binds the adenomatous polyposis coli protein (APC), a central β-catenin-binding region and a more *C*-terminal region that can interact with a similar region at the *N*-terminus of Dvl. Crucially, axin can also bind GSK-3 and the phosphatase PP-2A. GSK-3 phosphorylates axin, strengthening its binding; it also phosphorylates APC, which increases its ability to bind β-catenin[49]. PP-2A may reverse the phosphorylations of the scaffold if GSK-3 is inhibited or removed.

In resting cells, β-catenin is either sequestered at adherens junctions (bound to E-cadherin and α-catenin) or bound and processed by the axin-scaffolded APC/GSK-3[113]. Phosphorylation of β-catenin by scaffolded GSK-3 causes it to be recognised by an F-box protein (βTrCP) that delivers it to the ubiquitinating SCF complex and the consequence is that β-catenin is degraded by the proteasome. The GSK-3 phosphorylations occur at the *N*-terminus of β-catenin (Thr41 first, followed by Ser37, then Ser33), after a 'priming' phosphorylation by casein kinase I[109]. However, CK-Iϵ can also serve to stabilise β-catenin, possibly by causing release of Frat[112]. *In vitro*, β-catenin is a poor GSK-3 substrate, which probably explains why axin scaffolding is needed for

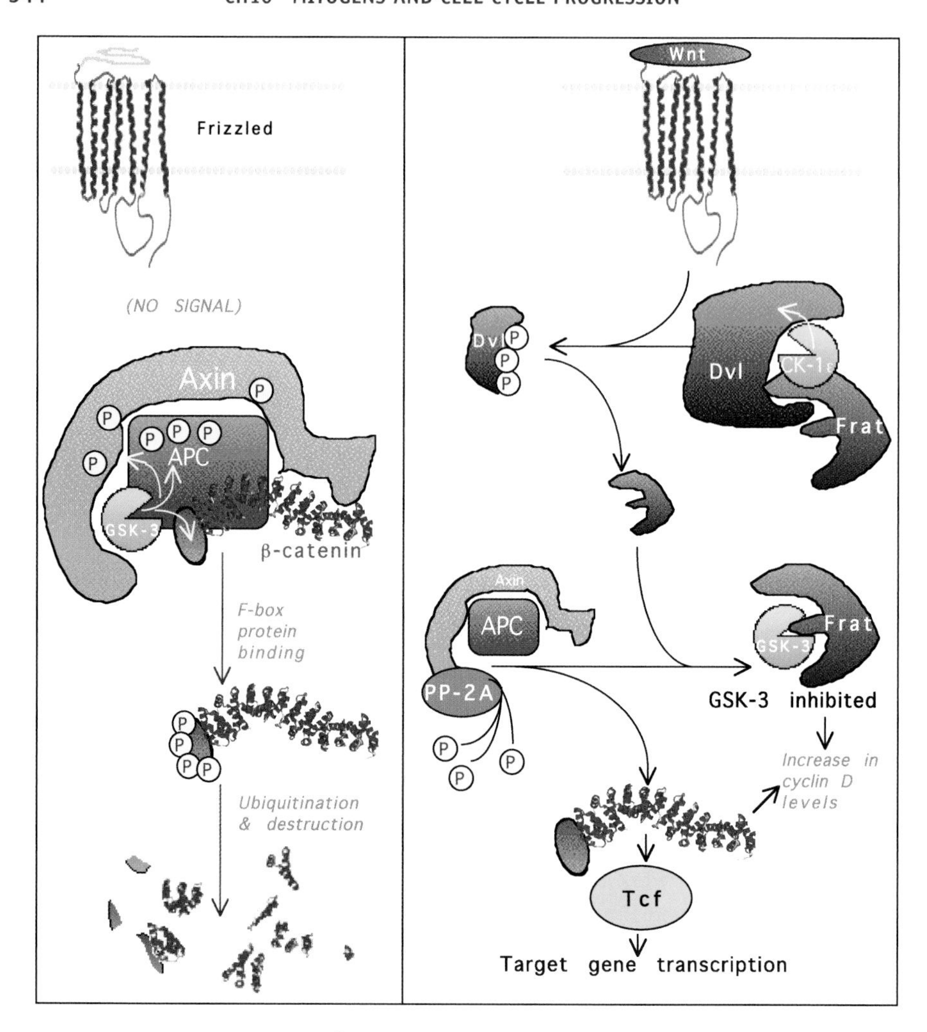

Figure 10.24 Wnt signalling

efficient phosphorylation. Axin-scaffolding also serves to segregate a discrete pool of GSK-3, insulating it from modulation by other pathways (like insulin, for example).

Wnt/frizzled activation leads to Dvl activation, possible via its phosphorylation by CK-Iε. This causes release of Frat, also known as GBP (GSK-3 binding protein), which inactivates GSK-3 by direct binding. Frat-bound GSK-3 can no longer associate with the axin scaffold. Frat binding also serves to transport GSK-3 out of the nucleus[49]. The downstream inactivation of GSK-3 by Wnt prevents β-catenin phosphorylation/destruction; instead β-catenin levels build-up. β-catenin can then translocate to the nucleus where it binds to DNA-bound members of the Tcf/LEF family. Cyclin D1 synthesis is both induced, and its destruction limited, by the inhibition of GSK-3 – GSK-3 would normally phosphorylate cyclin D1, which targets it for destruction by the proteasome[110].

10.18 Concluding remarks and caveats

Perhaps the most challenging recent developments have been the separate studies on cyclin D1,2,3- and Cdk-4,6-knockouts in mice[114,115], which together question an 'indispensable' role for D-type cyclin/Cdks in mitosis.

Mice embryos with a complete absence of Cdk-4 and Cdk-6 showed normal organ development, but died from embryonal day 14.5 onwards due to profound anaemia. Those with all D-type cyclins knocked out also displayed normal embryonic organogenesis but died before embryonal day 18.5, again through anaemia. Fibroblasts from these embryos were able to proliferate in culture (but with slower kinetics) and, if made quiescent, would re-enter the cell cycle in response to serum (although with lowered efficiency). Significantly, cyclin D1,2,3-knockout fibroblasts were resistant to transformation by Ras+Myc or Ras+dominant-negative-p53 oncogenic combinations and Cdk-4,6 knockouts were also resistant to transformation.

These findings should not be surprising because they were presaged by studies starting in 1989 when budding yeast was found to proliferate normally even with two out of the three G1 cyclins knocked out (Figure 10.2), and all G1 cyclins could be knocked out in fission yeast with no effect on cell cycle progression. These and the subsequent murine knockout studies through the 1990s and 2000s were recently reviewed[116] and are summarised in Table 10.5.

So what are we to make of these data? The studies were done in mice, and so the obvious question is 'do these findings also apply to human cell cycle progression?' There are no answers at present except to say that the classical human oncogenic axis (from growth factors, via Ras/MAPK, to R_B inactivation, E2F release and cyclin D/E induction) is not seriously in doubt. However, it does suggest that we are missing something. If human embryonic development and organogenesis does resemble the mouse – i.e., is not *critically* dependent upon D-type cyclins and kinases – then, it is argued that human cancer therapies specifically targetted at cyclin D/Cdk-4/6 might be *too* specific[16]. There are two overlapping explanations for the accumulated data.

- Cyclin and Cdk redundancy has been noted in many instances (reviewed in Reference 115) – in simple terms: if one cyclin is missing, another can take over (and the same applies to Cdks, except Cdk-1). In Cdk-4,6-null mice, Cdk-2 appears to take over their role as a G1 kinase partnered with cyclin D[115]. In these double knockouts, cyclin D is associated with Cdk-2 and the liaison changes Cdk-2's activity – complexed with cyclin A, Cdk-2 phosphorylates histones; with cyclin D, it does not, but does phosphorylate R_B. Similarly, cyclin E can replace cyclin D in D-type triple murine knockouts[114] or human cells over-expressing $p16^{INK}$ [117]. In these cases, cyclin E/ Cdk-2 phosphorylation of R_B appears to be enough to allow cell cycle progression. So a Cdk inhibitor might be better to have broad specificity for Cdks, rather than being directed at a single Cdk. What is more difficult to account for is the ability of serum to stimulate cell cycle entry and progression in D-type triple murine knockouts. It will be interesting to see if cyclin genes, other than D-type, have Myc or AP-1 sites on their promoters.

Table 10.5 Summary of murine cyclin/Cdk knockouts

Genes knocked out		Viability	Pathology
D-type cyclins*	cyclin D1	Viable	Retinopathy, neuropathy, defects in mammary gland development
	cyclin D2	Viable	Sterility, defects in B-lymphocyte proliferation and brain development
	cyclin D3	Viable	Thymus dysfunction, defective T-cell maturation
	cyclin D2,3	Lethal *in utero* (E18·5)	Megaloblastic anaemia
	cyclin D1,3	Death soon after birth	Neuropathy, retinopathy
	cyclin D1,2	Dead within three weeks of birth	Retarded growth and brain development
	cyclin D1,2,3	Lethal *in utero* (E16·5)	Profound anemia, heart defects
"D-type Cdk's"*	Cdk-4	Viable	Sterile, diabetic
	Cdk-6	Viable	Thymus and spleen dysfunction
	Cdk-4,6	Lethal *in utero* (from E14·5), a few die soon after birth	Severe megaloblastic anaemia
E-type cyclins*	cyclin E1	Viable	Normal
	cyclin E2	Viable	Male infertility
	cyclin E1,2	Lethal *in utero* (E11·5)	Heart defects
A-type cyclins§	cyclin A1	Viable	Normal
	cyclin A2	Lethal *in utero*	
	Cdk-2	Viable	Sterility
B-type cyclins†	cyclin B1	Lethal *in utero* (earlier than E10)	
	cyclin B2	Viable	Normal
	Cdk-1	Lethal *in utero*‡	

Information sources: -
*Sherr, C.J. and Roberts, J.M. (2004) Living with or without cyclins and cyclin-dependent kinases. *Genes & Development,* **18**: 2699-2711.
§Traganos, F. (2004) Cycling without cyclins. *Cell Cycle,* **3**: 32-34.
†Brandeis, M., Rosewell, I., Carrington, M., Crompton, T., Jacobs, M.A., Kirk, J., Gannon, J. and Hunt, T. (1998) Cyclin B2-null mice develop normally and are fertile whereas cyclin B1-null mice die *in utero*. *Proc. Natl. Acad. Sci. USA,* **95**: 4344-4349.
‡Malumbres, M. (2005) Revisiting the "Cdk-centric" view of the mammalian cell cycle. *Cell Cycle,* **4**: 206-210.

- Alternatively, another layer of serum-inducible, non-cyclin, control may exist that is capable of mediating serum induced cell cycle entry and progression. Gene products such as Polo and STK-15 ('Aurora-A'), for example, have the highest correlation with human tumour cell cycles, in a genome-wide analysis[118].

By the time you are reading this book, some answers to the questions discussed may have come along, others may take a lot longer. Equally, some hypotheses put forward may turn out to be wrong. But one thing is for sure, as new techniques come on-stream and protein-to-protein interactions become more easily studied, an exciting time is ahead for 21st century cell signalling researchers.

References

1. Evans, T., Rosenthal, E.T., Youngblom, J., Distel, D. and Hunt, T. (1983) Cyclin: a protein specified by maternal mRNA in sea urchin eggs that is destroyed at each cleavage division. *Cell*, **33**: 389–396.
2. Nurse, P. (1975) genetic control of cell size at cell division in yeast. *Nature*, **256**: 547–551.
3. Lohka, M.J., Hayes, M.K. and Maller, J.L. (1988) Purification of maturation-promoting factor, an intracellular regulator of early mitotic events. *Proc. Natl. Acad. Sci. USA*, **85**: 3009–3013.
4. Murray, A.W. (2004) Recycling the cell cycle: cyclins revisited. *Cell*, **116**: 221–234.
5. McGarry, T.J. (2002) Geminin deficiency causes a Chk1-dependent G2 arrest in *Xenopus. Molecular Biology of the Cell*, **13**: 3662–3671.
6. Campisi, J. and Pardee, A.B. (1984) Post-translational control of the inset of DNA synthesis by an insulin-like growth factor. *Mol. Cell. Biol.*, **4**: 1807–1814.
7. Crissman, H.A., Gadbois, D.M., Tobey, R.A. and Bradbury, E.M. (1991) Transformed mammalian cells are deficient in kinase-mediated control of progression through the G1 phase of the cell cycle. *Proc. Natl. Acad. Sci. USA*, **88**: 7580–7584.
8. Cooper, S. (2003) Reappraisal of serum starvation, the restriction point, G0, and G1 phase arrest points. *FASEB J.*, **17**: 333–340.
9. Stoeber, K., Tisty, T.D., *et al.* (2001) DNA replication licensing and human cell proliferation. *J. Cell Sci.*, **114**: 2027–2041.
10. Treisman, R. (1995) Journey to the surface of the cell: Fos regulation and the SRE. *EMBO Journal*, **14**: 4905–4913.
11. Yu, J., Deuel, T.F. and Choi Kim, H-R. (2000) Platelet-derived growth factor (PDGF) receptor-α activates c-Jun NH_2-terminal kinase-1 and antagonizes PDGF receptor-β-induced phenotypic transformation. *J. Biol. Chem.*, **275**: 19076–19082.
12. Olayioye, M.A., Neve, R.M., Lane, H.A. and Hynes, M.E. (2000) The ErbB signaling network: receptor heterodimerisation in development and cancer. *EMBO J.*, **19**: 3159–3167.
13. Nakaza, W.A.H., Aguelon, A.M. and Yamasaki, H. (1992) Identification and quantification of a carcinogen-induced molecular initiation event in cell-transformation. *Oncogene*, **7**: 2295–2301.
14. Lacal, J.C., Srivastava, S.K., Anderson, P.S. and Aaronson, S.A. (1986) Ras p21 proteins with high or low GTP'ase activity can efficiently transform NIH/3T3 cells. *Cell*, **44**: 609–617.
15. Iyer, V., Eisen, M.B., Ross, D.T., Schuler, G., Moore, T., Lee, J.C.F., Trent, J.M., Staudt, L.M., Hudson Jr., J.H., Boguski, M.S., Lashkari, D., Shalon, D., Botstein, D. and Brown, P.O. (1999) The transcriptional program in the response of human fibroblasts to serum. *Science*, **283**: 83–87.
16. Malumbres, M. (2005) Revisiting the 'Cdk-centric' view of the mammalian cell cycle. *Cell Cycle*, **4**:206–210.
17. Russo, A.A., Jeffrey, P.D. and Pavletich, N.P. (1996) Structural basis of cyclin-dependent kinase activation by phosphorylation. *Nature Structural Biology*, **3**: 696–700.
18. McGowan, C.H. and Russell, P. (1993) Human Wee1 kinase inhibits cell division by phosphorylating p34cdc2 exclusively on Tyr15. *EMBO J.*, **12**: 75–85.
19. Booher, R.N., Holman, P.S. and Fattaey, A. (1997) Human Myt1 is a cell cycle-regulated kinase that inhibits Cdc2 but not Cdk2 activity. *J. Biol. Chem.*, **272**: 22300–22306.
20. McGarry, T.J. (2002) Geminin deficiency causes a Chk1-dependent G2 arrest in *Xenopus. Mol. Biol. Cell*, **13**: 3662–3671.
21. Bloomberg, I. and Hoffman, I. (1999) Ectopic expression of Cdc25A accelerates the G1/S transition and leads to premature activation of cyclin E- and cyclin A-dependent kinases. *Mol. Cell. Biol.*, **19**: 6183–6194.

22. Cangi, M.G., Cukor, B., Soung, P., Signoretti, S., Moriera Jr., G., Ranashinge, M., Cady, B., Pagano, M. and Loda, M. (2000) Role of Cdc25A phosphatase in human breast cancer. *J. Clin. Invest.*, **106**: 753–761.

23. Xiao, Z., Chen, Z., Gunasekera, A.H., Sowin, T.J., Rosenberg, S.H., Fesik, S. and Zhang, H. (2003) Chk1 mediates S and G2 arrests through Cdc25A degradation in response to DNA-damaging agents. *J. Biol. Chem.*, **278**: 21767–21773.

24. Jackson, P.K. (1996) Cell cycle: cull and destroy. *Current Biology*, **6**: 1209–1212.

25. Dale, T.C. (1998) Signal transduction by the Wnt family of ligands. *Biochem. J.*, **329**: 209–223.

26. Shirayama, M., Zachariae, W., Ciosk, R. and Nasmyth, K. (1998) The polo-like kinase Cdc5p and the WD-repeat protein Cdc20p/fizzy are regulators and substrates of the anaphase promoting complex in *Saccharomyces cerevisiae*. *EMBO J.*, **17**: 1336–1349.

27. Kraft, C., Herzog, F., Gieffers, C., Mechtler, K., Hagting, A., Pines, J. and Peters, J-M. (2003) Mitotic regulation of the human anaphase-promoting complex by phosphorylation. *EMBO J.*, **22**: 6598–6609.

28. Ekholm, S.V. and Reed, S.I. (2000) Regulation of G1 cyclin-dependent kinases in the mammalian cell cycle. *Current Opinion in Cell Biology*, **12**: 676–684.

29. Nishitani, H., Sugimoto, N., Roukos, V., Nakanishsi, Y., Saijo, M., Obuse, C., Tsurimoto, T., Nakayama, K.I., Nakayama, K., Fujita, M., Lygerou, Z and Nishimoto, T. (2006) *EMBO J.*, **25**: 1126–1136.

30. Geley, S., Kramer, E., Gieffers, C., Gannon, J., Peters, J.M. and Hunt, T. (2001) Anaphase-promoting complex/cyclosome-dependent proteolysis of human cyclin A starts at the beginning of mitosis and is not subject to the spindle assembly checkpoint. *J. Cell Biol.*, **153**: 137–148.

31. Yam, C.H., Ng, R.W.M., Siu, W.Y., Lau, A.W.S. and Poon, R.Y.C. (1999) Regulation of cyclin A-Cdk2 by SCF component Skp1 and F-box protein Skp2. *Mol. Cell. Biol.*, **19**: 635–645.

32. Meyer, T.E. and Habener, J.F. (1993) Cyclic adenosine 3′,5′-monophosphate response element binding proteins (CREB) and related transcription-activating deoxyribonucleic acid binding proteins. *Endocr. Rev.*, **14**: 269–290.

33. Joyce, D., Albanese, C., Steer, J., Fu, M., Bouzahzah, B. and Pestell, R.G. (2001) NF-κB and cell-cycle regulation: the cyclin connection. *Growth and Growth Factor Reviews*, **12**: 73–90.

34. Shapiro, G.I. (2006) Cyclin-dependent kinase pathways as targets for cancer treatment. *J. Clin. Oncol.*, **24**: 1770–1783.

35. Bouchard, C., Thieke, K., *et al.* (1999) Direct induction of cyclin D2 by Myc contributes to cell cycle progression and sequestration of p27. *EMBO Journal*, **18**: 5321–5333.

36. Olayioye, M.A., Beuvink, I., Horsch, K., Daly, J.M. and Hynes, N.E. (1999) ErbB receptor-induced activation of Stat transcription factors is mediated by Src tyrosine kinases. *J. Biol. Chem.*, **274**: 17209–17218.

37. Galaktionov, K. and Beach, D. (1996) Cdc25 cell-cycle phosphatase as a target of c-Myc. *Nature*, **382**: 511–517.

38. Frame, S. and Cohen, P. (2001) GSK3 takes centre stage more than 20 years after its discovery. *Biochem. J.*, **359**: 1–16.

39. Whitmarsh, A.J., Yang, S-H., Su, M.S-S., Sharrocks, A.D. and Davis, R.J. (1997) Role of p38 and JNK mitogen-activated protein kinases in the activation of ternary complex factors. *Mol. Cell Biol.*, **17**: 2360–2371.

40. Bienz, M. and Clevers, H. (2000) Linking colorectal cancer to Wnt signalling. *Cell*, **103**: 311–320.

41. Yang, S-H., Shore, P., Willingham, N., Lakey, J.H. and Sharrocks, A.D. (1999) The mechanism of phosphorylation-inducible activation of the ETS-domain transcription factor Elk-1. *EMBO J.*, **18**: 5666–5674.

42. Soh, J-W., Lee, E.H., Prywes, R. and Weinstein, I.B. (1999) Novel roles of specific isoforms of protein kinase C in activation of the c-fos serum response element. *Mol. Cell Biol.*, **19**: 1313–1324.

43. Frödin, M. and Gammeltoft, S. (1999) Role and regulation of 90 kDa ribosomal S6 kinase (RSK) in signal transduction. *Molecular and Cellular Endocrinology*, **151**: 65–77.

44. Gille, H., Kortenjaan, M., Thomae, O., Moomaw, C., Slaughter, C., Cobb, M.H. and Davis, R.J. (1995) ERK phosphorylation potentiates Elk-1-mediated ternary complex formation and transactivation. *EMBO J.*, **14**: 951–962.

45. Hassler, M. and Richmond, T. (2001) The B-box dominates SAP-1-SRF interactions in the structure of the ternary complex. *EMBO J.*, **20**: 3018–3028.

46. Murphy, L.O., Smith, S., Chen, R-H., Fingar, D.C. and Blenis, J. (2002) Molecular interpretation of ERK signal duration by immediate early response genes. *Nature Cell Biology*, **4**: 556–564.

47. Shaulian, E. and Karin, M. (2001) AP-1 in cell proliferation and cell survival. *Oncogene*, **20**: 2390–2400.

48. Sears, R.C. and Nevins, J.R. (2002) Signaling networks that link cell proliferation and cell fate. *J. Biol. Chem.*, **277**: 11617–11620.

49. Doble, B.W. and Woodgett, J.R. (2003) GSK-3: tricks of the trade for a multi-tasking kinase. *J. Cell Science*, **116**: 1175–1186.

50. Lee, C., Chang, J.H., Lee, H.S. and Cho, Y. (2002) Structural basis for the recognition of the E2F transactivation domain by the retinoblastoma tumor suppressor. *Genes & Development*, **16**: 3199–3212.

51. Rubin, E., Tamrakar, S. and Ludlow, J.W. (1998) Protein phosphatase type 1, the product of the retinoblastoma susceptibility gene, and cell cycle control. *Frontiers in Bioscience*, **3**: d1209–d1219.

52. Rubin, S.M., Gall, A-L., Zheng, N. and Pavletich, N.P. (2005) Structure of the Rb C-terminal domain bound to E2F1-DP1: a mechanism for phosphorylation-induced E2F release. *Cell*, **123**: 1093–1106.

53. La Thangue, N.B. (1994) DRTF1/E2F: an expanding family of heterodimeric transcription factors implicated in cell-cycle control. *TIBS*, **19**: 108–114.

54. Johnson, D.G. and Schneider-Broussard, R (1998) Role of E2F in cell cycle control and cancer. *Frontiers in Bioscience*, **3**: d447–d458.

55. Cobrink, D. (2005) Pocket proteins and cell cycle control. *Oncogene*, **24**: 2796–2809.

56. Gao, G., Bracken, A.P., Burkard, K., Pasini, D., Classon, M., Attwooll, C., Sagara, M., Imai, T., Helin, K. and Zhao, J. (2003) NPAT expression is regulated by E2F and is essential for cell cycle progression. *Molecular and Cellular Biology*, **23**: 2821–2833.

57. Fujita, M. (2006) Cdt1 revisited: complex and tight regulation during the cell cycle and consequences of deregulation in mammalian cells. *Cell Division*, **1**: 22.

58. Giacinti, C. and Giordano, A (2006) RB and cell cycle progression. *Oncogene*, **25**: 5220–5227.

59. Harbour, J.W. and Dean, D.C. (2000) Rb functions in cell-cycle regulation and apoptosis. *Nature Cell Biology*, **2**: E65–E67.

60. Ewen, M.E. (2000) Where the cell cycle and histone meet. *Genes & Development*, **14**: 2265–2270.

61. Zhao, J., Dynlacht, B., Imai, T., Hori, T-A. and Harlow, E (1998) Expression of NPAT, a novel substrate of cyclin E-CDK2, promotes S-phase entry. *Genes & Development*, **12**: 456–461.

62. Ma, T., Van Tine, B.A., Wei, Y. Garrett, M.D., Nelson, D., Adams, P.D., Wang, J., Qin, J., Chow, L.T. and Harper, W. (2000) Cell cycle-regulated phosphorylation of $p220^{NPAT}$ by cyclin E-cdk2 promotes S-phase entry. *Genes & Development*, **14**: 2298–2313.

63. Barcaroli, D., Bongiorno-Borbone, L., Terrinoni, A., Hofmann, T.G., Rossi, M., Knight, R.A., Matera, A.G., Melino, G. and De Laurenzi, V. (2006) FLASH is required for histone transcription and S-phase progression. *Proc. Natl. Acad. Sci. USA*, **103**: 14808–14812.

64. Zhao, J., Kennedy, B.K., Lawrence, B.D., Barbie, D.A., Matera, A.G., Fletcher, J.A. and Harlow, E. (2000) NPAT links cyclin E-Cdk2 to the regulation of replication-dependent histone gene transcription. *Genes & Development*, **14**: 2283–2297.

65. Okuda, M., Horn, H.F., Tarapore, P. *et al.* (2000) Nucleophosmin/B23 is a target of CDK2/cyclin E in centrosome duplication. *Cell*, **103**: 127–140.

66. Coverley, D., Laman, H. and Laskey, R.A. (2002) Distinct roles for cyclins E and A during DNA replication complex assembly and activation. *Nature Cell Biology*, **4**: 523–528.

67. Alexandrow, M.G. and Hamlin, J.L. (2005) Chromatin decondensation in S-phase involves recruitment of Cdk2 by Cdc45 and histone phosphorylation. *J. Cell Biol.*, **168**: 875–886.

68. Voitenleitner, C., Fanning, E. and Nasheuer, H-P. (1997) Phosphorylation of DNA polymerase α-primase by Cyclin A-dependent kinases regulates initiation of DNA replication *in vitro*. *Oncogene*, **14**: 1611–1615.

69. Ishimi, Y., Komamura-Kohno, Y., You, Z., Omori, A. and Kitagawa, M. (2002) Inhibition of Mcm4,6,7 helicase activity by phosphorylation with cyclin A/Cdk2. *J. Biol. Chem.*, **275**: 16235–16241.

70. Li, C-J., Vassilev, A., and DePamphilis, M.L. (2004) Role for Cdk1 (cdc2)/cyclin A in preventing the mammalian origin recognition complex's largest subunit (Orc1) from binding to chromatin during mitosis. *Mol. Cell. Biol.*, **24**: 5875–5886.

71. Kelman, L.M. and Kelman, Z. (2003) Archaea: an archetype for replication initiation studies. *Molecular Microbiology*, **48**: 605–615.

72. Toyn, J.H., Toone, M., Morgan, B.A. and Johnston, L.H. (1995) The activation of DNA replication in yeast. *TIBS*, **20**: 70–73.

73. Lei, M. and Tye, B.K. (2001) Initiating DNA synthesis: from recruiting to activating the MCM complex. *J. Cell Science*, **114**: 1447–1454.

74. Forsburg, S.L. (2004) Eukaryotic MCM proteins: beyond replication initiation. *Microbiology and Molecular Biology Reviews*, **68**: 109–131.

75. Bandura, J.L. and Calvi, B.R. (2001) Duplication of the genome in normal and cancer cell cycles. *Cancer Biology and Therapy*, **1**: 8–13.

76. Nishitani, H. and Lygerou, Z. (2002) Control of DNA replication licensing in a cell cycle. *Genes to Cells*, **7**: 523–534.

77. McGarry, T.J. and Kirschner, M.W. (1998) Geminin, an inhibitor of DNA replication, is degraded during mitosis. *Cell*, **93**: 1043–1053.

78. Wohlschlegel, J.A., Dwyer, B.T., Dhar, S.K., Cvetic, C., Walter, J.C. and Dutta, A. (2000) Inhibition of eukaryotic DNA replication by geminin binding to Cdt1. *Science*, **290**: 2309–2312.

79. Sclafani, R.A. (2000) Cdc7p-Dbf4p becomes famous in the cell cycle. *J. Cell Science*, **113**: 2111–2117.

80. Oshiro, G., Owens, J.C., Shellman, Y., Sclafani, R.A. and Li, J. (1999) Cell cycle control of Cdc7p kinase activity through regulation of Dbf4p stability. *Mol. Cell. Biol.*, **19**: 4888–4896.

81. Blow, J.J. (2001) Control of chromosomal DNA replication in the early Xenopus embryo. *EMBO J.*, **20**: 3293–3297.

82. Moyer, S.E., Lewis, P.W. and Botchan, M.R. (2006) Isolation of the Cdc45/Mcm2-7/GINS (CMG) complex, a candidate for the eukarytotic DNA replication fork helicase. *Proc. Natl. Acad. Sci. USA*, **103**: 10236–10241.

83. Cogswell, J.P. Godlevski, M.M., Bonham, M., Bisi, J. and Babiss, L. (1995) Upstream stimulatory factor regulates expression of the cell cycle-dependent cyclin B1 gene promoter. *Mol. Cell. Biol.*, **15**: 2782–2790.

84. Alvarez, B., Martinez-A, C. Burgering, B.M.T. and Carrera, A.C. (2001) Forkhead transcription factors contribute to execution of the mitotic programme in mammals. *Nature*, **413**: 744–747.

85. Takizawa, C.G. and Morgan, D.O. (2000) Control of mitosis by changes in the subcellular location of cyclin-B1-Cdk1 and Cdc25C

86. Hagting, A., Jackman, M., Simpson, K. and Pines, J. (1999) Translocation of cyclin B1 to the nucleus at prophase requires a phosphorylation-dependent nuclear import signal. *Current Biology*, **9**: 680–689.

87. Qian, Y-W., Erikson, E., Taieb, F.E. and Maller, J.L. (2001) The polo-like kinase Plx1 is required for activation of the phosphatase Cdc25C and cyclin B-Cdc2 in Xenpous oocytes. *Molecular Biology of the Cell*, **12**: 1791–1799.

88. Toyoshima-Morimoto, F., Taniguchi, E., Shinya, N., Iwamatsu, A. and Nishida, E. (2001) Polo-like kinase 1 phosphorylates cyclin B1 and targets it to the nucleus during prophase. *Nature*, **410**: 215–220.

89. Brotherton, D.H., Dhanaraj, V., Wick, S., Brizuela, L., Domaille, P.J., Volyanik, E., Xu, X., Parsini, E., Smith, B.O., Archer, S.J., Serrano, M., Brenner, S.L., Blundell, TL. and Laue, E.D. (1998) Crystal structure of the complex of the cyclin D-dependent kinase Cdk6 bound to the cell cycle inhibitor pINK4d. *Nature*, **395**: 244–250.

90. Sherr, C.J. (2001) The *INK4a/ARF* network in tumour suppression. *Molecular Cell Biology*, **2**: 731–737.

91. Massague, J. (2004) G1 cell-cycle control and cancer. *Nature*, **432**: 298–305.

92. Pei, X-H. and Xiong, Y. (2005) Biochemical and cellular mechanisms of mammalian CDK inhibitors: a few unresolved issues. *Oncogene*, **24**: 2787–2795.

93. Sherr, C.J. and Roberts, J.M. (1999) CDK inhibitors: positive and negative regulators of G1-phase progression. *Genes & Development*, **13**: 1501–1512.

94. Kehn, K., Deng, L., de la Fuente, C., Strouss, K., Wu, K., Maddukuri, A., Baylor, S., Rufner, R., Pumfrey, A., Bottazzi, M.E. and Kashanchi, F. (2004) The role of cyclin D2 and p21/waf1 in human T-cell leukemia virus type I infected cells. *Retrovirology*, **1**: 6.

95. Poon, R.Y.C., Jiang, W., Toyoshima, H. and Hunter, T. (1996) Cyclin-dependent kinases are inactivated by a combination of p21 and Thr-14/Tyr-15 phosphorylation after UV-induced DNA damage. *J. Biol. Chem.*, **271**: 13283–13291.

96. Bai, L. and Zhu, W-G. (2006) p53: structure, function and therapeutic applications. *Journal of Cancer Molecules*, **2**: 141–153.

97. Rhind, N. and Russell, P. (2000) Chk1 and Cds1: linchpins of the DNA damage and replication checkpoint pathways. *J. Cell Biol.*, **113**: 3889–3896.

98. Ries, S., Bieder, C., Woods, D., Shifman, O., Shirasawa, S., Sasazuki, T., McMahon, M., Oren, M. and McCormick, F. (2000) Opposing effects of Ras on p53: transcriptional activation of *mdm2* and induction of p19$^{\mathbf{ARF}}$. *Cell*, **103**: 321–320.

99. Malumbres, M., Perez de Castro, I., Hernandez, M.I., Jimenez, M., Corral, T. and Pellicer, A. (2000) Cellular response to oncogenic Ras involves induction of the Cdk4 and Cdk6 inhibitor p15$^{\mathbf{INK4b}}$. *Mol. Cell. Biol.*, **20**: 2915–2925.

100. Kerr J.F., Wyllie A.H. and Currie A.R. (1972) Apoptosis: A basic biological phenomenon with wide-ranging implications in tissue kinetics. *Br J Cancer*, **26**: 239–257.

101. Green, D.R. and Evan, G.I. (2002) A matter of life and death. *Cancer Cell*, **1**: 19–30.

102. Cheng, E.H-Y., Wei, M.C., Weiler, S., Flavell, R.A., Mak, T.W., Lindsten, T. and Korsmeyer, S.J. (2001) BCL-2, BCL-XL sequester BH3 domain-only molecules preventing BAX- and BAK-mediated mitochondrial apoptosis. *Molecular Cell*, **8**: 705–711.

103. Rapp, U.R., Rennefahrt, U. and Troppmair, J. (2004) Bcl-2 proteins: master switches at the interaction of death signaling and the survival control of Raf kinases. *Biochim. Biophys. Acta.*, **1644**: 149–158.

104. Landis, C.A., Masters, S.B., Spada, A., Pace, A.M., Bourne, H.R. and Vallar, L. (1989) GTP'ase inhibiting mutations activate the α chain of Gs and stimulate adenylyl cyclase in human pituitary tumours. *Nature*, **340**: 692–696.

105. Rosenberg, D. Groussin, L., Jullian, E., Perlemoine, K., Bertagna, X. and Bertherat, J. (2002) Role of the PKA-regulated transcription factor CREB in development and tumorigenesis of endocrine tissues. *Ann. N.Y. Acd. Sci.*, **968**: 65–74.

106. Janknecht, R. and Hunter, T. (1996) Transcriptional control: versatile molecular glue. *Current Biology*, **6**: 951–954.

107. Taipale, J. and Beachy, P.A. (2001) The Hedgehog and Wnt signalling pathways in cancer. *Nature*, **411**: 349–354.

108. Huelsken, J. and Birchmeier, W. (2001) New aspects of Wnt signaling in higher vertebrates. *Current Opinion in Genetics & Development*, **11**: 547–553.

109. van Es, J.H., Barker, N. and Clevers, H. (2003) You Wnt some, you lose some: oncogenes in the Wnt signaling pathway. *Current Opinion in Genetics & Development*, **13**: 28–33.

110. Rimerman, R.A., Gellert-Randleman, A. and Diehl, J.A. (2000) Wnt and MEK1 cooperate to promote cyclin D1 accumulation and cellular transformation. *J. Biol. Chem.*, **275**: 14736–14742.

111. Poy, F., Lepourcelet, M., Shivdasani, R.A. and Eck, M.J. (2001) Structure of a human Tcf4-b-catenin complex. *Nature Structural Biology*, **8**: 1053–1057.

112. Peifer, M. and Polakis, P. (2000) Wnt signaling in oncogenesis and embryogenesis – a look outside the nucleus. *Science*, **287**: 1606–1609.

113. Willert, K. and Nusse, R. (1998) β-catenin: a key mediator of Wnt signalling. *Current Opinion in Genetics & Development*, **8**: 95–102.

114. Kozar, K., Ciermerych, M.A., Rebel, V.I., Shigematsu, H., Zagozdzon, A., Sicinska, E., Geng, Y., Yu, Q., Bhattacharya, S., Bronson, R.T., Akashi, K. and Sicinski, P. (2004) Mouse development and cell proliferation in the absence of D-cyclins. *Cell*, **118**: 477–491.

115. Malumbres, M., Sotillo, R., Santamaria, D., Galan, J., Cerezo, A., Ortega, S., Dubus, P. and Barbacid, M. (2004) Mammalian cells cycle without the D-type cyclin-dependent kinases Cdk4 and Cdk6. *Cell*, **118**: 493–504.

116. Sherr, C.J. and Roberts, J.M. (2004) Living with or without cyclins and cyclin-dependent kinases. *Genes & Development*, **18**: 2699–2711.

117. Gray-Bablin, J., Zalvide, J., Fox, M.P., Knickerbocker, C.J., DeCaprio, J.A. and Keyomarsi, K. (1996) Cyclin E, a redundant cyclin in breast cancer. *Proc. Natl. Acad. Sci. USA*, **93**: 15215–15220.

118. Whitfield, M.L., Sherlock, G., Saldanha, A.J., Murray, J.I., Ball, C.A., Alexander, K.E., Matese, J.C., Perou, C.M., Hurt, M.M., Brown, P.O. and Botstein, D. (2002) Identification of genes periodically expressed in the human cell cycle and their expression in tumors. *Molecular Biology of the Cell*, **13**: 1977–2000.

Appendix 1: Worked examples

A.1 Enzyme and receptor assays worked out from raw data examples

A.1.1 An alkaline phosphatase assay

Alkaline phosphatase (AP) is a good example of a Michaelian enzyme. As the name suggests, it works optimally at alkaline pH (8–10) and is fairly promiscuous in substrate preference. Calf intestinal AP is often used as a reporter in Western blotting because it is easily conjugated to secondary antibodies without loss of enzymic activity, allowing visualisation of primary antibodies bound to antigenic target proteins through production of coloured substrate.

Commercially available AP comes in a variety of forms, from relatively crude to highly purified. Purity is reflected in the units of activity per mg of solid quoted in the catalogue. This section uses data from an undergraduate practical class, which aims to analyse a sample of calf intestinal AP from the Sigma Chemical Company. The activity of the enzyme is given as 'DEA units per mg solid'. 'DEA' refers to the buffer diethanolamine (pH 9·8) and DEA units are the number of μmoles of para-nitrophenyl phosphate turned over per minute. 'Glycine' units are the number of μmoles of para-nitrophenyl phosphate turned over per minute in a pH 9·6 glycine buffer. One glycine unit is equivalent to about three DEA units, so you can see that even if pH is similar, the buffer composition used has an influence on the observed rates.

You are supplied with stock solutions of AP (0·04 mg/ml), and the (colourless) substrate para-nitrophenyl phosphate (3 mM in alkaline buffer (0·1 M Tris/HCl, pH 8·5) and are asked to obtain the K_M and *Vmax*. From this you can work out the units

Structure and Function in Cell Signalling John Nelson

of activity (μmole min^{-1} mg^{-1}) in Tris buffer and the number of katals in this AP preparation. *Katal* is the SI unit of enzyme activity and is defined as the number of moles of substrate turned over per second per kilogram of solid.

The reaction is followed by production of the yellow product (the nitrophenate ion), which is measured spectrophotometrically at 400 nm. The incubations are carried out in a 3 ml glass cuvette. Initially, you would do a few runs at different enzyme dilutions with fixed substrate concentration in a 'range-finding' exercise to establish an enzyme dilution that gives manageable initial rates that can be realistically measured. If enzyme is too concentrated, the reaction will be over before you get the cuvette into the spectrophotometer; too dilute and you will be there all day!

In this case, 0·14 ml of AP stock solution in a final volume of 3 ml is appropriate. The molar extinction coefficient (ε) of the para-nitrophenate ion = 18,300 M^{-1} cm^{-1} and the 3 ml cuvette has a pathlength of 1 cm.

Incubations of a range of substrate concentrations are performed separately with the same amount of enzyme in the same final volume and absorbance readings taken every 30 seconds. Table A.1 shows the raw data. To check for linearity and obtain the reaction rates for each incubation, the OD values are plotted *versus* time. This is shown in Figure A.1. Initial rates (ΔOD min^{-1}) are calculated from the individual

Table A.1 Alkaline phosphatase – product progress plots at varying substrate concentrations

TIME (s)	0·01 mM	0·0125 mM	0·015 mM	0·01875 mM
30	0·011	0·002	0·009	0
60	0·006	0·001	0·012	0·015
90	0·021	0·025	0·030	0·034
120	0·033	0·035	0·048	0·050
150	0·040	0·050	0·065	0·068
180	0·059	0·065	0·079	0·085
210	0·069	0·075	0·093	0·110
240	0·078	0·089	0·105	0·118
270	0·087	0·100	0·117	0·134
300	0·095	0·110	0·130	0·150
TIME (s)	0·025 mM	0·04 mM	0·10 mM	0·15 mM
30	0·003	0·020	0·037	0·055
60	0·022	0·045	0·066	0·082
90	0·046	0·065	0·094	0·111
120	0·068	0·090	0·120	0·139
150	0·091	0·110	0·150	0·169
180	0·111	0·133	0·175	0·197
210	0·131	0·157	0·205	0·226
240	0·148	0·180	0·235	0·253
270	0·166	0·205	0·260	0·281
300	0·183	0·225	0·290	0·309

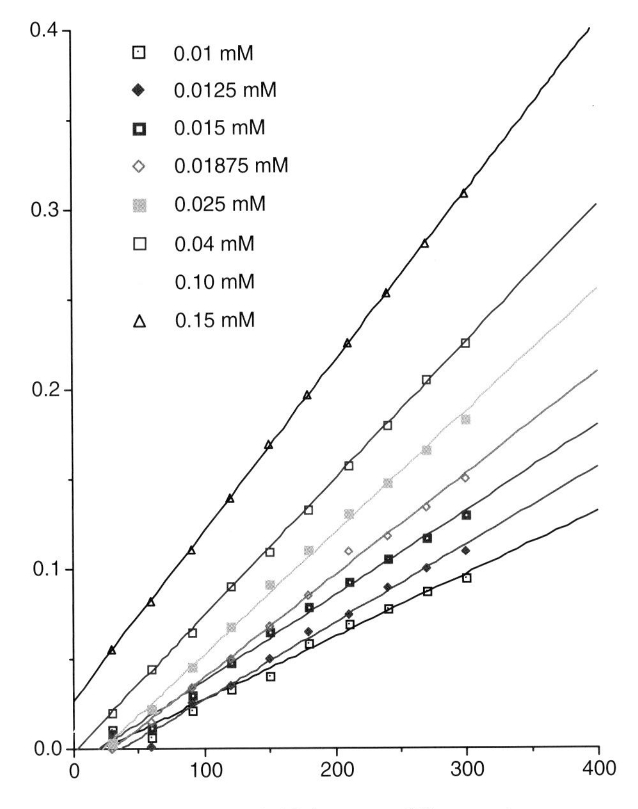

Figure A.1 Alkaline phosphatase initial rates at different substrate concentrations

slopes (shown in Table A.2). Now we are in a position to calculate K_M and *Vmax* values. A Michaelis-Menten plot shows you that we have not reached saturation (Figure 2.1, Chapter 2) and, because it is an asymptote, it is difficult to extrapolate the curve visually.

Now recalculate the data to obtain $1/v$ and $1/[S]$ (Table A.3) and this can be used to do a double-reciprocal Lineweaver–Burke plot (Figure 2.2, Chapter 2). From this graph, both

Table A.2 Initial rates for alkaline phosphatase

[S] (mM)	v (ΔOD per min)
0·01	0·023
0·0125	0·026
0·015	0·03
0·1875	0·034
0·025	0·042
0·04	0·046
0·1	0·056
0·15	0·057

Table A.3 Data prepared for Lineweaver-Burke or Eadie-Hoftsee plots

[S]	v (OD/min)	1/v	1/[S]	v/[S]
0·01000	0·023	43·478	100·000	2·300
0·01250	0·026	38·462	80·000	2·080
0·01500	0·030	33·333	66·667	2·000
0·01875	0·034	29·412	53·333	1·813
0·02500	0·042	23·810	40·000	1·680
0·04000	0·046	21·739	25·000	1·150
0·10000	0·056	17·857	10·000	0·560
0·15000	0·057	17·544	6·667	0·380

K_M and *Vmax* can be obtained. Note the data points are colour-coded so you can easily see the distribution and sequence of data in each plot.

The intersect on the 1/[S] axis is −50 and K_M is therefore equal to $1/50 = 0{\cdot}02$ mM.

The intersect on the $1/v$ axis is 15 and *Vmax* is therefore equal to $1/15 = 0{\cdot}067$ OD units per minute. You can work out the molar rate from the ε value:

$$\mathrm{OD} = \varepsilon \times (\mathrm{concentration}) \times (\mathrm{pathlength})$$

or

$$\mathrm{C} = \mathrm{OD} \div (\varepsilon \times 1).$$

$$Vmax = 0{\cdot}067 \div 18{,}300 = 3{\cdot}67 \times 10^{-6}\,\mathrm{M\ min}^{-1}$$

or

$$3{\cdot}67\,\mu\mathrm{M\ min}^{-1}.$$

In a volume of 3 ml, this is equal to

$$(3{\cdot}67 \div 1000) \times 3 = 0{\cdot}011\,\mu\mathrm{mole\ min}^{-1}.$$

The amount of enzyme producing this rate is 0·14 ml of 0·04 mg/ml stock = $5{\cdot}6 \times 10^{-3}$ mg. Thus,

$$Vmax = 0{\cdot}011 \div 5{\cdot}6 \times 10^{-3} = 1{\cdot}96\ \mu\mathrm{mole\ min}^{-1}\ \mathrm{mg}^{-1}.$$
$$= 0{\cdot}0327\,\mathrm{katals(i.e., mole\ s^{-1}\ kg^{-1})}$$

A.1.1.1 Hill plot

You can check the above data for any sign of cooperativity by construction of a Hill plot. This is a log-log plot of [S] versus (*v*/Vmax–*v*) and first requires conversion of the initial rates from OD units per minute into the same concentration terms as substrate – i.e., mM $\min^{-1}$. Plot log[S] versus log(v/Vmax–v) in arithmetic space (Figure A.2). The slope of this log-log plot is the **Hill coefficient** and is a measure of cooperativity. A Hill coefficient greater than 1 indicates positive cooperativity, less than 1 indicates negative cooperativity. You can see that our data for alkaline phosphatase displays a Hill coefficient $\approx$1 and is thus a non-cooperative Michaelian enzyme.

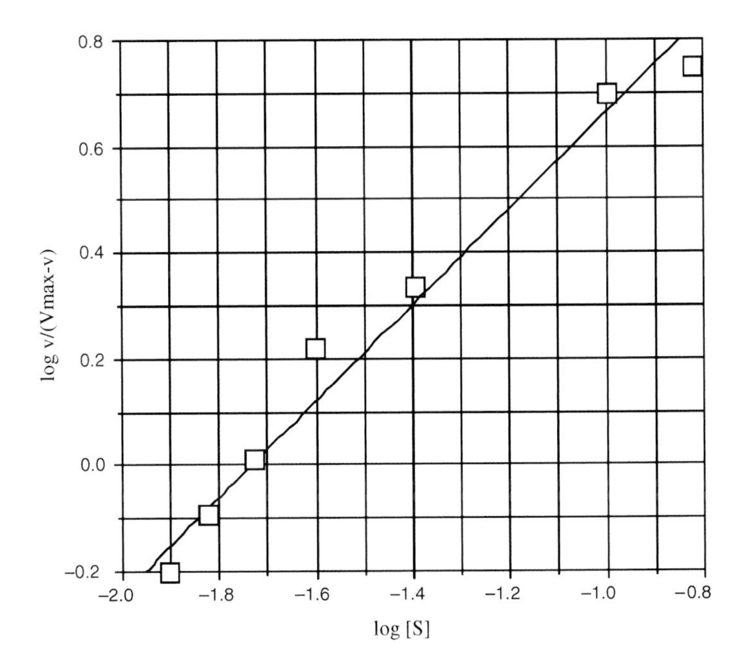

[S] (mM)	v (OD/min)	→ convert to mM	v (mM/min)
0.01000	0.023	18.300	0.001
0.01250	0.026	18.300	0.001
0.01500	0.030	18.300	0.002
0.01875	0.034	18.300	0.002
0.02500	0.042	18.300	0.002
0.04000	0.046	18.300	0.003
0.10000	0.056	18.300	0.003
0.15000	0.057	18.300	0.003

Vmax(mM/min)	Vmax-v	v/(Vmax-v)	log v/(Vmax-v)	log[S]
0.004	0.002	0.521	-0.283	-2.000
0.004	0.002	0.632	-0.199	-1.903
0.004	0.002	0.807	-0.093	-1.824
0.004	0.002	1.025	0.011	-1.727
0.004	0.001	1.669	0.222	-1.602
0.004	0.001	2.174	0.337	-1.398
0.004	0.001	5.017	0.700	-1.000
0.004	0.001	5.610	0.749	-0.824

Figure A.2 Hill plot of alkaline phosphatase data

A.1.1.2 An oestrogen receptor assay by Scatchard plot

Receptor assays appear simple in theory, but can be tricky in practice. This is because one encounters 'non-specific' binding of the ligand in all but the purest of receptor preparations. Although enzyme assays must also be corrected – for spontaneous hydrolysis of substrate – this is usually a very small component of measured rates in the presence of enzyme. Non-specific ligand binding, on the other hand, can contribute a large artefactual component to receptor assays.

Cytosolic extracts of a breast cancer cell line were subjected to whole cell oestrogen receptor assay, as follows.

Two sets of replicate incubation mixtures were set up. One set contained increasing concentrations of radioligand the other contained the same concentrations of radioligand plus an excess (1 μM) of cold oestradiol. The radioactivity in each solution was measured by scintillation counting, giving a figure for 'total ligand' (TL) at each point in the range. The ligand solutions were added to equal aliquots of cytocol at time zero and were incubated for one hour at 37°C. The unbound ligand was removed (using dextran-coated charcoal), and radioactivity in the supernatants was counted in a scintillation counter (Table A.4).

The 'total binding' (TB) data is from incubations with radioligand only; the 'non-specific binding' (NSB) data is from the parallel incubations containing radioligand plus cold excess ligand. The cold excess should ideally be at least 100-fold higher than the K_D of the receptor.

Given that each replicate incubation contained 0·45 mg of cellular protein in a final volume of 0·4 ml, and that the radioligand, [^{3}H]-oestradiol, was radiolabelled to a specific activity = 25 Curies per mmol, you are asked to derive the *Bmax* and K_D values for this receptor preparation.

In the presence of *HOT ligand only*, the observed binding is total binding:

$$TB = SB + NSB$$

In the presence of an *excess of COLD ligand*, the specific binding sites are saturated with cold ligand (plus a negligible amount of hot). In all NSB incubations the observed

Table A.4 Oestradiol receptor binding assay: TB, NSB, SB

1	2	3	4	5
F	SB	TL	TB	NSB
4.163	10692	4.177	13892	3200
1.870	10312	1.882	11749	1437
0.613	8270	0.640	8755	485
0.450	7423	0.458	7768	345
0.190	5060	0.195	5206	146
0.097	3212	0.100	3287	75
0.002	790	0.003	820	30

binding of radiolabel is overwhelmingly non-specific. *The NSB sites are considered infinite in number and, therefore, cannot be saturated.*

$$SB = TB - NSB$$

The specifically-bound ligand is in equilibrium with free ligand. Because specific sites have a very much higher affinity than non-specific sites, NSB is considered to be 'free'.

$$F = TL - SB \ldots (TL = \text{total ligand})$$

Note: In Scatchard plot analysis, SB and F should be expressed in concentration terms, but if a fixed incubation volume is used throughout, SB and F in disintegrations per minute (dpm) may be plotted.

Figure 2.3 (Chapter 2) shows TB, NSB and SB plotted against F. You can see that total binding is curvilinear, being made up of the linear non-saturable NSB component and the hyperbolic SB saturation curve. As with the Michaelis-Menten plot, the curve approaches saturation (at *Bmax*) but is often difficult to extrapolate visually. Plotting SB/F versus SB gives the linear Scatchard plot (Table A.5; Figure 2.4, Chapter 2).

Calculate *Bmax*. The intercept on the SB axis = 11,500 dpm per incubation.

Calculate K_D. The slope of the graph $= -(4{\cdot}1 \times 10^{-6}) = -(1/K_D).K_D = 244{,}000$ dpm per incubation.

A.1.1.3 An EGF receptor assay by displacement method

Using a 'tracer' amount of radioactivity is a more economical assay method and relies upon the ability of unlabelled ligand to 'displace' tracer from the specific binding sites. As cold ligand is increased, all tracer will eventually end up as non-specific binding. The following assay was performed on a human cancer cell line that over-expresses the EGF receptor:

Table A.5 Oestradiol receptor binding assay: Scatchard data

1	2	3
SB	SB/F	F
10692	0.003	4163000
10312	0.006	1943000
8270	0.013	765125
7423	0.017	555000
5060	0.027	277000
3212	0.033	97125
790	0.044	27750

Table A.6 EGF displacement receptor binding assay data

1 COLD EGF (nM)	2 dpm	3 Column 3
0.1	9697	1000
0.15	9510	800
0.25	8866	700
0.50	7260	500
1.00	5447	400
2.00	3843	300
5.00	1974	400
7.00	1580	100
10.00	1008	100
50.00	347	100
600.00	375	100

- Total hot ligand counts for each well were 10,000 dpm
- Final volume = 0·25 ml each
- Specific activity = 610 Curie/mmole
- Cold EGF added as indicated.

The raw data are shown in Table A.6 and the plot is shown in Figure 2.5 (Chapter 2).

Table A.7 Protein kinase classification scheme of Hanks & Hunter. Adapted (with kind permission) from: - Hanks, S.K. and Hunter, T. (1995) The eukaryotic protein kinase superfamily: kinase (catalytic) domain structure and classification. *FASEB J.*, **9**: 576–596

A. Protein serine/threonine kinases		
The **A-G-C** group	c*A*MP-dependent-, c*G*MP-dependent-, *C*a^{2+}-dependent-kinases	
AGC-I family	cyclic nucleotide-dependent	(A) PKA types (B) PKG types
AGC-II family	calcium-/DAG/phospholipid-dependent	(A) 'conventional' PKC, (B) 'novel' PKC (C) 'atypical' PKC
AGC-III family	***R**elated to PK**A**/PK**C***	RAC types
AGC-IV family	Kinases that phosphorylate 7-pass GPCRs	βARK
AGC-V family	Yeast-only kinases	
AGC-VI family	Kinases that phosphorylate ribosomal S6 protein	p70Rsk; p90Rsk
AGC-VII family	Yeast-only kinases	
AGC-VIII family	Plant-only kinases	
The **CaMK** group	Ca^{2+}/***cal**modulin*-dependent ***k**inases*	
CaMK-I	(A) CaMK-1,2,4 (B) phosphorylase kinase γ^{SUB} (C) myosin light chain kinase (MLCK)	
CaMK-II/AMPK	AMP-activated protein kinase (AMPK)	
The **C-M-G-C** group	***c**yclin-dependent kinase/**M**APK/**G**SK-3/**C**dk-like*	
CMGC-I family	*cyclin **d**ependent **k**inases*	Cdk-1, 2, 3, 4, 5, 6
CMGC-II family	Mitogen-activated protein kinase (MAPK) also known as extracellular signal-regulated kinase (Erk)	p44 MAPK (Erk1); p42 MAPK (Erk2); SAPK
CMGC-III family	***G**lycogen **s**ynthase **k**inase-3* (GSK-3) family	GSK-3α; GSK-3β; Casein kinase-II
CMGC-IV family	Cdc-like kinase	Clk
B. Conventional protein tyrosine kinase (PTK) group (and others)		
PTK families I-X: non-receptor		
PTK-I	the Src family	Src, Yes, Yrk, Fyn, Fgr, Lyn...
PTK-II	the Brk family	Brk
PTK-III	the Tec family	Tec
PTK-IV	the Csk family	***C**-terminal **S**rc **k**inase* (Csk)
PTK-V	the Fes family	Fes/Fps
PTK-VI	the Abl family	the ***Abel**son* kinase
PTK-VII	Syk/Zap70	***S**pleen tyrosine **k**inase*
PTK-VIII	the Janus kinase family	Jak1,2,3
PTK-IX	the Ack family	Cdc42Hs-associated kinase
PTK-X	the Fak family	***F**ocal **a**dhesion **k**inase* (Fak)

(*continued*)

Table A.7 (*Continued*)

PTK families XI-XIII: receptor- and membrane-spanning tyrosine kinases		
PTK-XI	the ***e***pidermal ***g***rowth ***f***actor ***r***eceptor (EGFR) family	ErbB1; ErbB2; ErbB3; ErbB4
PTK-XII	the Eph family	***e***rytropoetin-***p***roducing ***h***epatoma (Eph) kinase and Eph-like
PTK-XIII	the Axl family	Axl; Eyk
PTK-XIV	the Tie/Tek family	
PTK-XV(A)	the ***p***latelet ***d***erived ***g***rowth ***f***actor ***r***eceptor (PDGFR) family	PDGFR; Kit; FLT3 (also known as 'Flk-2') CSF-1R (5 IgG domains);
PTK-XV(B)	***v***ascular ***e***ndothelial ***g***rowth ***f***actor ***r***eceptor (VEGFR) family	VEGFR (7 IgG domains)
PTK-XVI	the FGFR family	***f***ibroblast ***g***rowth ***f***actor ***r***eceptor (FGFR)
PTK-XVII	the insulin receptor	InsR; InsR-related (IRR) and the insulin-like growth factor-1 receptor (IGF-I receptor)
PTK-XVIII	Ltk/Alk	
PTK-XIX	Ros/Sev	
PTK-XX	Trk/Ror	high MW NGF receptor
PTK-XXI	Ddr/Tkt	
PTK-XXII	the hepatocyte growth factor receptor family	MET
PTK-XXIII	*C. elegans* PTK's	
Other protein kinases		
O-I	Polo kinases	
O-II	Mixed function STY-kinases: MAPK kinase (also known as MAP Erk kinase, or Mek)	
O-III	MAPK kinase (also known as Mek kinase, MEKK)	
O-VI	Wee family (mixed function? — has STY activity *in vitro*)	
O-VIII	the Raf family: Raf-1; A-Raf; B-Raf	
Receptor serine/threonine kinases		
O-IX(A)	type-I activin/***t***ransforming ***g***rowth ***f***actor-β receptor: TGF-βRI)	
O-IX(B)	type-II activin/TGF-β receptor: TGF-βRII	
Casein kinase-I family		
O-XII	Casein kinase1α,β,γ,δ	
other kinases without grouping include Mos, Cdc7		

Appendix 2: RasMol: installation and use

All versions of RasMol are curated by Bernstein and Sons and can be downloaded from...
http://www.rasmol.org

Windows, Mac OS, Linux and Unix platforms are catered for and 8-, 16- and 32-bit versions are available. Most computers these days are perfectly happy running the 32-bit version.

A simple Windows download is available from...
http://www.umass.edu/microbio/rasmol/getras.htm

For Macintosh OSX users, there is a simple download (created by Mamoru Yamanishi), which runs as an Apple X11 application – these are labelled RasMacX and can be downloaded from...
http://blondie.dowling.edu/projects/rasmacx/
You will, of course need to install X11 from your system disk. In RasMacX, the command line runs in a Terminal window, otherwise it works exactly like the Mac Classic or OS9 version

The following simple instructions are based on Mac OS 9. RasMol on Windows behaves similarly. There are many more things you can do with RasMOL than can be covered here. There are numerous helpful tutorials on the Web to take you further.

Structure and Function in Cell Signalling John Nelson

- First create a folder called "Rasmol"
- Place the downloaded Rasmol software icon in this folder
- You can also download the Rasmol help file and place this in your Rasmol folder
- Downloaded structure files should ALWAYS be placed in the same Rasmol folder and all scripts must also be placed here
- To get a structure file, go to NCBI and, after selecting the structure file you want, download it and save
- You can configure your browser to open the file 'on the fly' as described in the tutorial web site (below)
- If the structure file needs to be RENAMED – call it something you'll recognise and add the file extension .pdb
- You can now open the file from the RasMol file menu
- If you write a script, make sure it ends with .spt and ensure it is placed in the RasMol folder
- You can also open the files in Word, edit them and then save as a text file
- The edited file should then be RENAMED with the. pdb file extension
- Again, make sure you put it in the RasMol folder

As mentioned in the preface, bear in mind that the sequence numbering in PDB files is not always the same as the native sequence numbering because proteins are often truncated to aid crystallisation. Always read the source paper before attempting to examine the structural model. Check the primary amino acid sequence from the paper or protein database against the RasMol file. The structure's sequence can be viewed using the 'show sequence' command.

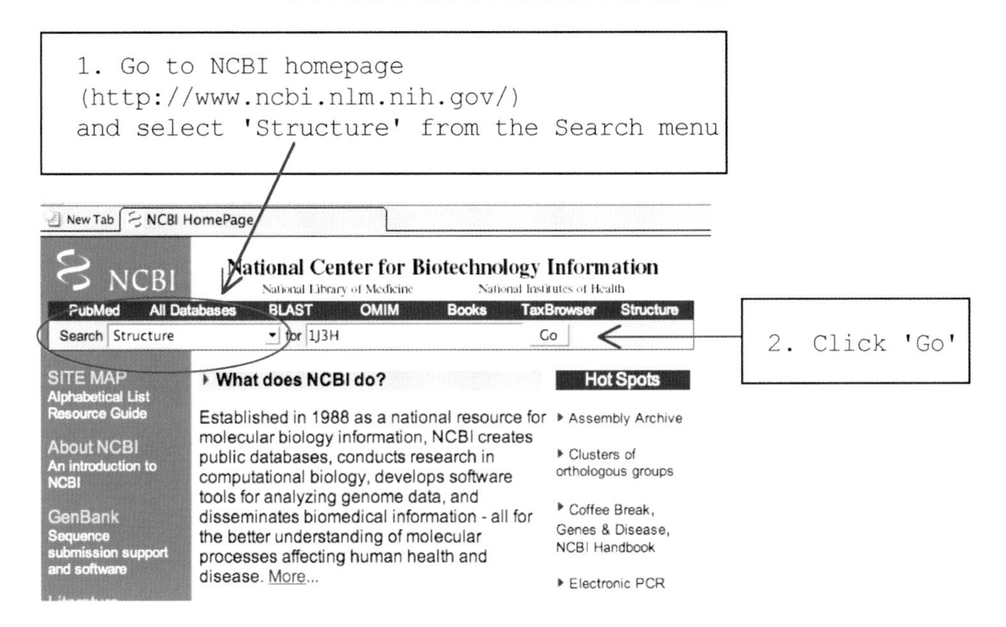

2. The unique PDB file code '1J3H' gives a single file (typing in 'Camp-dependent protein kinase' would give many entries). Double click on the thumbnail structure or the blue accession number

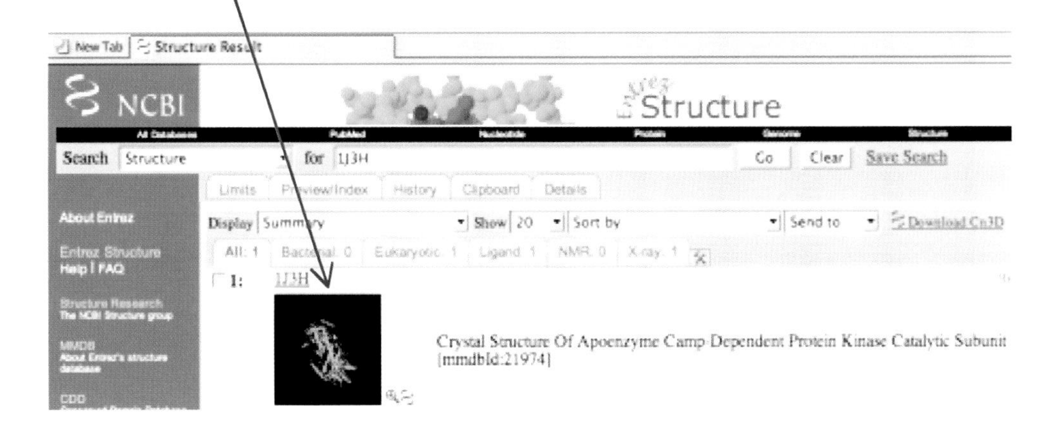

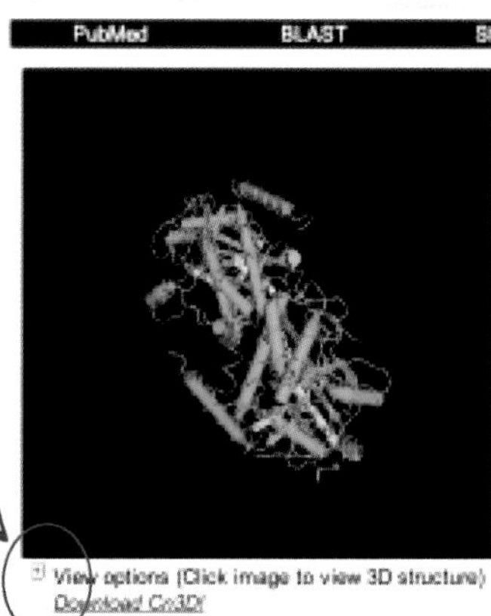

3. In the new window, the thumbnail is enlarged. Clicking in the clickbox beneath the picture allows you to select RasMol as your 'View Option'

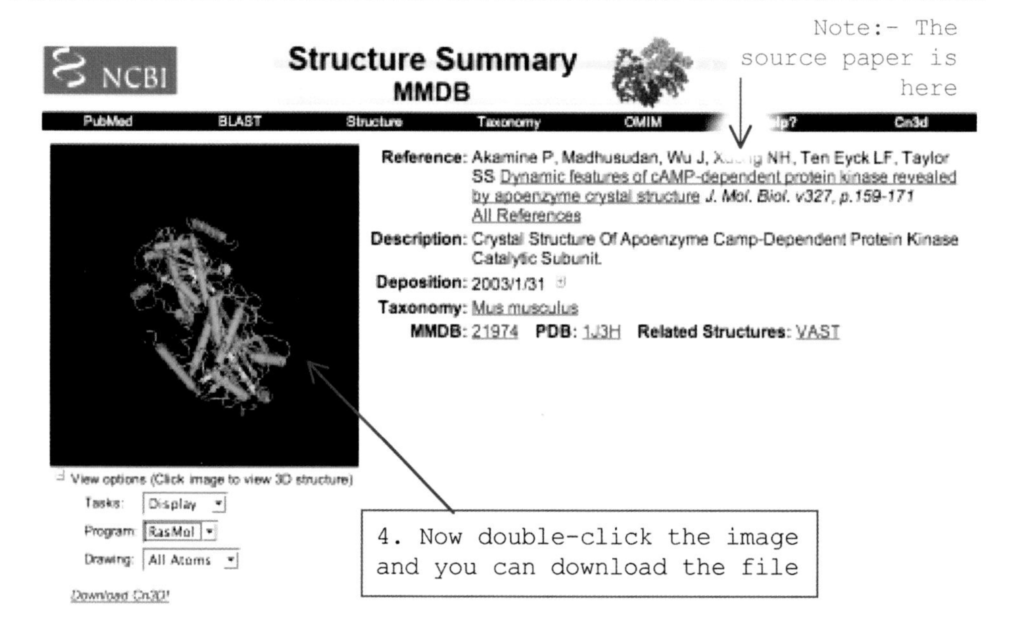

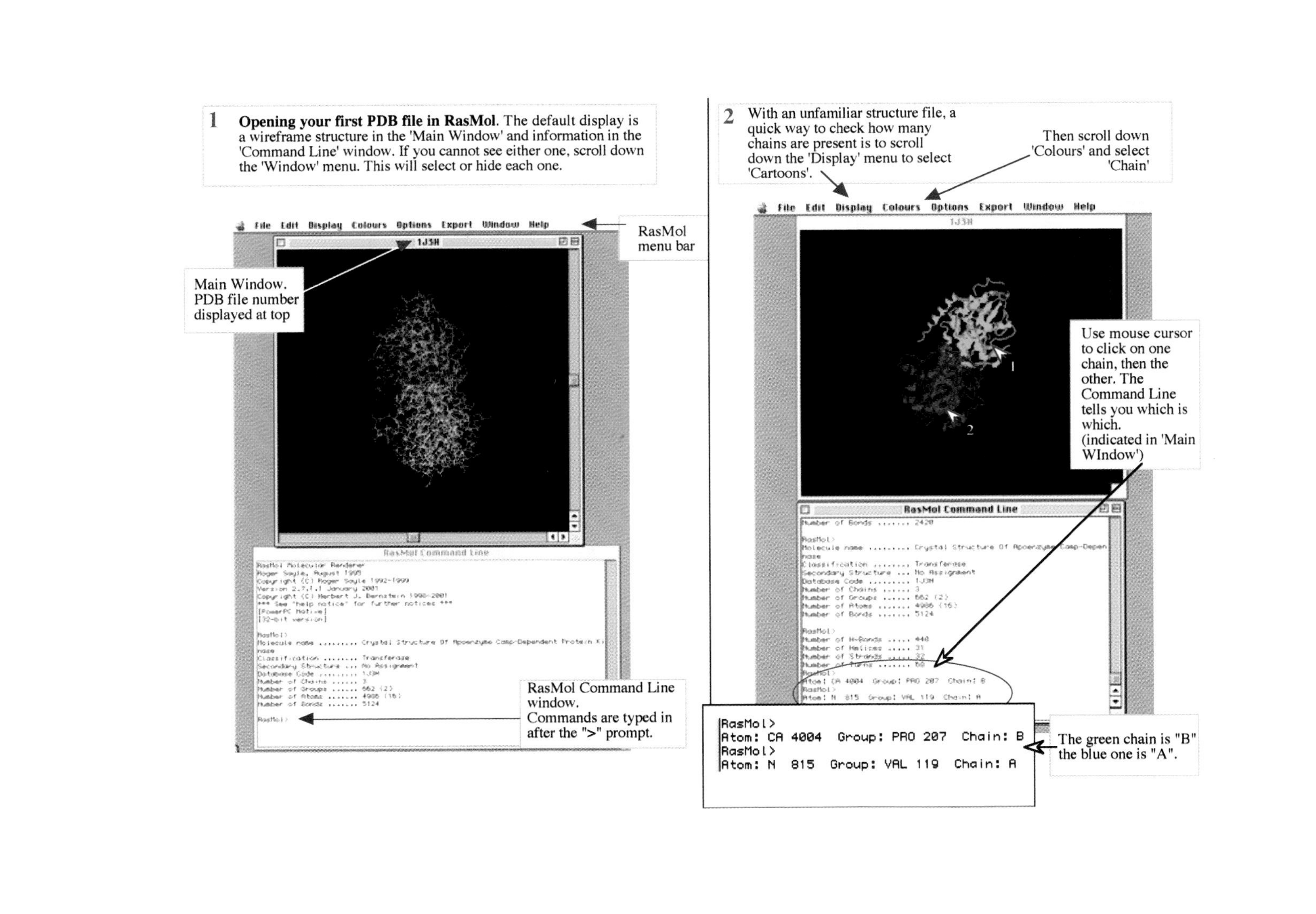

1 Opening your first PDB file in RasMol. The default display is a wireframe structure in the 'Main Window' and information in the 'Command Line' window. If you cannot see either one, scroll down the 'Window' menu. This will select or hide each one.
File Edit Display Colours Options Export Window Help
RasMol menu bar
1J3H
Main Window. PDB file number displayed at top
RasMol Command Line
RasMol Command Line window. Commands are typed in after the ">" prompt.
2 With an unfamiliar structure file, a quick way to check how many chains are present is to scroll down the 'Display' menu to select 'Cartoons'.
Then scroll down 'Colours' and select 'Chain'
1
2
Use mouse cursor to click on one chain, then the other. The Command Line tells you which is which. (indicated in 'Main WIndow')
RasMol>
Atom: CA 4004 Group: PRO 207 Chain: B
RasMol>
Atom: N 815 Group: VAL 119 Chain: A
The green chain is "B" the blue one is "A".

3 You know from the original paper that this is not a natural homodimer but two molecules that happened to crystalise in the same space. It is a good idea to create a file containing just one structure for clarity.

In many structures, 'hetero' atoms are included. Some of these may be important substrates or co-factors. In many instances there are ions and waters retained from the cryatalisation solution. In this apo-PKA structure, there is no adenine nucleotide and no waters. However, there are two molecules of MPD - one for each protein. MPD is '2-methyl-2,4- pentanediol', the preciptating agent in crystal growth, and it occupies a myristic acid binding site.

Imagining for one moment that you want to retain these, you need to identify which is associated with the chain you want to keep.

In the Command Line, type in...

`select hetero`

...press return. Then scroll down the 'Display'menu to select 'Spacefill'.

(In many structural files, this will reveal a blizzard of water molecules)

File Edit Display Colours Options Export Window Help

Wireframe
Backbone
Sticks
Spacefill
Ball & Stick
Ribbons
Strands
Cartoons

1J3H

```
Atom: CA 4004  Group: PRO 207  Chain: B
RasMol>
Atom: N  815  Group: VAL 119  Chain: A
RasMol> select hetero
16 atoms selected!
RasMol> |
```

Use mouse cursor to click on the hetero atoms to identify them. MPD 4 belongs with the green Chain B

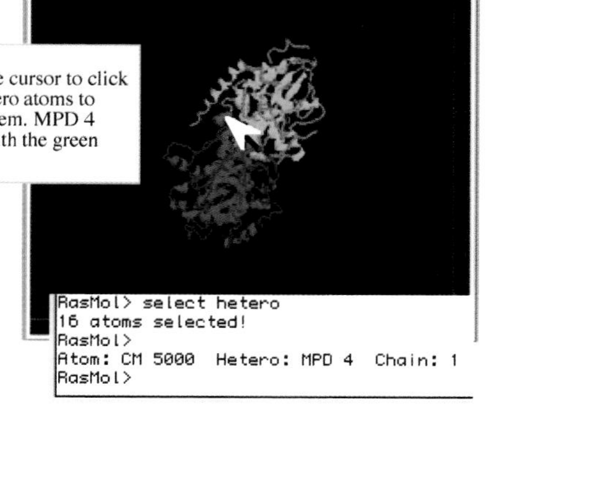

4 Quit RasMol. Now open the PDB file from the 'File' menu of Word or a text-editing application. The cooridnate file looks like this...

X-ray file has each (non-hydrogen atom) for each amino acid listed along with their x,y,z coordinates

Apo-PKA (1J3H)

```
HEADER    Transferase                                     13-JUL-93   1J3H
COMPND    Crystal Structure Of Apoenzyme Camp-Dependent Protein Kinase
ATOM      1  C    GLY A   9      12.288  22.380   5.336  1.00120.27
ATOM      2  CA   GLY A   9      11.484  21.096   5.436  1.00120.16
ATOM      3  N    GLY A   9      10.120  21.204   4.824  1.00119.84
ATOM      4  O    GLY A   9      11.708  23.468   5.252  1.00119.81
ATOM      5  C    SER A  10      14.160  24.540   4.324  1.00119.81
ATOM      6  CA   SER A  10      14.580  23.356   5.220  1.00120.19
ATOM      7  CB   SER A  10      15.960  22.824   4.788  1.00119.59
ATOM      8  N    SER A  10      13.620  22.240   5.300  1.00120.18
ATOM      9  O    SER A  10      13.372  25.400   4.744  1.00119.43
ATOM     10  C    GLU A  11      13.240  25.192   1.320  1.00116.30
ATOM     11  CA   GLU A  11      14.428  25.672   2.164  1.00117.68
ATOM     12  CB   GLU A  11      15.648  25.988   1.284  1.00118.23
ATOM     13  CD   GLU A  11      18.060  26.856   1.224  1.00119.02
ATOM     14  CG   GLU A  11      16.920  26.304   2.080  1.00118.87
ATOM     15  N    GLU A  11      14.748  24.616   3.124  1.00119.07
ATOM     16  O    GLU A  11      13.288  25.184   0.092  1.00115.59
ATOM     17  OE1  GLU A  11      18.324  28.084   1.304  1.00118.79
ATOM     18  OE2  GLU A  11      18.700  26.072   0.488  1.00118.47
ATOM     19  C    GLN A  12       9.940  25.372   1.228  1.00112.68
```

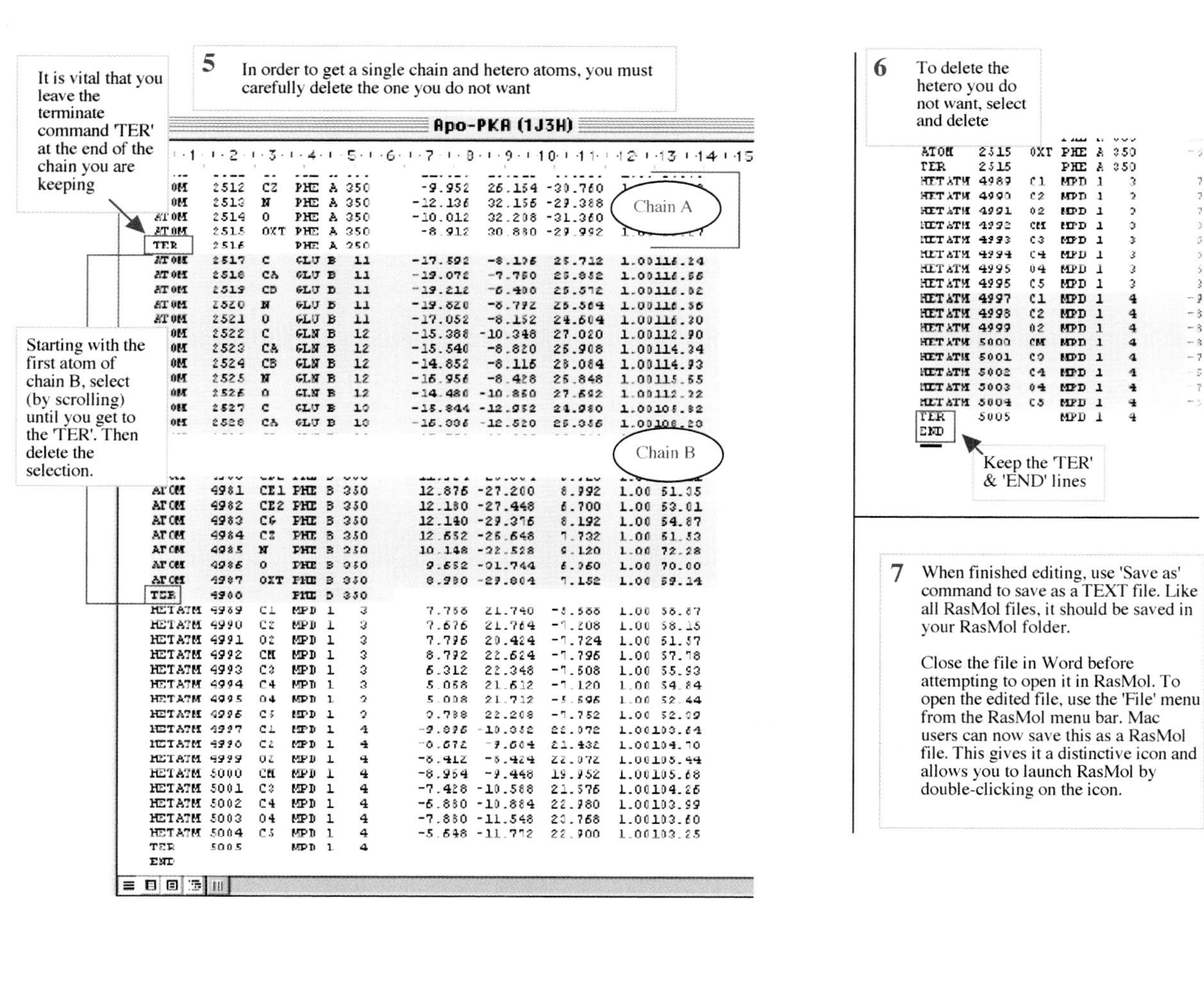
5
In order to get a single chain and hetero atoms, you must carefully delete the one you do not want
It is vital that you leave the terminate command 'TER' at the end of the chain you are keeping
Starting with the first atom of chain B, select (by scrolling) until you get to the 'TER'. Then delete the selection.
Apo-PKA (1J3H)
Chain A
Chain B
6
To delete the hetero you do not want, select and delete
Keep the 'TER' & 'END' lines
7
When finished editing, use 'Save as' command to save as a TEXT file. Like all RasMol files, it should be saved in your RasMol folder.
Close the file in Word before attempting to open it in RasMol. To open the edited file, use the 'File' menu from the RasMol menu bar. Mac users can now save this as a RasMol file. This gives it a distinctive icon and allows you to launch RasMol by double-clicking on the icon.

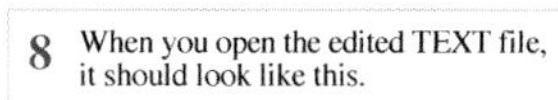

8 When you open the edited TEXT file, it should look like this.

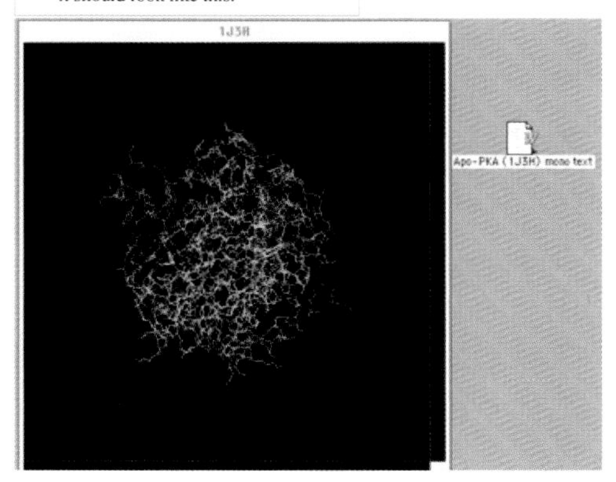

Use the RasMol 'Display' to select 'cartoons' rendering. Then scroll down 'Colour' menu to select 'structure'. The default colours alpha-helices magenta and beta-strands in yellow

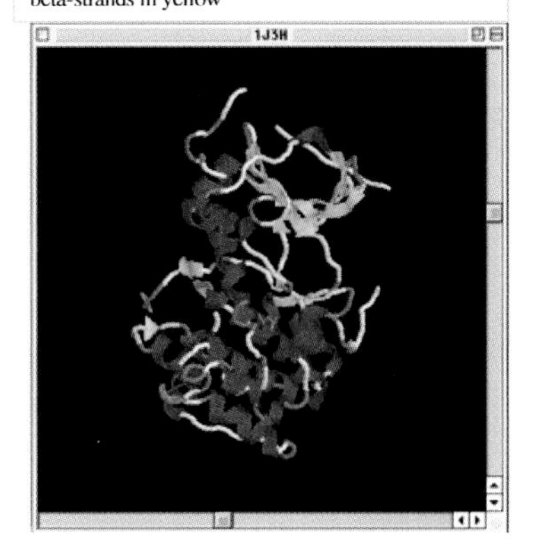

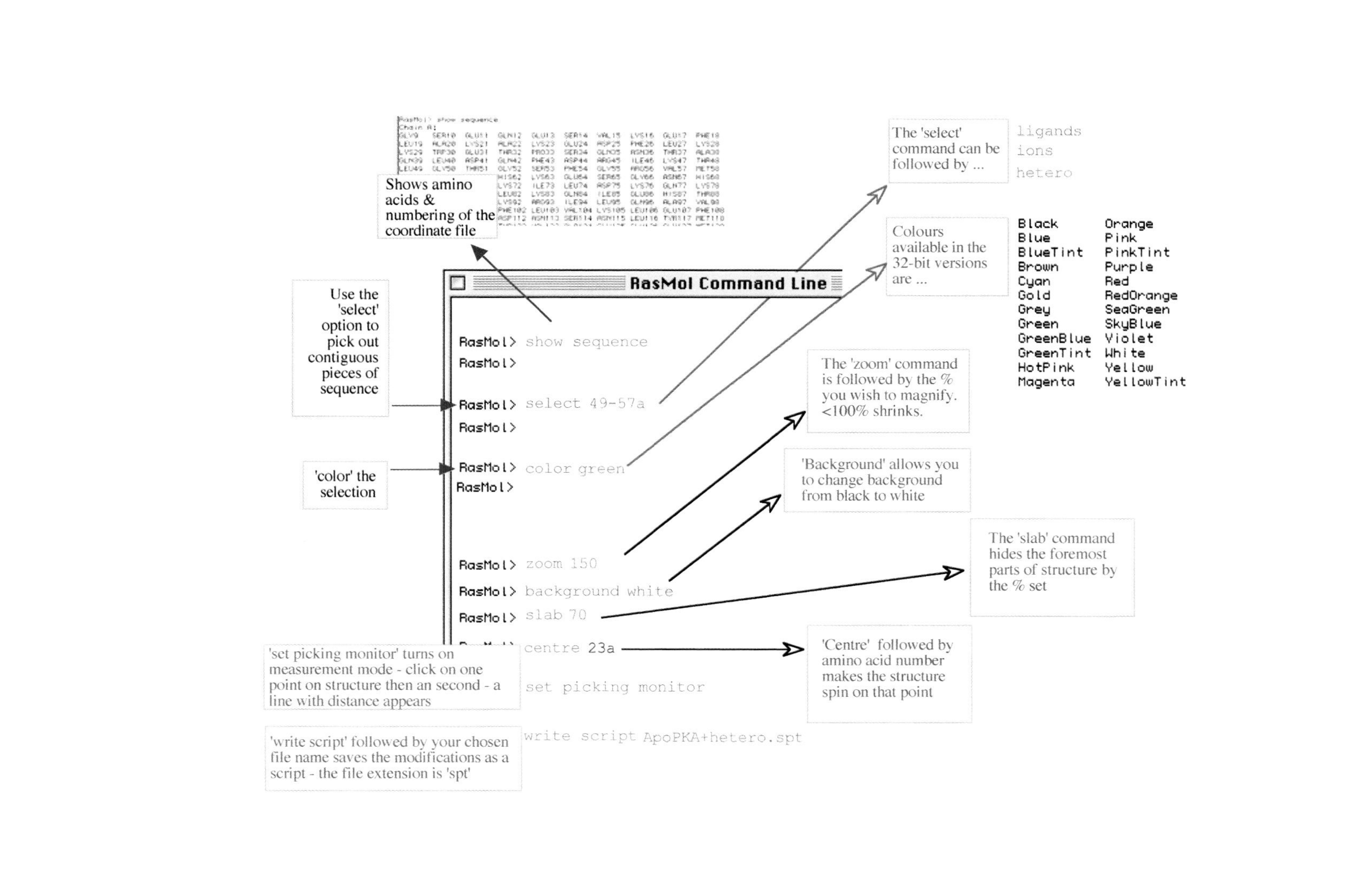

RasMol Command Line
RasMol> show sequence
RasMol>
RasMol> select 49-57a
RasMol>
RasMol> color green
RasMol>
RasMol> zoom 150
RasMol> background white
RasMol> slab 70
centre 23a
set picking monitor
write script ApoPKA+hetero.spt
Shows amino acids & numbering of the coordinate file
Use the 'select' option to pick out contiguous pieces of sequence
'color' the selection
The 'select' command can be followed by ...
ligands
ions
hetero
Colours available in the 32-bit versions are ...
Black
Blue
BlueTint
Brown
Cyan
Gold
Grey
Green
GreenBlue
GreenTint
HotPink
Magenta
Orange
Pink
PinkTint
Purple
Red
RedOrange
SeaGreen
SkyBlue
Violet
White
Yellow
YellowTint
The 'zoom' command is followed by the % you wish to magnify. <100% shrinks.
'Background' allows you to change background from black to white
The 'slab' command hides the foremost parts of structure by the % set
'Centre' followed by amino acid number makes the structure spin on that point
'set picking monitor' turns on measurement mode - click on one point on structure then an second - a line with distance appears
'write script' followed by your chosen file name saves the modifications as a script - the file extension is 'spt'

Bond	*Length* (Å)
O—H···O	2.70
O—H···O$^-$	2.63
O—H···N	2.88
N—H···O	3.04
N$^+$—H···O	2.93
N—H···N	3.10

Figure A2.1 Typical Hydrogen Bond Lengths. The distances are from the donor atom to the acceptor atom (the hydrogen is somewhere between them). Reproduced from *Biochemistry*, 3rd Edition by Stryer L. (1988). Copyright, W. H. Freeman & Co., New York

Index

Structure and Function in Cell Signalling John Nelson